CONCRETE

Microstructure, Properties, and Materials
Second Edition

P. Kumar Mehta
Paulo J.M. Monteiro
University of California, Berkeley

The McGraw-Hill Companies, Inc.
College Custom Series

New York St. Louis San Francisco Auckland Bogotá
Caracas Lisbon London Madrid Mexico Milan Montreal
New Delhi Paris San Juan Singapore Sydney Tokyo Toronto

McGraw·Hill
A Division of The McGraw·Hill Companies

CONCRETE: Microstructure, Properties, and Materials

Copyright © 1993 by The McGraw-Hill Companies, Inc. All rights reserved. Printed in the United States of America. Except as permitted under the United States Copyright Act of 1976, no part of this publication may be reproduced or distributed in any form or by any means, or stored in a data base retrieval system, without prior written permission of the publisher.

This book was previously published by Prentice-Hall, Inc.

1 2 3 4 5 6 7 8 9 0 BBC BBC 9 0 9 8 7 6

ISBN 0-07-041344-4

Editor: Todd Bull
Cover Designer: Ryan Sanzari
Printer/Binder: Braceland Brothers, Inc.

*This book is humbly dedicated
to pioneers in concrete
who are attempting
to extend the use of the material
to new frontiers,
and to make it more durable,
energy-efficient, and environment friendly*

Overview

PART I
STRUCTURE AND PROPERTIES OF HARDENED CONCRETE
---

1. Introduction to Concrete, *1*
2. The Structure of Concrete, *17*
3. Strength, *43*
4. Dimensional Stability, *78*
5. Durability, *113*

PART II
CONCRETE MATERIALS, MIX PROPORTIONING, AND EARLY-AGE PROPERTIES
---

6. Hydraulic Cements, *179*
7. Aggregates, *226*
8. Admixtures, *256*
9. Proportioning Concrete Mixtures, *290*
10. Concrete at Early Ages, *309*

PART III
RECENT ADVANCES AND CONCRETE IN THE FUTURE

11. Progress in Concrete Technology, *357*

12. Advances in Concrete Mechanics, *445*

13. The Future of Concrete, *525*

Contents

Foreword to the First Edition, *xix*

Preface to the First Edition, *xxi*

Preface to the Second Edition, *xxv*

Acknowledgments, *xxvii*

PART I
STRUCTURE AND PROPERTIES OF HARDENED CONCRETE

CHAPTER 1
___
Introduction to Concrete, *1*

    CONCRETE AS A STRUCTURAL MATERIAL, *1*
    COMPONENTS OF MODERN CONCRETE, *8*
    TYPES OF CONCRETE, *10*
    PROPERTIES OF HARDENED CONCRETE AND THEIR SIGNIFICANCE, *11*
    UNITS OF MEASUREMENT, *14*
    TEST YOUR KNOWLEDGE, *15*
    SUGGESTIONS FOR FURTHER STUDY, *16*

CHAPTER 2
___
The Structure of Concrete, *17*

    DEFINITIONS, *17*
    SIGNIFICANCE, *18*
    COMPLEXITIES, *18*
    STRUCTURE OF THE AGGREGATE PHASE, *21*
    STRUCTURE OF HYDRATED CEMENT PASTE, *22*
        Solids in Hydrated Cement Paste, *24*
            Calcium silicate hydrate, 24. Calcium hydroxide, 25.
            Calcium sulfoaluminates, 26. Unhydrated clinker grains, 26.

Voids in Hydrated Cement Paste, 26
  Interlayer space in C-S-H, 26. Capillary voids, 27.
  Air voids, 28.
Water in Hydrated Cement Paste, 29
  Capillary water, 29. Adsorbed water, 29.
  Interlayer water, 29. Chemically combined water, 30.
Structure-Property Relationships in Hydrated Cement Paste, 30
  Strength, 30. Dimensional stability, 32. Durability, 34.
TRANSITION ZONE IN CONCRETE, 36
  Significance of the Transition Zone, 36
  Structure of the Transition Zone, 37
  Strength of the Transition Zone, 37
  Influence of the Transition Zone on Properties of Concrete, 39
TEST YOUR KNOWLEDGE, 41
SUGGESTIONS FOR FURTHER STUDY, 42

CHAPTER 3

Strength, 43

DEFINITION, 43
SIGNIFICANCE, 44
STRENGTH-POROSITY RELATIONSHIP, 44
FAILURE MODES IN CONCRETE, 46
COMPRESSIVE STRENGTH AND FACTORS AFFECTING IT, 47
  Characteristics and Proportions of Materials, 47
    Water/cement ratio, 47. Air entrainment, 48.
    Cement type, 50. Aggregate, 51. Mixing water, 53.
    Admixtures, 56.
  Curing Conditions, 56
    Time, 56. Humidity, 57. Temperature, 59.
  Testing Parameters, 60
    Specimen parameters, 60. Loading conditions, 61.
BEHAVIOR OF CONCRETE UNDER VARIOUS STRESS STATES, 62
  Behavior of Concrete under Uniaxial Compression, 62
  Behavior of Concrete under Uniaxial Tension, 66
    Testing methods for tensile strength, 66.
  Relationship between the Compressive and Tensile Strengths, 69
  Tensile Strength of Mass Concrete, 71
  Behavior of Concrete under Shearing Stress, 72
  Behavior of Concrete under Biaxial and Multiaxial Stresses, 73
TEST YOUR KNOWLEDGE, 76
SUGGESTIONS FOR FURTHER STUDY, 77

CHAPTER 4

Dimensional Stability, 78

TYPES OF DEFORMATIONS AND THEIR SIGNIFICANCE, 78
ELASTIC BEHAVIOR, 80
  Nonlinearity of the Stress–Strain Relationship, 80
  Types of Elastic Moduli, 82
  Determination of the Static Elastic Modulus, 84
  Poisson's Ratio, 85

Factors Affecting Modulus of Elasticity, *86*
    Aggregate, *86*. Cement paste matrix, *87*.
    Transition zone, *87*. Testing parameters, *88*.
DRYING SHRINKAGE AND CREEP, *89*
    Causes, *89*
    Effect of Loading and Humidity Conditions on Drying Shrinkage and Viscoelastic Behavior, *90*
    Reversibility, *90*
    Factors Affecting Drying Shrinkage and Creep, *93*
        Materials and mix-proportions, *93*. Time and humidity, *97*.
        Geometry of the concrete element, *98*.
        Additional factors affecting creep, *100*.
THERMAL SHRINKAGE, *101*
    Factors Affecting Thermal Stresses, *103*
        Degree of restraint (K_r), *103*. Temperature change (ΔT), *103*. Heat losses, *107*.
THERMAL PROPERTIES OF CONCRETE, *108*
EXTENSIBILITY AND CRACKING, *109*
TEST YOUR KNOWLEDGE, *111*
SUGGESTIONS FOR FURTHER STUDY, *112*

CHAPTER 5

Durability, *113*

DEFINITION, *114*
SIGNIFICANCE, *114*
GENERAL OBSERVATIONS, *114*
WATER AS AN AGENT OF DETERIORATION, *115*
    Structure of Water, *116*
PERMEABILITY, *117*
    Permeability of Cement Paste, *118*
    Permeability of Aggregates, *119*
    Permeability of Concrete, *120*
CLASSIFICATION OF CAUSES OF CONCRETE DETERIORATION, *122*
DETERIORATION BY SURFACE WEAR, *123*
CRACKING BY CRYSTALLIZATION OF SALTS IN PORES, *126*
DETERIORATION BY FROST ACTION, *127*
    Frost Action on Hardened Cement Paste, *129*
    Frost Action on Aggregate, *132*
    Factors Controlling Frost Resistance of Concrete, *133*
        Air entrainment, *133*. Water/cement ratio and curing, *135*.
        Degree of saturation, *136*. Strength, *136*.
    Concrete Scaling, *136*
DETERIORATION BY FIRE, *138*
    Effect of High Temperature on Cement Paste, *139*
    Effect of High Temperature on Aggregate, *139*
    Effect of High Temperature on Concrete, *140*
DETERIORATION BY CHEMICAL REACTIONS, *140*
    Hydrolysis of Cement Paste Components, *142*
    Cation-Exchange Reactions, *144*
        Formation of soluble calcium salts, *144*.
        Formation of insoluble and nonexpansive calcium salts, *145*.
        Chemical attack by solutions containing magnesium salts, *145*.

REACTIONS INVOLVING FORMATION OF EXPANSIVE PRODUCTS, *146*
SULFATE ATTACK, *146*
    Chemical Reactions Involved in Sulfate Attack, *147*
    Selected Case Histories, *148*
    Control of Sulfate Attack, *152*
ALKALI-AGGREGATE REACTION, *154*
    Cements and Aggregate Types Contributing to the Reaction, *154*
    Mechanisms of Expansion, *155*
    Selected Case Histories, *157*
    Control of Expansion, *158*
HYDRATION OF CRYSTALLINE MgO AND CaO, *159*
CORROSION OF EMBEDDED STEEL IN CONCRETE, *160*
    Mechanisms Involved in Concrete Deterioration by Corrosion of Embedded Steel, *161*
    Selected Case Histories, *163*
    Control of Corrosion, *165*
CONCRETE IN SEAWATER, *167*
    Theoretical Aspects, *168*
    Case Histories of Deteriorated Concrete, *171*
    Lessons from Case Histories, *171*
TEST YOUR KNOWLEDGE, *176*
SUGGESTIONS FOR FURTHER STUDY, *177*

PART II
CONCRETE MATERIALS, MIX PROPORTIONING, AND EARLY-AGE PROPERTIES

CHAPTER 6

Hydraulic Cements, *179*

HYDRAULIC AND NONHYDRAULIC CEMENTS, *180*
    Definitions, and the Chemistry of Gypsum and Lime Cements, *180*
PORTLAND CEMENT, *180*
    Definition, *180*
    Manufacturing Process, *180*
    Chemical Composition, *182*
    Determination of Compound Composition from Chemical Analysis, *185*
    Crystal Structures and Reactivity of Compounds, *186*
        Calcium silicates, 187. Calcium aluminate and ferroaluminate, 187.
        Magnesium oxide and calcium oxide, 189.
        Alkali and sulfate compounds, 189.
    Fineness, *190*
HYDRATION OF PORTLAND CEMENT, *190*
    Significance, *190*
    Mechanism of Hydration, *191*
    Hydration of the Aluminates, *193*
    Hydration of the Silicates, *196*
HEAT OF HYDRATION, *198*
PHYSICAL ASPECTS OF THE SETTING AND HARDENING PROCESS, *199*
EFFECT OF CEMENT CHARACTERISTICS ON STRENGTH AND HEAT
  OF HYDRATION, *201*
TYPES OF PORTLAND CEMENT, *201*

Contents   xiii

SPECIAL HYDRAULIC CEMENTS, *205*
    Classification and Nomenclature, *205*
    Blended Portland Cements, *207*
        The pozzolanic reaction and its significance, 209.
        Heat of hydration, 212. Strength development, 212. Durability, 212.
    Expansive Cements, *214*
    Rapid Setting and Hardening Cements, *216*
        Very high early strength and high iron cements, 217.
    Oil-Well Cements, *217*
    White or Colored Cements, *219*
    Calcium Aluminate Cement, *220*
TEST YOUR KNOWLEDGE, *224*
SUGGESTIONS FOR FURTHER STUDY, *225*

CHAPTER 7

Aggregates, 226

SIGNIFICANCE, *226*
CLASSIFICATION AND NOMENCLATURE, *227*
NATURAL MINERAL AGGREGATES, *227*
    Description of Rocks, *228*
    Description of Minerals, *230*
        Silica minerals, 230. Silicate minerals, 230.
        Carbonate minerals, 231. Sulfide and sulfate minerals, 231.
LIGHTWEIGHT AGGREGATES, *231*
HEAVYWEIGHT AGGREGATES, *235*
BLAST-FURNACE SLAG AGGREGATES, *236*
AGGREGATE FROM FLY ASH, *237*
AGGREGATES FROM RECYCLED CONCRETE AND MUNICIPAL WASTES, *237*
PRODUCTION OF AGGREGATES, *238*
AGGREGATE CHARACTERISTICS AND THEIR SIGNIFICANCE, *241*
    Density and Apparent Specific Gravity, *242*
    Absorption and Surface Moisture, *242*
    Crushing Strength, Abrasion Resistance, and Elastic Modulus, *244*
    Soundness, *244*
    Size and Grading, *245*
    Shape and Surface Texture, *247*
    Deleterious Substances, *250*
METHODS OF TESTING AGGREGATE CHARACTERISTICS, *252*
TEST YOUR KNOWLEDGE, *253*
SUGGESTIONS FOR FURTHER STUDY, *254*

CHAPTER 8

Admixtures, 256

SIGNIFICANCE, *256*
NOMENCLATURE, SPECIFICATIONS, AND CLASSIFICATIONS, *257*
SURFACE-ACTIVE CHEMICALS, *259*
    Nomenclature and Composition, *259*
    Mechanism of Action, *259*
        Air-entraining surfactants, 259. Water-reducing surfactants, 260.

Applications, 262
　　Air-entraining admixtures, 262. Water-reducing admixtures, 262.
Superplasticizers, 263
SET-CONTROLLING CHEMICALS, 265
　Nomenclature and Composition, 265
　Mechanism of Action, 266
　Applications, 268
　　Accelerating admixtures, 268. Retarding admixtures, 268.
MINERAL ADMIXTURES, 271
　Significance, 271
　Classification, 272
　Natural Materials, 272
　　Volcanic glasses, 274. Volcanic tuffs, 276.
　　Calcined clays or shales, 276. Diatomaceous earth, 276.
　By-Product Materials, 276
　　Fly ash, 277. Blast-furnace slag, 278. Condensed silica fume, 279.
　　Rice husk ash, 280.
　Applications, 281
　　Improvement in workability, 281. Durability to thermal cracking, 281.
　　Durability to chemical attacks, 282.
　　Production of high-strength concrete, 284.
CONCLUDING REMARKS, 285
TEST YOUR KNOWLEDGE, 287
SUGGESTIONS FOR FURTHER STUDY, 288

CHAPTER 9
Proportioning Concrete Mixtures, 290

SIGNIFICANCE AND OBJECTIVES, 290
GENERAL CONSIDERATIONS, 291
　Cost, 292
　Workability, 293
　Strength and Durability, 293
　Ideal Aggregate Grading, 294
　Specific Principles, 294
　　Workability, 294. Strength, 295. Durability, 296.
PROCEDURES, 296
SAMPLE COMPUTATIONS, 302
APPENDIX: METHODS OF DETERMINING AVERAGE COMPRESSIVE
　STRENGTH FROM THE SPECIFIED STRENGTH, 305
TEST YOUR KNOWLEDGE, 307
SUGGESTIONS FOR FURTHER STUDY, 308

CHAPTER 10
Concrete at Early Ages, 309

DEFINITIONS AND SIGNIFICANCE, 310
BATCHING, MIXING, AND CONVEYING, 311
PLACING, COMPACTING, AND FINISHING, 317
CONCRETE CURING AND FORMWORK REMOVAL, 320
WORKABILITY, 322
　Definition and Significance, 322

Measurement, *322*
    Slump test, 323. Vebe test, 323. Compacting factor test, 323.
Factors Affecting Workability and Their Control, *324*
    Water content, 325. Cement content, 325.
    Aggregate characteristics, 326. Admixtures, 326.

SLUMP LOSS, *326*
    Definitions, *326*
    Significance, *327*
    Causes and Control, *328*

SEGREGATION AND BLEEDING, *330*
    Definitions and Significance, *330*
    Measurement, *331*
    Causes and Control, *332*

EARLY VOLUME CHANGES, *332*
    Definitions and Significance, *332*
    Causes and Control, *333*

SETTING TIME, *334*
    Definitions and Significance, *334*
    Measurement and Control, *335*

TEMPERATURE OF CONCRETE, *337*
    Significance, *337*
    Cold-Weather Concreting, *337*
        The maturity method, 339. Control of concrete temperature, 339.
    Hot-Weather Concreting, *340*
        Control of concrete temperature, 341.

TESTING AND CONTROL OF CONCRETE QUALITY, *344*
    Methods and Their Significance, *344*
    Accelerated Strength Testing, *345*
        Procedure A (warm-water method), 345.
        Procedure B (boiling-water method), 345.
        Procedure C (autogenous method), 345.
    In Situ and Nondestructive Testing, *346*
        Surface hardness methods, 347. Penetration resistance techniques, 347.
        Pullout tests, 347. Ultrasonic pulse velocity method, 348.
        Maturity meters, 348.
        Methods for assessing properties other than strength, 348.
    Core Tests, *348*
    Quality Control Charts, *349*

EARLY AGE CRACKING IN CONCRETE, *350*
CONCLUDING REMARKS, *354*
TEST YOUR KNOWLEDGE, *355*
SUGGESTIONS FOR FURTHER STUDY, *356*

PART III
RECENT ADVANCES AND CONCRETE IN THE FUTURE

CHAPTER 11

Progress in Concrete Technology, *357*

STRUCTURAL LIGHTWEIGHT CONCRETE, *358*
    Definitions and Specifications, *358*
    Mix-Proportioning Criteria, *359*

Properties, *361*
  Workability, *361*. Unit weight, *361*. Strength, *361*.
  Dimensional stability, *363*. Durability, *363*.
Applications, *365*

HIGH-STRENGTH CONCRETE, *367*
Definition, *367*
Significance, *367*
Materials and Mix Proportions, *369*
  General considerations, *369*. Typical concrete mixtures, *371*.
  Superplasticized concrete, *372*.
Properties, *374*
  Workability, *374*. Strength, *374*.
  Microstructure, stress-strain relation, fracture, drying shrinkage, and creep, *374*.
  Durability, *375*.
Applications, *377*

HIGH-WORKABILITY CONCRETE, *381*
Definition and Significance, *381*
Mix Proportions, *384*
Important Properties, *385*
Applications, *387*

SHRINKAGE-COMPENSATING CONCRETE, *392*
Definition and the Concept, *392*
Significance, *392*
Materials and Mix Proportions, *394*
Properties, *395*
  Workability, *395*. Slump loss, *395*. Plastic shrinkage, *396*.
  Strength, *396*. Volume changes, *396*. Durability, *398*.
Applications, *398*

FIBER-REINFORCED CONCRETE, *404*
Definition and Significance, *404*
Toughening Mechanism, *405*
Materials and Mix Proportioning, *407*
  Fibers, *407*. General considerations, *408*.
  Typical concrete mixtures, *410*.
Properties, *412*
  Workability, *412*. Strength, *412*. Toughness and impact resistance, *415*.
  Elastic modulus, creep, and drying shrinkage, *415*. Durability, *415*.
Applications, *416*

CONCRETES CONTAINING POLYMERS, *418*
Nomenclature and Significance, *418*
Polymer Concrete, *419*
Latex-Modified Concrete, *422*
Polymer-Impregnated Concrete, *423*

HEAVYWEIGHT CONCRETE FOR RADIATION SHIELDING, *425*
Significance, *425*
Concrete as a Shielding Material, *426*
Materials and Mix Proportions, *427*
Important Properties, *427*

MASS CONCRETE, *428*
Definition and Significance, *428*
General Considerations, *429*
Materials and Mix Proportions, *429*
  Cement, *429*. Admixtures, *430*. Aggregate, *430*. Mix design, *432*.
  Construction practices for controlling temperature rise, *434*.

Application of the Principles, 436
Roller-Compacted Concrete, 439
   Concept and significance, 439.
   Materials, mix proportions, and properties, 440.
TEST YOUR KNOWLEDGE, 442
SUGGESTIONS FOR FURTHER STUDY, 443

CHAPTER 12

Advances in Concrete Mechanics, 445

ELASTIC BEHAVIOR, 446
   Hashin-Shtrikman (H-S) Bonds, 453
   Transport Properties, 454
VISCOELASTICITY, 454
   Basic Rheological Models, 456
   Generalized Rheological Models, 466
   Time-Variable Rheological Models, 469
   Superposition Principle and Integral Representation, 471
   Mathematical Expressions for Creep, 473
   Methods for Predicting Creep and Shrinkage, 475
      CEB 1990, 476.
   Shrinkage, 477
      CEB 1978, 478. ACI 209, 479. Bazant-Panula Method, 480.
TEMPERATURE DISTRIBUTION IN MASS CONCRETE, 481
   Heat Transfer Analysis, 482
   Initial Condition, 484
   Boundary Conditions, 484
      I. Prescribed temperature boundary, 484. II. Prescribed heat flow boundary, 484.
      III. Convection boundary condition, 485. IV. Radiation boundary condition, 485.
   Finite Element Formulation, 485
   Examples of Application, 488
FRACTURE MECHANICS, 494
   Linear Elastic Fracture Mechanics, 494
   Finite Elements for Cracking Problems, 500
   Concrete Fracture Mechanics, 505
   Fracture Process Zone, 509
   Fictitious Crack Model, 511
   Fracture Resistance Curves, 517
   Two-Parameter Fracture Model, 519
   Size Effect, 519
TEST YOUR KNOWLEDGE, 522
SUGGESTIONS FOR FURTHER STUDY, 523

CHAPTER 13

The Future of Concrete, 525

FUTURE DEMAND FOR STRUCTURAL MATERIALS, 525
FUTURE SUPPLY OF CONCRETE, 527
ADVANTAGES OF CONCRETE OVER STEEL STRUCTURES, 528
   Engineering Considerations, 528
      Maintenance, 528. Fire resistance, 528. Resistance to cyclic loading, 529.

Vibration damping, 529. Control of deflections, 529.
Explosion resistance, 529. Resistance to cryogenic temperatures, 529.
Economic Considerations, *530*
Energy Considerations, *530*
Ecological Considerations, *532*
A BETTER PRODUCT IN THE FUTURE, *534*
CONCLUDING REMARKS, *536*

Index, 537

Foreword to the First Edition

Professor Mehta has presented the subject of concrete in a remarkably clear and logical manner. Actually, he has adopted a rather revolutionary approach, rejecting the dry and pedantic presentations of past texts, in order to address concrete as a living material, both in itself and in its application to structures and facilities built to serve society.

While this book accurately reflects the latest scientific advances in concrete structure and technology, it recognizes that working with concrete is an "art." Thus he has structured the book's arrangement and presentation from the point of view of the professional engineer charged with designing and building facilities of concrete.

He introduces not only the latest understanding of this complex material but the new and exciting techniques that enable dramatic improvements in the properties and performance of concrete.

The book is written primarily as an introductory text for Civil Engineering undergraduate students, but graduate students and professionals alike will find it useful for its lucid explanations and comprehensive treatment of the many interactive aspects.

Ben C. Gerwick, Jr.
Professor of Civil Engineering
University of California, Berkeley

Preface to the First Edition

Portland cement concrete is presently the most widely used manufactured material. Judging from world trends, the future of concrete looks even brighter because for most purposes it offers suitable engineering properties at low cost, combined with energy-saving and ecological benefits. It is therefore desirable that engineers know more about concrete than about other building materials.

There are several difficulties in preparing a scientific treatise on concrete as a material. First, in spite of concrete's apparent simplicity, it has a highly complex structure; therefore, the structure-property relations that are generally so helpful in the understanding and control of material properties cannot be easily applied. Concrete contains a heterogeneous distribution of many solid components as well as pores of varying shapes and sizes which may be completely or partially filled with alkaline solutions. Analytical methods of material science and solid mechanics which work well with manufactured materials that are relatively homogeneous and far less complex, such as steel, plastics, and ceramics, do not seem to be very effective with concrete.

Second, compared to other materials, the structure of concrete is not a static property of the material. This is because two of the three distinctly different components of the structure—the bulk cement paste and the transition zone between the aggregate and the bulk cement paste—continue to change with time. In this respect, concrete resembles wood and other living systems. In fact, the word *concrete* comes from the Latin term *concretus*, which means to grow. Strength and some other properties of concrete depend on the cement hydration products, which continue to form for several years. Although the products are relatively insoluble, they can slowly dissolve and recrystallize in moist environments, thus imparting to concrete the ability to heal microcracks.

Third, unlike other materials which are delivered in a ready-to-use form, concrete often has to be manufactured just before use at or near the job site. Typically, a book on concrete begins with a detailed account of the composition and properties of concrete-making materials, e.g., cements, aggregates, and admixtures.

This is followed by descriptions of methods for mix proportioning; equipment for batching, mixing, and transporting; and the technology of compacting, finishing, and curing concrete. The properties of concrete as a material and the principles governing them appear much later in the book, and are usually lost in a maze of nonscientific information, such as test methods, specifications, and applications.

This book is not intended to be an exhaustive treatise on concrete. Written primarily for the use of undergraduate students in civil engineering, it is proposed to present the art and science of concrete in a *simple, clear, and scientific manner*. The term *scientific manner* does not imply an emphasis on theoretical physics, chemistry, or mathematics. Because of the highly complex and dynamic nature of the material, theoretical models have produced only "theoretical concretes," and have proven to be of little value in practice. In fact, there is a popular joke in the concrete industry: What is abstract cannot be concrete. Most of our knowledge of the properties of concrete and the factors affecting it which forms the basis for current codes of concrete practice comes not from theoretical studies, but from laboratory and field experience. This experience provides adequate explanations for the properties of concrete and *how* and *why* they are influenced by various factors. By a scientific treatment of the subject, therefore, the author means that, as far as possible, structure–property relations are emphasized; that is, in addition to a presentation of the state of the art, rational explanations are provided for the observed behavior.

In regard to the organization of the subject matter, the author has taken a somewhat different than traditional approach. In many countries, since most of the concrete is ready-mixed and since the ready-mixed concrete industry has increasingly assumed the responsibility of selecting concrete-making materials and mix proportions, it is not essential to emphasize these topics in the beginning of a book. Most civil engineers involved with design, construction, and analysis of concrete structure are interested primarily in the properties of hardened concrete. The *first part* of this three-part book is therefore devoted to the properties of hardened concrete: for example, strength, elastic modulus, drying shrinkage, thermal shrinkage, creep, tensile strain capacity, permeability, and durability to physical and chemical processes of degradation. Definition of terms, the significance and origin of each property, and controlling factors are set forth in a clear and concise manner.

The *second part* of the book deals with the production of concrete. Separate chapters contain current information on the composition and properties of commonly used cements, aggregates, and admixtures. One chapter is devoted to the principles underlying the proportioning of concrete mixtures; another describes the properties of concrete at an early age and how they influence the operations to which freshly produced concrete is subjected. The latter also includes a brief discussion of quality assurance programs, such as accelerated tests, in situ tests, and statistical control charts.

In the *third part* of the book, advances in concrete technology resulting from innovations to adapt the material for special engineering applications are described. Current information on composition, properties, and applications of several types of special concrete is provided, including structural lightweight concrete, heavyweight concrete for nuclear shielding, high-strength concrete, high-workability concrete, shrinkage-compensating concrete, fiber-reinforced concrete, concretes containing polymers, and mass concrete. The final chapter includes some reflections on the future of concrete as a building material. These reflections are based on engineering properties, cost economy, energy savings, and ecological considerations.

Many unique diagrams, photographs, and summary tables are included to serve as teaching aids. New terms are indicated in boldface type and are defined when they appear first in the text. In the beginning of each chapter, a preview is given; at the end a self-test and a guide to further reading are provided. When the book is to be used as a text in a course in civil engineering materials, depending on the level at which instruction is being offered, individual instructors may wish to omit a part of the material from some chapters (e.g., chapters on cements and admixtures may be judged as too comprehensive for an undergraduate course), or supplement the others with additional reading (e.g., chapters on strength and dimensional changes may be judged as too elementary for a graduate course).

The field of concrete is vast and human effort is never perfect. Therefore, the readers may find shortcomings in this book. This author is conscious of some of the omissions. For instance, a large amount of excellent literature on concrete which comes from outside the United States has not been included in the list of references, in part because the author is not very familiar with these publications. Also limitations of space were a major constraint. It is hoped that this deficiency can be made up by referring to the books and reports that are listed for further study at the end of every chapter. Again, several important subjects are not covered. It is a good idea for civil engineers to know about architectural concrete, repair and maintenance of concrete structures, and methods of testing concrete-making materials (cement, aggregate, and admixtures). Regrettably, in a book of this size it was not possible to include all the material that was considered useful.

In the era of computers it might have been desirable to give more space to mathematical concepts developed for predicting the properties of concrete: for example, drying shrinkage, creep, cracking, and durability. Some of the work reported in the published literature is intellectually stimulating and indeed should be used for deeper and advanced study. On the other hand, a lot of the work is based on questionable assumptions about the microstructure of the material and is therefore of limited value. The author's failure to distinguish between the significant and the insignificant in this area of endeavor is largely responsible for its exclusion from the book. It is hoped that individual instructors and students can make up this deficiency on their own.

The author, however, would like to add a word of caution. Since concrete tends to behave like *living systems*, it cannot be left solely to mechanistic treatments. The nature of the material is such that as a whole it is different from the sum of its parts. Therefore, the properties of the material are destroyed when it is dissected into isolated elements, either physically or theoretically. In his book *The Turning Point*, F. Capra, commenting on the systems view of living systems, says that the reductionist description of organisms can be useful and may in some cases be necessary, but it is dangerous when taken to be the complete explanation. Several thousand years ago, the same view was expressed in Srimad Bhagvad Gita:

> That knowledge which clings to one single effect as if it were the whole, without reason, without foundation in truth, is narrow and therefore trivial.

My advice to students who will be tomorrow's engineers: In regard to models, mathematical abstractions, and computer programs developed to predict the properties of concrete, by all means keep an open mind. But never forget that like the human world, the world of concrete is nonlinear and has discontinuities within the nonlinearities. Therefore, empirical observations from laboratory and field experience will have to continue to supplement the theory.

P. Kumar Mehta
University of California, Berkeley

27 August, 1985

Preface to the Second Edition

Due to favorable reaction from the readers, especially from the education community, the key features of the first edition are retained in this revised edition.

In the *first part* of the three-part book, the microstructure and properties of hardened concrete are described. Only minor revisions are made to the chapters on microstructures (Chapter 2), strength (Chapter 3), and durability of concrete (Chapter 5). However, the chapter on dimensional stability (Chapter 4) has been rewritten to clarify further the viscoelastic behavior of concrete and to include a comprehensive treatment of thermal shrinkage and stresses, which are usually responsible for cracking in structures that are more than one meter thick.

The *second part* of this book, Chapters 6–10, deals with concrete production. Separate chapters contain state of the art reports on the composition and properties of concrete-making materials, namely cement, aggregates, and admixtures, followed by chapters on mix proportioning and early age properties of concrete. Again, only minor revisions have been necessary to the chapters in this section, except that in Chapter 10 a description of plastic settlement cracks and crazing is included in a section on general review of cracks in concrete.

The *third part* of the book, which contains a significant portion of the new material, should be of considerable interest to advanced students in modern concrete technology and concrete behavior. Chapter 11 contains up-to-date information on the technology of structural light-weight concrete, heavy-weight concrete, high-strength, flowing concrete, shrinkage-compensating concrete, fiber-reinforced concrete, mass concrete, and roller-compacted concrete.

A new chapter (Chapter 12) has been added to describe advances in concrete mechanics. In this chapter, theory of composite materials is used for modeling the elastic and the viscoelastic properties of concrete. The modern approach to composite modeling is based on the use of tensor notation which is a bit intimidating for students who may not have been exposed to tensor calculus. The authors decided to sacrifice elegance for simplicity by avoiding the use of tensor notation in this chapter. An effort is made to integrate rheological models and fundamental viscoelastic principles in order to develop methods for predicting drying shrinkage and creep. Advanced treatment of thermal shrinkage and thermal stresses is yet another key feature of this chapter. A finite element method is described for computing the temperature distribution in a mass concrete structure. To illustrate the application of the method, a number of examples simulating concrete construction are presented. Another subject covered by this chapter is the fracture mechanics of concrete, which is not yet a fully mature field but has been sufficiently developed to provide significant insights into size effect and crack propagation problems. A summary of linear elastic fracture mechanics, with its limitations, is presented, and some nonlinear fracture mechanics models are described. The last chapter (Chapter 13) on the future of concrete as a building material, presents an evaluation of commonly used construction materials from the standpoint of engineering properties, energy requirements, and ecological considerations.

Paulo Monteiro wishes to thank Rubens Bittencourt and Jose Thomas for examples of thermal analysis in Chapter 12. Others who reviewed portions of the material for this chapter and gave helpful suggestions include L. Biolzi, G. J. Creuss, M. Ferrari, P. Helene, J. Lubliner, R. Piltner, P. Papadopoulos, S. P. Shah, V. Souza Lima, and R. Zimmerman. Special thanks are due to Christine Human and Anne Robertson for their insightful comments.

Self-tests and references at the end of each chapter have been updated. The use of both U.S. customary (English units) and S.I. units is retained. However, as far as possible, conversions from the U.S. customary to S.I. units are provided in the second edition. Finally, as with the First Edition, the authors have made every effort in the Second Edition to uphold the goal of presenting the art and science of concrete in a simple, clear and scientific manner.

P. Kumar Mehta
Paulo Monteiro
University of California at Berkeley

Acknowledgments

For the information presented in this book, the authors are indebted to researchers from Vicat (early 19th century) and Powers (1940s), as well as colleagues and students at the University of California at Berkeley and at other institutions throughout the world. Since the list of individual names would be too long to present here, suffice it to say that the text represents the collective contribution of numerous minds, the authors' contribution being limited to the selection and organization of the subject matter. In handling the issue of individual acknowledgments in this unconventional manner, we take comfort in the thought that the real reward for those who produced the original information contained in this book, who furnished photographs and other materials for use, or who reviewed parts of the manuscript lies in making the study of concrete science a meaningful, educational experience for the reader.

Special recognition must be given to several organizations who have generously permitted the authors to reproduce material from their publications. Since these organizations continue to render invaluable service to the concrete profession, for the benefit of the reader their names and addresses are listed on the next page.

Organizations concerned with concrete are usually referred to by their initials. Listed below are initial designations, followed by the full names and addresses of these organizations.

ACI
American Concrete Institute
P.O. Box 19150
Redford Station
Detroit, Michigan 48219

ASTM
American Society for Testing and Materials
1916 Race Street
Philadelphia, PA 19103

BRE
Building Research Establishment
Garston, Watford WD2 7JR
United Kingdom

CANMET
Canada Center for Mineral Energy and Technology
405 Rochester Street
Ottawa, Ontario, KIA OG1
Canada

BCA
British Cement Association
Wexham Springs, Slough SL3 6PL
United Kingdom

CEB
Comite Euro-International du Beton
EPFL—Case Postale 88-CH
1015 Lausanne, Switzerland

CE
Corps of Engineers, U.S.
Waterways Experiment Station
Vicksburg, Mississippi 39180

NRMCA
National Ready Mixed Concrete Association
900 Spring Street
Silver Spring, Maryland 20910

PCA
Portland Cement Association
5420 Old Orchard Road
Skokie, Illinois 60076

RILEM
International Union of Testing and Research Laboratories for Materials and Structures
12, rue Brancion, 75015
Paris, France

USBR
U.S. Bureau of Reclamation
P.O. Box 25007
Denver Federal Center
Denver, Colorado 80225

CHAPTER 1

Introduction to Concrete

PREVIEW

In this chapter important applications of concrete are described, and the reasons that concrete is the most widely used structural material in the world today are examined. The principal components of modern concrete are identified and defined. A brief description of the major concrete types is given.

For the benefit of beginning students, an introduction to important properties of engineering materials, with special reference to concrete, is also included in this chapter. The properties discussed are strength, elastic modulus, toughness, dimensional stability, and durability. Finally, an introduction to the International System of Units (SI units) and multiplication factors for conversion of U.S. customary or English units to SI units are given.

CONCRETE AS A STRUCTURAL MATERIAL

In an article published by the *Scientific American* in April 1964, S. Brunauer and L. E. Copeland, two eminent scientists in the field of cement and concrete, wrote:

> The most widely used construction material is concrete, commonly made by mixing portland cement with sand, crushed rock, and water. Last year in the U.S. 63 million tons of portland cement were converted into 500 million tons of concrete, five times the consumption by weight of steel. In many countries the ratio of concrete consumption to steel consumption exceeds ten to one. The total world consumption of concrete last year is estimated at three billion tons, or one ton for every living human being. Man consumes no material except water in such tremendous quantities.

Today the rate at which concrete is used is not much different than it was 30 years ago. It is estimated that the present consumption of concrete in the world is of the order of 5.5 billion tons every year.

Concrete is neither as strong nor as tough as steel, *so why is it the most widely used engineering material?* There are a number of reasons. First, concrete[1] possesses *excellent resistance to water.* Unlike wood and ordinary steel, the ability of concrete to withstand the action of water without serious deterioration makes it an ideal material for building structures to control, store, and transport water. In fact, some of the earliest known applications of the material consisted of aqueducts and waterfront retaining walls constructed by the Romans. The use of concrete in dams, canals, water pipes, and storage tanks is now a common sight almost everywhere in the world (Figs. 1–1 to 1–4). The durability of concrete to some aggressive waters

Figure 1-1 Central Arizona project pipeline. (Photograph courtesy of Ameron Pipe Division.)

The largest circular precast concrete structure ever built for the transportation of water is part of the Central Arizona Project—a $1.2 billion U.S. Bureau of Reclamation development, which will provide water from the Colorado River for agricultural, industrial, and municipal use in Arizona, including the metropolitan areas of Phoenix and Tucson. The system contains 1560 pipe sections, each 22 ft (6.7 m) long, 24-1/2 ft (7.5 m) outside diameter (equivalent to the height of a two-story building), 21 ft (6.4 m) inside diameter, and weighing up to 225 tons.

---

[1] In this book, the term **concrete** refers to portland cement concrete, unless stated otherwise.

Concrete as a Structural Material

Figure 1-2 Itaipu Dam, Brazil. (Photograph courtesy of Promon Eng., Brazil.)

This spectacular 12,600-MW hydroelectric project at Itaipu, estimated to cost $18.5 billion, includes a 180-m-high hollow-gravity concrete dam at the Paraná River on the Brazil-Paraguay border. By 1982 twelve types of concrete, totaling 12.5 million cubic meters, had been used for the construction of the dam, piers of diversion structure, and precast beams, slabs, and other structural concrete elements for the power plant.

The designed compressive strengths of concrete ranged from as low as 14 MPa at 1 year for mass concrete for the dam to as high as 35 MPa at 28 days for precast concrete members. All coarse aggregate and about 70 percent of the fine aggregate was obtained by crushing basalt rock available at the site. The coarse aggregates were separately stockpiled into gradations of 6 in. (150 mm), 3 in. (75 mm), 1-1/2 in. (38 mm), and 3/4 in. (19 mm) maximum size aggregate. A combination of several aggregates containing different size fractions was necessary to reduce the void content and therefore the cement content of the mass concrete mixtures. As a result, the cement content of mass concrete was limited to as low as 108 kg/m³, and the adiabatic temperature rise to 19°C at 28 days. Furthermore, to prevent thermal cracking, it was specified that the temperature of freshly cooled concrete would be limited to 7°C by precooling the constituent materials.

---

is responsible for the fact that its use has been extended to many hostile industrial and natural environments (Fig. 1–5).

Structural elements exposed to moisture, such as piles, foundations, footings, floors, beams, columns, roofs, exterior walls, and pavements, are frequently built with concrete, which is reinforced with steel. **Reinforced concrete**[2] is a concrete

---

[2] It should be noted that the design and behavior of both reinforced and prestressed concrete structures are beyond the scope of this book.

Figure 1–3 California aqueduct. (Photograph courtesy of the State of California, Department of Water Resources.)

In California, about three-fourths of the fresh water in the form of rain and snowfall is found in the northern one-third of the state; however, three-fourths of the total water is needed in the lower two-thirds, where some major centers of population, industry, and agriculture are located. Therefore, in the 1960s, at an estimated cost of $4 billion, California undertook to build a water system capable of handling 4.23 million acre-feet of water annually. Eventually extending more than 600 miles (900 km) from north to south to provide supplemental water, flood control, hydroelectric power, and recreational facilities, this project called for the construction of 23 dams and reservoirs, 22 pumping plants, 473 miles of canal (California Aqueduct), 175 miles of pipelines, and 20 miles of tunnels.

An awesome task before the project was to transport water from an elevation near the seafloor in the San Joaquin Delta across the Tehachapi mountains over to the Los Angeles metropolitan area. This is accomplished by pumping the large body of water in a single 1926-ft (587 m) lift. At its ultimate capacity, this pumping plant will consume nearly 6 billion kilowatt-hours a year.

Approximately 4 million cubic yards (3 million m^3) of concrete is used for the construction of tunnels, pipelines, pumping plants, and canal lining. One of the early design decisions for the California Aqueduct was to build a concrete rather than compacted-earth lined canal, because concrete-lined canals have relatively lower head loss, pumping and maintenance cost, and seepage loss. Depending on the side slope of the canal section, 2- to 4-in (50- to 100-mm)-thick unreinforced concrete lining is provided. Concrete, containing 380 to 400 lb/yd^3 (225 to 237 kg/m^3) portland cement and 70 lb/yd^3 (42 kg/m^3) pozzolan, showed 2000 psi (14 MPa), 3500 psi (24 MPa), and 4500 psi (31 MPa) compressive strength in test cylinders cured for 7, 28, and 91 days, respectively. Adequate speed of construction of the concrete lining was assured by the slip-forming operation shown in the photograph.

Figure 1–4 Country club reservoir for water storage. (Photograph courtesy of East Bay Municipal Utility District, Oakland, California.)

This buried prestressed concrete reservoir for water storage, located in the Oakland-Berkeley hills, was constructed in 1976–1977 and has 2.5 million gallons ($132 \times 10^3 m^3$) capacity. It is 121 ft (37 m) in diameter, 30 ft (9 m) high, and has a 10- to 14-in.-thick (250- to 350-mm) wall.

usually containing steel bars, which is designed on the assumption that the two materials act together in resisting forces. **Prestressed concrete** is a concrete in which by tensioning steel tendons, prestress of such magnitude and distribution is introduced that the tensile stresses resulting from the service loads are counteracted to a desired degree. It is believed that a large amount of concrete finds its way to reinforced or prestressed concrete elements.

The second reason for the widespread use of concrete is the *ease with which structural concrete elements can be formed into a variety of shapes and sizes* (Figs. 1–6 to 1–8). This is because freshly made concrete is of a plastic consistency, which permits the material to flow into prefabricated formwork. After a number of hours, the formwork can be removed for reuse when the concrete has solidified and hardened to a strong mass.

The third reason for the popularity of concrete with engineers is that it is usually the *cheapest and most readily available material on the job*. The principal ingredients for making concrete—portland cement and aggregates—are relatively inexpensive and are more commonly available in most areas of the world. Although in certain geographical locations the cost of concrete may be as high as $80 per ton, at others it is as low as $20 per ton, which amounts to only 1 cent per pound.

Compared to most other engineering materials, the production of concrete requires considerably less energy input. In addition, large amounts of many industrial wastes can be recycled as a substitute for the cementitious material or aggregates

Figure 1-5 Statfjord B offshore concrete platform, Norway. (Photograph courtesy of Norwegian Contractors, Inc.)

Since 1971, 15 concrete platforms requiring about 1.3 million cubic meters of concrete have been installed in the British and the Norwegian sectors of the North Sea. Statfjord B, the largest concrete platform, built in 1981, has a base area of 18,000 m^2, 24 oil storage cells with about 2 million barrels of storage capacity, four prestressed concrete shafts between the storage cells and the deck frame, and 42 drilling slots on the deck. The structure was built and assembled at a dry dock in Stavanger; then the entire assembly, weighing 40,000 tons, was towed to the site of the oil well, where it was submerged to a water depth of about 145 m.

The prestressed and heavily reinforced concrete elements of the structure are exposed to the corrosive action of seawater and are designed to withstand 31-m-high waves. Therefore, the selection and proportioning of materials for the concrete mixture was governed primarily by consideration of the speed of construction by slip-forming and the durability of hardened concrete to the hostile environment. A free-flowing concrete mixture (220-mm slump), containing 380 kg of finely ground portland cement, 20 mm of maximum-size coarse aggregate, an 0.42 water-cement ratio, and a superplasticizing admixture was found satisfactory for the job. The tapered shafts under the slip-forming operation are shown in the Figure.

in concrete. Therefore, in the future, ***considerations of energy and resource conservation*** are likely to make the choice of concrete as a structural material even more attractive.

In his 1961 presidential address to the ACI convention, calling concrete a ***universal material*** and emphasizing that all engineers need to know more about concrete, J. W. Kelly said:

> One would not think of using wood for a dam, steel for pavement, or asphalt for a building frame, but concrete is used for each of these and for many other uses than other construction materials. Even where another material is the principal component of a structure, concrete is usually used with it for certain portions of the work. It is used to support, to enclose, to surface, and to fill. More people need to know more about concrete than about other specialized materials.

Concrete as a Structural Material

Figure 1-6 Fountain of Time: A sculpture in concrete. (Photograph courtesy of R. W. Steiger, *Concr. Construct.*, Vol. 29, pp. 797–802, September 1984. By permission of Concrete Construction Publications, Inc., 426 South Westgate, Addison, Illinois 60101.)

"Time goes, you say? Ah, no.
Alas, time stays; we go."

Concrete is an extraordinary material because it can be not only cast into a variety of complex shapes, but also given special surface effects. Aesthetically pleasing sculpture, murals, and architectural ornaments can be created by suitable choice of concrete-making materials, formwork, and texturing techniques. Fountain of Time is a massive 120- by 18- by 14-ft (36 by 5 by 4 m) work of art in concrete on the south side of the University of Chicago campus. The sculpture is a larger-than-life representation of 100 individual human figures, all cast in place in the exposed aggregate finish. In the words of Steiger, the central figure is Time the conqueror, seated on an armored horse and surrounded by young and old, soldiers, lovers, religious practitioners, and many more participants in the diversity of human life, finally embracing death with outstretched arms. Lorado Taft made the model for this sculpture in 1920 after 7 years of work. About the choice of concrete as a medium of art, the builder of the sculpture, John J. Earley, had this to say: "Concrete as an artistic medium becomes doubly interesting when we realize that in addition to its economy it possesses those properties which are the most desirable of both metal and stone. Metal is cast, it is an exact mechanical reproduction of the artist's work, as is concrete. . . . Stone (sculpture) is an interpretation of an original work and more often than not is carried out by another artist. But stone has the advantage of color and texture which enable it to fit easily into varied surroundings, a capability lacking in metal. Concrete, treated as in the Foundation of Time, presents a surface almost entirely of stone with all its visual advantages while at the same time offering the precision of casting that would otherwise only be attained in metal."

Figure 1-7 Candlestick Park Stadium, San Francisco, California. (Photograph courtesy of Interactive Resources, Inc., Structural Engineers, Point Richmond, California.)

Cast-in-place and precast concrete elements can be assembled to produce large structures of different shapes. The photograph shows Candlestick Park Stadium, San Francisco, California, which was constructed in 1958 with about 60,000 seating capacity. The roof canopy is supported by 24-ft (7.3 m) cantilevered precast concrete girders. Through a roof girder connection the cantilevered concrete member is supported by joining it to a cast-in-place concrete bleacher girder.

COMPONENTS OF MODERN CONCRETE

Although the composition and properties of the materials used for making concrete are discussed in Part II, at this stage it is useful to define concrete and the principal concrete-making components. The following definitions are based on ASTM C 125[3] (*Standard Definition of Terms Relating to Concrete and Concrete Aggregates*), and ACI Committee 116 (A Glossary of Terms in the Field of Cement and Concrete Technology):

Concrete is a composite material that consists essentially of a binding medium within which are embedded particles or fragments of aggregates. In hydraulic-cement concrete, the binder is formed from a mixture of hydraulic cement and water.

Aggregate is the granular material, such as sand, gravel, crushed stone, or iron blast-furnace slag, used with a cementing medium to form hydraulic-cement concrete or mortar. The term **coarse aggregate** refers to aggregate particles larger than 4.75 mm (No. 4 sieve), and the term **fine aggregate** refers to aggregate particles smaller than 4.75 mm but larger than 75 mm (No. 200 sieve). **Gravel** is the coarse aggregate resulting from natural disintegration and abrasion of rock or processing

---

[3] The ACI committee reports and the ASTM (American Society for Testing and Materials) standards are updated from time to time. The definitions given here are from the ASTM standard approved in 1982.

Figure 1-8 Baha'i Temple, Wilmette, Illinois. (Photograph courtesy of R. W. Steiger, Trimedia Studios, Farmington Hills, Michigan.)

The Baha'i Temple is an example of exceedingly beautiful, ornamental architecture that can be created in concrete. Describing the concrete materials and the temple, F. W. Cron (Concrete Construction, V. 28, No. 2, 1983) wrote: "The architect had wanted the building and especially the great dome, 90 ft. (27m) diameter, to be as white as possible, but not with a dull and chalky appearance. To achieve the desired effect Earley proposed an opaque white quartz found in South Carolina to reflect light from its broken face. This would be combined with a small amount of translucent quartz to provide brilliance and life. Puerto Rican sand and white portland cement were used to create a combination that reflected light and imparted a bright glow to the exposed-aggregate concrete surface.... On a visit to the Temple of Light one can marvel at its brilliance in sunlight. If one returns at night the lights from within and the floodlights that play on its surface turn the building into a shimmering jewel. The creativity of Louis Bourgeois and the superbly crafted concrete from the Earley Studios have acted in concert to produce this great performance."

of weakly bound conglomerate. The term **sand** is commonly used for fine aggregate resulting from natural disintegration and abrasion of rock or processing of friable sandstone. **Crushed stone** is the product resulting from industrial crushing of rocks, boulders, or large cobblestones. **Iron blast-furnace slag,** a by-product of the iron industry, is the material obtained by crushing blast-furnace slag that solidified under atmospheric conditions.

Mortar is a mixture of sand, cement, and water. It is essentially concrete without a coarse aggregate. **Grout** is a mixture of cementitious material and aggregate, usually fine aggregate, to which sufficient water is added to produce a pouring consistency without segregation of the constituents. **Shotcrete** refers to a mortar or concrete that is pneumatically transported through a hose and projected onto a surface at a high velocity.

Cement is a finely pulverized material which by itself is not a binder, but develops the binding property as a result of hydration (i.e., from chemical reactions between cement minerals and water). A cement is called **hydraulic** when the hydration products are stable in an aqueous environment. The most commonly used hydraulic cement for making concrete is **portland cement,** which consists essentially of hydraulic calcium silicates. The calcium silicate hydrates formed on the hydration of portland cement are primarily responsible for its adhesive characteristic, and are stable in aqueous environments.

The foregoing definition of concrete as a mixture of hydraulic cement, aggregates, and water does not include a fourth component, admixtures, which are almost always used in the modern practice. **Admixtures** are defined as materials other than aggregates, cement, and water, which are added to the concrete batch immediately before or during mixing. The use of admixtures in concrete is now widespread due to many benefits which are possible by their application. For instance, chemical admixtures can modify the setting and hardening characteristic of the cement paste by influencing the rate of cement hydration. Water-reducing admixtures can plasticize fresh concrete mixtures by reducing the surface tension of water, air-entraining admixtures can improve the durability of concrete exposed to cold weather, and mineral admixtures such as pozzolans (materials containing reactive silica) can reduce thermal cracking in mass concrete. A detailed description of the types of admixtures, their composition, and mechanism of action in concrete is given in Chapter 8.

TYPES OF CONCRETE

Based on unit weight, concrete can be classified into three broad categories. Concrete containing natural sand and gravel or crushed-rock aggregates, generally weighing about 2400 kg/m^3 (4000 lb/yd^3), is called **normal-weight concrete,** and is the most commonly used concrete for structural purposes. For applications where a higher strength-to-weight ratio is desired, it is possible to reduce the unit weight of concrete by using certain natural or pyro-processed aggregates having lower bulk density. The term **lightweight concrete** is used for concrete that weighs less than about 1800 kg/m^3 (3000 lb/yd^3). On the other hand, **heavyweight concrete,** used at times for radiation shielding, is concrete produced from high-density aggregates, and generally weighs more than 3200 kg/m^3 (5300 lb/yd^3).

Strength grading of concrete, which is prevalent in Europe and many other countries, is not practiced in the United States. However, from the standpoint of distinct differences in structure–property relationships, which will be discussed later, it is useful to divide concrete into three general categories based on compressive strength:

- **Low-strength concrete:** less than 20 MPa (3000 psi) compressive strength
- **Moderate-strength concrete:** 20 to 40 MPa (3000 to 6000 psi) compressive strength
- **High-strength concrete:** more than 40 MPa (6000 psi) compressive strength.

Moderate-strength concrete is ordinary or normal concrete, which is used for most structural work. High-strength concrete is used for special applications, described in Chapter 11.

Typical proportions of materials for producing low-strength, moderate-strength, and high-strength concretes with normal-weight aggregates are shown in Table 1–1. The relationships between the cement paste content and strength, and the water/cement ratio of the cement paste and strength should be noted from the data.

It is not possible here to list all concrete types. There are numerous modified concretes which are appropriately named: for example, fiber-reinforced concrete, expansive-cement concrete, and latex-modified concrete. The composition and properties of special concretes are described in Chapter 11.

PROPERTIES OF HARDENED CONCRETE AND THEIR SIGNIFICANCE

The selection of an engineering material for a particular application has to take into account its ability to withstand the applied force. Traditionally, the deformation occurring as a result of applied load is expressed as **strain,** which is defined as the change in length per unit length; the load is expressed as **stress,** which is defined as the force per unit area. Depending on how the stress is acting on the material, the stresses are further distinguished from each other: for example, compression, tension, flexure, shear, and torsion. The stress-strain relationships in materials are generally expressed in terms of strength, elastic modulus, ductility, and toughness.

Strength is a measure of the amount of stress required to fail a material. The working stress theory for concrete design considers concrete as mostly suitable for bearing compressive load; this is why it is the compressive strength of the material

TABLE 1–1 TYPICAL PROPORTIONS OF MATERIALS IN CONCRETES OF DIFFERENT STRENGTH

	Low-strength		Moderate-strength		High-strength	
	lb/yd^3	ft^3/yd^3	lb/yd^3	ft^3/yd^3	lb/yd^3	ft^3/yd^3
Cement	430	2.19	600	3.06	860	4.39
Water	300	4.80	300	4.80	300	4.80
Fine aggregate	1350	8.08	1430	8.56	1500	8.82
Coarse aggregate	1970	11.80	1740	10.42	1470	8.80
Cement paste proportion						
percent by mass		18		22.1		28.1
percent by volume		26		29.3		34.3
Water/cement						
ratio by mass		0.70		0.50		0.35
Strength[a]						
psi		2650		4520		9010
MPa		18		30		60

[a] In American practice, unless otherwise specified, the strength of concrete is measured by crushing 6- by 12-in. cylindrical specimens under compression after 28 days of standard curing (73.4 ± 3°F, 100% relative humidity).

that is generally specified. Since the strength of concrete is a function of the cement hydration process, which is relatively slow, traditionally the specifications and tests for concrete strength are based on specimens cured under standard temperature-humidity conditions for a period of 28 days. As stated earlier, for most structural applications, moderate strength concrete (20 to 40 MPa compressive strength) is used, although recently high-strength concretes of up to 130 MPa or 20,000 psi strength have been produced commercially.

It may be mentioned here that typically the tensile and flexural strengths of concrete are of the order of 10 and 15 percent, respectively, of the compressive strength. The reason for such a large difference between the tensile and compressive strength is attributed to the heterogeneous and complex structure of concrete.

For many engineering materials, such as steel, the observed stress-strain behavior when a specimen is subjected to incremental loads can be divided into two parts (Fig. 1–9). Initially, when the strain is proportional to the applied stress and is reversible upon unloading the specimen, it is called **elastic strain.** The **modulus of elasticity** is defined as the ratio between the stress and this reversible strain. In homogeneous materials, the elastic modulus is a measure of the interatomic bonding forces and is unaffected by microstructural changes. This is not true of the heterogeneous multiphase materials such as concrete. The elastic modulus of concrete in compression varies from 14×10^3 to 40×10^3 MPa (2×10^6 to 6×10^6 psi). The significance of the elastic limit in structural design lies in the fact that it represents the maximum allowable stress before the material undergoes permanent deformation. Therefore, the engineer must know the elastic modulus of the material because it influences the rigidity of a design.

At a high stress level (Fig. 1–9), the strain no longer remains proportional to the applied stress, and also becomes permanent (i.e., it will not be reversed if the specimen is unloaded). This strain is called **plastic** or **inelastic strain.** The amount of inelastic strain that can occur before failure is a measure of the **ductility** of the material. The energy required to break the material, the product of force times distance, is represented by the area under the stress–strain curve. The term **toughness** is used as a measure of this energy. The contrast between toughness and strength should be noted; the former is a measure of energy, whereas the latter is a measure of the stress required to fracture the material. Thus two materials may have identical strength but different values of toughness. In general, however, when the strength of a material goes up, the ductility and the toughness go down; also, very high-strength materials usually fail in a brittle manner (i.e., without undergoing any significant inelastic strain).

Although under compression concrete appears to show some inelastic strain before failure, typically the strain at fracture is of the order of 2000×10^{-6}, which is considerably lower than the failure strain in structural metals. For practical purposes, therefore, designers do not treat concrete as a ductile material and do not recommend it for use under conditions where it is subjected to heavy impact unless reinforced with steel.

From observation of the elastic–plastic behavior, concrete seems to be a

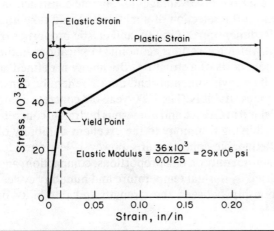

Figure 1–9 Stress–strain behavior of a steel specimen subjected to incremental loads.

complex material. Many characteristics of this composite material do not follow the laws of mixture of two components. For instance, under compressive loading both the aggregate and the hydrated cement paste, if separately tested, would fail elastically, whereas concrete itself shows inelastic behavior before fracture. Also, the strength of concrete is lower than the individual strengths of the two components. Such anomalies in the behavior of concrete will be explained on the basis of its structure, in which the transition zone between coarse aggregate and cement paste plays an important part.

The stress–strain behavior of materials shown in Fig. 1–9 is typical of specimens loaded to failure in a short time in the laboratory. For some materials the relationship between stress and strain is independent of the loading time; for others it is not. Concrete belongs to the latter category. If a concrete specimen is held for a long period under a constant stress which, for instance, is 50 percent of the ultimate strength of the material, it will exhibit plastic strain. The phenomenon of gradual increase in strain with time under a sustained stress is called **creep.** When creep in concrete is restrained, it is manifested as a progressive decrease of stress with time. The stress relief associated with creep has important implications for behavior of both plain and prestressed concretes.

Strains can arise even in unloaded concrete as a result of changes in the environmental humidity and temperature. Freshly formed concrete is moist; it undergoes **drying shrinkage** when exposed to the ambient humidity. Similarly, shrinkage strains result when hot concrete is cooled to the ambient temperature. Massive concrete elements can register considerable rise in temperature due to poor dissipation of heat evolved by cement hydration, and **thermal shrinkage** would occur on cooling of the hot concrete. Shrinkage strains are critical to concrete because, when restrained, they manifest themselves into tensile stresses. Since the tensile strength of concrete is low, concrete structures often crack as a result of restrained shrinkage, caused by either moisture or temperature changes. In fact, the cracking tendency of the material is one of the serious disadvantages in concrete construction.

Finally, professional judgment in the selection of a material should take into consideration not only the strength, dimensional stability, and elastic properties of the material, but also its durability, which has serious economic implications in the form of maintenance and replacement costs of a structure. **Durability** is defined as the service life of a material under given environmental conditions. Generally, dense or watertight concretes enjoy long-time durability. The 2700-year-old concrete lining of a water storage tank on Rodhos Island in Greece and numerous hydraulic concrete structures built by the Romans are a living testimony to the excellent durability of concrete in moist environments. Permeable concretes are, however, less durable. The permeability of concrete depends not only on mix proportions, compaction, and curing, but also on microcracks caused by normal temperature and humidity cycles. In general, there is a close relationship between the strength and durability of concrete.

UNITS OF MEASUREMENT

The metric system of measurement, which is prevalent in most countries of the world, uses millimeters, centimeters, and meters for length, grams and kilograms for mass, liters for volume, kilogram force per unit area for stress, and degrees Celsius for temperature. The United States is almost alone in the world in using English units of measurement such as inches, feet, and yards for length, pounds or tons for mass, gallons for volume, pounds per square inch (psi) for stress, and degrees Fahrenheit for temperature. Multinational activity in design and construction of large engineering projects is commonplace in the modern world. Therefore, it is becoming increasingly important that scientists and engineers throughout the world speak the same language of measurement.

The metric system is simpler than the English system and has recently been modernized in an effort to make it universally acceptable. The modern version of the metric system, called the **International System of Units** (Système International d'Unités), abbreviated **SI,** was approved in 1960 by 30 participating nations in the General Conference on Weights and Measures.

In *SI measurements,* meter and kilogram are the only units permitted for length and mass, respectively. A series of approved prefixes, shown in Table 1–2, are used for the formation of multiples and submultiples of various units. The force required to accelerate a mass of 1 kilogram by 1 meter is expressed as 1 **newton** (N), and a stress of 1 newton per square meter is expressed as 1 **pascal** (Pa). The ASTM Standard E 380-70 contains a comprehensive guide to the use of SI units.

In 1975, the U.S. Congress passed the Metric Conversion Act, which declares that it will be the policy of the United States to coordinate and plan the increasing use of the metric system of measurement (SI units). Meanwhile, a bilinguality in the units of measurement is being practiced so that engineers should become fully

TABLE 1–2 MULTIPLE AND SUBMULTIPLE SI UNITS AND SYMBOLS

Multiplication factor	Prefix	SI symbol
$1\,000\,000\,000 = 10^9$	giga	G
$1\,000\,000 = 10^6$	mega	M
$1\,000 = 10^3$	kilo	k
$100 = 10^2$	hecto[a]	h
$10 = 10^1$	deka[a]	da
$0.1 = 10^{-1}$	deci[a]	d
$0.01 = 10^{-2}$	centi[a]	c
$0.001 = 10^{-3}$	milli	m
$0.000\,001 = 10^{-6}$	micro	μ
$0.000\,000\,001 = 10^{-9}$	nano[b]	n

[a] Not recommended but occasionally used.
[b] 0.1 nanometer (nm) = 1 angstrom (Å) is a non-SI unit which is commonly used.

conversant with both systems. This is why, as far as possible, data in this book are presented in both units. To aid quick conversion from the U.S. customary or English units to SI units, a list of the commonly needed multiplication factors is given below.

To convert from:	To:	Multiply by:
yards (yd)	meters (m)	0.9144
feet (ft)	meters (m)	0.3048
inches (in.)	millimeter (mm)	25.4
cubic yards (yd^3)	cubic meters (m^3)	0.7646
U.S. gallons (gal)	cubic meters (m^3)	0.003785
U.S. gallons (gal)	liters	3.785
pounds, mass (lb)	kilograms (kg)	0.4536
U.S. tons (t)	tonnes (T)	0.9072
pounds/cubic yard (lb/yd^3)	kilograms/cubic meter (kg/m^3)	0.5933
kilogram force (kgf)	newtons (N)	9.807
pounds force (lbf)	newtons (N)	4.448
kips per square inch (ksi)	megapascal (MPa or N/mm^2)	6.895
degrees Fahrenheit (°F)	degrees Celsius (°C)	(°F − 32)/1.8

TEST YOUR KNOWLEDGE

1. Why is concrete the most widely used engineering material?
2. What are reinforced concrete and prestressed concrete?
3. Define the following terms: fine aggregate, coarse aggregate, gravel, grout, shotcrete, hydraulic cement.

4. What are the typical unit weights for normal-weight, lightweight, and heavyweight concretes? How would you define high-strength concrete?
5. What is the significance of elastic limit in structural design?
6. What is the difference between strength and toughness? Why is it the compressive strength of concrete at 28 days that is generally specified?
7. Discuss the significance of drying shrinkage, thermal shrinkage, and creep in concrete.
8. How would you define durability? In general, what concrete types are expected to show better long-time durability?

SUGGESTIONS FOR FURTHER STUDY

American Concrete Institute Committee Report by Committee 116, *Cement and Concrete Terminology,* SP-19, 1985.

American Society for Testing and Materials, *Annual Book of ASTM Standards,* Vol. 04.01 (Cement, Lime, and Gypsum), 1991.

American Society for Testing and Materials, *Annual Book of ASTM Standards,* Vol. 04.02 (Concrete and Mineral Aggregates), 1991.

Van Vlack, L. H., *Elements of Material Science and Engineering,* 6th ed., Addison-Wesley Publishing Company, Inc., Reading, Mass., 1989.

CHAPTER 2

The Structure of Concrete

PREVIEW

Structure–property relationships are at the heart of modern material science. Concrete has a highly heterogeneous and complex structure. Therefore, it is very difficult to constitute exact models of the concrete structure from which the behavior of the material can be reliably predicted. However, a knowledge of the structure and properties of the individual components of concrete and their relationship to each other is useful for exercising some control on the properties of the material. In this chapter three components of concrete structure—the hydrated cement paste, the aggregate, and the transition zone between the cement paste and the aggregate—are described. The structure–property relationships are discussed from the standpoint of selected characteristics of concrete, such as strength, dimensional stability, and durability.

DEFINITIONS

The type, amount, size, shape, and distribution of phases present in a solid constitute its **structure.** The gross elements of the structure of a material can readily be seen, whereas the finer elements are usually resolved with the help of a microscope. The term **macrostructure** is generally used for the gross structure, visible to the human eye. The limit of resolution of the unaided human eye is approximately one-fifth of a millimeter (200 μm). The term **microstructure** is used for the microscopically magnified portion of a macrostructure. The magnification capability of modern electron optical microscopes is of the order of 10^5 times; thus the application of transmission and scanning electron optical microscopy techniques has made it possible to resolve the structure of materials to a fraction of a micrometer.

SIGNIFICANCE

Progress in the field of materials has resulted primarily from recognition of the principle that the properties of a material originate from its internal structure; in other words, the properties can be modified by making suitable changes in the structure of a material. Although concrete is the most widely used structural material, its structure is heterogeneous and highly complex. The structure–property relationships in concrete are not yet well developed; however, an understanding of some of the elements of the concrete structure is essential before we discuss the factors influencing the important engineering properties of concrete, such as strength (Chapter 3), elasticity, shrinkage, creep, and cracking (Chapter 4), and durability (Chapter 5).

COMPLEXITIES

From examination of a cross section of concrete (Fig. 2–1), the two phases that can easily be distinguished are aggregate particles of varying size and shape and the binding medium, composed of an incoherent mass of the hydrated cement paste (henceforth abbreviated **hcp**). At the macroscopic level, therefore, concrete may be considered to be a two-phase material, consisting of aggregate particles dispersed in a matrix of the cement paste.

At the microscopic level, the complexities of the concrete structure begin to show up. It becomes obvious that the two phases of the structure are neither homogeneously distributed with respect to each other, nor are they themselves homogeneous. For instance, in some areas the *hcp* mass appears to be as dense as the aggregate while in others it is highly porous (Fig. 2–2). Also, if several specimens of concrete containing the same amount of cement but different amounts of water are examined at various time intervals it will be seen that, in general, the volume of capillary voids in the *hcp* would decrease with decreasing water/cement ratio or with increasing age of hydration. For a well-hydrated cement paste, the inhomogeneous distribution of solids and voids alone can perhaps be ignored when modeling the behavior of the material. However, microstructural studies have shown that this cannot be done for the *hcp* present in concrete. In the presence of aggregate, the structure of *hcp* in the vicinity of large aggregate particles is usually very different from the structure of bulk paste or mortar in the system. In fact, many aspects of concrete behavior under stress can be explained only when the cement paste-aggregate interface is treated as a third phase of the concrete structure. Thus the *unique features of the concrete structure* can be summarized as follows.

First, there is a third phase, the **transition zone**, which represents the interfacial region between the particles of coarse aggregate and the *hcp*. Existing as a thin shell, typically 10 to 50 μm thick around large aggregate, the transition zone is generally weaker than either of the two main components of concrete, and therefore it

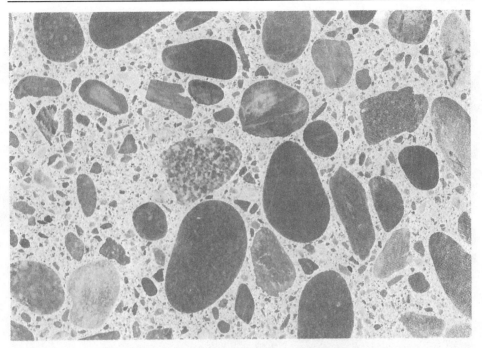

Figure 2-1 Polished section from a concrete specimen.

Macrostructure is the gross structure of a material that is visible to the unaided human eye. In the macrostructure of concrete two phases are readily distinguished: aggregates of varying shapes and size, and the binding medium, which consists of an incoherent mass of the hydrated cement paste.

exercises a far greater influence on the mechanical behavior of concrete than is reflected by its size. Second, each of the three phases is itself multiphase in nature. For instance, each aggregate particle may contain several minerals, in addition to microcracks and voids. Similarly, both the bulk *hcp* and the transition zone generally contain a heterogeneous distribution of different types and amounts of solid phases, pores, and microcracks which will be described below. Third, unlike other engineering materials, the structure of concrete does not remain stable (i.e., it is not an intrinsic characteristic of the material). This is because the two components of the structure—the *hcp* and the transition zone—are subject to change with time, environmental humidity, and temperature.

The highly heterogeneous and dynamic nature of the structure of concrete are the primary reasons why the theoretical structure–property relationship models, generally so helpful for predicting the behavior of engineering materials, are of little use in the case of concrete. A broad knowledge of the important features of the structure of individual components of concrete, as furnished below, is nevertheless essential for understanding and controlling the properties of the composite material.

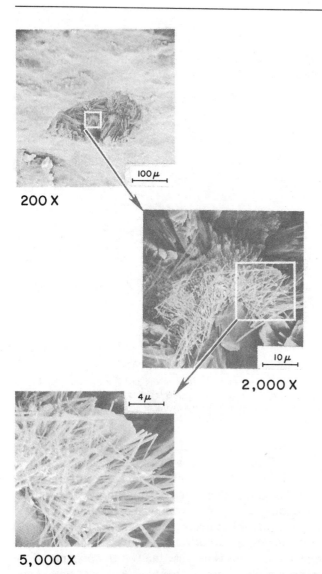

Figure 2-2 Microstructure of a hydrated cement paste.

Microstructure is the subtle structure of a material that is resolved with the help of a microscope. A low-magnification (200 ×) electron micrograph of a hydrated cement paste shows that the structure is not homogeneous; while some areas are dense, the others are highly porous. In the porous areas, it is possible to resolve the individual hydrated phases by using higher magnifications. For example, massive crystals of calcium hydroxide, long and slender needles of ettringite, and aggregation of small fibrous crystals of calcium silicate hydrate can be seen at 2000 × and 5000 × magnifications.

STRUCTURE OF THE AGGREGATE PHASE

The composition and properties of different types of concrete aggregates are described in detail in Chapter 7. Hence only a brief description of the general elements of the aggregate structure, which exercise a major influence on properties of concrete, will be given here.

The aggregate phase is predominantly responsible for the unit weight, elastic modulus, and dimensional stability of concrete. These properties of concrete depend to a large extent on the bulk density and strength of the aggregate, which, in turn, are determined by the physical rather than chemical characteristics of the aggregate structure. In other words, the chemical or mineralogical composition of the solid phases in aggregate is usually less important than the physical characteristics such as the volume, size, and distribution of pores.

In addition to porosity, the shape and texture of the coarse aggregate also affect the properties of concrete. Some typical aggregate particles are shown in Fig. 2-3. Generally, natural gravel has a rounded shape and a smooth surface texture. Crushed rocks have a rough texture; depending on the rock type and the choice of crushing equipment, the crushed aggregate may contain a considerable proportion of flat or elongated particles, which adversely affect many properties of concrete. Lightweight aggregate particles from pumice, which is highly cellular, are also angular and have a rough texture, but those from expanded clay or shale are generally rounded and smooth.

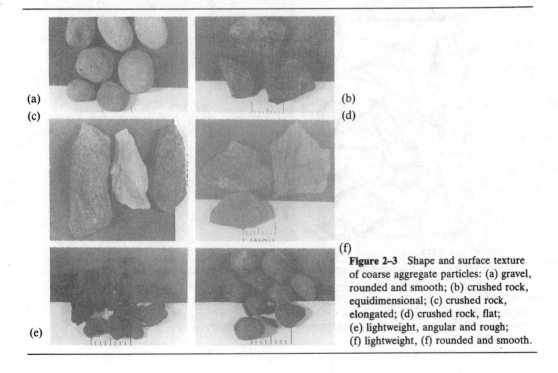

Figure 2-3 Shape and surface texture of coarse aggregate particles: (a) gravel, rounded and smooth; (b) crushed rock, equidimensional; (c) crushed rock, elongated; (d) crushed rock, flat; (e) lightweight, angular and rough; (f) lightweight, (f) rounded and smooth.

Being generally stronger than the other two phases of concrete, the aggregate phase has no direct influence on the strength of concrete except in the case of some highly porous and weak aggregates, such as the pumice aggregate described above. The size and the shape of coarse aggregate can, however, affect the strength of concrete in an indirect way. It is obvious from Fig. 2–4 that the larger the size of aggregate in concrete and the higher the proportion of elongated and flat particles, the greater will be the tendency for water films to accumulate next to the aggregate surface, thus weakening the cement paste–aggregate transition zone. This phenomenon, known as *internal bleeding,* is discussed in detail in Chapter 10.

STRUCTURE OF HYDRATED CEMENT PASTE

It should be understood that the term hydrated cement paste (*hcp*) as used in this text refers generally to pastes made from portland cement. Although the composition and properties of portland cement are discussed in detail in Chapter 6, a summary of the composition will be given here before we discuss how the structure of the *hcp* develops as a result of chemical reactions between the portland cement minerals and water.

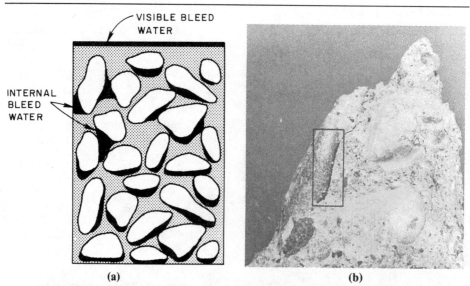

Figure 2–4 (a) Diagrammatic representation of bleeding in freshly deposited concrete; (b) shear-bond failure in a concrete specimen tested in uniaxial compression.

Internal bleed water tends to accumulate in the vicinity of elongated, flat, and large pieces of aggregate. In these locations, the aggregate-cement paste transition zone tends to be weak and easily prone to microcracking. This phenomenon is responsible for the shear-bond failure at the surface of the aggregate particle marked in the photograph.

Anhydrous portland cement is a gray powder that consists of angular particles typically in the size range 1 to 50 μm. It is produced by pulverizing a clinker with a small amount of calcium sulfate, the clinker being a heterogeneous mixture of several minerals produced by high-temperature reactions between calcium oxide and silica, alumina, and iron oxide. The chemical composition of the principal clinker minerals corresponds approximately to C_3S,[1] C_2S, C_3A, and C_4AF; in ordinary portland cement their respective amounts usually range between 45 and 60, 15 and 30, 6 and 12, and 6 and 8 percent.

When portland cement is dispersed in water, the calcium sulfate and the high-temperature compounds of calcium tend to go into solution, and the liquid phase gets rapidly saturated with various ionic species. As a result of combinations between calcium, sulfate, aluminate, and hydroxyl ions within a few minutes of cement hydration, first the needle-shaped crystals of a calcium sulfoaluminate hydrate called ettringite make their appearance; a few hours later large prismatic crystals of calcium hydroxide and very small fibrous crystals of calcium silicate hydrates begin to fill the empty space formerly occupied by water and the dissolving cement particles. After some days, depending on the alumina-to-sulfate ratio of the portland cement, ettringite may become unstable and decompose to form the monosulfate hydrate, which has hexagonal-plate morphology. Hexagonal-plate morphology is also the characteristic of calcium aluminate hydrates, which are formed in the hydrated pastes of either undersulfated or high-C_3A portland cements. A scanning electron micrograph illustrating the typical morphology of phases prepared by mixing a calcium aluminate solution with calcium sulfate solution is shown in Fig. 2–5. A model of the essential phases present in the microstructure of a well-hydrated portland cement paste is shown in Fig. 2–6.

From the microstructure model of the *hcp* shown in Fig. 2–6, it may be noted that the various phases are neither uniformly distributed nor they are uniform in size and morphology. In solids, microstructural inhomogeneities can lead to serious effects on strength and other related mechanical properties because these properties are controlled by the microstructural extremes, not by the average microstructure. Thus, in addition to the evolution of the microstructure as a result of the chemical changes, which occur after cement comes in contact with water, attention has to be paid to certain rheological properties of freshly mixed cement paste which are also influential in determining the microstructure of the hardened paste. For instance, as will be discussed later (see Fig. 8–2c), the anhydrous particles of cement have a tendency to attract each other and form flocs, which entrap large quantities of mixing water. Obviously, local variations in water-to-cement ratio would be the primary source of evolution of the heterogeneous pore structure. With highly flocculated cement paste systems not only the size and shape of pores but also the crystalline products of hydration are known to be different when compared to well-dispersed systems.

---

[1] It is convenient to follow the abbreviations used by cement chemists: $C = CaO$; $S = SiO_2$; $A = Al_2O_3$; $F = Fe_2O_3$; $\bar{S} = SO_3$; $H = H_2O$.

Figure 2-5 Scanning electron micrograph of typical hexagonal crystals of monosulfate hydrate and needlelike crystals of ettringite formed by mixing calcium aluminate and calcium sulfate solutions. (Courtesy of F. W. Locher, Research Institute for Cement Industry, Dusseldorf, Federal Republic of Germany.)

Solids in Hydrated Cement Paste

The types, amounts, and characteristics of the four principal solid phases generally present in a *hcp*, that can be resolved by an electron microscope, are as follows.

Calcium silicate hydrate. The calcium silicate hydrate phase, abbreviated **C-S-H,** makes up 50 to 60 percent of the volume of solids in a completely hydrated portland cement paste and is, therefore, the most important in determining the properties of the paste. The fact that the term C-S-H is hyphenated signifies that C-S-H is not a well-defined compound; the C/S ratio varies between 1.5 to 2.0 and the structural water content varies even more. The morphology of C-S-H also varies from poorly crystalline fibers to reticular network. Due to their colloidal dimensions and a tendency to cluster, C-S-H crystals could only be resolved with the advent of electron optical microscopy. The material is often referred to as **C-S-H gel** in older literature. The internal crystal structure of C-S-H also remains unresolved. Previously, it was assumed to resemble the natural mineral tobermorite; this is why C-S-H was sometimes called **tobermorite gel.**

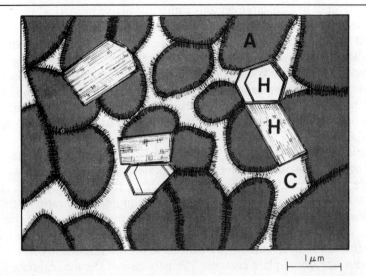

Figure 2-6 Model of a well-hydrated portland cement paste. A represents aggregation of poorly crystalline C-S-H particles which have at least one colloidal dimension (1 to 100 nm). Inter-particle spacing within an aggregation is 0.5 to 3.0 nm (avg. 1.5 nm). H represents hexagonal crystalline products such as CH, $C_4A\bar{S}H_{18}$, C_4AH_{19}. They form large crystals, typically 1 μm wide. C represents capillary cavities or voids which exist when the spaces originally occupied with water do not get completely filled with the hydration products of cement. The size of capillary voids ranges from 10 nm to 1 μm, but in well-hydrated, low water/cement ratio pastes, they are < 100 nm.

Although the exact structure of C-S-H is not known, several models have been proposed to explain the properties of the materials. According to the *Powers-Brunauer model*,[2] the material has a layer structure with a very high surface area. Depending on the measurement technique, surface areas on the order of 100 to 700 m²/g have been proposed for C-S-H. The strength of the material is attributed mainly to van der Waals forces, the size of **gel pores** or the solid-to-solid distance[3] being about 18 Å. The *Feldman-Sereda model*[4] visualizes the C-S-H structure as being composed of an irregular or kinked array of layers which are randomly arranged to create interlayer spaces of different shapes and sizes (5 to 25 Å).

Calcium hydroxide. Calcium hydroxide crystals (also called portlandite) constitute 20 to 25 percent of the volume of solids in the hydrated paste. In contrast to the C-S-H, the calcium hydroxide is a compound with a definite stoichiometry,

---

[2] T. C. Powers, *J. Am. Ceram. Soc.*, Vol. 61, No. 1, pp. 1–5, 1958; and S. Brunauer, *American Scientist*, Vol. 50, No. 1, pp. 210–29, 1962.

[3] In the older literature, the solid-to-solid distances between C-S-H layers were called gel pores. In modern literature it is customary to call them interlayer spaces.

[4] R. F. Feldman and P. J. Sereda, *Engineering Journal (Canada)*, Vol. 53, No. 8/9, pp. 53–59, 1970.

$Ca(OH)_2$. It tends to form large crystals with a distinctive hexagonal-prism morphology. The morphology usually varies from nondescript to stacks of large plates, and is affected by the available space, temperature of hydration, and impurities present in the system. Compared with C-S-H, the strength-contributing potential of calcium hydroxide due to van der Waals forces is limited as a result of a considerably lower surface area. Also, the presence of a considerable amount of calcium hydroxide in hydrated portland cement has an adverse effect on chemical durability to acidic solutions because of the higher solubility of calcium hydroxide than C-S-H.

Calcium sulfoaluminates. Calcium sulfoaluminate compounds occupy 15 to 20 percent of the solids volume in the hydrated paste and therefore play only a minor role in the structure–property relationships. It has already been stated that during the early stages of hydration the sulfate/alumina ionic ratio of the solution phase generally favors the formation of trisulfate hydrate, $C_6A\bar{S}_3H_{32}$, also called **ettringite,** which forms needle-shaped prismatic crystals. In pastes of ordinary portland cement, ettringite eventually transforms to the monosulfate hydrate, $C_4A\bar{S}H_{18}$, which forms hexagonal-plate crystals. The presence of the monosulfate hydrate in portland cement concrete makes the concrete vulnerable to sulfate attack. It should be noted that both ettringite and the monosulfate contain small amounts of iron oxide, which can substitute for the aluminum oxide in the crystal structures.

Unhydrated clinker grains. Depending on the particle size distribution of the anhydrous cement and the degree of hydration, some unhydrated clinker grains may be found in the microstructure of hydrated cement pastes, even long after hydration. As stated earlier, the clinker particles in modern portland cement generally conform to the size range 1 to 50 μm. With the progress of the hydration process, first the smaller particles get dissolved (i.e., disappear from the system) and then the larger particles appear to grow smaller. Because of the limited available space between the particles, the hydration products tend to crystallize in close proximity to the hydrating clinker particles, which gives the appearance of a coating formation around them. At later ages, due to a lack of available space, in situ hydration of clinker particles results in the formation of a very dense hydration product, which at times resembles the original clinker particle in morphology.

Voids in Hydrated Cement Paste

In addition to the above-described solids, *hcp* contains several types of voids which have an important influence on its properties. The typical sizes of both the solid phases and the voids in *hcp* are shown in Fig. 2–7. The various types of voids, and their amount and significance, are discussed next.

Interlayer space in C-S-H. Powers assumed the width of the interlayer space within the C-S-H structure to be 18 Å and determined that it accounts for 28 percent porosity in solid C-S-H; however, Feldman and Sereda suggest that the space may

Structure of Hydrated Cement Paste

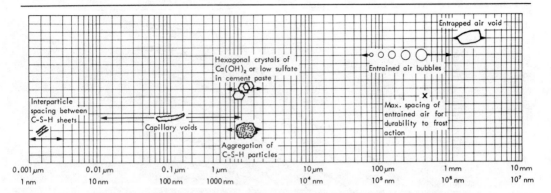

Figure 2–7 Dimensional range of solids and pores in a hydrated cement paste.

vary from 5 to 25 Å. This void size is too small to have an adverse affect on the strength and permeability of the *hcp*. However, as discussed below, water in these small voids can be held by hydrogen bonding, and its removal under certain conditions may contribute to drying shrinkage and creep.

Capillary voids. Capillary voids represent the space not filled by the solid components of the *hcp*. The total volume of a cement–water mixture remains essentially unchanged during the hydration process. The average bulk density of the hydration products[5] is considerably lower than the density of anhydrous portland cement; it is estimated that 1 cm^3 of cement, on complete hydration, requires about 2 cm^3 of space to accommodate the products of hydration. Thus cement hydration may be looked upon as a process during which the space originally occupied by cement and water is being replaced more and more by the space filled by hydration products. The space not taken up by the cement or the hydration products consists of capillary voids, the volume and size of the capillary voids being determined by the original distance between the anhydrous cement particles in the freshly mixed cement paste (i.e., water/cement ratio) and the degree of cement hydration. A method of calculating the total volume of capillary voids, popularly known as **porosity,** in portland cement pastes having either different water/cement ratios or different degrees of hydration will be described later.

In well-hydrated, low water/cement ratio pastes, the capillary voids may range from 10 to 50 nm; in high water/cement ratio pastes, at early ages of hydration the capillary voids may be as large as 3 to 5 μm. Typical pore size distribution plots of several *hcp* specimens tested by the mercury intrusion technique are shown in Fig. 2–8. It has been suggested that the pore size distribution, not the total capillary porosity, is a better criterion for evaluating the characteristics of a *hcp* capillary voids larger than 50 nm, referred to as *macropores* in the modern literature, are assumed

---

[5] It should be noted that the interlayer space within the C-S-H phase is considered as a part of the solids in the hcp.

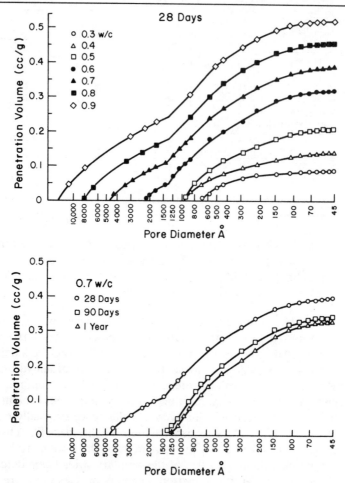

Figure 2-8 Pore size distribution in hydrated cement pastes. (From P. K. Mehta and D. Manmohan, *Proc. 7th Int. Congr. on Chemistry of Cements,* Paris, 1980.)

It is not the total porosity but the pore size distribution that actually controls the strength, permeability, and volume changes in a hardened cement paste. Pore size distributions are affected by the water/cement ratio, and the age (degree) of cement hydration. Large pores influence mostly the compressive strength and permeability; small pores influence mostly the drying shrinkage and creep.

to be detrimental to strength and impermeability, while voids smaller than 50 nm, referred to as *micropores* are assumed to be more important to drying shrinkage and creep.

Air voids. Whereas the capillary voids are irregular in shape, the air voids are generally spherical. For various reasons, which are discussed in Chapter 8, admixtures may be added to concrete purposely to entrain very small air voids in the

Structure of Hydrated Cement Paste

cement paste. Air can be entrapped in the fresh cement paste during the mixing operation. Entrapped air voids may be as large as 3 nm; entrained air voids usually range from 50 to 200 μm. Therefore, both the entrapped and entrained air voids in the *hcp* are much bigger than the capillary voids, and are capable of adversely affecting its strength and impermeability.

Water in Hydrated Cement Paste

Under electron microscopic examination, the voids in the *hcp* appear to be empty. This is because the specimen preparation technique calls for drying the specimen under high vacuum. Actually, depending on the environmental humidity and the porosity of the paste, the untreated cement paste is capable of holding a large amount of water. Like the solid and void phases discussed before, water can exist in the *hcp* in many forms. The classification of water into several types is based on the degree of difficulty or ease with which it can be removed from the *hcp*. Since there is a continuous loss of water from a saturated cement paste as the relative humidity is reduced, the dividing line between the different states of water is not rigid. In spite of this, the classification is useful for understanding the properties of the *hcp*. In addition to vapor in empty or partially water-filled voids, water exists in the *hcp* in the following states.

Capillary water. This is the water present in voids larger than about 50 Å. It may be pictured as the bulk water which is free from the influence of the attractive forces exerted by the solid surface. Actually, from the standpoint of the behavior of the capillary water in the *hcp*, it is desirable to divide the capillary water into two categories: the water in large voids of the order of > 50 nm (0.05 μm), which may be considered as **free water** because its removal does not cause any volume change, and the water held by capillary tension in small capillaries (5 to 50 nm) which on removal may cause shrinkage of the system.

Adsorbed water. This is the water that is close to the solid surface; that is, under the influence of attractive forces water molecules are physically adsorbed onto the surface of solids in the *hcp*. It has been suggested that up to six molecular layers of water (15 Å) can be physically held by hydrogen bonding. Since the bond energies of the individual water molecules decrease with distance from the solid surface, a major portion of the adsorbed water can be lost by drying the *hcp* to 30 percent relative humidity. The loss of adsorbed water is mainly responsible for the shrinkage of the *hcp* on drying.

Interlayer water. This is the water associated with the C-S-H structure. It has been suggested that a monomolecular water layer between the layers of C-S-H is strongly held by hydrogen bonding. The interlayer water is lost only on strong drying (i.e., below 11 percent relative humidity). The C-S-H structure shrinks considerably when the interlayer water is lost.

Chemically combined water. This is the water that is an integral part of the structure of various cement hydration products. This water is not lost on drying; it is evolved when the hydrates decompose on heating. Based on the Feldman–Sereda model, different types of water associated with the C-S-H are illustrated in Fig. 2–9.

Structure–Property Relationships in Hydrated Cement Paste

The desirable engineering characteristics of hardened concrete—strength, dimensional stability, and durability—are influenced not only by the proportion but also by the properties of the *hcp*, which, in turn, depend on the microstructural features (i.e., the type, amount, and distribution of solids and voids). The structure–property relationships of the *hcp* are discussed briefly next.

Strength. It should be noted that the principal source of strength in the solid products of the *hcp* is the existence of the van der Waals forces of attraction. Adhesion between two solid surfaces can be attributed to these physical forces, the degree of the adhesive action being dependent on the extent and the nature of the surfaces involved. The small crystals of C-S-H, calcium sulfoaluminate hydrates, and hexagonal calcium aluminate hydrates possess enormous surface areas and adhesive capability. These hydration products of portland cement tend to adhere strongly not only to each other, but also to low surface-area solids such as calcium hydroxide, anhydrous clinker grains, and fine and coarse aggregate particles.

It is a well-known fact that in solids there is an inverse relationship between porosity and strength. Strength resides in the solid part of a material; therefore, voids are detrimental to strength. In the *hcp*, the interlayer space within the C-S-H

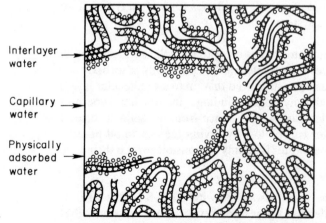

Figure 2–9 Types of water associated with the calcium silicate hydrate. [Based on R. F. Feldman and P. J. Sereda, *Eng. J.* (Canada), Vol. 53, No. 8/9, 1970.)

In the hydrated cement paste, water can exist in many forms; these can be classified depending on the degree of ease with which water can be removed. This classification is useful in understanding the volume changes in cement paste that are associated with the water held by small pores.

structure and the small voids which are within the influence of the van der Waals forces of attraction cannot be considered detrimental to strength, because stress concentration and subsequent rupture on application of load begin at large capillary voids and micro-cracks that are invariably present. As stated earlier, the volume of capillary voids in a *hcp* depends on the amount of water mixed with the cement at the start of hydration and the degree of cement hydration. When the paste sets, it acquires a stable volume that is approximately equal to the volume of the cement plus the volume of the water. Assuming that 1 cm^3 of cement produces 2 cm^3 of the hydration product, Powers made simple calculations to demonstrate the changes in capillary porosity with varying degrees of hydration in cement pastes of different water/cement ratios. Based on his work, two illustrations of the process of progressive reduction in the capillary porosity, either with increasing degrees of hydration (case A) or with decreasing water/cement ratios (case B), are shown in Fig. 2–10. Since the water/cement ratio is generally given by mass, it is necessary to know the specific gravity of portland cement (approximately 3.14) in order to calculate the volume of water and the total available space, which is equal to the sum of the volumes of water and cement.

In case A, a 0.63 water/cement-ratio paste containing 100 cm^3 of the cement will require 200 cm^3 of water; this adds to 300 cm^3 of paste volume or total available space. The degree of hydration of the cement will depend on the curing conditions (duration of hydration, temperature, and humidity). Assuming that under the ASTM standard curing conditions,[6] the volume of cement hydrated at 7, 28, and 365 days is 50, 75, and 100 percent, respectively, the calculated volume of solids (anhydrous cement plus the hydration product) is 150, 175, and 200 cm^3. The volume of capillary voids can be found by the difference between the total available space and the total volume of solids. This is 50, 42, and 33 percent, respectively, at 7, 28, and 365 days of hydration.

In case B, a 100 percent degree of hydration is assumed for four cement pastes made with water corresponding to water/cement ratios 0.7, 0.6, 0.5, or 0.4. For a given volume of cement, the paste with the largest amount of water will have the greatest total volume of available space. However, after complete hydration, all the pastes would contain the same quantity of the solid hydration product. Therefore, the paste with the greatest total space would end up with a correspondingly larger volume of capillary voids. Thus 100 cm^3 of cement at full hydration would produce 200 cm^3 of solid hydration products in every case; however, since the total space in the 0.7, 0.6, 0.5, or 0.4 water/cement-ratio pastes was 320, 288, 257, and 225 cm^3, the calculated capillary voids are 37, 30, 22, and 11 percent, respectively. Under the assumptions made here, with a 0.32 water/cement-ratio paste there would be no capillary porosity when the cement had completely hydrated.

For normally hydrated portland cement mortars, Powers showed that there is an exponential relationship of the type $S = kx^3$ between the compressive strength (S) and the solids-to-space ratio (x), where k is a constant equal to 34,000 psi. Assuming a given degree of hydration, such as 25, 50, 75, and 100 percent, it is

---

[6] ASTM C 31 requires moist curing at 73.4 ± 3°F until the age of testing.

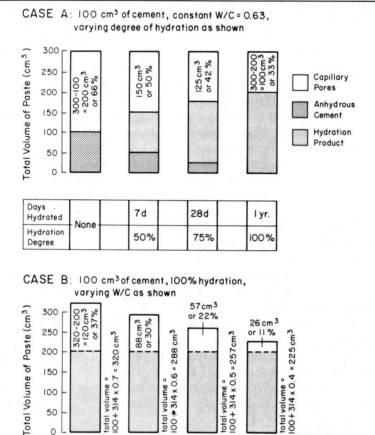

Figure 2-10 Changes in the capillary porosity with varying water/cement ratio and degree of hydration.

By making certain assumptions, calculations can be made to show how, with a given water/cement ratio, the capillary porosity of a hydrated cement paste would vary with varying degrees of hydration. Alternatively, capillary porosity variations, for a given degree of hydration but variable water/cement ratios, can be determined.

possible to calculate the effect of increasing the water/cement ratio, first on porosity, and subsequently, on strength, by using Powers's formula. The results are plotted in Fig. 2-11a. The permeability curve of this figure will be discussed later.

Dimensional stability. Saturated *hcp* is not dimensionally stable. As long as it is held at 100 percent relative humidity (RH), practically no dimensional change will occur. However, when exposed to environmental humidity, which normally is much lower than 100 percent, the material will begin to lose water and shrink. How the water loss from saturated *hcp* is related to RH on one hand, and to drying

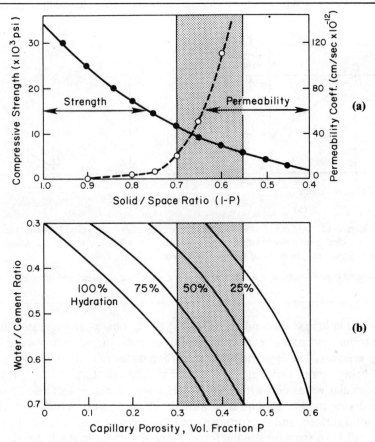

Figure 2–11 Influence of water/cement ratio and degree of hydration on strength and permeability.

A combination of water/cement ratio and degree of hydration determines the porosity of hydrated cement paste. The porosity and the opposite of porosity (solid/space ratio) are exponentially related to both the strength and permeability of the material. The shaded area shows the typical capillary porosity range in hydrated cement pastes.

shrinkage on the other, is described by L'Hermite (Fig. 2–12). As soon as the RH drops below 100 percent, the free water held in large cavities (e.g., >50 nm) begins to escape to the environment. Since the free water is not attached to the structure of the hydration products by any physical–chemical bonds, its loss would not be accompanied by shrinkage. This is shown by curve '*A-B*' in Fig. 2–12. Thus a saturated *hcp* exposed to slightly less than 100 percent RH can lose a considerable amount of total evaporable water before undergoing any shrinkage.

When most of the free water has been lost, on continued drying it is found that further loss of water begins to result in considerable shrinkage. This phenomenon, shown by curve '*B-C*' in Fig. 2–12, is attributed mainly to the loss of adsorbed water

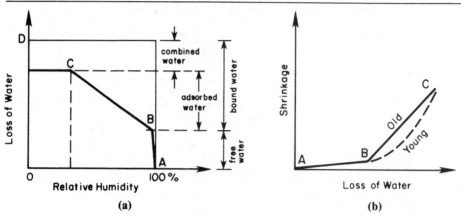

Figure 2-12 (a) Loss of water as a function of the relative humidity; (b) shrinkage of a cement mortar as a function of the water loss. (From R. L'Hermite, *Proc. Fourth Int. Symp. on Chemistry of Cements,* Washington, D.C., 1960.)

From a saturated cement paste, it is the loss of adsorbed water that is mainly responsible for the drying shrinkage.

and the water held in small capillaries (see Fig. 2-9). It has been suggested that when confined to narrow spaces between two solid surfaces, the adsorbed water causes a ***disjoining pressure.*** The removal of the adsorbed water reduces the disjoining pressure and brings about shrinkage of the system. The interlayer water, present as a mono-molecular water film within the C-S-H layer structure, can also be removed by severe drying conditions. This is because the closer contact of the interlayer water with the solid surface, and the tortuosity of the transport path through the capillary network, call for a stronger driving force. Since the water in small capillaries (5 to 50 nm) exerts ***hydrostatic tension,*** its removal tends to induce a compressive stress on the solid walls of the capillary pore, thus also causing contraction of the system.

It should be pointed out here that the mechanisms which are responsible for drying shrinkage are also responsible for creep of the *hcp*. In the case of creep, a sustained external stress becomes the driving force for the movement of the physically adsorbed water and the water held in small capillaries. Thus creep strain can occur even at 100 percent RH.

Durability. The term **durability** of a material relates to its service life under given environmental conditions. The *hcp* is alkaline; therefore, exposure to acidic waters is detrimental to the material. Under these conditions, **impermeability,** also called **watertightness,** becomes a primary factor in determining durability. The impermeability of the *hcp* is a highly prized characteristic because it is assumed that an impermeable *hcp* would result in an impermeable concrete (the aggregate in concrete is generally assumed to be impermeable). **Permeability** is defined as the ease with which a fluid can flow through a solid. It should be obvious that the size and continuity of the pores in the structure of the solid determine its permeability.

Strength and permeability of the *hcp* are two sides of the same coin in the sense that both are closely related to the capillary porosity or the solid/space ratio. This is evident from the permeability curve in Fig. 2–11 which is based on the experimentally determined values of permeability by Powers.

The exponential relationship between permeability and porosity shown in Fig. 2–11 can be understood from the influence that various pore types exert on permeability. As hydration proceeds, the void space between the originally discrete cement particles gradually begins to fill up with the hydration products. It has been shown (Fig. 2–10) that the water/cement ratio (i.e., original capillary space between cement particles) and the degree of hydration determine the total capillary porosity, which decreases with the decreasing water/cement ratio and/or increasing degree of hydration. Mercury-intrusion porosimetric studies on the cement pastes of Fig. 2–8, hydrated with different water/cement ratios and to various ages, have shown that the decrease in total capillary porosity was associated with reduction of large pores in the *hcp* (Fig. 2–13). From the data in Fig. 2–11 it is obvious that the coefficient of permeability registered an exponential drop when the fractional volume of capillary pores was reduced from 0.4 to 0.3. This range of capillary porosity, therefore, seems to correspond to the point when both the volume and size of capillary pores in the *hcp* are so reduced that the interconnections between them have become difficult. As a result, the permeability of a fully hydrated cement paste may be of the order

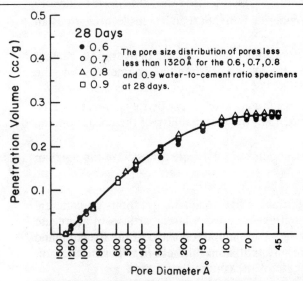

Figure 2–13 Distribution plots of small pores in cement pastes of varying water/cement ratios. (From P. K. Mehta and D. Manmohan, *Proc. 7th Int. Congr. on Chemistry of Cements*, Paris, 1980.)

When the data of Fig. 2–8 are replotted after omitting the large pores (i.e., > 1320 Å), it was found that a single curve could fit the pore size distributions in the 28-day-old pastes made with four different water/cement ratios. This shows that in hardened cement pastes, the increase in total porosity resulting from increasing water/cement ratios manifests itself in the form of large pores only. This observation has great significance from the standpoint of the effect of water/cement ratio on strength and permeability, which are controlled by large pores.

of 10^6 times less than that of a young paste. Powers showed that even an 0.6-water/cement-ratio paste, on complete hydration, can become as impermeable as a dense rock such as basalt or marble.

It should be noted that the porosities represented by the C-S-H interlayer space and small capillaries do not contribute to permeability of *hcp*. On the contrary, with increasing degree of hydration, although there is a considerable increase in the volume of pores due to the C-S-H interlayer space and small capillaries, the permeability is greatly reduced. In *hcp* a direct relationship was noted between the permeability and the volume of pores larger than about 100 nm.[7] This is probably because the pore systems, comprised mainly of small pores, tend to become discontinuous.

TRANSITION ZONE IN CONCRETE

Significance of the Transition Zone

Have you ever wondered why:

- Concrete is brittle in tension but relatively tough in compression?
- The components of concrete when tested separately in a uniaxial compression remain elastic until fracture, whereas concrete itself shows inelastic behavior?
- The compressive strength of a concrete is higher than its tensile strength by an order of magnitude?
- At a given cement content, water/cement ratio, and age of hydration, cement mortar will always be stronger than the corresponding concrete? Also, the strength of concrete goes down as the coarse aggregate size is increased.
- The permeability of a concrete containing even a very dense aggregate will be higher by an order of magnitude than the permeability of the corresponding cement paste?
- On exposure to fire, the elastic modulus of a concrete drops more rapidly than its compressive strength?

The answers to the above and many other enigmatic questions on concrete behavior lie in the transition zone that exists between large particles of aggregate and the *hcp*. Although composed of the same elements as the *hcp*, the structure and properties of the transition zone are different from the bulk *hcp*. It is, therefore, desirable to treat it as a separate phase of the concrete structure.

---

[7] P. K. Mehta and D. Manmohan, *Proceedings of the Seventh International Congress on the Chemistry of Cements,* Editions Septima, Vol. III, Paris, 1980.

Structure of the Transition Zone

Due to experimental difficulties, the information about the transition zone in concrete is scarce; however, based on a description given by Maso[8] some understanding of its structural characteristics can be obtained by following the sequence of its development from the time concrete is placed.

> First, in freshly compacted concrete, water films form around the large aggregate particles. This would account for a higher water/cement ratio closer to the larger aggregate than away from it (i.e., in the bulk mortar). Next, as in the bulk paste, calcium, sulfate, hydroxyl, and aluminate ions, produced by the dissolution of calcium sulfate and calcium aluminate compounds, combine to form ettringite and calcium hydroxide. Owing to the high water/cement ratio, these crystalline products in the vicinity of the coarse aggregate consist of relatively larger crystals, and therefore form a more porous framework than in the bulk cement paste or mortar matrix. The platelike calcium hydroxide crystals tend to form in oriented layers, for instance, with the c-axis perpendicular to the aggregate surface. Finally, with the progress of hydration, poorly crystalline C-S-H and a second generation of smaller crystals of ettringite and calcium hydroxide start filling the empty space that exists between the framework created by the large ettringite and calcium hydroxide crystals. This helps to improve the density and hence the strength of the transition zone.

A diagrammatic representation and scanning electron micrograph of the transition zone in concrete are shown in Fig. 2–14.

Strength of the Transition Zone

As in the case of the *hcp,* the cause of adhesion between the hydration products and the aggregate particle is the van der Waals force of attraction; therefore, the strength of the transition zone at any point depends on the volume and size of voids present. Even for low water/cement ratio concrete, at early ages the volume and size of voids in the transition zone will be larger than in the bulk mortar; consequently, the former is weaker in strength (Fig. 2–15). However, with increasing age the strength of the transition zone may become equal to or even greater than the strength of the bulk mortar. This could happen as a result of crystallization of new products in the voids of the transition zone by slow chemical reactions between the cement paste constituents and the aggregate, formation of calcium silicate hydrates in the case of siliceous aggregates, or formation of carboaluminate hydrates in the case of limestone. Such interactions are strength contributing because they also tend to reduce the concentration of the calcium hydroxide in the transition zone. The large calcium

---

[8] J. C. Maso, *Proceedings of the Seventh International Congress on the Chemistry of Cements,* Vol. 1, Editions Septima, Paris, 1980.

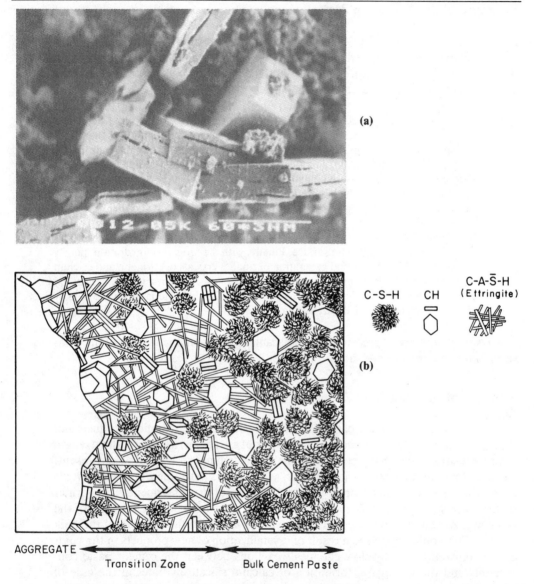

Figure 2–14 (a) Scanning electron micrograph of the calcium hydroxide crystals in the transition zone. (b) Diagrammatic representation of the transition zone and bulk cement paste in concrete.

At early ages, especially when a considerable internal bleeding has occurred, the volume and size of voids in the transition zone are larger than in the bulk cement paste or mortar. The size and concentration of crystalline compounds such as calcium hydroxide and ettringite are also larger in the transition zone. The cracks are formed easily in the direction perpendicular to the c-axis. Such effects account for the lower strength of the transition zone than the bulk cement paste in concrete.

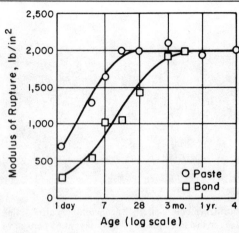

Figure 2-15 Effect of age on the bond (transition zone) strength and bulk cement paste strength. (From K. M. Alexander, J. Wardlaw, and D. J. Gilbert, *The Structure of Concrete,* Cement and Concrete Association, London, 1968, p. 65.)

As a result of slow chemical interaction between the cement paste and aggregate, at later ages the transition-zone strength improves more than the bulk cement paste strength.

hydroxide crystals possess less adhesion capacity, not only because of the lower surface area and correspondingly weak van der Waals forces of attraction, but also because they serve as preferred cleavage sites owing to their oriented structure.

In addition to the large volume of capillary voids and oriented calcium hydroxide crystals, a major factor responsible for the poor strength of the transition zone in concrete is the presence of microcracks. The amount of microcracks depends on numerous parameters, including aggregate size and grading, cement content, water/cement ratio, degree of consolidation of fresh concrete, curing conditions, environmental humidity, and thermal history of concrete. For instance, a concrete mixture containing poorly graded aggregate is more prone to segregation in compacting; thus thick water films can form around the coarse aggregate, especially beneath the particle. Under identical conditions, the larger the aggregate size the thicker would be the water film. The transition zone formed under these conditions will be susceptible to cracking when subjected to the influence of tensile stresses induced by differential movements between the aggregate and the *hcp*. Such differential movements commonly arise either on drying or on cooling of concrete. In other words, concrete has microcracks in the transition zone even before a structure is loaded. Obviously, short-term impact loads, drying shrinkage, and sustained loads at high stress levels will have the effect of increasing the size and number of microcracks (Fig. 2-16).

Influence of the Transition Zone on Properties of Concrete

The transition zone, generally the *weakest link of the chain,* is considered the strength-limiting phase in concrete. It is because of the presence of the transition zone that concrete fails at a considerably lower stress level than the strength of either

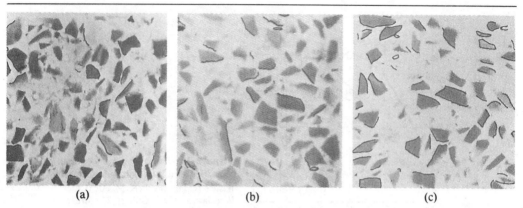

Figure 2–16 Typical cracking maps for normal (medium-strength) concrete: (a) after drying shrinkage; (b) after short-term loading; (c) for sustained loading for 60 days at 65 percent of the 28-day compressive strength. (From A. J. Ngab, F. O. Slate, and A. M. Nilson, *J. ACI*, Proc., Vol. 78, No. 4, 1981.)

As a result of short-term loading, drying shrinkage, and creep, the transition zone in concrete contains microcracks.

of the two main components. Since it does not take very high energy levels to extend the cracks already existing in the transition zone, even at 40 to 70 percent of the ultimate strength, higher incremental strains are obtained per unit of applied stress. This explains the phenomenon that the components of concrete (i.e., aggregate and *hcp* or mortar) usually remain elastic until fracture in a uniaxial compression test, whereas concrete itself shows inelastic behavior.

At stress levels higher than about 70 percent of the ultimate strength, the stress concentrations at large voids in the mortar matrix become large enough to initiate cracking there. With increasing stress, the matrix cracks gradually spread until they join the cracks originating from the transition zone. The crack system then becomes continuous and the material ruptures. Considerable energy is needed for the formation and extension of matrix cracks under a compressive load. On the other hand, under tensile loading cracks propagate rapidly and at a much lower stress level. This is why concrete fails in a brittle manner in tension but is relatively tough in compression. This is also the reason why the tensile strength is much lower than the compressive strength of a concrete. This subject is discussed in greater length in Chapters 3 and 4.

The structure of the transition zone, especially the volume of voids and microcracks present, have a great influence on the stiffness or the elastic modulus of concrete. In the composite material, the transition zone serves as a bridge between the two components: the mortar matrix and the coarse aggregate particles. Even when the individual components are of high stiffness, the stiffness of the composite may be low because of the **broken bridges** (i.e., voids and microcracks in the transition zone), which do not permit stress transfer. Thus due to microcracking on exposure to fire, the elastic modulus of concrete drops faster than the compressive strength.

The characteristics of the transition zone also influence the durability of concrete. Prestressed and reinforced concrete elements often fail due to corrosion of the embedded steel. The rate of corrosion of steel is greatly influenced by the permeability of concrete. The existence of microcracks in the transition zone at the interface with steel and coarse aggregate is the primary reason that concrete is more permeable than the corresponding *hcp* or mortar. It should be noted that the permeation of air and water is a necessary prerequisite to corrosion of the steel in concrete.

The effect of the water/cement ratio on permeability and strength of concrete is generally attributed to the relationship that exists between the water/cement ratio and the porosity of the *hcp* in concrete. The foregoing discussion on the influence of structure and properties of the transition zone on concrete shows that, in fact, it is more appropriate to think in terms of the effect of the water/cement ratio on the concrete mixture as a whole. This is because depending on aggregate characteristics, such as the maximum size and grading, it is possible to have large differences in the water/cement ratio between the mortar matrix and the transition zone. In general, everything else remaining the same, the larger the aggregate the higher will be the local water/cement ratio in the transition zone and, consequently, the weaker and more permeable will be the concrete.

TEST YOUR KNOWLEDGE

1. What is the significance of the structure of a material? How do you define structure?
2. Describe some of the unique features of the concrete structure that make it difficult to predict the behavior of the material from its structure.
3. Discuss the physical–chemical characteristics of the C-S-H, calcium hydroxide, and calcium sulfoaluminates present in a well-hydrated portland cement paste.
4. How many types of voids are present in a hydrated cement paste? What are their typical dimensions? Discuss the significance of the C-S-H interlayer space with respect to properties of the hydrated cement paste.
5. How many types of water are associated with a saturated cement paste? Discuss the significance of each. Why is it desirable to distinguish between the free water in large capillaries and the water held in small capillaries?
6. What would be the volume of capillary voids in an 0.2-water/cement ratio paste that is only 50 percent hydrated? Also calculate the water/cement ratio needed to obtain zero porosity in a fully hydrated cement paste.
7. When a saturated cement paste is dried, the loss of water is not directly proportional to the drying shrinkage. Explain why.
8. In a hydrating cement paste the relationship between porosity and impermeability is exponential. Explain why.
9. Draw a typical sketch showing how the structure of hydration products in the aggregate-cement paste transition zone is different from the bulk cement paste in concrete.

10. Discuss why the strength of the transition zone is generally lower than the strength of the bulk hydrated cement paste. Explain why concrete fails in a brittle manner in tension but not in compression.
11. Everything else remaining the same, the strength and impermeability of a mortar will decrease as coarse aggregate of increasing size is introduced. Explain why.

SUGGESTIONS FOR FURTHER STUDY

DIAMOND, S., *Proceedings of the Conference on Hydraulic Cement Pastes,* Cement and Concrete Association, Wexham Springs, Slough, U.K., pp. 2–30, 1976.

LEA, F. M., *The Chemistry of Cement and Concrete,* Chemical Publishing Company, Inc., New York, 1971, Chap. 10, The Setting and Hardening of Portland Cement.

MINDESS, S., and J. F. YOUNG, *Concrete,* Prentice Hall, Inc., Englewood Cliffs, N.J., 1981, Chap. 4, Hydration of Portland Cement.

POWERS, T. C., Properties of Fresh Concrete, John Wiley and Sons, Inc., 1968, Chapters 2, 9, and 11.

Proceedings of the Seventh International Congress on the Chemistry of Cement (Paris, 1980), Eighth Congress (Rio de Janeiro, 1986), and Ninth Congress (New Delhi, 1992).

RAMACHANDRAN, V. S., R. F. FELDMAN, and J. J. BEAUDOIN, *Concrete Science,* Heyden & Son Ltd., London, 1981, Chaps. 1 to 3, Microstructure of Cement Paste.

SKALNY, J. P., ED., *Material Science of Concrete,* Vol. 1, The American Ceramic Society Inc., 1989.

CHAPTER 3
Strength

PREVIEW

The strength of concrete is the property most valued by designers and quality control engineers. In solids, there exists a fundamental inverse relationship between porosity (volume fraction of voids) and strength. Consequently, in multiphase materials such as concrete, the porosity of each component of the structure can become strength limiting. Natural aggregates are generally dense and strong; therefore, it is the porosity of the cement paste matrix as well as the transition zone between the matrix and coarse aggregate which usually determines the strength characteristic of normal-weight concrete.

Although the water/cement ratio is important in determining the porosity of both the matrix and the transition zone and hence the strength of concrete, factors such as compaction and curing conditions (degree of cement hydration), aggregate size and mineralogy, admixtures, specimen geometry and moisture condition, type of stress, and rate of loading can also have an important effect on strength. In this chapter the influence of various factors on concrete strength is examined in detail. Since the uniaxial strength in compression is commonly accepted as a general index of concrete strength, the relationships between the uniaxial compressive strength and other strength types such as tensile, flexural, shear, and biaxial strength are discussed.

DEFINITION

The **strength** of a material is defined as the ability to resist stress without failure. Failure is sometimes identified with the appearance of cracks. However, it should be noted that unlike most structural materials, concrete contains fine cracks even before it is subjected to external stresses. In concrete, therefore, strength is related to the stress required to cause fracture and is synonymous with the degree of failure at which the applied stress reaches its maximum value. In tension tests fracture of the test piece usually signifies failure; in compression the test piece is considered to have failed when no sign of external fracture is visible, yet internal cracking is so advanced that the specimen is unable to carry a higher load without fracture.

SIGNIFICANCE

In concrete design and quality control, strength is the property generally specified. This is because, compared to most other properties, testing of strength is relatively easy. Furthermore, many properties of concrete, such as elastic modulus, watertightness or impermeability, and resistance to weathering agents including aggressive waters, are directly related to strength and can therefore be deduced from the strength data. It was pointed out earlier (Chapter 1) that the compressive strength of concrete is many times greater than other types of strength, and a majority of concrete elements are designed to take advantage of the higher compressive strength of the material. Although in practice most concrete is subjected simultaneously to a combination of compressive, shearing, and tensile stresses in two or more directions, the uniaxial compression tests are the easiest to perform in the laboratory, and the 28-day compressive strength of concrete determined by a standard uniaxial compression test is accepted universally as a general index of concrete strength.

STRENGTH–POROSITY RELATIONSHIP

In general, there exists a ***fundamental inverse relationship between porosity and strength*** of solids which, for simple homogeneous materials, can be described by the expression

$$S = S_0 e^{-kp} \tag{3-1}$$

where S is the strength of the material which has a given porosity p; S_0 is the intrinsic strength at zero porosity; and k is a constant. For many materials the ratio S/S_0 plotted against porosity follows the same curve. For instance, the data in Fig. 3–1a represent normally cured cements, autoclaved cements, and a variety of aggregates. Actually, the strength–porosity relationship is applicable to a very wide range of materials, such as iron, stainless steel, plaster of paris, sintered alumina, and zirconia (Fig. 3–1b).

Powers[1] found that the 28-day compressive strength f_c of three different mortar

---

[1] T. C. Powers, *J. Am. Ceram. Soc.*, Vol. 41, No. 1, pp. 1–6, 1958.

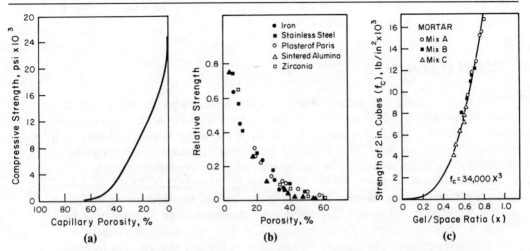

Figure 3-1 Porosity-strength relation in solids: (a) normally cured cements, autoclaved cements, and aggregates; (b) iron, stainless steel, plaster of Paris, sintered alumina, and zirconia; (c) portland cement mortars with different mix proportions. [(a), From G. J. Verbeck and R. A. Helmuth, *Proc. Fifth Int. Symp. on Chemistry of Cements*, Tokyo, Vol. 3, pp. 1–32, 1968; (b), from A. M. Neville, *Properties of Concrete*, Pitman Publishing, Inc., Marshfield, Mass., p. 271, 1981; (c), from T. C. Powers, *J. Am. Ceram. Soc.*, Vol. 41, No. 1, pp. 1–6, 1958.]

The inverse relationship between porosity and strength is not limited to cementitious products; it is generally applicable to a very wide variety of materials.

mixtures was related to the gel/space ratio, or the ratio between the solid hydration products in the system and the total space:

$$f_c = ax^3 \qquad (3\text{–}2)$$

where (a) is the intrinsic strength of the material at zero porosity (p) and (x) the solid/space ratio or the amount of solid fraction in the system, which is therefore equal to $(1 - p)$. Powers's data are shown in Fig. 3–1c; he found the value of (a) to be 34,000 psi (234 MPa). The similarity of the three curves in Fig. 3–1 confirms the general validity of the strength–porosity relationship in solids.

Whereas in hardened cement paste or mortar the porosity can be related to strength, with concrete the situation is not simple. The presence of microcracks in the transition zone between the coarse aggregate and the cement paste matrix makes concrete too complex a material for prediction of strength by precise strength–porosity relations. The general validity of strength–porosity relation, however, must be respected because porosities of the component phases of concrete, including the transition zone, indeed become strength limiting. With concrete containing the conventional low-porosity or high-strength aggregates, the strength of the material will be governed both by the strength of the cement paste matrix and the strength of the transition zone. Typically, at early ages the transition zone is weaker than the matrix, but at later ages the reverse seems to be the case.

FAILURE MODES IN CONCRETE

With a material such as concrete, which contains void spaces of various size and shape in the matrix and microcracks at the transition zone between the matrix and coarse aggregates, the failure modes under stress are very complex and vary with the type of stress. A brief review of the failure modes, however, will be useful in understanding and controlling the factors that influence concrete strength.

Under uniaxial tension, relatively less energy is needed for the initiation and growth of cracks in the matrix. Rapid propagation and interlinkage of the crack system, consisting of preexisting cracks at the transition zone and newly formed cracks in the matrix, account for the brittle failure. In compression, the failure mode is less brittle because considerably more energy is needed to form and to extend cracks in the matrix. It is generally agreed that in a uniaxial compression test on medium- or low-strength concrete, no cracks are initiated in the matrix up to about 50 percent of the failure stress; at this stage a stable system of cracks, called shear-bond cracks, already exists in the vicinity of coarse aggregate. At higher stress levels, cracks are initiated within the matrix; their number and size increases progressively with increasing stress levels. The cracks in the matrix and the transition zone (shear-bond cracks) eventually join up, and generally a failure surface develops at about 20 to 30° from the direction of the load, as shown in Fig. 3-2.

Figure 3-2 Typical failure mode of concrete in compression.

COMPRESSIVE STRENGTH AND FACTORS AFFECTING IT

The response of concrete to applied stress depends not only on the stress type but also on how a combination of various factors affects porosity of the different structural components of concrete. The factors include properties and proportions of materials that make up the concrete mixture, degree of compaction, and conditions of curing. From the standpoint of strength, the **water/cement ratio–porosity relation** is undoubtedly the most important factor because, independent of other factors, it affects the porosity of both the cement paste matrix and the transition zone between the matrix and the coarse aggregate.

Direct determination of porosity of the individual structural components of concrete—the matrix and the transition zone—is impractical, and therefore precise models of predicting concrete strength cannot be developed. However, over a period of time many useful empirical relations have been found which for practical use provide enough indirect information about the influence of numerous factors on compressive strength (compressive strength being widely used as an index of all other types of strength). Although the actual response of concrete to applied stress is a result of complex interactions between various factors, to simplify an understanding of these factors they are discussed separately under three categories: (1) characteristics and proportions of materials, (2) curing conditions, and (3) testing parameters.

Characteristics and Proportions of Materials

Before making a concrete mixture, the selection of proper constituent materials and determination of their proportions is the first step toward obtaining a concrete that will meet the specified strength. Although the composition and properties of concrete-making materials are discussed in detail in Chapters 6–8, the aspects that are important from the standpoint of concrete strength will be considered here. It should be noted that in practice many mix design parameters are interdependent, therefore their influences cannot really be separated.

Water/cement ratio. In 1918, as a result of extensive testing at the Lewis Institute, University of Illinois, Duff Abrams found that a relation existed between water/cement ratio and concrete strength. Popularly known as **Abram's water/cement ratio rule**, this inverse relation is represented by the expression

$$f_c = \frac{k_1}{k_2^{w/c}} \tag{3-3}$$

where w/c represents the water/cement ratio of the concrete mixture and k_1 and k_2 are empirical constants. Typical curves illustrating the relationship between water/cement ratio and strength at a given moist curing age are shown in Fig. 3–3.

From an understanding of the factors responsible for the strength of the hydrated cement paste and the effect of increasing water/cement ratio on porosity at a given degree of cement hydration (Fig. 2–10, case B), the water/cement

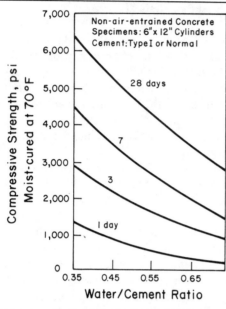

Figure 3-3 Influence of the water/cement ratio and moist curing age on concrete strength. (From *Design and Control of Concrete Mixtures, 13th Edition,* Portland Cement Association, Skokie, Ill., 1988. p. 6.)

Compressive strength of concrete is a function of the water/cement ratio and degree of cement hydration. At a given temperature of hydration, the degree of hydration is time dependent and so is the strength.

ratio–strength relationship in concrete can easily be explained as the natural consequence of a progressive weakening of the matrix caused by increasing porosity with increase in the water/cement ratio. This explanation, however, does not consider the influence of the water/cement ratio on the strength of the transition zone. In low- and medium-strength concrete made with normal aggregate, both the transition zone porosity and the matrix porosity determine the strength, and a direct relation between the water/cement ratio and the concrete strength holds. This seems no longer to be the case in high-strength (i.e., very low water/cement ratio) concretes. For water/cement ratios under 0.3, disproportionately high increases in the compressive strength can be achieved for very small reductions in water/cement ratio. The phenomenon is attributed mainly to a significant improvement in the strength of the transition zone at very low water/cement ratios. One of the explanations is that the size of calcium hydroxide crystals become smaller with decreasing water/cement ratios.

Air entrainment. For the most part, it is the water/cement ratio that determines the porosity of the cement paste matrix at a given degree of hydration; however, when air voids are incorporated into the system, either as a result of inadequate compaction or through the use of an air-entraining admixture, they also have the effect of increasing the porosity and decreasing the strength of the system. At a given water/cement ratio, the effect on the compressive strength of concrete of increasing the volume of entrained air is shown by the curves in Fig. 3–4a.

Compressive Strength and Factors Affecting It

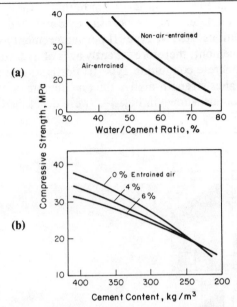

Figure 3-4 Influence of the water/cement ratio, entrained air, and cement content on concrete strength. (From *Concrete Manual,* U.S. Bureau of Reclamation, 1981, and W. A. Cordon, *Properties, Evaluation, and Control of Engineering Materials,* McGraw-Hill Book Company, New York, 1979.)

At a given water/cement ratio or cement content, entrained air generally reduces the strength of concrete. For very low cement contents, entrained air may actually increase the strength.

It has been observed that the extent of strength loss as a result of entrained air depends not only on the water/cement ratio of the concrete mixture (Fig. 3-4a), but also on the cement content. In short, as a first approximation, the strength loss due to air entrainment can be related to the general level of concrete strength. The data in Fig. 3-4b show that at a given water/cement ratio, high-strength concretes (containing a high cement content) suffer a considerable strength loss with increasing amounts of entrained air, whereas low-strength concretes (containing a low cement content) tend to suffer only a little strength loss or may actually gain some strength as a result of air entrainment. This point is of great significance in the design of mass-concrete mixtures (Chapter 11).

The influence of the water/cement ratio and cement content on the response of concrete to applied stress can be explained from the two opposing effects caused by incorporation of air into concrete. By increasing the porosity of the matrix, entrained air will have an adverse affect on the strength of the composite material. On the other hand, by improving workability and compactibility of the mixture, entrained air tends to improve the strength of the transition zone (especially in mixtures with low water and cement contents) and thus improves the strength of concrete. It seems that with a concrete having a low cement content, when air entrainment is accompanied by a substantial reduction in the water content, the adverse effect of air entrainment on the strength of the matrix is more than compensated by the beneficial effect on the transition zone.

Cement type. It may be recalled from Fig. 2–10 that both the water/cement ratio and degree of cement hydration determine the porosity of a hydrated cement paste. Under standard curing conditions, ASTM Type III portland cement hydrates more rapidly than Type I portland cement; therefore, at early ages of hydration and at a given water/cement ratio, a concrete containing Type III portland cement will have lower porosity and have a higher-strength matrix than a concrete containing Type I portland cement. Strength bands, shown in Fig. 3–5, were developed by the

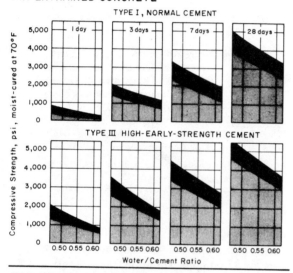

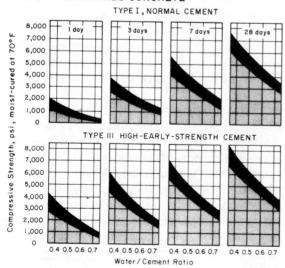

Figure 3–5 Influence of the water/cement ratio, moist curing age, cement type, and air entrainment on concrete strength. (From *Design and Control of Concrete Mixtures, 11th Edition,* Portland Cement Association, Skokie, Ill., 1968, p. 44.)

Portland Cement Association, that take into account the effect of the water/cement ratio and the cement type on both non-air-entrained and air-entrained concrete mixtures. It is stated that a majority of the strength data from laboratories using a variety of materials falls within the strength band curves. For job materials, if laboratory test data or field experience records concerning the relationship between water/cement ratio and strength are not available, it is recommended that the water/cement ratio be estimated from these curves, using the lower edge of the applicable strength band.

It should be noted that at normal temperature the rates of hydration and strength development of ASTM Types II, IV, and V portland cements, Type IS (portland blastfurnace slag cement), and Type IP (portland pozzolan cement) are somewhat slower than ASTM Type I portland cement. At ordinary temperatures, for different types of portland and blended portland cements, the degree of hydration at 90 days and above is usually similar; therefore, the influence of cement composition on porosity of the matrix and strength of concrete is limited to early ages. The effect of portland cement type on the relative strength of concrete at 1, 7, 28, and 90 days is shown by the data in Table 3–1.

Aggregate. In concrete technology, an overemphasis on the relationship between water/cement ratio and strength has caused some problems. For instance, the influence of aggregate on concrete strength is not generally appreciated. It is true that aggregate strength is usually not a factor in normal concrete strength because, with the exception of lightweight aggregates, the aggregate particle is several times stronger than the matrix and the transition zone in concrete. In other words, with most natural aggregates the strength of the aggregate is hardly utilized because the failure is determined by the other two phases.

There are, however, aggregate characteristics other than strength, such as the size, shape, surface texture, grading (particle size distribution), and mineralogy which are known to affect concrete strength in varying degrees. Frequently, the

TABLE 3–1 APPROXIMATE RELATIVE STRENGTH OF CONCRETE AS AFFECTED BY TYPE OF CEMENT

Type of portland cement		Compressive strength (percent of Type I or normal portland cement concrete)			
ASTM	Description	1 day	7 days	28 days	90 days
I	Normal or general purpose	100	100	100	100
II	Moderate heat of hydration and moderate sulfate resisting	75	85	90	100
III	High early strength	190	120	110	100
IV	Low heat of hydration	55	65	75	100
V	Sulfate resisting	65	75	85	100

Source: Adapted from *Design and Control of Concrete Mixtures, 11th Edition,* Portland Cement Association, Skokie, Ill., 1968.

effect of aggregate characteristics on concrete strength can be traced to a change of water/cement ratio. But there is sufficient evidence in the published literature that this is not always the case. Also, from theoretical considerations it may be anticipated that, independent of the water/cement ratio, the size, shape, surface texture, and mineralogy of aggregate particles would influence the characteristics of the transition zone and therefore affect concrete strength.

A change in the *maximum size* of well-graded coarse aggregate of a given mineralogy can have two opposing effects on the strength of concrete. With the same cement content and consistency, concrete mixtures containing larger aggregate particles require less mixing water than those containing smaller aggregate. On the contrary, larger aggregates tend to form weaker transition zones containing more microcracks. The net effect will vary with the water/cement ratio of the concrete and the applied stress. Cordon and Gillispie[2] (Fig. 3–6) showed that in the No. 4 mesh to 3 in. (5 mm to 75 mm) range the effect of increasing maximum aggregate size on the 28-day compressive strengths of the concrete was more pronounced with a high-strength (0.4 water/cement ratio) and a moderate-strength (0.55 water/cement ratio) concrete than with a low-strength concrete (0.7 water/cement ratio). This is because at low water/cement ratios the reduced porosity of the transition zone also begins to play an important role in concrete strength. Furthermore, since the transition zone characteristics appear to affect the tensile strength of concrete more than the compressive strength, it is to be expected that in a given concrete mixture, at a constant water/cement ratio, the tensile/compressive strength ratio would increase with the decreasing size of coarse aggregate.

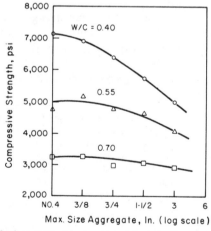

Figure 3–6 Influence of the aggregate size and the water/cement ratio on concrete strength. (From W. A. Cordon and H. A. Gillespie, *J. ACI*, Proc., Vol. 60, No. 8, 1963.)

Generally, high-strength, i.e., low-water/cement ratio, concrete is adversely affected by increasing the size of aggregate. For a given water/cement ratio, the aggregate size does not seem to have much effect on strength in the case of low-strength, i.e. high-water/cement-ratio, concretes.

---

[2] W. A. Cordon and H. A. Gillispie, *J. ACI*, Proc., Vol. 60, No. 8, pp. 1029–50, 1963.

A change in the *aggregate grading* without any change in the maximum size of coarse aggregate, and with water/cement ratio held constant, can influence the concrete strength when this change causes a corresponding change in consistency and bleeding characteristics of the concrete mixture. In a laboratory experiment, with a constant water/cement ratio of 0.6, when the coarse/fine aggregate proportion and the cement content of a concrete mixture were progressively raised to increase the consistency from 2 to 6 in. (50 to 150 mm) of slump, there was about 12 percent decrease in the average 7-day compressive strength (from 3350 psi to 2950 psi). The effects of increased consistency on the strength and cost of the concrete mixtures are shown in Fig. 3-7. The data demonstrate the economic significance of making concrete mixtures at the stiffest possible consistency that is workable from the standpoint of proper consolidation.

It has been observed that a concrete mixture containing a *rough-textured* or crushed aggregate would show somewhat higher strength (especially tensile strength) at early ages than a corresponding concrete containing smooth or naturally weathered aggregate of similar mineralogy. A stronger physical bond between the aggregate and the hydrated cement paste is assumed to be responsible for this. At later ages, when chemical interaction between the aggregate and the cement paste begins to take effect, the influence of the surface texture of aggregate on strength may be reduced. From the standpoint of the physical bond with the cement paste, it may be noted that a smooth-looking particle of weathered gravel will be found possessing adequate roughness and surface area when observed under a microscope. Also, with a given cement content, somewhat more mixing water is usually needed to obtain the desired workability in a concrete mixture containing rough-textured aggregates; thus the small advantage due to better physical bonding may be lost as far as the overall strength is concerned.

Differences in the *mineralogical composition* of aggregates are also known to affect concrete strength. There are numerous reports in the published literature showing that, under identical conditions, the substitution of calcareous for siliceous aggregate resulted in substantial improvement in concrete strength. This was also confirmed from the results of a recent study at the University of California at Berkeley. Not only a decrease in the maximum size of coarse aggregate (Fig. 3-8a), but also a substitution of limestone for sandstone (Fig. 3-8b), improved the ultimate strength (e.g., 56-day) of concrete significantly. The concrete mixtures shown in Fig. 3-8 contained 800 lb/yd^3 (475 kg/m^3) ASTM Type I portland cement, 200 lb/yd^3 (119 kg/m^3) Class F fly ash, 330 lb/yd^3 water (196 kg/m^3), 1750 lb/yd^3 (1038 kg/m^3) crushed coarse aggregate, 1000 lb/yd^3 (590 kg/m^3) natural silica sand, and 32 oz (1 liter) of a water-reducing admixture.

Mixing water. Impurities in water used for mixing concrete, when excessive, may affect not only the concrete strength but also setting time, **efflorescence** (deposits of white salts on the surface of concrete), and corrosion of reinforcing or prestressing steel. In general, mixing water is rarely a factor in concrete strength, because in may specifications for making concrete mixtures the quality of water is

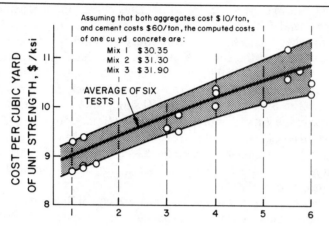
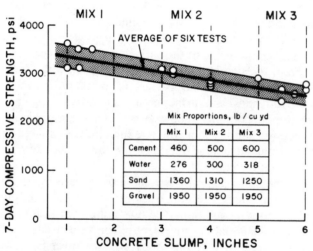

Figure 3-7 Influence of the concrete slump on compressive strength and cost. (Data from students' experiments, University of California at Berkeley.)

For a given water/cement ratio, concrete mixtures with higher slumps tend to bleed and therefore give lower strength. It is not cost-effective to produce concrete mixtures with slumps higher than needed.

protected by a clause stating that the water should be fit for drinking. Municipal drinking water seldom contains dissolved solids in excess of 1000 ppm (parts per million).

As a rule, water unsuitable for drinking may not necessarily be unfit for mixing concrete. From the standpoint of concrete strength, acidic, alkaline, salty, brackish, colored, or foul-smelling water should not be rejected outright. This is important because recycled waters from mining and many other industrial operations can be

Compressive Strength and Factors Affecting It

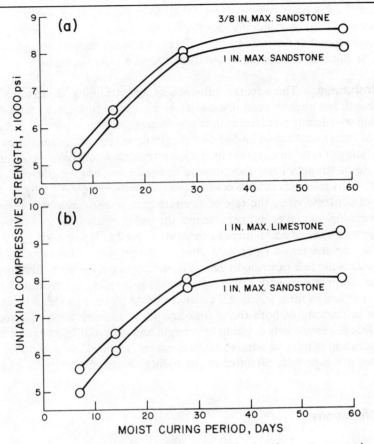

Figure 3-8 Influence of the aggregate size and mineralogy on compressive strength of concrete. (Data from students' experiments, University of California at Berkeley.)

For a given water/cement ratio and cement content, the strength of concrete can be significantly affected by the choice of aggregate size and type.

---

safely used as mixing waters for concrete. The best way to determine the suitability of a water of unknown performance for making concrete is to compare the setting time of cement and the strength of mortar cubes made with the unknown water and a reference water that is clean. The cubes made with the questionable water should have 7- and 28-day compressive strengths equal to or at least 90 percent of the strength of reference specimens made with a clean water; also, the quality of mixing water should not affect the setting time of cement to an unacceptable degree.

Seawater, which contains about 35,000 ppm dissolved salts, is not harmful to the strength of plain concrete. However, with reinforced and prestressed concrete it increases the risk of corrosion of steel; therefore, the use of seawater as concrete-mixing water should be avoided under these circumstances. As a general guideline, from the standpoint of concrete strength the presence of excessive amounts of algae, oil, salt, or sugar in the mixing water should send a warning signal.

Admixtures. The adverse influence of air-entraining admixtures on concrete strength has already been discussed. At a given water/cement ratio the presence of water-reducing admixtures in concrete generally has a positive influence on the rate of cement hydration and early strength development. Admixtures capable of accelerating or retarding cement hydration obviously would have a great influence on the rate of strength gain; the ultimate strengths are not significantly affected. However, many researchers have pointed out the tendency toward a higher ultimate strength of concrete when the rate of strength gain at early ages was retarded.

For ecological and economic reasons, the use of pozzolanic and cementitious by-products as mineral admixtures in concrete is gradually increasing. When used as a partial replacement for portland cement, the presence of mineral admixtures usually retards the rate of strength gain. The ability of a mineral admixture to react at normal temperatures with calcium hydroxide present in the hydrated portland cement paste and to form additional calcium silicate hydrate can lead to significant reduction in porosity of both the matrix and the transition zone. Consequently, considerable improvements in ultimate strength and watertightness can be achieved by incorporation of mineral admixtures in concrete. It should be noted that mineral admixtures are especially effective in increasing the tensile strength of concrete (p. 70).

Curing Conditions

The term **curing of concrete** stands for procedures devoted to promote cement hydration, consisting of control of time, temperature, and humidity conditions immediately after the placement of a concrete mixture into formwork.

At a given water/cement ratio, the porosity of a hydrated cement paste is determined by the degree of cement hydration (Fig. 2–10, case A). Under normal temperature conditions some of the constituent compounds of portland cement begin to hydrate as soon as water is added, but the hydration reactions slow down considerably when the products of hydration coat the anhydrous cement grains. This is because hydration can proceed satisfactorily only under conditions of saturation; it almost stops when the vapor pressure of water in capillaries falls below 80 percent of the saturation humidity. Time and humidity are therefore important factors in the hydration processes controlled by water diffusion. Also, like all chemical reactions, temperature has an accelerating effect on the hydration reactions.

Time. It should be noted that the time-strength relations in concrete technology generally assume moist curing conditions and normal temperatures. At a given water/cement ratio, the longer the moist curing period the higher the strength (Fig. 3–3) assuming that the hydration of anhydrous cement particles is still going

Compressive Strength and Factors Affecting It

on. In thin concrete elements, if water is lost by evaporation from the capillaries, air-curing conditions prevail, and strength will not increase with time (Fig. 3–9).

The evaluation of compressive strength with time is of great concern for structural engineers. ACI Committee 209 recommends the following relationship for moist-cured concrete made with normal portland cement (ASTM Type I):

$$f_{cm}(t) = f_{c28}\left(\frac{t}{4 + 0.85t}\right) \tag{3-4}$$

For concrete specimens cured at 20°C, the CEB-FIP Models Code (1990) suggests the following relationship:

$$f_{cm}(t) = \exp\left[s\left(1 - \left(\frac{28}{t/t_1}\right)^{1/2}\right)\right]f_{cm} \tag{3-5}$$

where $f_{cm}(t)$ = mean compressive strength at age t days
f_{cm} = mean 28-day compressive strength
s = coefficient depending on the cement type, such as $s = 0.20$ for high early strength cements; $s = 0.25$ for normal hardening cements; $s = 0.38$ for slow hardening cements
t_1 = 1 day

Humidity. The influence of the curing humidity on concrete strength is obvious from the data in Fig. 3–9, which show that after 180 days at a given water/cement ratio, the strength of the continuously moist-cured concrete was three times greater than the strength of the continuously air-cured concrete. Furthermore, probably as a result of microcracking in the transition zone caused by drying shrinkage, a slight retrogression of strength occurs in thin members of moist-cured concrete when they are subjected to air drying. The rate of water loss from concrete soon after placement depends not only on the surface/volume ratio of the concrete

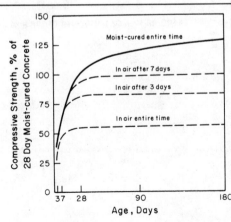

Figure 3–9 Influence of curing conditions on strength. (From *Concrete Manual*, 8th Edition, U.S. Bureau of Reclamation, 1981.)

The curing age would not have any beneficial effect on the concrete strength unless curing is carried out in the presence of moisture.

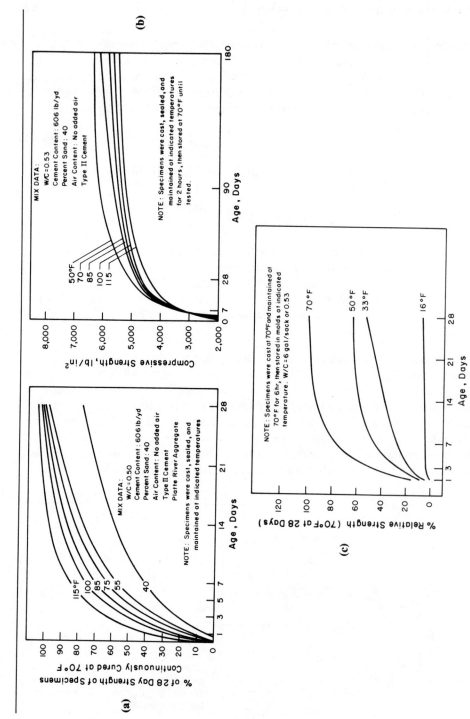

Figure 3-10 Influence of casting and curing temperatures on concrete strength. (From *Concrete Manual*, U.S. Bureau of Reclamation, 1975.)

Concrete casting (placement) and curing temperatures control the degree of cement hydration and thus have a profound influence on the rate of strength development as well as the ultimate strength.

element but also on the temperature, relative humidity, and velocity of the surrounding air.

A minimum period of 7 days of moist curing is generally recommended for concrete containing normal portland cement; obviously, for concretes containing either a blended portland cement or a mineral admixture, a longer curing period would be desirable to ensure the strength contribution from the pozzolanic reaction. Moist curing is provided by spraying or ponding or by covering the concrete surface with wet sand, sawdust, or cotton mats. Since the amount of mixing water used in a concrete mixture is usually more than needed for portland cement hydration (estimated to be about 30 percent by weight of cement), the application of an impermeable membrane soon after concrete placement provides an acceptable way to maintain the process of cement hydration and to ensure a satisfactory rate of development of concrete strength.

Temperature. For moist-cured concrete the influence of temperature on strength depends on the time-temperature history of *casting and curing*. This can be illustrated with the help of three cases: concrete cast (placed) and cured at the same temperature, concrete cast at different temperatures but cured at a normal temperature, and concrete cast at a normal temperature but cured at different temperatures.

In the temperature range 40 to 115°F, when concrete is cast and cured at a specific constant temperature, it is generally observed that up to 28 days, the higher the temperature the more rapid is the cement hydration and the strength gain resulting from it. From the data in Fig. 3–10a, it is evident that the 28-day strength of specimens cast and cured at 40°F was about 80 percent of those cast and cured at 70 to 115°F. At later ages, when the differences in the degree of cement hydration are reduced, such differences in concrete strength are not sustained. On the other hand, as explained below, it has been observed that the higher the casting and curing temperature, the lower will be the ultimate strength.

The data in Fig. 3–10b represent a different time–temperature history of casting and curing. The casting temperature (i.e., the temperature during the first 2 hours after making concrete) was varied between 40 and 115°F; thereafter all concretes were moist cured at a constant temperature of 70°F. The data show that ultimate strengths (180-day) of the concrete cast at 40 or 55° were higher than those cast at 70, 85, 100, or 115°F. From microscopic studies many researchers have concluded that, with low temperature curing, a relatively more uniform microstructure of the hydrated cement paste (especially the pore size distribution) accounts for the higher strength.

For concretes cast at 70°F and subsequently cured at different temperatures from below freezing to 70°F, the effect of curing temperature on strength is shown in Fig. 3–10c. In general, the lower the curing temperature, the lower the strengths up to 28 days. At a curing temperature near freezing (33°F), the 28-day strength was about one-half of the strength of the concrete cured at 70°F; hardly any strength developed at the below-freezing curing temperature (16°F). Since the hydration reactions of portland cement compounds are slow, it appears that adequate temperature levels must be maintained for a sufficient time to provide the needed activation energy for the reactions. This allows the strength development process, which is associated with the progressive filling of voids with hydration products, to proceed unhindered.

The influence of time–temperature history on concrete strength has several important applications in the concrete construction practice. Since the curing temperature is far more important to strength than the placement temperature, ordinary concrete placed in cold weather must be maintained above a certain minimum temperature for a sufficient length of time. Concrete cured in summer or in a tropical climate can be expected to have a higher early strength but a lower ultimate strength than the same concrete cured in winter or in a colder climate. In the precast concrete products industry, steam curing is used to accelerate strength development to achieve quicker mold release. In massive elements, when no measures for temperature control are taken, for a long time the temperature of concrete will remain at a much higher level than the environmental temperature. Therefore, compared to the strength of specimens cured at the normal laboratory temperature, the in situ concrete strength will be higher at early ages and lower at later ages.

Testing Parameters

It is not always appreciated that the results of concrete strength tests are significantly affected by parameters involving the test specimen and loading conditions. Specimen parameters include the influence of size, geometry, and the moisture state of concrete; loading parameters include stress level and duration, and the rate at which stress is applied.

Specimen parameters. In the United States, the standard specimen for testing the compressive strength of concrete is a 6- by 12-in. cylinder. While maintaining the height/diameter ratio equal to 2, if a concrete mixture is tested in compression with cylindrical specimens of varying diameter, the larger the diameter the lower will be the strength. The data in Fig. 3–11 show that, compared to the

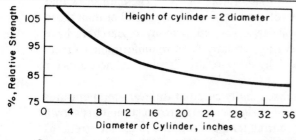

Figure 3–11 Influence of the specimen diameter on concrete strength when the length/diamteter ratio is equal to 2. (From *Concrete Manual*, U.S. Bureau of Reclamation, 1975, pp. 574–75.

Specimen geometry can affect the laboratory test data on concrete strength. The strengths of cylindrical specimens with a slenderness ratio (L/D) above 2 or a diameter above 12 in. are not much influenced by the size effects.

standard specimens, the average strength of 2- by 4-in. and 3- by 6-in. cylindrical specimens was 106 and 108 percent, respectively. When the diameter is increased beyond 18 in., a much smaller reduction in strength is observed. Such variations in strength with variation of the specimen size are to be expected due to the increasing degree of statistical homogeneity in large specimens.

The effect of change in specimen geometry (height/diameter ratio) on the compressive strength of concrete is shown in Fig. 3–12. In general, the greater the ratio of the specimen height to diameter, the lower will be the strength. For instance, compared to the strength of the standard specimens (height/diameter ratio equal to 2), the specimens with the height/diameter ratio of 1 showed about 15 percent higher strength. It may be of interest to point out that concrete strength based on the 6-in. standard cube test, which is prevalent in Europe, is reported to be 10 to 15 percent higher than the strength of the same concrete tested in accordance with standard U.S. practice (6- by 12-in. cylinders).

Because of the effect of moisture state on concrete strength, the standard procedure requires that the specimens be in a moist condition at the time of testing. In compression tests it has been observed that air-dried specimens show 20 to 25 percent higher strength than corresponding specimens tested in a saturated condition. The lower strength of the saturated concrete is probably due to the existence of disjoining pressure within the cement paste.

Loading conditions. The compressive strength of concrete is measured in the laboratory by a uniaxial compression test (ASTM C 469) in which the load is progressively increased to fail the specimen within 2 to 3 min. In practice, most structural elements are subjected to a dead load for an indefinite period and, at times, to repeated loads or to impact loads. It is, therefore, desirable to know the relationship between the concrete strength under laboratory testing conditions and actual loading conditions. The behavior of concrete under various stress states is

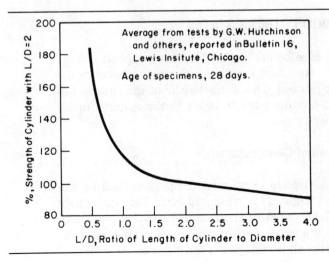

Figure 3–12 Influence of varying the length/diameter ratio on concrete strength. (From *Concrete Manual*, U.S. Bureau of Reclamation, 1975, pp. 574–75.)

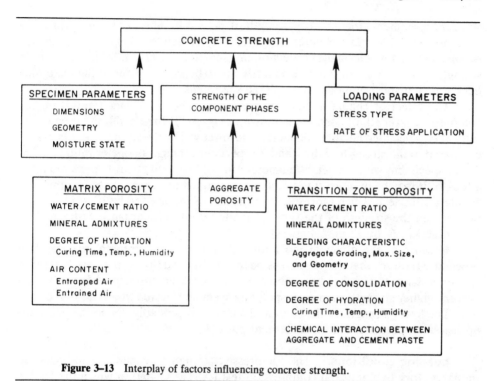

Figure 3–13 Interplay of factors influencing concrete strength.

described in the next section. From this description it can be concluded that the loading condition has an important influence on strength.

To appreciate at a glance the complex web of numerous factors that are capable of influencing the strength of concrete, a summary is presented in Fig. 3–13.

BEHAVIOR OF CONCRETE UNDER VARIOUS STRESS STATES

It was described in Chapter 2 that, even before any load has been applied, a large number of microcracks exist in the transition zone (i.e., the region between the cement paste matrix and coarse aggregate). This characteristic of the structure of concrete plays a decisive role in determining the behavior of the material under various stress states that are discussed next.

Behavior of Concrete under Uniaxial Compression

Stress-strain behavior of concrete subjected to uniaxial compression will be discussed in detail in Chapter 4; only a summary is presented here. The stress-strain curve (Fig. 3–14a) shows a linear-elastic behavior up to about 30 percent of the ultimate strength (f'_c), because under **short-term loading** the microcracks in the transition zone remain undisturbed. For stresses above this point, the curve shows a gradual

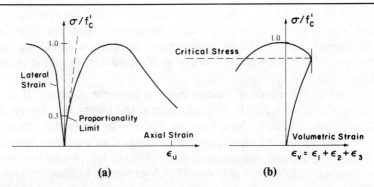

Figure 3-14 Typical plots of compressive stress versus (a) axial and lateral strains, and (b) volumetric strain. (From W. F. Chen, *Plasticity in Reinforced Concrete,* McGraw-Hill Book Company, 1982, p. 20.)

increase in curvature up to about 0.75 f'_c to 0.9 f'_c, then it bends sharply (almost becoming flat at the top) and, finally, descends until the specimen is fractured.

From the shape of the stress-strain curve it seems that, for a stress between 30 to 50 percent of f'_c the microcracks in the transition zone show some extension due to stress concentrations at crack tips; however, no cracking occurs in the mortar matrix. Until this point, crack propagation is assumed to be *stable* in the sense that crack lengths rapidly reach their final values if the applied stress is held constant. For a stress between 50 to 75 percent of f'_c, increasingly the crack system tends to be unstable as the transition zone cracks begin to grow again. When the available internal energy exceeds the required crack-release energy, the rate of crack propagation will increase and the system becomes *unstable*. This happens at compressive stresses above 75 percent of f'_c, when complete fracture of the test specimen can occur by bridging of the mortar and transition zone cracks.

The stress level of about 75 percent of f'_c, which represents the onset of unstable crack propagation, is termed *critical stress*[3]; critical stress also corresponds to the maximum value of volumetric strain (Fig. 3-14b). From the figure it may be noted that when volumetric strain $\epsilon_v = \epsilon_1 + \epsilon_2 + \epsilon_3$ is plotted against stress, the initial change in volume is almost linear up to about 0.75 f'_c; at this point the direction of the volume change is reversed, resulting in a volumetric expansion near or at f'_c.

Above the critical stress level, concrete shows a time-dependent fracture; that is, under *sustained stress conditions* crack bridging between the transition zone and the matrix would lead to failure at a stress level that is lower than the short-term loading strength f'_c. In an investigation reported by Price,[4] when the sustained stress was 90 percent of the ultimate short-time stress, the failure occurred in 1 hr; however, when the sustained stress was about 75 percent of the ultimate short-time stress, it

---

[3] W. F. Chen, *Plasticity in Reinforced Concrete,* McGraw-Hill Book Co., pp. 20-21, 1982.
[4] W. H. Price, *J. ACI,* Proc., Vol. 47, pp. 417-32, 1951.

took 30 years to fail. As the value of the sustained stress approaches that of the ultimate short-time stress, the time to failure decreases. Rusch[5] confirmed this in his tests on 56-day-old, 5000 psi (34MPa) compressive strength, concrete specimens. The long-time failure limit was found to be about 80 percent of the ultimate short-time stress (Fig. 3–15).

In regard to the *effect of loading rate* on concrete strength, it is generally reported that the more rapid the rate of loading, the higher the observed strength value. However, Jones and Richart[6] found that within the range of customary testing, the effect of rate of loading on strength is not large. For example, compared with the data from the standard compression test (ASTM C 469), which requires the rate of uniaxial compression loading to be 35 ± 5 psi/sec, a loading rate of 1 psi/sec reduced the indicated strength of 6- by 12-in. concrete cylinders by about 12 percent; on the other hand, a loading rate of 1000 psi/sec increased the indicated strength by a similar amount.

It is interesting to point out here that the *impact strength* of concrete increases greatly with the rate at which the impact stress is applied. It is generally assumed that the impact strength is directly related to the compressive strength, since both are adversely affected by the presence of microcracks and voids. This assumption

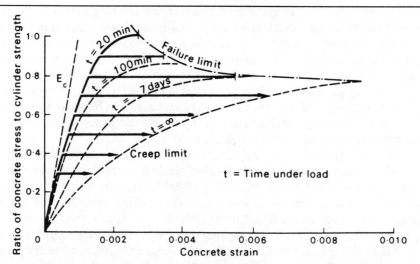

Figure 3–15 Relationship between the short-term and long-term loading strengths. (From H. Rusch. *J. ACI*, Proc., Vol. 57, No. 1, 1960.)

The ultimate strength of concrete is also affected by the rate of loading. Due to progressive microcracking at sustained loads, a concrete will fail at a lower stress than that induced by instantaneous or short-time loading normally used in the laboratory.

---

[5] H. Rusch, *J. ACI*, Proc., Vol. 57, pp. 1–28, 1960.
[6] P. G. Jones and F. E. Richart, *ASTM Proc.*, Vol. 36, pp. 380–91, 1936.

is not completely correct; for the same compressive strength, Green[7] found that the impact strength increased substantially with the angularity and surface roughness of coarse aggregate, and decreased with the increasing size of aggregate. It seems that the impact strength is more influenced than the compressive strength by the transition zone characteristics and thus is more closely related to the tensile strength.

The CEB-FIP Model Code (1990) recommends that the increase in compressive strength due to impact, with rates of loading less than 10^6 MPa/sec., can be computed using the relationship:

$$f_{c,\text{imp}}/f_{cm} = (\dot{\sigma}/\dot{\sigma}_0)_s^\alpha \tag{3-6}$$

where $f_{c,\text{imp}}$ is the impact compressive strength, f_{cm} the compressive strength of concrete $\dot{\sigma}_0 = -1.0$ MPa/sec., $\dot{\sigma}$ the impact stress rate, and $\alpha_s = 1/(5 + 9 f_{cm}/f_{cmo})$, and $f_{cmo} = 10$ MPa.

Ople and Hulsbos[8] reported that, **repeated** or **cyclic loading** has an adverse effect on concrete strength at stress levels greater than 50 percent of f'_c. For instance, in 5000 cycles of repeated loading, concrete failed at 70 percent of the ultimate monotonic loading strength. Progressive microcracking in the transition zone and the matrix are responsible for this phenomenon.

Typical behavior of plain concrete subjected to cyclic compressive loading is shown in Fig. 3–16. For stress levels between 50 and 75 percent of f'_c, a gradual degradation occurs in both the elastic modulus and the compressive strength. As the number of loading cycles increases, the unloading curves show nonlinearities and a characteristic hysteresis loop is formed on reloading. For stress levels at about 75 percent of f'_c, the unloading-reloading curves exhibit strong nonlinearities (i.e., the elastic property of the material has greatly deteriorated). In the beginning, the area of the hysteresis loop decreases with each successive cycle but eventually increases before fatigue failure.[9] Fig. 3–16 shows that the stress-strain curve for monotonic loading serves as a reasonable envelope for the peak values of stress for concrete under cyclic loading.

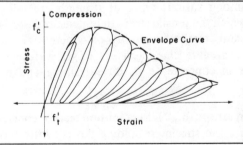

Figure 3–16 Response of concrete to repeated uniaxial loading. (Adapted from P. Karson and J. O. Jirsa, *ASCE Jour. Str. Div.*, Vol. 95, No. ST12, Paper 6935, 1969.)

---

[7] H. Green, *Proceedings, Inst. of Civil Engineers (London)*, Vol. 28, No. 3, pp. 383–96, 1964.
[8] F. S. Ople and C. L. Hulsbos, *J. ACI*, Proc., Vol. 63, pp. 59–81, 1966.
[9] W. F. Chen, *Plasticity in Reinforced Concrete*, McGraw-Hill Book Co., p. 23, 1982.

Behavior of Concrete under Uniaxial Tension

The shape of the stress-strain curve, the elastic modulus, and the Poisson's ratio of concrete under uniaxial tension are similar to those under uniaxial compression. However, there are some important differences in behavior. As the uniaxial tension state of stress tends to arrest cracks much less frequently than the compressive states of stress, the interval of stable crack propagation is expected to be short. Explaining the relatively brittle fracture behavior of concrete in tension tests, Chen states:

> The direction of crack propagation in uniaxial tension is transverse to the stress direction. The initiation and growth of every new crack will reduce the available load-carrying area, and this reduction causes an increase in the stresses at critical crack tips. The decreased frequency of crack arrests means that the failure in tension is caused by a few bridging cracks rather than by numerous cracks, as it is for compressive states of stress. As a consequence of rapid crack propagation, it is difficult to follow the descending part of the stress-strain curve in an experimental test.[10]

The ratio between uniaxial tensile and compressive strengths is generally in the range 0.07 to 0.11. Owing to the ease with which cracks can propagate under a tensile stress, this is not surprising. Most concrete elements are therefore designed under the assumption that the concrete would resist the compressive but not the tensile stresses. However, tensile stresses cannot be ignored altogether because cracking of concrete is frequently the outcome of a tensile failure caused by restrained shrinkage; the shrinkage is usually due either to lowering of concrete temperature or to drying of moist concrete. Also, a combination of tensile, compressive, and shear stresses usually determines the strength when concrete is subjected to flexural or bending loads, such as in highway pavements.

In the preceding discussion on factors affecting the compressive strength of concrete, it was assumed that the compressive strength is an adequate index for all types of strength, and therefore a direct relationship ought to exist between the compressive and the tensile or flexural strength of a given concrete. While as a first approximation the assumption is valid, this may not always be the case. It has been observed that the relationships among various types of strength are influenced by factors such as the methods by which the tensile strength is measured (i.e., direct tension test, splitting test, or flexure test), the quality of concrete (i.e., low-, medium-, or high-strength), the aggregate characteristics (e.g., surface texture and mineralogy), and admixtures in concrete (e.g., air-entraining and mineral admixtures).

Testing methods for tensile strength. Direct tension tests of concrete are seldom carried out, mainly because the specimen holding devices introduce secondary stresses that cannot be ignored. The most commonly used tests for estimating the tensile strength of concrete are the ASTM C 496 splitting tension test and the ASTM C 78 third-point flexural loading test (Fig. 3–17).

[10] Ibid., p. 25.

Behavior of Concrete Under Various Stress States

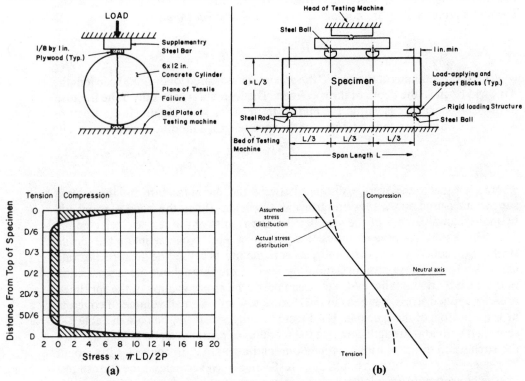

Figure 3–17 (a) Splitting tension test (ASTM C 496): top, diagrammatic arrangement of the test; bottom, stress distribution across the loaded diameter of a cylinder compressed between two plates. (b) Flexural test by third-point loading (ASTM C 78): top, diagrammatic arrangement of the test; bottom, stress distribution across the depth of a concrete beam under flexure.

In the splitting tension test a 6- by 12-in. concrete cylinder is subjected to compression loads along two axial lines which are diametrically opposite. The load is applied continuously at a constant rate within the splitting tension stress range of 100 to 200 psi until the specimen fails. The compressive stress produces a transverse tensile stress which is uniform along the vertical diameter. The splitting tension strength is computed from the formula

$$T = \frac{2P}{\pi l d} \tag{3-7}$$

where T is the tensile strength, P the failure load, l the length, and d the diameter of the specimen. Compared to direct tension, the splitting tension test is known to overestimate the tensile strength of concrete by 10 to 15 percent.

In the third-point flexural loading test, a 6- by 6- by 20-in. (150- by 150- by 500 mm) concrete beam is loaded at a rate of 125 to 175 psi/min. (0.8 to 1.2 MPa/min.).

Flexural strength is expressed in terms of the modulus of rupture, which is the maximum stress at rupture computed from the flexure formula

$$R = \frac{PL}{bd^2} \quad (3\text{--}8)$$

where R is the modulus of rupture, P the maximum indicated load, L the span length, b the width, and d the depth of the specimen. The formula is valid only if the fracture in the tension surface is within the middle third of the span length. If the fracture is outside by not more than 5 percent of the span length, a modified formula is used:

$$R = \frac{3Pa}{bd^2} \quad (3\text{--}9)$$

where a is equal to the average distance between the line of fracture and the nearest support measured on the tension surface of the beam. When the fracture is outside by more than 5 percent of the span length, the results of the test are rejected.

The results from the modulus of rupture test tend to overestimate the tensile strength of concrete by 50 to 100 percent, mainly because the flexure formula assumes a linear stress–strain relationship in concrete throughout the cross section of the beam. Additionally, in direct tension tests the entire volume of the specimen is under applied stress, whereas in the flexure test only a small volume of concrete near the bottom of the specimen is subjected to high stresses. The data in Table 3–2 show that with low-strength concrete the modulus of rupture can be as high as twice the strength in direct tension; for medium- and high-strength concretes it is about 70 percent and 50 to 60 percent higher, respectively. Nevertheless, the flexure test is usually preferred for quality control of concrete for highway and airport pavements, where the concrete is loaded in bending rather than in axial tension.

TABLE 3–2 RELATION BETWEEN COMPRESSIVE, FLEXURAL, AND TENSILE STRENGTH OF CONCRETE

Strength of concrete (psi)			Ratio (%)		
Compressive	Modulus of rupture	Tensile	Modulus of rupture to compressive strength	Tensile strength to compressive strength	Tensile strength to modulus of rupture
1000	230	110	23.0	11.0	48
2000	375	200	18.8	10.0	53
3000	485	275	16.2	9.2	57
4000	580	340	14.5	8.5	59
5000	675	400	13.5	8.0	59
6000	765	460	12.8	7.7	60
7000	855	520	12.2	7.4	61
8000	930	580	11.6	7.2	62
9000	1010	630	11.2	7.0	63

Source: W. H. Price, *J. ACI,* Proc., Vol. 47, p. 429, 1951.

The CEB-FIP Model Code (1990) suggests the following relationship between direct tension strength (f_{ctm}) and flexural strength ($f_{ct,fl}$)

$$f_{ctm} = f_{ct,fl} \frac{2.0(h/h_0)^{0.7}}{1 + 2.0(h/h_0)^{0.7}} \qquad (3\text{--}10)$$

where h is the depth of the beam in mm, $h_0 = 100$ mm and strengths are expressed in MPa units.

Relationship between the Compressive and Tensile Strengths

It has been pointed out before that the compressive and tensile strengths are closely related; however, there is no direct proportionality. As the compressive strength of concrete increases, the tensile strength also increases but at a decreasing rate (Fig. 3–18). In other words, the tensile/compressive strength ratio depends on the general level of the compressive strength; the higher the compressive strength, the lower the ratio. Relationships between compressive and tensile strengths in the f'_c range 1000 to 9000 psi are shown in Table 3–2. It appears that the direct tensile/compressive strength ratio is 10 to 11 percent for low-strength, 8 to 9 percent for medium-strength, and 7 percent for high-strength concrete.

The CEP-FIP Model Code (1990) recommends that the lower and upper bound

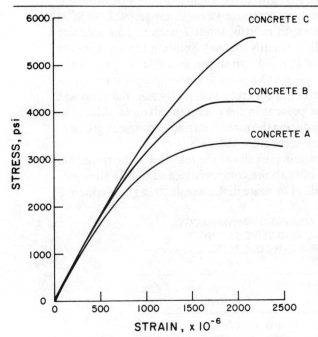

Figure 3–18 Influence of the water/cement ratio on tensile and compressive strengths.

values of the characteristic tensile strength, $f_{ctk,\,max}$ and $f_{ctk,\,min}$, may be estimated from the characteristic strength f_{ck} (in MPa units):

$$f_{ctk,\,min} = 0.95\left(\frac{f_{ck}}{f_{cko}}\right)^{2/3} \quad \text{and} \quad f_{ctk,\,max} = 1.85\left(\frac{f_{ck}}{f_{cko}}\right)^{2/3} \quad (3\text{–}11)$$

where $f_{cko} = 10$ MPa

The mean value of the tensile strength is given by the relationship:

$$f_{ctm} = 1.40\left(\frac{f_{ck}}{f_{cko}}\right)^{2/3} \quad (3\text{–}12)$$

The relationship between the compressive's strength and the tensile/compressive strength ratio seems to be determined by the effect of various factors on properties of both the matrix and the transition zone in concrete. It is observed that not only the curing age but also the characteristics of the concrete mixture, such as water/cement ratio, type of aggregate, and admixtures affect the tensile/compressive strength ratio to varying degrees. For example, after about 1 month of curing the tensile strength of concrete is known to increase more slowly than the compressive strength; that is, the tensile/compressive strength ratio decreases with the curing age. At a given curing age, the tensile/compressive ratio also decreases with the decrease in water/cement ratio.

In concrete containing calcareous aggregate or mineral admixtures, it is possible to obtain after adequate curing a relatively high tensile/compressive strength ratio even at high levels of compressive strength. From Table 3–2 it may be observed that with ordinary concrete, in the high compressive strength range (8000 to 9000 psi) the direct-tensile/compressive strength ratio is about 7 percent (the splitting tensile/compressive strength ratio will be slightly higher). Splitting tension data for the high-strength concrete mixtures of Fig. 3–8 are shown in Table 3–3. From the data it is evident that compared to a typical 7 to 8 percent splitting tension/compressive strength ratio (f_{st}/f_c) for a high-strength concrete with no fly ash, the ratio was considerably higher when fly ash was present in the concrete mixtures. Also, the beneficial effects on the f_{st}/f_c ratio of reducing the maximum size of coarse aggregate, or of changing the aggregate type, are clear from the data.

Whereas factors causing a decrease in porosity of the matrix and the transition zone lead to a general improvement of both the compressive and tensile strengths of concrete, it seems that the magnitude of increase in the tensile strength of concrete

TABLE 3–3 EFFECT OF AGGREGATE MINERALOGY AND SIZE ON TENSILE/COMPRESSIVE STRENGTH RELATIONS IN HIGH-STRENGTH CONCRETES (60-DAYS MOIST CURED)

	f_c (psi)	f_{st} (psi)	f_{st}/f_c (psi)
Sandstone, 1-in. max.	8100	760	0.09
Limestone, 1-in. max.	9270	1010	0.11
Sandstone, 3/8-in. max.	8550	860	0.10

remains relatively small unless the intrinsic strength of hydration products comprising the transition zone is improved at the same time. That is, the tensile strength of concrete with a low-porosity transition zone will continue to be weak as long as large numbers of oriented crystals of calcium hydroxide are present there (see Fig. 2–14). The size and concentration of calcium hydroxide crystals in the transition zone can be reduced as a result of chemical reactions when either a pozzolanic admixture (see Fig. 6–14) or a reactive aggregate is present. For example, an interaction between calcium hydroxide and the aggregate resulting in the formation of recrystallized calcium carbonate in the transition zone was probably the reason for the relatively large increase in the tensile strength of concrete containing the calcareous aggregate, as shown by the data in Table 3–3.

Tensile Strength of Mass Concrete

Engineers working with reinforced concrete ignore the low value of the tensile strength of concrete and use steel to pick up tensile loads. With massive concrete structures, such as dams, it is impractical to use steel reinforcement. Therefore, a reliable estimate of the tensile strength of concrete is necessary, especially for judging the safety of a dam under seismic loading. Raphael[11] recommends the values obtained by the splitting test of the modulus of rupture test, augmented by the multiplier found appropriate by dynamic tensile tests, or about 1.5. Alternatively, depending on the loading conditions, the plots of tensile strength as a function of compressive strength (Fig. 3–19) may be used for this purpose. The lowest plot,

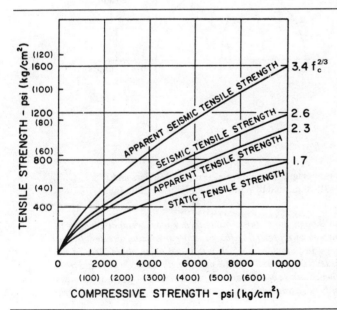

Figure 3–19 *Design chart for tensile strength* (ref. 11)

[11] J. Raphael, *J. ACI Proc.* Vol. 81, No. 2, pp. 158–64, 1984.

$f_t = 1.7 f_c^{2/3}$ represents actual tensile strength under long-time or static loading. The second plot, $f_r = 2.3 f_c^{2/3}$ is also for static loading but takes into account the nonlinearity of concrete and is to be used with finite element analyses. The third plot, $f_t = 2.6 f_c^{2/3}$ is the actual tensile strength of concrete under seismic loading, and the highest plot $f_r = 3.4 f_c^{2/3}$ is the apparent tensile strength under seismic loading that should be used with linear finite element analyses.

Behavior of Concrete under Shearing Stress

Although pure shear is not encountered in concrete structures, an element may be subject to the simultaneous action of compressive, tensile, and shearing stresses. Therefore, the failure analysis under multiaxial stresses is carried out from a phenomenological rather than a material standpoint. Although the Coulomb-Mohr theory is not exactly applicable to concrete, the Mohr rupture diagram (Fig. 3–20)

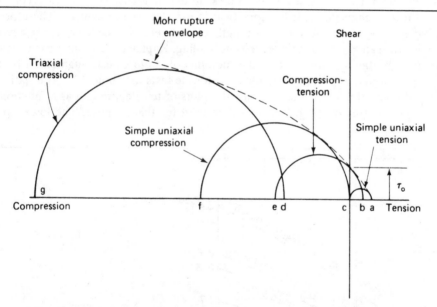

Figure 3–20 Typical Mohr rupture diagram for concrete. (From S. Mindess and J. F. Young, *Concrete*, © 1981, p. 401. Reprinted by permission of Prentice-Hall, Inc., Englewood Cliffs, N.J.)

According to Mindess and Young: "Even though the Columb-Mohr theory may not apply exactly to concrete, this remains the most convenient way of representing failure under multiaxial stresses. In this figure, τ_o represents the strength of concrete in pure shear, which was stated to be about 20% of the compressive strength. The distance c–f represents the uniaxial compressive strength (as might be found from the standard cylinder test). It may be seen, then, that an applied transverse compressive stress (c–d) increases the compressive strength (c–g). On the other hand, a transverse tensile stress (c–b) will decrease the apparent compressive strength to (c–e)."

offers a way of representing the failure under combined stress states from which an estimate of the shear strength can be obtained.

In Fig. 3–20, the strength of concrete in pure shear is represented by the point at which the failure envelope intersects the vertical axis, τ_o. By this method it has been found that the shear strength is approximately 20 percent of the uniaxial compressive strength.

Behavior of Concrete under Biaxial and Multiaxial Stresses

Biaxial compressive stresses $\sigma_1 = \sigma_2$ can be generated by subjecting a cylindrical specimen to hydrostatic pressure in radial directions. To develop a truly biaxial stress state, the friction between the concrete cylinder and the steel platens must be avoided. Also penetration of the pressure fluid into microcracks and pores on the surface of concrete must be prevented by placing the specimen into a suitable membrane.

Kupfer, Hilsdorf, and Rusch[12] investigated the biaxial strength of three types of concrete (2700, 4450, and 8350 psi unconfined uniaxial compressive strengths), when the specimens were loaded without longitudinal restraint by replacing the solid bearing platens of a conventional testing machine with *brush bearing platens*. These platens consisted of a series of closely spaced small steel bars that were flexible enough to follow the concrete deformations without generating appreciable restraint of the test piece. Figure 3–21 shows the typical stress-strain curves for concrete under (a) biaxial compression, (b) combined tension-compression, and (c) biaxial tension. Biaxial stress interaction curves are shown in Fig. 3–22.

The test data show that the strength of concrete subjected to biaxial compression (Fig. 3–21a) may be up to 27 percent higher than the uniaxial strength. For equal compressive stresses in two principal directions, the strength increase is approximately 16 percent. Under biaxial compression-tension (Fig. 3–21b), the compressive strength decreased almost linearly as the applied tensile strength increased. From the biaxial strength envelope of concrete (Fig. 3–22a) it can be seen that the strength of concrete under biaxial tension is approximately equal to the uniaxial tensile strength.

Chen[13] points out that concrete *ductility* under biaxial stresses has different values depending on whether the stress states are compressive or tensile. For instance, in biaxial compression (Fig. 3–21a) the average maximum compressive microstrain is about 3000 and the average maximum tensile microstrain varies from 2000 to 4000. The tensile ductility is greater in biaxial compression than in uniaxial compression. In biaxial tension-compression (Fig. 3–21b), the magnitude at failure of both the principal compressive and tensile strains decreases as the tensile stress increases. In biaxial tension (Fig. 3–21c), the average value of the maximum principal tensile microstrain is only about 80.

---

[12] H. Kupfer, H. K. Hilsdorf, and H. Rusch, *J. ACI*, Proc., Vol. 66, pp. 656–66, 1969.

[13] W. F. Chen, *Plasticity in Reinforced Concrete*, McGraw-Hill Book Co., p. 27, 1982.

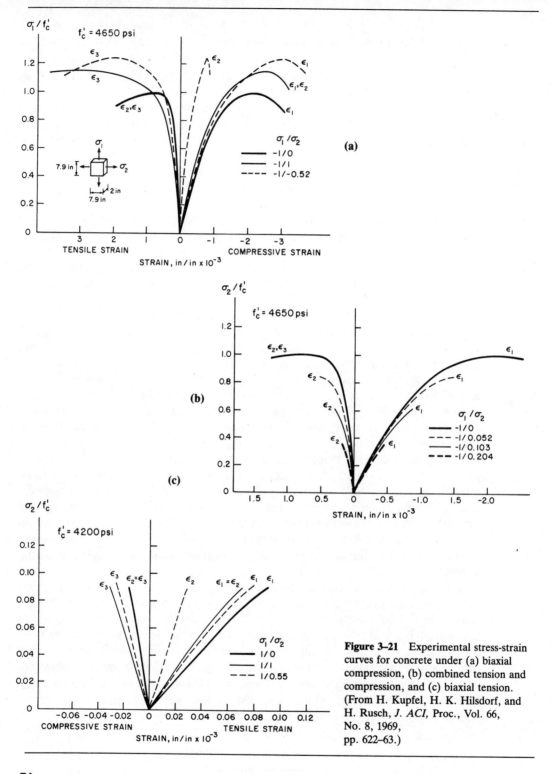

Figure 3-21 Experimental stress-strain curves for concrete under (a) biaxial compression, (b) combined tension and compression, and (c) biaxial tension. (From H. Kupfel, H. K. Hilsdorf, and H. Rusch, *J. ACI,* Proc., Vol. 66, No. 8, 1969, pp. 622-63.)

Behavior of Concrete Under Various Stress States

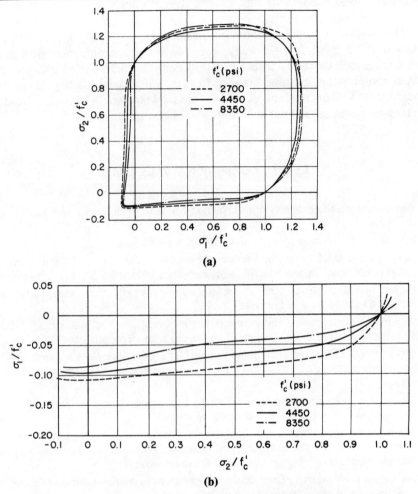

Figure 3–22 Biaxial stress interaction curves: (a) strength envelope; (b) strength under combined tension and compression, and under biaxial tension. (From H. Kupfel, H. K. Hilsdorf, and H. Rusch, *J. ACI*, Proc., Vol. 66, No. 8, 1969, pp. 662–63.)

The data in Fig. 3–22a show that the level of uniaxial compressive strength of concrete virtually does not affect the shape of the biaxial stress interaction curves or the magnitude of values (the uniaxial compressive strength of concretes tested was in the range 2700 to 8300 psi). However, in compression-tension and in biaxial tension (Fig. 3–22b), it is observed that the relative strength at any particular biaxial stress combination decreases as the level of uniaxial compressive strength increases. Neville[14] suggests that this is in accord with the general observation that the ratio

[14] A. Neville, *Hardened Concrete: Physical and Mechanical Aspects*, ACI Monograph No. 6, pp. 48–53, 1971.

of uniaxial tensile strength to compressive strength decreases as the compressive strength level rises (see Table 3–2).

The behavior of concrete under multiaxial stresses is very complex and, as was explained in Fig. 3–20, it is generally described from a phenomenological point of view. Unlike the laboratory tests for determining the behavior of concrete under uniaxial compression, splitting tension, flexure, and biaxial loading, there are no standard tests for concrete subjected to multiaxial stresses. Moreover, there is no general agreement as to what should be the failure criterion.

TEST YOUR KNOWLEDGE

1. Why is strength the property most valued in concrete by designers and quality control engineers?
2. In general, discuss how strength and porosity are related to each other.
3. Abrams established a rule which relates the water-cement ratio to strength of concrete. List two additional factors which have a significant influence on the concrete strength.
4. Explain how the water/cement ratio influences the strength of the cement paste matrix and the transition zone in concrete.
5. Why may air entrainment reduce the strength of medium- and high-strength concrete mixtures, but increase the strength of low-strength concrete mixtures?
6. For the five ASTM types of portland cements, at a given water/cement ratio would the ultimate strengths be different? Would the early-age strengths be different? Explain your answer.
7. Discuss the two opposing effects on strength that are caused by an increase in the maximum size of aggregate in a concrete mixture.
8. At a given water/cement ratio, a change in the cement content or aggregate grading can be made to increase the consistency of concrete. Why is it not desirable to produce concrete mixtures of a higher consistency than necessary?
9. Can we use recycled water from industrial operations as mixing water in concrete? What about the use of seawater for this purpose?
10. What do you understand by the term *curing of concrete?* What is the significance of curing?
11. From the standpoint of concrete strength, which of the two options is undesirable, and why?
    (a) Concrete cast at 40°F and cured at 70°F.
    (b) Concrete cast at 70°F and cured at 40°F.
12. Many factors have an influence on the compressive strength of concrete. Briefly explain which one of the two options listed below will result in higher strength at 28 days:
    (a) Water/cement ratio of 0.5 vs 0.4.
    (b) Moist curing temperature of 25°C vs 10°C.
    (c) Using test cylinders of size 150 by 300 mm vs 75 by 150 mm.
    (d) For the compression test using a loading rate of 250 psi/sec vs 50 psi/sec.
    (e) Testing the specimens in a saturated condition vs air-dry condition.

13. The temperature during the placement of concrete is known to have an effect on later age strength. What would be the effect on the 6-month strength when a concrete mixture is placed at (a) 10°C and (b) 35°C.
14. In general, how are the compressive and tensile strengths of concrete related? Is this relationship independent of concrete strength? If not, why? Discuss how admixtures and aggregate mineralogy can affect the relationship.

SUGGESTIONS FOR FURTHER STUDY

BROOKS, A. E., and K. NEWMAN, eds., *The Structure of Concrete,* Proc. Int. Conf., London, Cement and Concrete Association, Wexham Springs, Slough, U.K., pp. 49–318, 1968.

MINDESS, S., and J. F. YOUNG, *Concrete,* Prentice Hall, Inc., Englewood Cliffs, N.J., 1981, Chap. 15, pp. 381–406.

NEVILLE, A. M., *Properties of Concrete,* Pitman Publishing, Inc., Marshfield, Mass., 1981, Chap. 5, pp. 268–358.

SWAMY, R. N., "On the Nature of Strength in Concrete," in *Progress in Concrete Technology,* ed. V. M. Malhotra, CANMET, Ottawa, pp. 189–228, 1980.

CHAPTER 4

Dimensional Stability

PREVIEW

Concrete shows elastic as well as inelastic strains on loading, and shrinkage strains on drying or cooling. When restrained, shrinkage strains result in complex stress patterns that often lead to cracking.

In this chapter causes of nonlinearity in the stress–strain relation of concrete are discussed, and different types of elastic moduli and the methods of determining them are described. Explanations are provided as to why and how aggregate, cement paste, transition zone, and testing parameters affect the modulus of elasticity.

The stress effects resulting from drying shrinkage and viscoelastic strains in concrete are not the same; however, with both phenomena the underlying causes and the controlling factors have much in common. Important parameters that influence drying shrinkage and creep are discussed, such as aggregate content, aggregate stiffness, water content, cement content, time of exposure, relative humidity, and size and shape of a concrete member.

Thermal shrinkage is of great importance in large concrete elements. Its magnitude can be controlled by the coefficient of thermal expansion of aggregate, the cement content and type, and the temperature of concrete-making materials. The concepts of extensibility and tensile strain capacity and their significance to concrete cracking are introduced.

TYPES OF DEFORMATIONS AND THEIR SIGNIFICANCE

Deformations in concrete, which often lead to cracking, occur as a result of material's response to external load and environment. When freshly hardened concrete (whether loaded or unloaded) is exposed to the ambient temperature and humidity, it generally undergoes **thermal shrinkage** (shrinkage strain associated with cooling)[1] and **drying shrinkage** (shrinkage strain associated with the moisture loss). Which one of the two shrinkage strains will be dominant under a given condition depends, among other factors, on the size of the member, characteristics of concrete-making materials, and mix proportions. Generally in thick members (e.g., larger than a few meters) the drying shrinkage is less important a factor than the thermal shrinkage.

It should be noted that structural elements of hardened concrete are always under restraint, usually from subgrade friction, end members, reinforcing steel, or even from differential strains between the exterior and the interior of concrete. When the shrinkage strain in an elastic material is fully restrained, it results in elastic tensile stress; the magnitude of the induced stress σ is determined by the product of the strain ϵ and the elastic modulus E of the material ($\sigma = E\epsilon$). The elastic modulus of concrete is also dependent on the characteristics of concrete-making materials and mix proportions, but not necessarily to the same degree or even in the same way as the shrinkage strains. The material is expected to crack when a combination of the elastic modulus and the shrinkage strain induces a stress level that reaches its tensile strength (Fig. 4–1, curve a). Given the low tensile strength

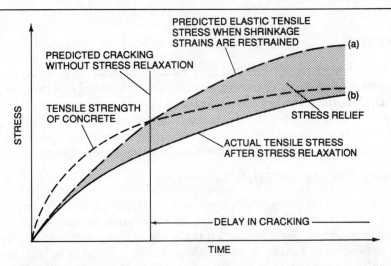

Figure 4–1 Influence of shrinkage and creep on concrete cracking. (Adapted from a presentation by J. W. Kelly at the Associated General Contractor's meeting in San Francisco, June 20, 1963.)

Under restraining conditions in concrete, the interplay between the elastic tensile stresses induced by shrinkage strains and the stress relief due to viscoelastic behavior is at the heart of deformations and cracking in most structures.

---

[1] Exothermic reactions between cement compounds and water tend to raise the temperature of concrete (see Chapter 6).

of concrete, this does happen in practice but, fortunately, not exactly as predicted by the theoretically computed values of the induced elastic tensile stress.

To understand the reason why a concrete element may not crack at all or may crack but not soon after exposure to the environment, we have to consider how concrete would respond to sustained stress or to sustained strain. The phenomenon of a gradual increase in strain with time under a given level of sustained stress is called **creep.** The phenomenon of gradual decrease in stress with time under a given level of sustained strain is called *stress relaxation.* Both manifestations are typical of viscoelastic materials. When a concrete element is restrained, the viscoelasticity of concrete will manifest into a progressive decrease of stress with time (Fig. 4-1, curve b). Thus under the restraining conditions present in concrete, the interplay between elastic tensile stresses induced by shrinkage strains and stress relief due to viscoelastic behavior is at the heart of deformations and cracking in most structures.

In practice the stress–strain relations in concrete are much more complex than indicated by Fig. 4-1. First, concrete is not a truly elastic material; second, neither the strains nor the restraints are uniform throughout a concrete member; therefore, the resulting stress distributions tend to vary from point to point. Nevertheless, it is important to know the elastic, drying shrinkage, thermal shrinkage, and viscoelastic properties of concrete, and the factors affecting these properties.

ELASTIC BEHAVIOR

The elastic characteristics of a material are a measure of its stiffness. In spite of the nonlinear behavior of concrete, an estimate of the elastic modulus (the ratio between the applied stress and instantaneous strain within an assumed proportional limit) is necessary for determining the stresses induced by strains associated with environmental effects. It is also needed for computing the design stresses under load in simple elements, and moments and deflections in complicated structures.

Nonlinearity of the Stress–Strain Relationship

From typical σ–ϵ curves for aggregate, hydrated cement paste, and concrete loaded in uniaxial compression (Fig. 4-2), it becomes immediately apparent that relative to aggregate and cement paste, concrete is really not an elastic material. Neither the strain on instantaneous loading of a concrete specimen is found to be truly directly proportional to the applied stress, nor is it fully recovered upon unloading. The cause for nonlinearity of the stress–strain relationship has been explained from studies on the process of progressive microcracking in concrete under load by researchers, including those from Cornell University.[2] Figure 4-3 is based on their work and a review of the subject by Glucklich.[3]

[2] T. C. Hsu, F. O. Slate, G. M. Sturman, and G. Winter, *J. ACI*, Proc., Vol. 60, No. 2, pp. 209–23, 1963; S. P. Shah and F. O. Slate, *Proceedings of a Conference on Structure of Concrete,* Cement and Concrete Association, Wexham Springs, Slough, U.K.; editors: A. E. Brooks and K. Newman, pp. 82–92, 1968.

[3] J. Glucklich, *Proceedings of a Conference on the Structure of Concrete,* Cement and Concrete Association, Wexham Springs, Slough, U.K., pp. 176–89, 1968.

Elastic Behavior

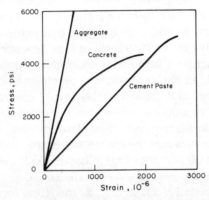

Figure 4–2 Typical stress-strain behaviors of cement paste, aggregate, and concrete. (Based on T. C. Hsu, ACI Monograph 6, 1971, p. 100.)

The properties of complex composite materials need not be equal to the sum of the properties of the components. Thus both hydrated cement paste and aggregates show linear elastic properties, whereas concrete does not.

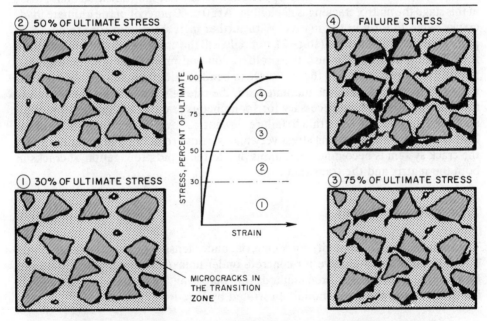

Figure 4–3 Diagrammatic representation of the stress-strain behavior of concrete under uniaxial compression. (Based on J. Glucklich, *Proc. Int. Conf. on the Structure of Concrete*, Cement and Concrete Association, Wexham Springs, Slough, U.K., 1968, pp. 176–85.)

The progress of internal microcracking in concrete goes through various stages, which depend on the level of applied stress.

From the standpoint of the relationship between the stress level (expressed as percent of the ultimate load) and microcracking in concrete, Figure 4-3 reflects four stages of concrete behavior. It is now well known that even before the application of external load, microcracks already exist in the transition zone between the matrix mortar and coarse aggregate in concrete. The number and width of these cracks in a concrete specimen would depend, among other factors, on bleeding characteristics, strength of the transition zone, and the curing history of concrete. Under ordinary curing conditions (when a concrete element is subjected to drying or thermal shrinkage effects), due to the differences in their elastic moduli, differential strains will be set up between the matrix and the coarse aggregate, causing cracks in the transition zone. Below about 30 percent of the ultimate load, the transition zone cracks remain stable; therefore, the σ–ϵ curve remains linear (stage 1, Fig. 4-3).

Above 30 percent of the ultimate load (stage 2), as the stress increases, the transition zone microcracks begin to increase in length, width, and numbers. Thus with increasing stress, the ϵ/σ ratio increases and the curve begins to deviate appreciably from a straight line. However, until about 50 percent of the ultimate stress, a **stable system of microcracks** may be assumed to exist in the transition zone; also, at this stage the matrix cracking is negligible. At 50 to 60 percent of the ultimate load, cracks begin to form in the matrix. With further increase in stress up to about 75 percent of the ultimate load (stage 3), not only will the crack system in the transition zone become **unstable** but also the proliferation and propagation of cracks in the matrix will increase, causing the σ–ϵ curve to bend considerably toward the horizontal. At 75 to 80 percent of the ultimate load, the rate of strain energy release seems to reach the critical level necessary for spontaneous crack growth under sustained stress and the material will strain to failure. In short, above 75 percent of the ultimate load (stage 4), with increasing stress very high strains are developed, indicating that the crack system is becoming continuous due to the rapid propagation of cracks in both the matrix and the transition zone.

Types of Elastic Moduli

The **static modulus of elasticity** for a material under tension or compression is given by the slope of the σ–ϵ curve for concrete under uniaxial loading. Since the curve for concrete is nonlinear, three methods for computing the modulus are used. This has given rise to the three moduli illustrated by Fig. 4-4:

1. The **tangent modulus** is given by the slope of a line drawn tangent to the stress–strain curve at any point on the curve.
2. The **secant modulus** is given by the slope of a line drawn from the origin to a point on the curve corresponding to a 40 percent stress of the failure load.
3. The **chord modulus** is given by the slope of a line drawn between two points on the stress-strain curve. Compared to the secant modulus, instead of the origin the line is drawn from a point representing a longitudinal strain of 50

Elastic Behavior

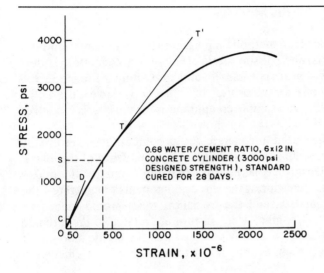

Figure 4–4 Different types of elastic moduli and the method by which these are determined.

μin./in. to the point that corresponds to 40 percent of the ultimate load. Shifting the base line by 50 microstrain is recommended to correct for the slight concavity that is often observed at the beginning of the stress-strain curve.

The **dynamic modulus of elasticity,** corresponding to a very small instantaneous strain, is approximately given by the **initial tangent modulus,** which is the tangent modulus for a line drawn at the origin. It is generally 20, 30, and 40 percent higher than the static modulus of elasticity for high-, medium-, and low-strength concretes, respectively. For stress analysis of structures subjected to earthquake or impact loading it is more appropriate to use the dynamic modulus of elasticity, which can be determined more accurately by a sonic test.

The **flexural modulus of elasticity** may be determined from the deflection test on a loaded beam. For a beam simply supported at the ends and loaded at midspan, ignoring the shear deflection, the approximate value of the modulus is calculated from:

$$E = \frac{PL^3}{48Iy}$$

where y is the midspan deflection due to load P, L the span length, and I the moment of inertia. The flexural modulus is commonly used for design and analysis of pavements.

Determination of the Static Elastic Modulus

ASTM C 469 describes a standard test method for measurement of the static modulus of elasticity (the chord modulus) and Poisson's ratio of 6- by 12-in. concrete cylinders loaded in longitudinal compression at a constant loading rate within the range 35 ± 5 psi. Normally, the deformations are measured by a linear variable differential transformer. Typical σ–ϵ curves, with sample computations for the elastic moduli of the three concrete mixtures of Fig. 3–18, are shown in Fig. 4–5.

The elastic modulus values used in concrete design computations are usually estimated from empirical expressions that assume direct dependence of the elastic modulus on the strength and density of concrete. As a first approximation this makes sense because the stress–strain behavior of the three components of concrete—the aggregate, the cement paste matrix, and the transition zone—would indeed be determined by their individual strengths, which in turn are related to the ultimate strength of the concrete. Furthermore, it may be noted that the elastic modulus of the aggregate (which controls the aggregate's ability to restrain volume changes in

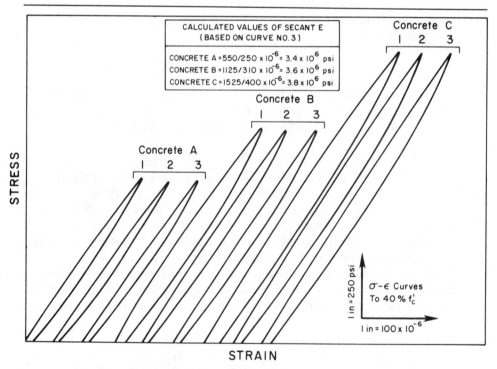

Figure 4–5 Determination of the secant modulus in the laboratory (ASTM C469). See Fig. 3–18 for the composition and strength characteristics of concrete mixtures. (Unpublished data from students' experiments, University of California at Berkeley.)

Elastic Behavior

the matrix) is directly related to its porosity, and the measurement of the unit weight of concrete happens to be the easiest way of obtaining an estimate of the porosity of aggregate in concrete.

According to ACI Building Code 318, with a concrete unit weight between 90 and 155 lb/ft³, the modulus of elasticity can be determined from

$$E_c = w_c^{1.5} \times 33 f_c'^{1/2}$$

where E_c is the static modulus of elasticity (psi), w_c the unit weight (lb/ft³), and f_c' the 28-day compressive strength of standard cylinders. In the CEB-FIP Model Code (1990), the modulus of elasticity of normal-weight concrete can be estimated from

$$E_c = 2.15 \times 10^4 (f_{cm}/10)^{1/3}$$

where E_c is the 28-day modulus of elasticity of concrete (MPa) and f_{cm} the average 28-day compressive strength. If the the actual compressive strength is not known, f_{cm} should be replaced by $f_{ck} + 8$, where f_{ck} is the characteristic compressive strength. The elastic modulus-strength relationship was developed for quartizitic aggregate concretes. For other types of aggregates, the modulus of elasticity can be obtained by multiplying E_c with factors α_e from Table 4-1. It should be mentioned that the CEB-FIP expression is valid for characteristic strengths up to 80 MPa (11,600 psi), while the ACI equation is valid up to 41 MPa (6000 psi). Extensions to the ACI formulation are presented in Chapter 11 (see high-strength concrete). Assuming concrete density to be 145 lb/ft³ (2320 kg/m³), the computed values of the modulus of elasticity for normal-weight concrete according to both the ACI Building Code and CEB-FIP Model Code (1990) are shown in Table 4-2.

From the following discussion of the factors affecting the modulus of elasticity of concrete, it will be apparent that the computed values shown in Table 4-2, which are based on strength and density of concrete, should be treated as approximate only. This is because the transition-zone characteristics and the moisture state of the specimen at the time of testing do not have similar effects on strength and elastic modulus.

TABLE 4-1 EFFECT OF TYPE OF AGGREGATE ON MODULUS OF ELASTICITY

Aggregate type	α_e
Basalt, dense limestone	1.2
Quartizitic	1.0
Limestone	0.9
Sandstone	0.7

Poisson's Ratio

For a material subjected to simple axial load, the ratio of the lateral strain to axial strain *within the elastic range* is called **Poisson's ratio.** Poisson's ratio is not generally needed for most concrete design computations; however, it is needed for structural analysis of tunnels, arch dams, and other statically indeterminate structures.

TABLE 4-2 ELASTIC MODULI FOR NORMAL-WEIGHT CONCRETES (QUARTIZITIC AGGREGATE)

ACI Building Code		CEB-FIP Model Code	
f'_{cm}	E_c	f'_{cm}	E_c
psi (MPa)	$\times 10^6$ (GPa)	psi (MPa)	$\times 10^6$ psi (GPa)
3,000 (21)	3.1 (21)	3,000 (21)	4.0 (28)
4,000 (27)	3.6 (25)	4,000 (27)	4.3 (30)
5,000 (34)	4.1 (28)	5,000 (34)	4.7 (32)
6,000 (41)	4.4 (30)	6,000 (41)	5.0 (34)

With concrete the values of Poisson's ratio generally vary between 0.15 and 0.20. There appears to be no consistent relationship between Poisson's ratio and concrete characteristics such as water/cement ratio, curing age, and aggregate gradation. However, Poisson's ratio is generally lower in high-strength concrete, and higher for saturated concrete and for dynamically loaded concrete.

Factors Affecting Modulus of Elasticity

In homogeneous materials a direct relationship exists between density and modulus of elasticity. In heterogeneous, multiphase materials such as concrete, the volume fraction, the density and the modulus of elasticity of the principal constituents and the characteristics of the transition zone determine the elastic behavior of the composite. Since density is oppositely related to porosity, obviously the factors that affect the porosity of aggregate, cement paste matrix, and the transition zone would be important. For concrete the direct relation between strength and elastic modulus arises from the fact that both are affected by the porosity of the constituent phases, although not to the same degree.

Aggregate. Among the coarse aggregate characteristics that affect the elastic modulus of concrete, porosity seems to be the most important. This is because aggregate porosity determines its stiffness, which in turn controls the ability of aggregate to restrain matrix strains. Dense aggregates have a high elastic modulus. In general, the larger the amount of coarse aggregate with a high elastic modulus in a concrete mixture, the greater would be the modulus of elasticity of concrete. Since in low-, or medium-strength concretes, the strength of concrete is not affected by the aggregate porosity, this shows that all variables may not control the strength and the elastic modulus in the same way.

Rock core tests have shown that the elastic modulus of natural aggregates of low porosity such as granite, trap rock, and basalt is in the range 10 to 20×10^6 psi, while with sandstones, limestones, and gravels of the porous variety it varies from 3 to 7×10^6 psi. Lightweight aggregates are highly porous; depending on the porosity the elastic modulus of a lightweight aggregate may be as low as 1×10^6 psi or as high as 4×10^6 psi. Generally, the elastic modulus of lightweight-aggregate concrete ranges from 2.0 to 3.0 × 10⁶ psi, which is between 50 and 75 percent of the modulus for normal-weight concrete of the same strength.

Other properties of aggregate also influence the modulus of elasticity of concrete. For example, the maximum size, shape, surface texture, grading, and mineralogical composition can influence microcracking in the transition zone and thus affect the shape of the stress–strain curve.

Cement paste matrix. The elastic modulus of the cement paste matrix is determined by its porosity. The factors controlling the porosity of the cement paste matrix, such as water/cement ratio, air content, mineral admixtures, and degree of cement hydration are listed in Fig. 3–13. Values in the range 1 to 4 × 10⁶ psi as the elastic moduli of hydrated portland cement pastes of varying porosity have been reported. It should be noted that these values are similar to the elastic moduli for lightweight aggregates.

Transition zone. In general, void spaces, microcracks, and oriented calcium hydroxide crystals are relatively more common in the transition zone than in the bulk cement paste; therefore, they play a very important part in determining the stress–strain relations in concrete. The factors controlling the porosity of the transition zone are listed in Fig. 3–13.

It has been reported that the strength and elastic modulus of concrete are not influenced to the same degree by curing age. With different concrete mixtures of varying strength, it was found that at later ages (i.e., 3 months to 1 year) the elastic modulus increased at a higher rate than the compressive strength (Fig. 4–6). It is possible that the beneficial effect of improvement in the density of the transition zone, as a result of slow chemical interaction between the alkaline cement paste and

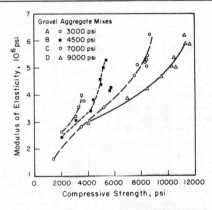

Figure 4–6 Relationship between the compressive strength and elastic modulus. (Based on J. J. Shideler, *J. ACI*, Proc., Vol. 54, No. 4, 1957.)

The upward tendency of the $E - f'_c$ curves from different-strength concrete mixtures tested at regular intervals up to 1 year shows that at later ages the elastic modulus increases at a faster rate than the compressive strength.

aggregate, is more pronounced for the stress–strain relationship than for compressive strength of concrete.

Testing parameters. It is observed that regardless of mix proportions or curing age, concrete specimens that are tested in wet conditions show about 15 percent higher elastic modulus than the corresponding specimens tested in dry condition. Interestingly, the compressive strength of the specimen behaves in the opposite manner; that is, the strength is higher by about 15 percent when the specimens are tested in dry condition. It seems that drying of concrete produces a different effect on the cement paste matrix than on the transition zone; while the former gains in strength owing to an increase in the van der Waals force of attraction in the hydration products, the latter loses strength due to microcracking. The compressive strength of the concrete increases when the matrix is strength determining; however, the elastic modulus is reduced because increase in the transition-zone microcracking greatly affects the stress–strain behavior. There is yet another explanation for the phenomenon. In a saturated cement paste the adsorbed water in the C-S-H is load-bearing, therefore its presence contributes to the elastic modulus; on the other hand, the disjoining pressure (see Chapter 2) in the C-S-H tends to reduce the van der Waals force of attraction, thus lowering the strength.

The advent and degree of nonlinearity in the stress–strain curve obviously would depend on the rate of application of load. At a given stress level the rate of crack propagation and hence the modulus of elasticity is dependent on the rate at which load is applied. Under instantaneous loading, only a little strain can occur prior to failure, and the elastic modulus is very high. In the time range normally required to test the specimens (2 to 5 min) the strain is increased by 15 to 20 percent, hence the elastic modulus decreases correspondingly. For very slow loading rates, the elastic and the creep strains would be superimposed, thus lowering the elastic modulus further.

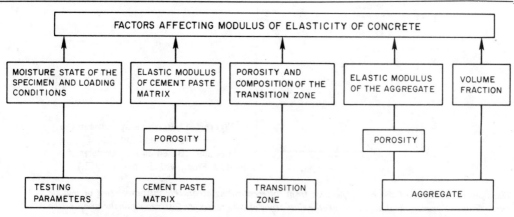

Figure 4-7 Various parameters that influence the modulus of elasticity of concrete.

Figure 4–7 presents a summary chart showing all the factors discussed above which affect the modulus of elasticity of concrete.

DRYING SHRINKAGE AND CREEP

For a variety of reasons it is desirable to discuss both the drying shrinkage and viscoelastic phenomena together. First, both drying shrinkage and creep originate from the same source, the hydrated cement paste; second, the strain–time curves are very similar; third, the factors that influence the drying shrinkage also influence the creep and generally in the same way; fourth, in concrete the microstrain of each, 400 to 1000×10^{-6}, is large and cannot be ignored in structural design; and fifth, both are partially reversible.

Causes

As described in Chapter 2, a saturated cement paste will not remain dimensionally stable when exposed to ambient humidities that are below saturation, mainly because the loss of physically adsorbed water from C-S-H results in a shrinkage strain. Similarly, when a hydrated cement paste is subjected to a sustained stress, depending on the magnitude and duration of applied stress, the C-S-H will lose a large amount of the physically adsorbed water and the paste will show a creep strain. This is not to suggest that there are no other causes contributing to creep in concrete; however, the loss of adsorbed water under sustained pressure appears to be the most important cause. In short, both the drying shrinkage and creep strains in concrete are assumed to be related mainly to the ***removal of adsorbed water*** from the hydrated cement paste. The difference is that in one case a differential relative humidity between the concrete and the environment is the driving force, while in the other it is a sustained applied stress. Again as stated in Chapter 2, a minor cause of the contraction of the system, either as a result of drying or an applied stress, is the removal of water held by hydrostatic tension in small capillaries (< 50 nm) of the hydrated cement paste.

The causes of creep in concrete are more complex. It is generally agreed that in addition to moisture movements there are other causes that contribute to the creep phenomenon. The nonlinearity of the stress–strain relation in concrete, especially at stress levels greater than 30 to 40 percent of the ultimate stress, clearly shows the contribution of the transition zone microcracks to creep. Increase in creep strain, which invariably occurs when concrete is simultaneously exposed to drying conditions, is caused by additional microcracking in the transition zone owing to drying shrinkage.

The occurrence of delayed elastic response in aggregate is yet another cause of creep in concrete. Since the cement paste and the aggregate are bonded together, the stress on the former gradually declines as load is transferred to the latter, which with increasing load transfer deforms elastically. Thus the delayed elastic strain in aggregate contributes to total creep.

Effect of Loading and Humidity Conditions on Drying Shrinkage and Viscoelastic Behavior

In practice, drying shrinkage and viscoelastic behavior usually take place simultaneously. Consider the various combinations of loading, restraining, and humidity conditions presented in Table 4-3. Application of a constant stress on a concrete specimen under conditions of 100 percent relative humidity, leads to an increase of strain over time, which is called **basic creep**. This condition often arises in massive concrete structures where drying shrinkage can be neglected. Now, instead of applying a constant stress let us analyze the case where a constant strain is imposed on the concrete specimen. When the strain is applied, the concrete specimen will have an instantaneous elastic stress; however the stress will decrease over time by the phenomenon of stress relaxation. Both creep and stress relaxation can be visualized as resulting from the application of stress to a classical spring and dashpot model (springs and dashpots connected either in series or in parallel are discussed in Chapter 12).

Exposing an unrestrained concrete specimen to low relative humidity conditions causes drying shrinkage which increases over time. However, if the specimen is restrained, i.e. if it is not free to move, the strain will be zero but tensile stresses will develop over time. This condition is the reason for most of the cracks due to drying shrinkage.

It has been observed that when concrete is under load and simultaneously exposed to low relative humidity environments, the total strain is higher than the sum of elastic strain, free shrinkage strain (drying shrinkage strain of unloaded concrete), and basic creep strain (without drying). The additional creep that occurs when the specimen under load is also drying is called **drying creep**. Total creep is the sum of basic and drying creep, however, it is a common practice to ignore the distribution between the basic and the drying creep, and creep is simply considered as the deformation under load in excess of the sum of the elastic strain and free-drying shrinkage strain.

The interaction between the restrained drying shrinkage strain and stress relaxation due to the viscoelastic behavior of concrete was illustrated in Fig. 4-1 and is also shown in Table 4-3. Because of the boundary conditions, the strain is zero and the magnitude of tensile stresses caused by drying shrinkage is reduced by stress relaxation. Note that presentation of creep data can be done in different ways which have given rise to special terminology. For instance, **specific creep** is the creep strain per unit of applied stress and **creep coefficient** is the ratio of creep strain to elastic strain.

Reversibility

Typical behavior of concrete on drying and rewetting or on loading and unloading is shown in Fig. 4-8. Both the drying shrinkage and the creep phenomena in concrete exhibit a degree of irreversibility that has practical significance. Fig. 4-8 shows that

Drying Shrinkage and Creep

TABLE 4–3 COMBINATIONS OF LOADING, RESTRAINING, AND HUMIDITY CONDITIONS

MECHANISM	DIAGRAM	STRAIN VERSUS TIME	STRESS VERSUS TIME	NOTES
BASIC CREEP	σ_0 applied	Elastic + Creep curve	Constant σ_0	• NO MOISTURE MOVEMENT BETWEEN CONCRETE AND AMBIENT (NO DRYING SHRINKAGE) • CONSTANT STRESS OVER TIME
STRESS RELAXATION	Initial configuration, ϵ_0 imposed	Constant ϵ_0	Elastic then Relaxation decay	• CONSTANT STRAIN OVER TIME
DRYING SHRINKAGE (UNRESTRAINED)	RH < 100%	Increasing shrinkage strain	Zero stress	• THE MEMBER IS FREE TO MOVE • NO STRESSES ARE GENERATED
DRYING SHRINKAGE (RESTRAINED)	l_0, fully restrained	Zero strain	Tensile stress develops	• DEVELOPMENT OF TENSILE STRESS
DRYING SHRINKAGE (UNDER CONSTANT STRAIN)	Initial configuration, ϵ_0, RH < 100%	Constant ϵ_0	Elastic then Shrinkage decay	• THE PREVIOUS EXAMPLE IS A PARTICULAR CASE WITH $\xi = 0$
CREEP + DRYING SHRINKAGE	σ_0 applied, RH < 100%	Elastic + Basic Creep + Drying Shrinkage + Drying Creep	Constant σ_0	• THE TOTAL STRAIN IS NOT THE SUM OF THE ELASTIC, BASIC CREEP AND DRYING SHRINKAGE STRAIN. THE STRAIN DUE TO DRYING CREEP SHOULD BE INCLUDED
DRYING SHRINKAGE + STRESS RELAXATION (RESTRAINED)	l_0, restrained, RH < 100%	Zero strain	Relaxation and Shrinkage → Resulting Stress	• THE RELAXATION STRESS OPPOSES THE STRESS DUE TO DRYING SHRINKAGE
DRYING SHRINKAGE + STRESS RELAXATION (UNDER CONSTANT STRAIN)	Initial configuration, ϵ_0, RH < 100%	Constant ϵ_0	Elastic + Shrinkage + Relaxation → Resulting Stress	• SHRINKAGE AND RELAXATION STRESS ACT IN THE SAME DIRECTION

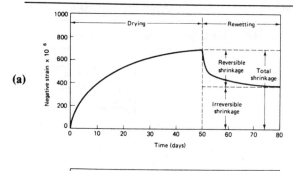

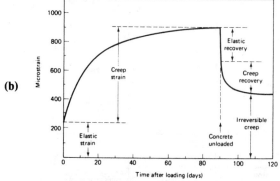

Figure 4–8 Reversibility of drying shrinkage and creep. (From S. Mindess and J. F. Young, *Concrete*, © 1981, pp. 486, 501. Reprinted by permission of Prentice-Hall, Inc., Englewood Cliffs, N.J.)

There is a remarkable similarity of concrete behavior on (a) drying and rewetting, and (b) loading (uniaxial compression) and unloading.

after the first drying, concrete did not return to the original dimension on rewetting. Drying shrinkage has therefore been categorized into **reversible shrinkage,** which is the part of total shrinkage that is reproducible on wet–dry cycles; and **irreversible shrinkage,** which is the part of total shrinkage on first drying that cannot be reproduced on subsequent wet–dry cycles. The irreversible drying shrinkage is probably due to the development of chemical bonds within the C-S-H structure as a consequence of drying. The improvement in the dimensional stability of concrete as a result of first drying has been used to advantage in the manufacture of precast concrete products.

The creep curve for plain concrete subjected to a sustained uniaxial compression for 90 days and thereafter unloaded is shown in Fig. 4-8b. When the specimen is unloaded the instantaneous or elastic recovery is approximately of the same order as the elastic strain on first application of the load. The instantaneous recovery is followed by a gradual decrease in strain called **creep recovery.** Although the creep recovery occurs more rapidly than creep, the reversal of creep strain is not total. Similar to the drying shrinkage (Fig. 4–8a), this phenomenon is defined by the corresponding terms, **reversible** and **irreversible creep.** A part of the reversible creep may be attributed to the delayed elastic strain in aggregate, which is fully recoverable.

Factors Affecting Drying Shrinkage and Creep

In practice, moisture movements in hydrated cement paste, which essentially control the drying shrinkage and creep strains in concrete, are influenced by numerous simultaneously interacting factors. The interrelations among these factors are quite complex and not easily understood. The factors are categorized and discussed below individually, solely for the purpose of understanding their relative significance.

Materials and mix-proportions. The main source of moisture-related deformations in concrete is the hydrated cement paste. Therefore, many attempts have been made to obtain expressions relating the drying shrinkage or creep strains to the volume fraction of hydrated cement paste in concrete (which is determined by the cement content and degree of hydration). Although both the drying shrinkage strain and the creep strain are a function of the hydrated cement paste content, a direct proportionality does not exist because the restraint against deformation has a major influence on the magnitude of the deformation.

Most theoretical expressions for predicting the drying shrinkage or creep of concrete assume that the elastic modulus of concrete can provide an adequate measure of the degree of restraint against deformation and that, as a first approximation, the elastic modulus of the aggregate determines the elastic modulus of the concrete. When the elastic modulus of the aggregate becomes part of a mathematical expression, it is convenient to relate the drying shrinkage or the creep strain to the aggregate fraction rather than the cement paste fraction in concrete. This is easily done because the sum of the two is constant.

Powers[4] investigated the drying shrinkage of concretes containing two different aggregates and water/cement ratios of 0.35 or 0.50. From the data shown in Fig. 4–9a) the ratio of shrinkage of concrete (S_c) to shrinkage of the cement paste (S_p) can be related exponentially to the volume fraction of the aggregate (g) in concrete

$$\frac{S_c}{S_p} = (1 - g)^n \tag{4-1}$$

L'Hermite[5] found that values of (n) varied between 1.2 and 1.7 depending on the elastic modulus of aggregate. From the standpoint of shrinkage-causing and shrinkage-restraining constituents in concrete, Powers suggested that any unhydrated cement present may be considered a part of the aggregate (Fig. 4–9a).

Figure 4–9b shows that a similar relationship exists between the volume concentration of aggregate and creep of concrete. Neville[6] suggested that creep of

---

[4] T. C. Powers, *Rev. Mater. Construct.* (Paris), No. 545, pp. 79–85, 1961.

[5] R. L'Hermite, *Proc. Fourth Int. Symp. Chemistry of Cements,* National Bureau of Standards, Washington, D.C., 1962, pp. 659–94.

[6] A. M. Neville, *Mag. Concr. Res.* (London), Vol. 16, No. 46, pp. 21–30, 1964.

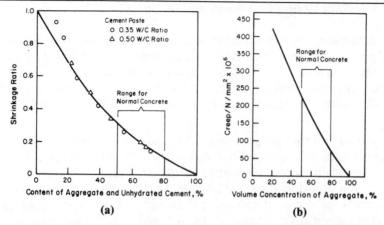

Figure 4-9 Influence of aggregate content on (a) drying shrinkage and (b) creep. [(a), From ACI Monograph 6, 1971, p. 126; (b), from Concrete Society (London), Technical Paper 101, 1973.]

Aggregate content in concrete is the most important factor affecting drying shrinkage and creep. Unhydrated cement does not shrink and therefore may be included in the aggregate.

concrete (C_c) and cement paste (C_p) can be related to the sum of the aggregate (g) and unhydrated cement (μ) contents:

$$\log \frac{C_p}{C_c} = \alpha \log \frac{1}{1 - g - \mu} \tag{4-2}$$

In well-cured concrete, neglecting the small fraction of unhydrated cement (μ), the expression can be rewritten as

$$\frac{C_c}{C_p} = (1 - g)^\alpha \tag{4-3}$$

Thus the expressions for creep and drying shrinkage are similar.

The grading, maximum size, shape, and texture of the aggregate have also been suggested as factors influencing the drying shrinkage and creep. It is generally agreed that the ***modulus of elasticity of the aggregate*** is the most important; the influence of other aggregate characteristics may be indirect, that is, through their effect on the ***aggregate content*** of concrete or on the compactability of the concrete mixture.

The influence of aggregate characteristics, primarily the elastic modulus, was confirmed by Troxell et al.'s study[7] (Fig. 4-10) of creep and shrinkage of concrete over a period of 23 years. Since the modulus of elasticity of aggregate affects the elastic deformation of concrete, a good correlation was found between the elastic deformation of concrete and the drying shrinkage or creep values. Using fixed mix

---

[7] G. E. Troxell, J. M. Raphael, and R. E. Davis, *Proc. ASTM,* Vol. 58, pp. 1101–20, 1958.

Drying Shrinkage and Creep

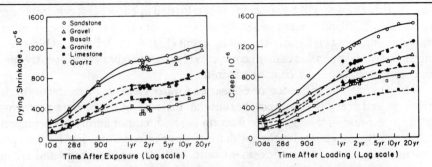

Figure 4-10 Influence of aggregate type on drying shrinkage and creep. (From G. E. Troxell et al., *Proc. ASTM,* Vol. 58, 1958; and ACI Monograph 6, 1971, pp. 128, 151. Reprinted with permission from ASTM Copyright, ASTM, 1916 Race Street, Philadelphia, PA 19103.

The modulus of elasticity of aggregate can affect the magnitude of ultimate drying shrinkage and creep up to 2-1/2 times. Generally, dense limestones and quartz have higher elastic moduli than sandstones and gravel.

proportions, it was found that the 23-year drying shrinkage values of concretes containing quartz and limestone aggregates were 550 and 650 × 10^{-6}, respectively; concretes containing gravel and sandstone showed 1140 and 1260 × 10^{-6} drying shrinkage, respectively. The elastic deformation of the concretes containing either quartz or limestone aggregate was approximately 220 × 10^{-6}; and the elastic deformation of the concretes containing either gravel or sandstone was approximately 280 × 10^{-6}. The corresponding creep values were 600, 800, 1070, and 1500 × 10^{-6} for concretes containing limestone, quartz, gravel, and sandstone, respectively. The importance of the aggregate modulus in controlling concrete deformations is obvious from Troxell et al.'s data, which show that both the drying shrinkage and creep of concrete increased 2.5 times when an aggregate with a high elastic modulus was substituted by an aggregate with low elastic modulus.

Although the influence of aggregate type on creep and drying shrinkage is similar, a closer examination of the curves in Fig. 4–10 shows subtle differences. For example, compared to the drying shrinkage strain, the creep of concretes containing basalt and quartz aggregates was relatively higher. A possible explanation is the higher degree of microcracking in the transition zone when a relatively nonreactive aggregate is present in the system. This underscores the point that creep in concrete is controlled by more than one mechanism.

Within limits, variations in the ***fineness and composition of portland cement*** affect the rate of hydration, but not the volume and the characteristics of the hydration products. Therefore, many researchers have observed that normal changes in cement fineness or composition, which tend to influence the drying shrinkage behavior of small specimens of cement paste or mortar, have a negligible effect on concrete. Obviously, with a given aggregate and mix proportions, if the type of cement influences the strength of concrete at the time of application of load, the creep of concrete will be affected. When loaded at early ages, concrete containing ordinary portland cement generally shows higher creep than the corresponding concrete containing high-early-strength cement (Fig. 4–11b). Portland blast-furnace slag and portland pozzolan cement concretes also show higher creep at early age than the corresponding Type I cement concrete.

In general, the influence of **cement and water contents** in concrete on drying shrinkage and creep is not direct, because an increase in the cement paste volume means a decrease in the aggregate fraction (g) and, consequently, a corresponding increase in the moisture-dependent deformations in concrete. For a given cement content, with increasing water/cement ratio both drying shrinkage and creep are known to increase. A decrease in strength (therefore, the elastic modulus) and increase in permeability of the system are probably responsible for this. The data in Fig. 4–11a) show that, for a given water/cement ratio, both drying shrinkage and creep increased with increasing cement content. This is expected due to an increase in the volume of the cement paste; however, in actual practice it does not always happen.

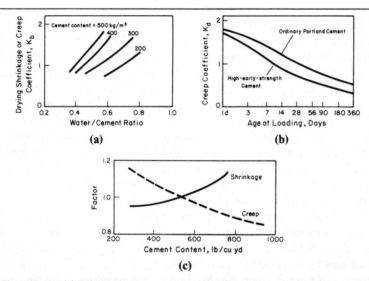

Figure 4–11 (a) Effect of water content on shrinkage or creep; (b) Effect of cement type on creep; (c) influence of cement content on drying shrinkage and creep. [(a)-(b), From *International Recommendations for the Design and Construction of Concrete Structures*, CEB/FIP, 1970; (c) From T. R. Jones, T. J. Hirsch, and H. K. Stephenson, Texas Transportation Institute Report E52, 1959; and ACI Monograph 6, 1971, p. 178.]

Probably due to a decrease in strength and an increase in the permeability of concrete, both the drying shrinkage and creep are increased with increasing water/cement ratios. Compared to normal or ordinary portland cement, creep is slightly reduced by using a high-early-strength cement. Not all parameters affect the drying shrinkage and creep in the same way; with a given water/cement ratio, a higher cement content will tend to increase the drying shrinkage but reduce the creep.

Drying Shrinkage and Creep

The results from many experimental investigations have shown that the foregoing theoretical analysis holds good for drying shrinkage but not always for creep. Experimental data show that within a wide range of concrete strengths, creep is inversely proportional to the *strength of concrete* at the time of application of load. It appears, therefore, that the effect of lowering the aggregate content on possible increase in creep is more than compensated by a reduction in creep, which is associated with the increase in concrete strength. Curves illustrating the effect of cement content on both drying shrinkage and creep at a constant water/cement ratio are shown in Fig. 4–11c.

Concrete admixtures such as calcium chloride, granulated slag, and pozzolans, tend to increase the volume of fine pores in the cement hydration product. Since drying shrinkage and creep in concrete are directly associated with the water held by small pores in the range 3 to 20 nm, concretes containing admixtures capable of pore refinement usually show higher drying shrinkage and creep. Water-reducing and set-retarding admixtures, which are capable of effecting better dispersion of anhydrous cement particles in water, also lead to pore refinement in the hydration product. In general, it is to be expected that admixtures which increase the drying shrinkage will increase the creep.

Time and humidity. Diffusion of the adsorbed water and the water held by capillary tension in small pores (under 50 nm) of hydrated cement paste to large capillary voids within the system or to the atmosphere is a time-dependent process that takes place over long periods. From long-time creep and drying shrinkage tests lasting for more than 20 years, Troxell et al. found that for a wide range of concrete mix proportions, aggregate types, and environmental and loading conditions, only 20 to 25 percent of the 20-year drying shrinkage was realized in 2 weeks, 50 to 60 percent in 3 months, and 75 to 80 percent in 1 year (Fig. 4–12a). Surprisingly, similar results were found for the creep strains shown in Fig. 4–12b.

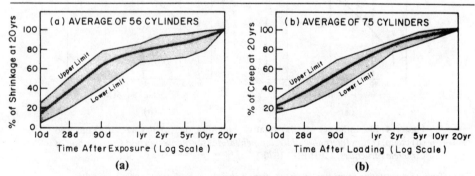

Figure 4–12 The time dependency of (a) drying shrinkage and (b) creep. (From G. E. Troxell et al., *Proc. ASTM,* Vol. 58, 1958. Reprinted with permission from ASTM. Copyright, ASTM, 1916 Race Street, Philadelphia PA 19103.)

For a wide range of concrete mixtures, drying shrinkage and creep show a similar time dependency.

An increase in the atmospheric humidity is expected to slow down the relative rate of moisture flow from the interior to the outer surfaces of concrete. For a given condition of exposure, the effects of relative humidity of air on drying shrinkage strain (Fig. 4-13a) and creep coefficient (Fig. 4-13b) are illustrated in the charts published by the Comité Euro-International du Béton (CEB).[8] At 100 percent RH, the drying shrinkage (E_c) is assumed to be zero, rising to about 200 microstrain at 80 percent RH, and 400 microstrain at 45 percent RH. Similarly, the creep coefficient which is one of the five partial coefficients contributing to total creep is assumed to be 1 at 100 percent RH, rising to about 2 at 80 percent RH, and 3 at 45 percent RH. Recent data from the Comité, showing the effect of humidity conditions and thickness of concrete structure on the ultimate drying shrinkage and creep, is presented in Fig. 4-14.

Geometry of the concrete element. Because of resistance to water transport from the interior of the concrete to the atmosphere, the rate of water loss would obviously be controlled by the length of the path traveled by the water, which is being expelled during drying shrinkage and/or creep. At a constant RH, both the size and the shape of a concrete element determine the magnitude of drying shrinkage and creep. It is convenient to express the size and shape parameters by a single quantity expressed in terms of effective or **theoretical thickness,** which is equal to the area of the section divided by the semiperimeter in contact with the atmosphere. The relations between theoretical thickness and the drying shrinkage and creep coefficient, as given in the CEB charts, are illustrated by Fig. 4-14a and b.

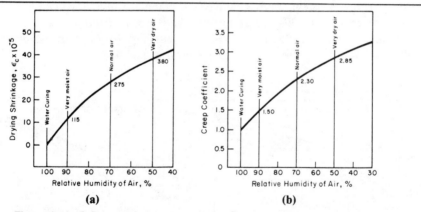

Figure 4-13 Influence of the relative humidity on (a) drying shrinkage and (b) creep. (From *International Recommendations for the Design and Construction of Concrete Structures,* CEB/FIP, 1970.)

The effects of the relative humidity of air on drying shrinkage and creep are also similar.

---

[8] *International Recommendations for the Design and Construction of Concrete Structures.* CEB/FIP, 1976.

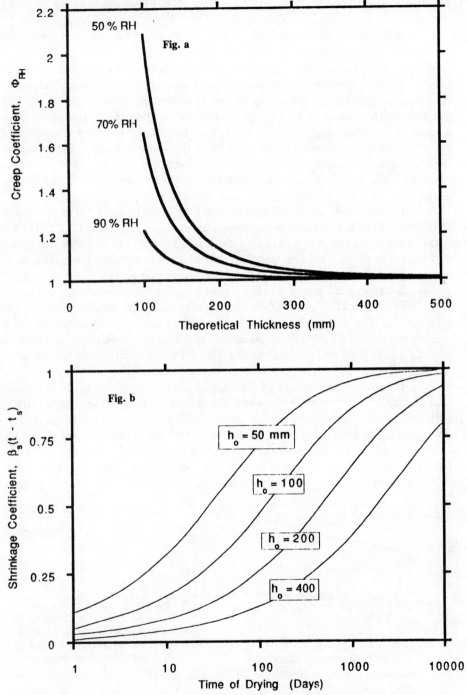

Figure 4–14 (a) Influence of specimen size and relative humidity on the creep coefficient. (b) Influence of exposure time and specimen size on the drying shrinkage coefficient (data from equations given by CEB-FIP Model Code, 1990).

Additional factors affecting creep. The curing history of concrete, temperature of exposure, and magnitude of applied stress are known to affect drying creep more than drying shrinkage, probably because of a greater influence of these factors on the transition-zone characteristics (i.e., porosity, microcracking, and strength). Depending on the *curing history* of a concrete element, creep strains in practice may be significantly different from those in a laboratory test carried out at a constant humidity. For instance, drying cycles can enhance microcracking in the transition zone and thus increase creep. For the same reason, it has often been observed that alternating the environmental humidity between two limits would result in a higher creep than that obtained at a constant humidity (within those limits).

The *temperature* to which concrete is exposed can have two counteracting effects on creep. If a concrete member is exposed to a higher than normal temperature as a part of the curing process before it is loaded, the strength will increase and the creep strain would be less than that of a corresponding concrete stored at a lower temperature. On the other hand, exposure to high temperature during the period under load can increase creep. Nasser and Neville[9] found that in the range 70 to 160°F (21° to 71° C), the 350-day creep increased approximately 3.5 times with temperature (Fig. 4–15). The influence of temperature on creep is of considerable interest to nuclear PCRV (prestressed concrete reactor vessel) structures because neutron attenuation and gamma-ray absorption causes the concrete temperature to rise (see Chapter 11).

In regard to the *intensity of the applied stress,* Troxell et al. found a direct proportionality between the magnitude of sustained stress and the creep of concrete with a 0.69-water/cement ratio (3000 psi or 20 MPa nominal compressive strength). For example, specimens cured for 90 days and then loaded for 21 years showed 680, 1000, and 1450×10^{-6} creep strains, corresponding to sustained stress levels of 600 (4 MPa), 900 (6 MPa), and 1200 psi (8 MPa), respectively (Fig. 4–16).

The proportionality is valid as long as the applied stress is in the linear domain

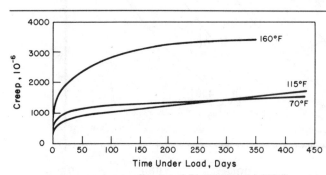

Figure 4–15 Effect of concrete temperature on creep. (From K. W. Nasser and A. M. Neville, *J. ACI, Proc.,* Vol. 64, No. 2, 1967; and ACI Monograph 6, 1971, p. 162.)

At a stress-strength ratio of 70 percent, the 350-day creep can increase 3-5 times if the surrounding temperature is raised from 70°F to 160°F.

---

[9] K. W. Nasser and A. M. Neville, *J. ACI, Proc.,* Vol. 64, No. 2, pp. 97–103, 1967.

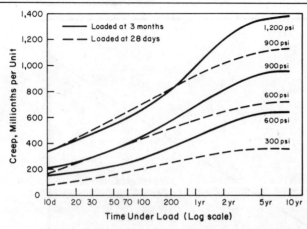

Figure 4–16 Effect of magnitude of sustained stress on creep. (From G. E. Troxell et al., *Proc. ASTM*, Vol. 58, 1958. Reprinted with permission, from ASTM, 1916 Race Street, Philadelphia PA 19103.)

Creep is directly proportional to the magnitude of sustained stress. With 90-day-old concrete specimens, the amount of ultimate creep doubled when the loading stress was increased from 600 psi to 1200 psi. Because of the effect of strength on creep, the figure shows that, at a given stress level, lower creep values were obtained for the longer period of curing before application of the load.

of stress-to-strain relationship (e.g., 0.4 stress-strength ratio for a compressive stress). At high stress-strength ratios a correction factor should be used from the data in Fig. 4–17.

THERMAL SHRINKAGE

In general, solids expand on heating and contract on cooling. The strain associated with change in temperature will depend on the coefficient of thermal expansion of the material and the magnitude of temperature drop or rise. Except under extreme climatic conditions, ordinary concrete structures suffer little or no distress from changes in ambient temperature. However, in massive structures, the combination of heat produced by cement hydration and relatively poor heat dissipation conditions results in a large rise in concrete temperature within a few days after placement. Subsequently, cooling to the ambient temperature often causes the concrete to crack. Since the primary concern in the design and construction of mass concrete structures is that the completed structure remain a monolith, free of cracks, every effort to control the temperature rise is made through selection of proper materials, mix proportions, curing conditions, and construction practices (see Chapter 11).

With low tensile strength materials, such as concrete, it is the shrinkage strain from cooling that is more important than the expansion from heat generated by cement hydration. This is because, depending on the elastic modulus, the degree of restraint, and stress relaxation due to creep, the resulting tensile stresses can be large enough to cause cracking. For instance, assuming that the coefficient of thermal expansion (α) of concrete is 10×10^{-6} per °C, and the temperature rise above the ambient (ΔT) from heat of hydration is 15 °C, then the thermal shrinkage (ϵ) caused

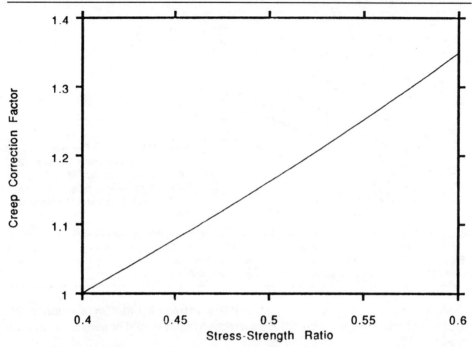

Figure 4-17 Correction factor for computing the creep coefficient at high stress levels (data from equations given by CEB-FIP Model Code, 1990).

by the 15 °C temperature drop will be $\alpha \Delta T$ or 150×10^{-6}. The elastic modulus (E) of ordinary concrete may be assumed as 3×10^6 psi. If the concrete member is fully restrained ($K_r = 1$), the cooling would produce a tensile stress of $\epsilon E = 450$ psi. Since the elastic tensile strength of ordinary concrete is usually less than 450 psi, it is likely to crack if there is no relief due to stress relaxation (Fig. 4-1).

However, there is always some relaxation of stress due to creep. When the creep coefficient is known, the resulting tensile stress (σ_t) can be calculated by the expression:

$$\sigma_t = K_r \frac{E}{1 + \varphi} \alpha \Delta T \tag{4-4}$$

where σ_t = tensile stress
  K_r = degree of restraint
  E = elastic modulus
  α = coefficient of thermal expansion
  ΔT = temperature change
  φ = creep coefficient

The factors influencing the modulus of elasticity and creep of concrete are

Thermal Shrinkage

described in the previous sections. An analysis of other factors in equation (4–4) which affect thermal stresses is presented next.

Factors Affecting Thermal Stresses

Degree of restraint (K_r). A concrete element, if free to move, would have no stress development associated with thermal deformation on cooling. However, in practice, the concrete mass will be restrained either externally by the rock foundation or internally by differential deformations within different areas of concrete due to the presence of temperature gradients. For example, assuming a rigid foundation, there will be full restraint at the concrete-rock interface ($K_r = 1.0$), however, as the distance from the interface increases, the restraint will decrease, as shown in Fig. 4–18. The same reasoning can be applied to determine the restraint between different concrete lifts. If the foundation is not rigid, the degree of restraint will decrease. When dealing with a nonrigid foundation, ACI-207.2R recommends the following multipliers for K_r:

$$\text{multiplier} = \frac{1}{1 + \frac{A_g E}{A_f E_f}} \quad (4\text{--}5)$$

where A_g = gross area of concrete cross section
A_f = area of foundation or other restraining element (For mass concrete on rock, A_f can be assumed as 2.5 A_g)
E_f = modulus of elasticity of foundation or restraining element
E = modulus of elasticity of concrete

Temperature change (ΔT). The hydration of cement compounds involves exothermic reactions which generate heat, and increase the temperature of concrete mass. Heating causes expansion, and expansion under restraint results in compressive stress. However, at early ages, the elastic modulus of concrete is low and the stress relaxation is high, therefore, the compressive stress will be very small, even in areas of full restraint. In design, to be conservative, it is assumed that a condition of no initial compression exists. The change of temperature (ΔT) in equation (4–4) is the difference between the peak temperature of concrete and the service temperature of the structure, as shown in Fig. 4–19. The change of temperature can also be expressed as:

ΔT = placement temperature of fresh concrete + adiabatic temperature rise
  − ambient or service temperature − temperature drop due to heat losses.

Controlling the *placement temperature* is one of the best ways to avoid thermal cracks in concrete. *Precooling of fresh concrete* is a commonly used method of controlling the subsequent temperature drop. Often, chilled aggregates and/or ice shavings are specified for making mass concrete mixtures in which the temperature of fresh

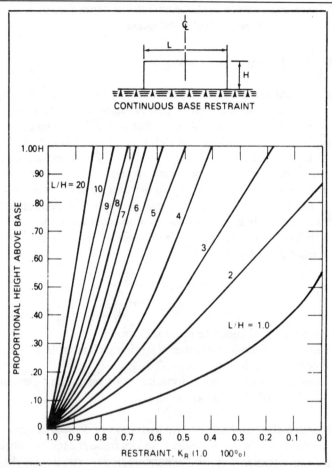

Figure 4–18 Degree of tensile restraint at center section. (Source: ACI Committee 207, Cooling Mass Concrete, 1986).

concrete is limited to 10 °C or less. During the mixing operation the latent heat needed for fusion of ice is withdrawn from other components of the concrete mixture, providing a very effective way to lower the temperature. ACI 207.4R suggests a placement temperature such that the tensile strain caused by the temperature drop should not exceed the tensile strain capacity of the concrete. This is expressed by the relationship:

$$T_i = T_f + \frac{C}{\alpha K_r} - T_r \qquad (4\text{–}6)$$

where T_i = placement temperature of concrete
T_f = final stable temperature of concrete
C = tensile strain capacity of concrete

Thermal Shrinkage

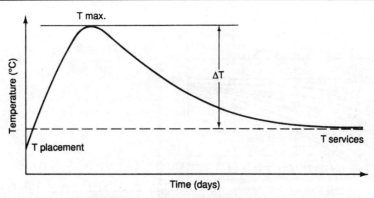

Figure 4-19 Temperature change with time.

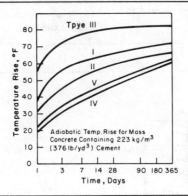

Figure 4-20 Adiabatic temperature rise in mass concretes containing 376 lb/yd³ (223 kg/m³) cement of different types. From W. H. Price, *Concr. Int.*, Vol. 4, No. 10, 1982.)

K_r = degree of restraint
α = coefficient of thermal expansion
T_r = initial temperature rise of concrete

The rate and magnitude of the ***adiabatic temperature rise*** is a function of the amount, composition and fineness of cement, and its temperature during hydration. Finely ground portland cements, or cements with relatively high C_3A and C_3S contents show higher heats of hydration than coarser cements or cements with low C_3A and C_3S (see Chapter 6). The adiabatic temperature rise curves for a concrete containing 223 kg/m³ cement and any one of the five types of portland cements are shown in Fig. 4-20. It can be seen that between a normal cement (Type I) and a low-heat cement (Type IV) the difference in temperature rise is 13 °C in 7 days and 9 °C in 90 days. Note that at this cement content the total adiabatic temperature rise was still above 30 °C even with the ASTM Type IV, low-heat, cement. Also, as shown in Figs. 4-20 and 4-21, the composition of cement and the placement temperature appear to affect mainly the rate of heat generation rather than the total heat

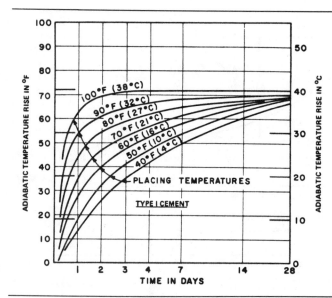

Figure 4–21 Effect of placing temperature on temperature rise of mass concrete containing 376 pcy (223 kg/m³) of Type 1 cement. (ACI Committee 207. Effect of restraint, volume changes and reinforcement on cracking of massive concrete, 1986.)

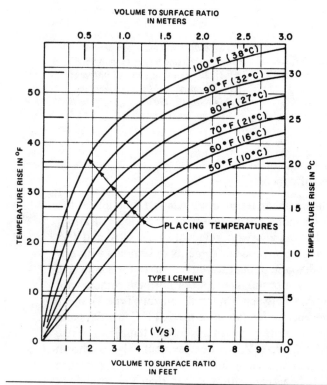

Figure 4–22 Temperature rise of concrete members containing 376 pcy (223 kg/m³) of cement. (ACI Committee 207. Effect of restraint, volume changes and reinforcement on cracking of massive concrete, 1986.)

Thermal Shrinkage

produced. Figure 4–22 shows the affect of volume-to-surface ratio of concrete on the adiabatic temperature rise at different placement temperatures.

Another effective means of reducing the magnitude of the adiabatic temperature rise is the inclusion of a pozzolan as a partial replacement for cement. Typical data given by Carlson et al.[10] on adiabatic temperature rise in mass concrete containing different types and amounts of cementitious materials are shown in Fig. 4–23. In a concrete containing 223 kg/m^3 cement, the replacement of ASTM Type I cement by Type II cement reduced the 28-day adiabatic temperature rise from 37 °C to 32 °C; a partial replacement of the Type II cement by 30 volume percent of pozzolan (25 weight percent) further reduced the temperature rise to 28 °C.

Heat losses depend on the thermal properties of concrete, and the construction technology adopted. A concrete structure can lose heat through its surface, and the magnitude of heat loss is a function of the type of environment in immediate contact with the concrete surface. Table 4–4 shows surface transmission coefficients for different isolation environments. Numerical methods of computing the temperature distribution in mass of concrete will be presented in Chapter 12.

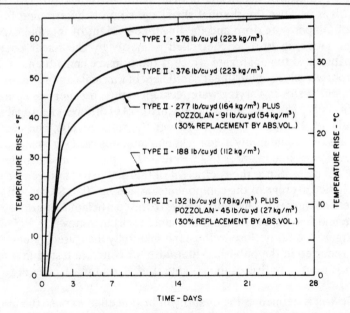

Figure 4–23 Effect of cement and pozzolan contents on temperature rise in concrete. (From R. W. Carlson et al., *J. ACI*, Proc., Vol. 76, No. 7, 1979.)

The use of a low cement content, an ASTM Type II portland cement instead of Type I, and a partial substitution of the portland cement by a pozzolan, are effective means by which the adiabatic temperature rise in mass concrete can be significantly reduced.

---

[10] R. W. Carlson, D. L. Houghton, and M. Polivka, *J. ACI*, Proc., Vol. 76, No. 7, pp. 821–37, 1979.

TABLE 4-4 COEFFICIENT OF HEAT TRANSMISSION OF DIFFERENT ISOLATION ENVIRONMENTS

Type of isolation	Surface transmission coefficient (kcal/m2.h.C)
Concrete-air	11.6
Concrete-curing water	300.0
Concrete-wood-air	2.6
Concrete-metal-air	11.6
Concrete-isolant-air	2.0

THERMAL PROPERTIES OF CONCRETE

Coefficient of thermal expansion (α) is defined as the change in unit length per degree of temperature change. Selecting an *aggregate with a low coefficient of thermal expansion* when it is economically feasible and technologically acceptable, may, under certain conditions, become a critical factor for crack prevention in mass concrete. This is because the thermal shrinkage strain is determined both by the magnitude of temperature drop and the linear coefficient of thermal expansion of concrete; the latter, in turn, is controlled primarily by the linear coefficient of thermal expansion of the aggregate which is the primary constituent of concrete.

The reported values of the linear coefficient of thermal expansion for saturated portland cement pastes of varying water/cement ratios, for mortars containing 1:6 cement/natural silica sand, and for concrete mixtures of different aggregate types are approximately 18, 12, and $6 - 12 \times 10^{-6}$ per °C, respectively. The coefficient of thermal expansion of commonly used rocks and minerals varies from about 5×10^{-6} per °C for limestones and gabbros to $11 - 12 \times 10^{-6}$ per °C for sandstones, natural gravels, and quartzite. Since the coefficient of thermal expansion can be estimated from the weighted average of the components, assuming 70 to 80 percent aggregate in the concrete mixture, the calculated values of the coefficient for various rock types (both coarse and fine aggregate from the same rock) are shown in Fig. 4-24. The data in the figure are fairly close to the experimentally measured values of thermal coefficients reported in the published literature for concrete tested in moist conditions, which is representative of the condition of typical mass concrete.

Specific heat is defined as the quantity of heat needed to raise the temperature of a unit mass of a material by one degree. The specific heat of normal weight concrete is not very much affected by the type of aggregate, temperature and other parameters. Typically the values of specific heat are in the range of 0.22 to 0.25 Btu/lb.F.

Thermal conductivity gives the heat flux transmitted through a unit area of a material under a unit temperature gradient. The thermal conductivity of concrete

Thermal Shrinkage

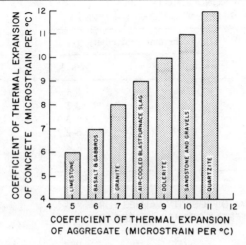

Figure 4-24 Influence of the aggregate type on the coefficient of thermal expansion of concrete.

Since the coefficient of thermal expansion of concrete is directly related to the coefficient of expansion of the aggregate present, in mass concrete the selection of an aggregate with a lower coefficient provides another approach toward lowering the thermal strain.

is influenced by the mineralogical characteristics of aggregate, and by the moisture content, density, and temperature of concrete. Table 4–5a shows typical values of thermal conductivity for concretes containing different aggregate types.

Thermal Diffusivity is defined as:

$$\kappa = \frac{K}{c\rho} \qquad (4-7)$$

where K = diffusivity, ft²/h (m²/h)
 K = conductivity, Btu/ft.h.F (J/m.h.K)
 c = specific heat, Btu/lb.F (J/kg.K)
 ρ = density of concrete, lb/ft³ (kg/m³)

Heat will move more readily through a concrete with a higher thermal diffusivity. For normal-weight concrete, the conductivity usually controls the thermal diffusivity because the density and specific heat do not vary much. Table 4–5b shows typical values of thermal diffusivity for concretes made with different types of coarse aggregate.

EXTENSIBILITY AND CRACKING

As stated earlier, the primary significance of deformations caused by applied stress and by thermal and moisture-related effects in concrete is whether or not their interaction would lead to cracking. Thus the magnitude of the shrinkage strain is only

TABLE 4–5a THERMAL CONDUCTIVITY VALUES FOR CONCRETE WITH DIFFERENT AGGREGATE TYPES

Aggregate type	Thermal conductivity	
	Btu in./h. ft^{2}F	W/m.K
Quartzite	24	3.5
Dolomite	22	3.2
Limestone	18–23	2.6–3.3
Granite	18–19	2.6–2.7
Rhyolite	15	2.2
Basalt	13–15	1.9–2.2

TABLE 4–5b THERMAL DIFFUSIVITY VALUES FOR CONCRETE WITH DIFFERENT COARSE AGGREGATES

Coarse Aggregate	ft^{2}h	m^{2}h
Quartzite	0.058	0.0054
Limestone	0.051	0.0047
Dolomite	0.050	0.0046
Granite	0.043	0.0040
Rhyolite	0.035	0.0033
Basalt	0.032	0.0030

Source: ACI Committee 207—Cooling and Insulating Systems for mass concrete, 1986.

one of the factors governing the cracking of concrete. From Fig. 4–1 it is clear that the other factors are:

- **Modulus of elasticity.** The lower the modulus of elasticity, the lower will be the amount of the induced elastic tensile stress for a given magnitude of shrinkage.
- **Creep.** The higher the creep, the higher is the amount of stress relaxation and lower the net tensile stress.
- **Tensile strength.** The higher the tensile strength, the lower is the risk that the tensile stress will exceed the strength and crack the material.

The combination of factors that are desirable to reduce the advent of cracking in concrete can be described by a single term called **extensibility**. Concrete is said to have a high degree of extensibility when it can be subjected to large deformations without cracking. Obviously, for a minimum risk of cracking, the concrete should undergo not only less shrinkage but also should have a high degree of extensibility (i.e., low elastic modulus, high creep, and high tensile strength). In general, high-strength concretes may be more prone to cracking because of greater thermal shrinkage and lower stress relaxation; on the other hand, low-strength concretes tend to crack less, because of lower thermal shrinkage and higher stress relaxation. Note that the preceding statement is applicable to massive concrete members; with thin sections the effect of drying shrinkage strain would be more important.

It may be of interest to point out that many factors which reduce the drying shrinkage of concrete will also tend to reduce the extensibility. For instance, an increase in the aggregate content or stiffness will reduce the drying shrinkage but at the same time reduce stress relaxation and extensibility. This example demonstrates the difficulty of practicing concrete technology from purely theoretical considerations.

The cracking behavior of concrete in the field can be more complex than indicated by Fig. 4–1, that is, the rates at which the shrinkage and stress relaxation develop may not be similar to those shown in the figure. For example, in mass concrete compressive stresses are developed during the very early period when temperatures are rising, and the tensile stresses do not develop until at a later age when the temperature begins to decline. However, due to the low strength of concrete at early ages, most of the stress relaxation takes place during the first week after placement. In this way, concrete loses most of the stress-relaxing capacity before this is needed for prevention of cracking induced by tensile stresses.

For thermal-shrinkage cracking, whether related to internal temperature effects in mass concrete or to external temperature effects in extreme climates, the significance of **tensile strain capacity,** which is defined as the failure strain under tension, is noteworthy. It is generally agreed that the failure of concrete loaded in uniaxial compression is mainly a tensile failure. Also, there are indications that it is not a limiting tensile strength, but a limiting tensile strain that determines the fracture strength of concrete under static loading. Accordingly, Houghton[11] has described a simple method to determine the ultimate tensile strain for quick loading by taking a ratio of the modulus of rupture to the elastic modulus in compression. Since the modulus of rupture is 20 to 40 percent higher than the true tensile strength, and the modulus of elasticity in compression is higher than the stress/strain ratio by a similar order of magnitude, it is claimed that the method gives a true value of the ultimate elastic strain for quick loading. By adding to this strain the creep strain due to slow loading, an estimate of the tensile strain capacity can be obtained. For the purpose of risk analysis against thermal cracking, it is suggested that the determination of the tensile strain capacity is a better criterion than the practice of converting the thermal strain to induced elastic stress. A general method of computing stress in viscoelastic materials is presented in Chapter 12, which also contains a finite element method to compute temperature distributions in mass concrete.

TEST YOUR KNOWLEDGE

1. What is a truly elastic material? Is concrete truly elastic? If not, why? Describe the various stages of microcracking when a concrete specimen is loaded to failure.

[11] D. L. Houghton, *J. ACI, Proc.,* Vol. 73, No. 12, pp. 691–700, 1976.

2. Draw a typical stress–strain curve for concrete. From this, how would you determine the dynamic modulus of elasticity and the different types of the static elastic moduli? Typically, what are their magnitudes for a medium-strength concrete?
3. What are the assumptions underlying the formulas used by the ACI Building Code and the CEB-FIP Model Code for predicting the static elastic modulus of concrete? Can you point out any limitations of these formulas?
4. How does the moisture state of a concrete test specimen affect the elastic modulus and strength values? Explain why both properties are not affected in the same manner.
5. What is the significance of adiabatic temperature in concrete? How much adiabatic temperature rise can occur in a typical low-strength concrete containing ASTM Type II cement? How can this be reduced?
6. Can we control the coefficient of thermal expansion of concrete? If so, how?
7. What are the typical ranges of drying shrinkage strain and creep strain in concrete; what is their significance? How are the two phenomena similar to each other?
8. What do you understand by the terms *basic creep, specific creep, drying creep,* and *creep coefficient*?
9. List the most important factors that affect drying shrinkage and creep, and discuss when the effects are similar or opposite.
10. Which factors affect creep only, and why?
11. What is the significance of the term *theoretical thickness*?
12. Beside the magnitude of shrinkage strain, which other factors determine the risk of cracking in a concrete element?
13. What is the usefulness of the *extensibility* concept? Why would high-strength concrete be more prone to cracking than low-strength concrete?
14. Ideally, from the standpoint of crack resistance, a concrete should have low shrinkage and high extensibility. Give examples to show why this may not be possible to achieve in practice.
15. What is the significance of tensile strain capacity? How can you determine it?

SUGGESTIONS FOR FURTHER STUDY

ACI, *Designing for Creep and Shrinkage in Concrete Structures*, SP-76, 1983.

BROOKS, A. E., and K. NEWMAN, eds., *The Structure of Concrete*, Proc. Int. Conf., London, Cement and Concrete Association, Wexham Springs, Slough, U. K., pp. 82–92, 176–89, 319–447, 1968.

CARLSON, R. W., D. L. HOUGHTON, and M. POLIVKA, "Causes and Control of Cracking in Unreinforced Mass Concrete," *J. ACI*, Proc., Vol. 76, No. 7, pp. 821–37, 1979.

NEVILLE, A. M., and BROOKS, J. J., *Concrete Technology*. Longman Scientific and Technical Publ., Chapters 12 and 13, 1987.

CHAPTER 5

Durability

PREVIEW

Designers of concrete structures have been mostly interested in the strength characteristics of the material; for a variety of reasons, they must now become durability conscious. Whereas properly constituted, placed, and cured concrete enjoys a long service life under most natural and industrial environments, premature failures of concrete structures do occur and they provide valuable lessons for control of factors responsible for lack of durability.

Water is generally involved in every form of deterioration, and in porous solids permeability of the material to water usually determines the rate of deterioration. Therefore, at the beginning of this chapter the structure and properties of water are described with special reference to its destructive effect on porous materials; then factors controlling the permeability of cement paste, aggregates, and concrete, are presented.

Physical effects that adversely influence the durability of concrete include surface wear, cracking due to crystallization pressure of salts in pores, and exposure to temperature extremes such as frost or fire. Deleterious chemical effects include leaching of the cement paste by acidic solutions, and expansive reactions involving sulfate attack, alkali-aggregate attack, and corrosion of the embedded steel in concrete. The significance, physical manifestations, mechanisms, and control of various causes of concrete deterioration are discussed in detail.

In the end, special attention is given to performance of concrete in seawater. Since numerous physical and chemical causes of deterioration are simultaneously at work, a study of the behavior of concrete in seawater provides an excellent opportunity to appreciate the complexity of durability problems affecting concrete structures in field practice.

DEFINITION

A long service life is considered synonymous with durability. Since durability under one set of conditions does not necessarily mean durability under another, it is customary to include a general reference to the environment when defining durability. According to ACI Committee 201, **durability** of portland cement concrete is defined as its ability to resist weathering action, chemical attack, abrasion, or any other process of deterioration; that is, durable concrete will retain its original form, quality, and serviceability when exposed to its environment.

No material is inherently durable; as a result of environmental interactions the microstructure and, consequently, the properties of materials change with time. A material is assumed to reach the end of service life when its properties under given conditions of use have deteriorated to an extent that the continuing use of the material is ruled either unsafe or uneconomical.

SIGNIFICANCE

It is generally accepted now that in designing structures the durability characteristics of the materials under consideration should be evaluated as carefully as other aspects such as mechanical properties and initial cost. First, there is a better appreciation of the socioeconomic implications of durability. Increasingly, repair and replacement costs of structures arising from material failures have become a substantial portion of the total construction budget. For example, it is estimated that in industrially developed countries over 40 percent of the total resources of the building industry are applied to repair and maintenance of existing structures, and less than 60 percent to new installations. The escalation in replacement costs of structures and the growing emphasis on life-cycle cost rather than first cost are forcing engineers to become durability conscious. Next, there is a realization that a close relation exists between durability of materials and ecology. Conservation of natural resources by making materials last longer is, after all, an ecological step. Also, the uses of concrete are being extended to increasingly hostile environments, such as offshore platforms in the North Sea, containers for handling liquefied gases at cryogenic temperatures, and high-pressure reaction vessels in the nuclear industry. Recent failures of offshore steel structures in Norway and Newfoundland showed that both the human and economic costs associated with premature and sudden failure of the material of construction can be very high.[1]

GENERAL OBSERVATIONS

Before a discussion of important aspects of durability of concrete, a few general remarks on the subject will be helpful. First, *water*, which is the primary agent of

---

[1] On March 27, 1980, Alexander Kjeland, a steel structure drilling platform off the coast of Stavanger (North Sea) failed suddenly, resulting in the death of 123 persons. Shortly after this incident an offshore oil-drilling steel structure collapsed near Newfoundland, causing the death of 64 persons.

both creation and destruction of many natural materials, happens to be central to most durability problems in concrete. In porous solids, water is known to be the cause of many types of *physical processes of degradation*. As a vehicle for transport of aggressive ions, water can also be a source of *chemical processes of degradation*. Second, the physical-chemical phenomena associated with water movements in porous solids are controlled by the *permeability* of the solid. For instance, the rate of chemical deterioration would depend on whether the chemical attack is confined to the surface of concrete, or whether it is also at work inside the material. Third, the rate of deterioration is affected by the type of concentration of ions in water and by the chemical composition of the solid. Unlike natural rocks and minerals, *concrete is a basic material* (because alkaline calcium compounds constitute the hydration products of portland cement paste); therefore, acidic waters are expected to be particularly harmful to concrete.

Most of our knowledge of physical-chemical processes responsible for concrete deterioration comes from case histories of structures in the field, because it is difficult in the laboratory to simulate the combination of long-term conditions normally present in real life. However, in practice, deterioration of concrete is seldom due to a single cause; usually, at advanced stages of material degradation more than one deleterious phenomena are found at work. In general, the physical and chemical causes of deterioration are so closely intertwined and mutually reinforcing that even separation of the cause from the effect often becomes impossible. Therefore, a classification of concrete deterioration processes into neat categories should be treated with some care. Since the purpose of such classifications is to explain, systematically and individually, the various phenomena involved, there is a tendency to overlook the interactions when several phenomena are present simultaneously.

WATER AS AN AGENT OF DETERIORATION

Concrete is not the only material that is vulnerable to physical and chemical processes of deterioration associated with water. Therefore it will be desirable to review, in general, the characteristics of water that make it the principal agent of destruction of materials.

In the form of seawater, groundwater, rivers, lakes, rain, snow, and vapor, water is undoubtedly the most abundant fluid in nature. Being small, water molecules are capable of penetrating extremely fine pores or cavities. As a solvent, water is noted for its ability to dissolve more substances than any other known liquid. This property accounts for the presence of many ions and gases in some waters, which, in turn, become instrumental in causing chemical decomposition of solid materials. It may also be noted that water has the highest heat of vaporization among the common liquids; therefore, at ordinary temperatures it has a tendency to remain in a material in the liquid state, rather than to vaporize and leave the material dry.

In porous solids, internal movements and changes of structure of water are known to cause disruptive volume changes of many types. For example, freezing of water into ice, formation of ordered structure of water inside fine pores, development of osmotic pressure due to differences in ionic concentration, and hydrostatic pressure buildup by differential vapor pressures can lead to high internal stresses within a moist solid. A brief review of the structure of water will be useful for understanding these phenomena.

Structure of Water

The H-O-H molecule is covalent bonded. Due to differences in the charge centers of hydrogen and oxygen, the positively charged proton of the hydrogen ion belonging to a water molecule attracts the negatively charged electrons of the neighboring water molecules. This relatively weak force of attraction, called the *hydrogen bond,* is responsible for the *ordered structure of water.*

The highest manifestation of the long-range order in the structure of water due to hydrogen bonding is seen in ice (Fig. 5–1a). Each molecule of water in ice is surrounded by four molecules such that the group has one molecule at the center and the other four at the corners of a tetrahedron. In all three directions the molecules and groups of molecules are held together by hydrogen bonding. Ice melts at 0°C when approximately 15 percent of the hydrogen bonds break up. As a result of the partial breakdown in directionality of the tetrahedral bond, each water

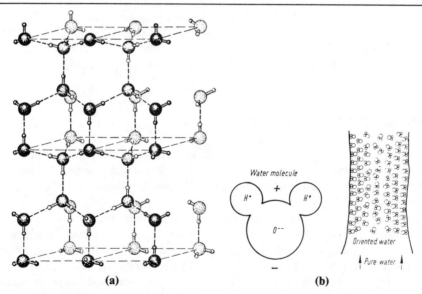

Figure 5–1 (a) Structure of ice; (b) structure of oriented water molecules in a micropore. [(a), Reprinted from Linus Pauling, *The Nature of the Chemical Bond, Third Edition.* Copyright © 1960 by Cornell University. Used by permission of the publisher, Cornell University Press, 1960; (b), from E. M. Winkler, *Stone: Properties, Durability in Man's Environment,* Springer-Verlag, New York, 1973.]

The structure and properties of water are affected by temperature and by the size of pores in a solid.

molecule can acquire more than four nearest neighbors, the density thus rising from 0.917 to 1. The reversibility of the process accounts for the phenomenon that liquid water, on solidification, expands rather than shrinks.

Compared to the structure of ice, water at room temperature has approximately 50 percent of the hydrogen bonds broken. Materials in the broken-bond state have unsatisfied surface charges, which give rise to surface energy. The surface energy in liquids causes surface tension, which accounts for the tendency of a large number of molecules to adhere together. It is the *high surface tension of water* (defined as the force required to pull the water molecules apart) which prevents it from acting as an efficient plasticizing agent in concrete mixtures until suitable admixtures are added (p. 253).

Formation of *oriented structure* of water by hydrogen bonding in micropores is known to cause expansion in many systems. In solids the surface energy due to unsatisfied charges depends on the surface area; therefore, the surface energy is high when numerous fine pores are present. If water is able to permeate such micropores, and if the forces of attraction at the surface of the pores are strong enough to break down the surface tension of bulk water and orient the molecules to an ordered structure (analogous to the structure of ice), this oriented or ordered water, being less dense than bulk water, will require more space and will therefore tend to cause expansion (Fig. 5–1b).

PERMEABILITY

In concrete, the role of water has to be seen in a proper perspective because, as a necessary ingredient for the cement hydration reactions and as a plasticizing agent for the components of concrete mixtures, water is present from the beginning. Gradually, depending on the ambient conditions and the thickness of a concrete element, most of the evaporable water in concrete (all the capillary water and a part of the absorbed water, p. 29) will be lost, leaving the pores empty or unsaturated. Since it is the evaporable water which is freezable and which is also free for internal movement, a concrete will not be vulnerable to water-related destructive phenomena, provided that there is a little or no evaporable water left after drying and provided that the subsequent exposure of the concrete to the environment does not lead to resaturation of the pores. The latter, to a large extent, depends on the hydraulic conductivity, which is also known as the coefficient of permeability (K). Note that in concrete technology it is a common practice to drop the adjective and refer to K simply as the permeability.

Garboczi[2] has reviewed several theories which attempt to relate the microstructural parameters of cement products with either diffusivity (the rate of diffusion of ions through water-filled pores) or permeability (the rate of viscous flow of fluids through the pore structure). For materials like concrete, with numerous microcracks,

---

[2] E. J. Garboczi: *Cement and Concrete Research,* Vol. 20, No. 4, pp. 591–601, 1990.

a satisfactory *pore structure transport property factor* is difficult to determine due to unpredictable changes in the pore structure on penetration of an external fluid. Note that the pore structure transport property of the material is changing continuously because of ongoing cycles of narrowing and widening of pores and microcracks from physical-chemical interactions between the penetrating fluid and the minerals of the cement paste. According to Garboczi, for a variety of reasons the diffusivity predictions need more development and validation before their practical usefulness can be proven. For practical purposes, therefore, in this text only permeability is discussed. However, it is implied that the term, in a crude sense, covers the overall fluid transport property of the material.

Permeability is defined as the property that governs the rate of flow of a fluid into a porous solid. For *steady-state flow,* the coefficient of permeability (K) is determined from Darcy's expression:

$$\frac{dq}{dt} = K \frac{\Delta H A}{L \mu}$$

where dq/dt is the rate of fluid flow, μ the viscosity of the fluid, ΔH the pressure gradient, A the surface area, and L the thickness of the solid. The coefficient of permeability of a concrete to gases or water vapor is much lower than the coefficient for liquid water; therefore, tests for measurement of permeability are generally carried out using water that has no dissolved air. Unless otherwise stated, the data in this chapter pertain to permeability of concrete to pure water. It may also be noted that due to their interaction with cement paste the permeabilities of solutions containing ions would be different from the water permeability.

Permeability of Cement Paste

In a hydrated cement paste, the size and continuity of the pores at any point during the hydration process would control the coefficient of permeability. As discussed earlier (p. 33), the mixing water is indirectly responsible for permeability of the hydrated cement paste because its content determines first the total space and subsequently the unfilled space after the water is consumed by either cement hydration reactions or evaporation to the environment. The coefficient of permeability of freshly mixed cement paste is of the order of 10^{-4} to 10^{-5} cm/sec; with the progress of hydration as the capillary porosity decreases, so does the coefficient of permeability (Table 5–1), but there is no direct proportionality between the two. For instance, when the capillary porosity decreases from 40 percent to 30 percent (Fig. 2–11), the coefficient of permeability drops by a much greater amount (i.e., from about 110 to 20×10^{-12} cm/sec). However, a further decrease in the porosity from 30 percent to 20 percent would bring about only a small drop in permeability. This is because, in the beginning, as the cement hydration process progresses even a small decrease in the total capillary porosity is associated with considerable segmentation of large pores, thus greatly reducing the size and number of channels of flow in the cement paste. Typically, about 30 percent capillary porosity represents a point when the

TABLE 5–1 REDUCTION IN PERMEABILITY OF CEMENT PASTE (WATER/CEMENT RATIO = 0.7) WITH THE PROGRESS OF HYDRATION

Age days	Coefficient of permeability (cm/sec $\times 10^{-11}$)
Fresh	20,000,000
5	4,000
6	1,000
8	400
13	50
24	10
Ultimate	6

Source: T. C. Powers, L. E. Copeland, J. C. Hayes, and H. M. Mann, *J. ACI*, Proc., Vol. 5, pp. 285–98, 1954.

interconnections between the pores have already become so tortuous that a further decrease in the porosity of the paste is not accompanied by a substantial decrease in the permeability coefficient.

In general, when the water/cement ratio is high and the degree of hydration is low, the cement paste will have high capillary porosity; it will contain a relatively large number of big and well-connected pores and, therefore, its coefficient of permeability will be high. As hydration progresses, most of the pores will be reduced to small size (e.g., 100 nm or less) and will also lose their interconnections; thus the permeability drops. The coefficient of permeability of cement paste when most of the capillary voids are small and not interconnected is of the order of 10^{-12} cm/sec. It is observed that in normal cement pastes the discontinuity in the capillary network is generally reached when the capillary porosity is about 30 percent. With 0.4, 0.5, 0.6, and 0.7 water/cement ratio pastes this generally happens in 3, 14, 180, and 365 days of moist curing, respectively. Since the water/cement ratio in most concrete mixtures seldom exceeds 0.7, it should be obvious that in well-cured concrete the cement paste is not the principal contributing factor to the coefficient of permeability.

Permeability of Aggregates

Compared to 30 to 40 percent capillary porosity of typical cement pastes in hardened concrete, the volume of pores in most natural aggregates is usually under 3 percent and rarely exceeds 10 percent. It is expected, therefore, that the permeability of aggregate would be much lower than that of the typical cement paste. This may not necessarily be the case. From the permeability data of some natural rocks and cement pastes (Table 5–2) it appears that the coefficients of permeability of aggregates are as variable as those of hydrated cement pastes of water/cement ratios in the range 0.38 to 0.71.

TABLE 5-2 COMPARISON BETWEEN PERMEABILITIES OF ROCKS AND CEMENT PASTES

Type of rock	Coefficient of permeability (cm/sec)	Water/cement ratio of mature paste with the same coefficient of permeability
Dense trap	2.47×10^{-12}	0.38
Quartz diorite	8.24×10^{-12}	0.42
Marble	2.39×10^{-11}	0.48
Marble	5.77×10^{-10}	0.66
Granite	5.35×10^{-9}	0.70
Sandstone	1.23×10^{-8}	0.71
Granite	1.56×10^{-8}	0.71

Source: T. C. Powers, *J. Am. Ceram. Soc.*, Vol. 4, No. 1, pp. 1–5, 1958.

Whereas the coefficient of permeability of most marble, traprock, diorite, basalt, and dense granite may be of the order of 1 to 10×10^{-12} cm/sec, some varieties of granite, limestone, sandstones, and cherts show values that are higher by two orders of magnitude. The reason some aggregates, with as low as 10 percent porosity, may have much higher permeability than cement pastes is because the size of capillary pores in aggregates is usually much larger. Most of the capillary porosity in a mature cement paste lies in the range 10 to 100 nm, while pores in aggregates are, on the average, larger than 10 μm. With some cherts and limestones the pore size distribution involves a considerable content of finer pores; therefore, permeability is low but the aggregates are subject to expansion and cracking associated with sluggish moisture movement and resulting hydrostatic pressure.

Permeability of Concrete

Theoretically, the introduction of aggregate particles of low permeability into a cement paste is expected to reduce the permeability of the system (especially with high-water/cement-ratio pastes at early ages when the capillary porosity is high) because the aggregate particles should intercept the channels of flow within the cement paste matrix. Compared to the neat cement paste, therefore, mortar or concrete with the same water/cement ratio and degree of maturity should give a lower coefficient of permeability. Test data indicate that, in practice, this is not the case. The two sets of data[3] in Fig. 5-2 clearly show that the addition of aggregate to a cement paste or a mortar increases the permeability considerably; in fact, the larger the aggregate size, the greater the coefficient of permeability. Typically,

---

[3] The permeability coefficient in SI units is expressed as kg/Pa · m · sec, which is approximately 10^{-3} times smaller than the coefficient expressed in cm/sec.

Permeability

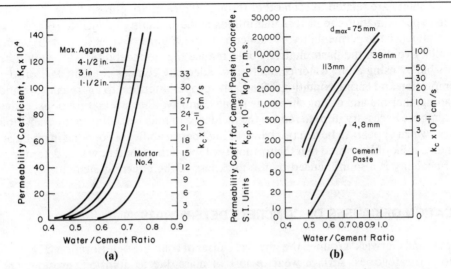

Figure 5–2 Influence of water/cement ratio and maximum aggregate size on concrete permeability: (a) K_q is a relative measure of the flow of water through concrete in cubic feet per year per square foot of area for a unit hydraulic gradient. [(a), From *Concrete Manual, 8th Edition,* U.S. Bureau of Reclamation, 1975, p. 37, (b), adapted from *Beton-Bogen,* Aalborg Cement Co., Aalborg, Denmark, 1979.]

The permeability of concrete to water depends mainly on the water/cement ratio (which determines the size, volume, and continuity of capillary voids) and maximum aggregate size (which influences the microcracks in the transition zone between the coarse aggregate and the cement paste).

permeability coefficients for moderate-strength concrete (containing 38 mm aggregate and 356 kg/m³ cement and an 0.5 water/cement ratio), and low-strength concrete used in dams (75 to 150 mm aggregate, 148 kg/m³ cement, and an 0.75 water/cement ratio) are of the order of 1×10^{-10} and 30×10^{-10} cm/sec, respectively.

The explanation as to **why the permeability of mortar or concrete is higher than the permeability of the corresponding cement paste** lies in microcracks that are present in the transition zone between the aggregate and the cement paste. As stated earlier (p. 22), the aggregate size and grading affect the bleeding characteristics of a concrete mixture which, in turn, influence the strength of the transition zone. During early hydration periods the transition zone is weak and vulnerable to cracking due to differential strains between the cement paste and the aggregate induced generally by drying shrinkage, thermal shrinkage, and externally applied load. The cracks in the transition zone are too small to be seen by the naked eye, but they are larger in width than most capillary cavities present in the cement paste matrix, and thus instrumental in establishing the interconnections, which increase the permeability of the system.

Owing to the significance of the permeability to physical and chemical processes of deterioration of concrete, which will be described below, a brief review of the *factors controlling permeability* of concrete should be useful. Since strength and permeability are related to each other through the capillary porosity (Fig. 2–11), as a first approximation the factors that influence the strength of concrete (Fig. 3–14) also influence the permeability. A reduction in the volume of large (e.g., > 100 nm) capillary voids in the paste matrix would reduce the permeability. This should be possible by using a low water/cement ratio, adequate cement content, and proper compaction and curing conditions. Similarly, proper attention to aggregate size and grading, thermal and drying shrinkage strains, and avoiding premature or excessive loading are necessary steps to reduce the incidence of microcracking in the transition zone, which appears to be a major cause of high permeability of concrete in practice. Finally, it should be noted that the tortuosity of the path of fluid flow that determines permeability is also influenced by the thickness of the concrete element.

CLASSIFICATION OF CAUSES OF CONCRETE DETERIORATION

Mehta and Gerwick[4] grouped the physical causes of concrete deterioration (Fig. 5–3) into two categories: surface wear or loss of mass due to abrasion, erosion, and cavitation; and cracking due to normal temperature and humidity gradients, crystallization pressures of salts in pores, structural loading, and exposure to temperature extremes such as freezing or fire. Similarly, the authors grouped the chemical causes of deterioration (p. 134) into three categories: (1) hydrolysis of the cement paste components by soft water; (2) cation-exchange reactions between aggressive fluids and the cement paste; and (3) reactions leading to formation of expansive products, such as in sulfate expansion, alkali-aggregate expansion, and corrosion of steel in concrete.

It needs to be emphasized again that the distinction between the physical and

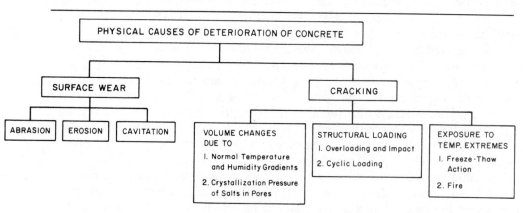

Figure 5–3 Physical causes of concrete deterioration. (From P. K. Mehta and B. C. Gerwick, Jr., *Concr. Int.*, Vol. 4, pp. 45–51, 1982.)

---

[4] P. K. Mehta and B. C. Gerwick, Jr., *Concr. Int.*, Vol. 4, No. 10, pp. 45–51, 1982.

chemical causes of deterioration is purely arbitrary; in practice, the two are frequently superimposed on each other. For example, loss of mass by surface wear and cracking increases the permeability of concrete, which then becomes the primary cause of one or more processes of chemical deterioration. Similarly, the detrimental effects of the chemical phenomena are physical; for instance, leaching of the components of hardened cement paste by soft water or acidic fluids would increase the porosity of concrete, thus making the material more vulnerable to abrasion and erosion.

An excellent review of the causes, mechanisms, and control of all types of cracking in concrete is published by the ACI Committee 224.[5] Cracking of concrete due to normal temperature and humidity gradients was discussed in Chapter 4; deterioration by surface wear, pressure of crystallization salts in pores, freeze-thaw cycles, fire, and various chemical processes will be discussed here.

DETERIORATION BY SURFACE WEAR

Progressive loss of mass from a concrete surface can occur due to abrasion, erosion, and cavitation. The term **abrasion** generally refers to dry attrition, such as in the case of wear on pavements and industrial floors by vehicular traffic. The term **erosion** is normally used to describe wear by the abrasive action of fluids containing solid particles in suspension. Erosion takes place in hydraulic structures, for instance on canal lining, spillways, and pipes for water or sewage transport. Another possibility of damage to hydraulic structures is by **cavitation,** which relates to loss of mass by formation of vapor bubbles and their subsequent collapse due to sudden change of direction in rapidly flowing water.

Hardened cement paste does not possess a high resistance to attrition. Service life of concrete can be seriously shortened under conditions of repeated attrition cycles, especially when cement paste in the concrete is of high porosity or low strength, and is inadequately protected by an aggregate which itself lacks wear resistance. Using a special test method, Liu[6] found a good correlation between water/cement ratio and abrasion resistance of concrete (Fig. 5–4a). Accordingly, *for obtaining abrasion resistance concrete surfaces,* ACI Committee 201 recommends that in no case should the compressive strength of concrete be less than 4000 psi (28MPa). Suitable strengths may be attained by a low water/cement ratio, proper grading of fine and coarse aggregate (limit the maximum size to 25 mm), lowest consistency practicable for proper placing and consolidation (maximum slump 75 mm; for toppings 25 mm), and minimum air content consistent with exposure conditions.

When a fluid containing suspended solid particles is in contact with concrete, the impinging, sliding, or rolling action of particles will cause surface wear. The rate

---

[5] ACI Report 224R-90, Manual of Concrete Practice, Part 3, 1991.
[6] T. C. Liu, *J. ACI,* Proc., Vol. 78, No. 5, p. 346, 1981.

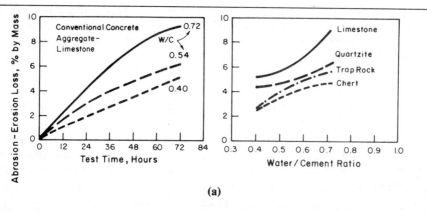

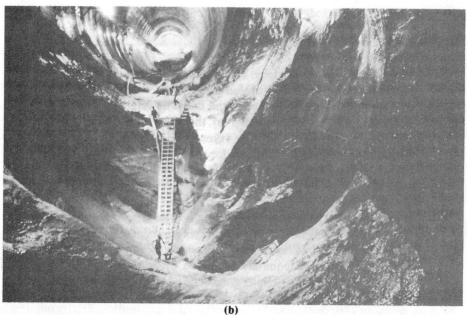

Figure 5-4 (a) Influence of water/cement ratio and aggregate type on abrasion-erosion damage in concrete; (b) cavitation damage to concrete lining in a 41-ft-diameter (12.5m) tunnel of the Glen Canyon Dam. [(a), From T. C. Liu, *J. ACI*, Proc., Vol. 78, No. 5, p. 346, 1981; (b) photograph courtesy of U.S. Bureau of Reclamation and William Scharf of Guy F. Atkinson Construction Co.]

of surface erosion will depend on the porosity or strength of concrete, and on the amount, size, shape, density, hardness, and velocity of the moving particles. It is reported that if the quantity and size of solids is small, for example, silt in an irrigation canal, erosion will be negligible at bottom velocities up to 1.8 m/sec (velocity at or above which a given particle can be transported). When **severe erosion** *or abrasion* conditions exist, it is recommended that, in addition to the use of hard aggregates, the concrete should be proportioned to develop at least 6000 psi (41MPa) compressive strength at 28 days and adequately cured before exposure to the aggressive environment. ACI Committee 201 recommends at least 7 days of continuous moist curing after the concrete has been finished.

As to *additional measures for improving the durability of concrete to abrasion or erosion,* it should be noted that the process of physical attrition of concrete occurs at the surface; hence particular attention should be paid to ensure that at least the concrete at the surface is of high quality. To reduce the formation of a weak surface called *laitance* (the term is used for a layer of fines from cement and aggregate), it is recommended to delay floating and troweling until the concrete has lost its surface bleed water. Heavy-duty industrial floors or pavements may be designed to have a 25- to 75-mm-thick topping, consisting of a low water/cement ratio concrete containing hard aggregate of 12.5 mm maximum size. Because of very low water/cement ratio, concrete toppings containing latex admixtures or superplasticizing admixtures are becoming increasingly popular for abrasion or erosion resistance. Also, the use of mineral admixtures, such as condensed silica fume, presents interesting possibilities. Besides causing a substantial reduction in the porosity of concrete after moist curing, the fresh concrete containing mineral admixtures is less prone to bleeding. Resistance to deterioration by permeating fluids and reduction in dusting due to attrition can also be achieved by application of surface-hardening solutions to well-cured new floors or abraded old floors. Solutions most commonly used for this purpose are magnesium or zinc fluosilicate or sodium silicate, which react with calcium hydroxide present in portland cement paste to form insoluble reaction products, thus sealing the capillary pores at or near the surface.

While good-quality concrete shows excellent resistance to steady high-velocity flow of clear water, nonlinear flow at velocities exceeding 12 m/sec (7 m/sec in closed conduits) may cause severe erosion of concrete through *cavitation.* In flowing water, vapor bubbles form when the local absolute pressure at a given point in the water is reduced to ambient vapor pressure of water corresponding to the ambient temperature. As the vapor bubbles flowing downstream with water enter a region of higher pressure, they implode with great impact because of the entry of high-velocity water into the previously vapor-occupied space, thus causing severe local pitting. Therefore, the concrete surface affected by cavitation is irregular or pitted, in contrast to the smoothly worn surface by erosion from suspended solids. Also, in contrast to erosion or abrasion, a strong concrete may not necessarily be effective in preventing damage due to cavitation; the best solution lies in removal of the causes of cavitation, such as surface misalignments or abrupt changes of slope. In 1984, extensive repairs were needed for the concrete lining of a tunnel of the Glen Canyon Dam (Fig. 5–4b); the damage was caused by cavitation attributable to surface irregularities in the lining.

Test methods for the evaluation of wear resistance of concrete are not always satisfactory, because simulation of the field conditions of wear is not easy in the laboratory. Therefore, laboratory methods are not intended to provide a quantitative measurement of the length of service that may be expected from a given concrete surface; they can be used to evaluate the effects of concrete materials and curing or finishing procedures on the abrasion resistance of concrete.

ASTM C 779 describes three optional methods for testing the relative abrasion resistance of horizontal concrete surfaces. In the steel-ball abrasion test, load is applied to a rotating head containing steel balls while the abraded material is removed by water circulation; in the dressing wheel test, load is applied through rotating dressing wheels of steel; and in the revolving-disk test, revolving disks of steel are used in conjunction with a silicon carbide abrasive. In each of the tests, the degree of wear can be measured in terms of weight loss after a specified time. ASTM C 418 describes the sandblast test, which covers determination of the abrasion resistance characteristics of concrete by subjecting it to the impingement of air-driven silica sand. There are no satisfactory tests for erosion resistance. Due to a direct relationship between the abrasion and erosion resistance, the abrasion resistance data can be used as a guide for erosion resistance.

CRACKING BY CRYSTALLIZATION OF SALTS IN PORES

ACI Committee 201 cites evidence that a purely physical action (not involving chemical attack on the cement) of crystallization of the sulfate salts in the pores of concrete can account for considerable damage. For instance, when one side of a retaining wall or slab of a permeable concrete is in contact with a salt solution and the other sides are subject to evaporation, the material can deteriorate by stresses resulting from the pressure of salts crystallizing in the pores. In many porous materials the crystallization of salts from supersaturated solutions is known to produce pressures that are large enough to cause cracking. In fact, moisture effects and salt crystallization are believed to be the two most damaging factors in the decay of historic stone monuments.

Crystallization from a salt solution can occur only when the concentration of the solute (C) exceeds the saturation concentration (C_s) at a given temperature. As a rule, the higher the C/C_s ratio (or degree of supersaturation), the greater the crystallization pressure. Winkler[7] determined the crystallization pressures for salts that are commonly found in pores of rocks, stones, and concrete; these pressures, calculated from the density, molecular weight, and molecular volume for a C/C_s ratio of 2, are shown in Table 5–3. At this degree of supersaturation, NaCl (Halite) crystallizing at 0°C, 25°C, and 50°C produces 554, 605, and 654 atm pressure, respectively. The stress is strong enough to disrupt most rocks. When the degree of supersaturation is 10, the calculated crystallization pressure is 1835 atm at 0°C, and 2190 atm at 50°C.

---

[7] E. M. Winkler, *Stone: Properties, Durability in Man's Environment,* Springer-Verlag, New York, 1975, p. 120.

TABLE 5–3 CRYSTALLIZATION PRESSURES FOR SALTS

Salt	Chemical formula	Density (g/cm³)	Molecular weight (g/mol)	Molar volume (cm³/mol)	Pressure (atm) $C/C_s = 2$	
					0°C	50°C
Anhydrite	$CaSO_4$	2.96	136	46	335	398
Bischofite	$MgCl_2 \cdot 6H_2O$	1.57	203	129	119	142
Dodekahydrate	$MgSO_4 \cdot 12H_2O$	1.45	336	232	67	80
Epsomite	$MgSO_4 \cdot 7H_2O$	1.68	246	147	105	125
Gypsum	$CaSO_4 \cdot 2H_2O$	2.32	127	55	282	334
Halite	$NaCl$	2.17	59	28	554	654
Heptahydrite	$Na_2CO_3 \cdot 7H_2O$	1.51	232	154	100	119
Hexahydrite	$MgSO_4 \cdot 6H_2O$	1.75	228	130	118	141
Kieserite	$MgSO_4 \cdot H_2O$	2.45	138	57	272	324
Mirabilite	$Na_2SO_4 \cdot 10H_2O$	1.46	322	220	72	83
Natron	$Na_2CO_3 \cdot 10H_2O$	1.44	286	199	78	92
Tachhydrite	$2MgCl_2 \cdot CaCl_2 \cdot 12H_2O$	1.66	514	310	50	59
Thenardite	Na_2SO_4	2.68	142	53	292	345
Thermonatrite	$Na_2CO_3 \cdot H_2O$	2.25	124	55	280	333

Source: E. M. Winkler, *Stone: Properties, Durability in Man's Environment,* Springer-Verlag, New York, 1975, p. 120.

DETERIORATION BY FROST ACTION

In cold climates, damage to concrete pavements, retaining walls, bridge decks, and railings attributable to **frost action** (freeze-thaw cycles) is one of the major problems requiring heavy expenditures for repair and replacement. The causes of deterioration of hardened concrete by frost action can be related to the complex microstructure of the material; however, the deleterious effect depends not only on characteristics of the concrete but also on specific environmental conditions. Thus a concrete that is frost resistant under a given freeze-thaw condition can be destroyed under a different condition.

The frost damage in concrete can take several forms. The most common is *cracking and spalling* of concrete that is caused by progressive expansion of cement paste matrix from repeated freeze-thaw cycles. Concrete slabs exposed to freezing and thawing in the presence of moisture and deicing chemicals are susceptible to **scaling** (i.e., the finished surface flakes or peels off). Certain coarse aggregates in concrete slabs are known to cause cracking, usually parallel to joints and edges, which eventually acquires a pattern resembling a large capital letter D (cracks curving around two of the four corners of the slab). This type of cracking is described by the term **D-cracking.** The different types of concrete deteriorations due to frost action are shown by the photographs in Fig. 5–5.

Air entrainment has proved to be an effective means of reducing the risk of

(a)

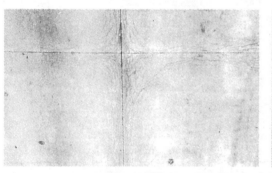

(b)

(c)

Figure 5–5 Types of frost action damage in concrete: (a) deterioration of a non-air-entrained concrete retaining wall along the saturation line (Lock and Dam No. 3, Monongahela River, Pittsburgh, Pa.); (b) severe D-cracking along longitudinal and transverse joints of a 9-year-old pavement; (c) scaling of a concrete surface. [(a), Photography courtesy of J. M. Scanlon, U.S. Army Corps of Engineers, Vicksburg, Miss.); (b), photograph courtesy of D. Stark, from *Report RD 023.01* P, Portland Cement Association, Skokie, Ill., 1974; (c), photograph courtesy of R. C. Meininger, from *Concrete in Practice,* Publ. 2, National Ready Mixed Concrete Association, Silver Springs, Md.]

(a) Progressive expansion of unprotected (non-air-entrained) cement paste by repeated freeze-thaw cycles leads to deterioration of concrete by cracking and spalling. Many Corps of Engineers lock walls which were built prior to the use of air entrainment in concrete suffer from deterioration from freezing and thawing in a saturated environment. Standard operating procedures normally require the water in the locks to remain at upper pool level during the winter so that the concrete is protected from the damaging environment. All Corps hydraulic projects built since the 1940s have been constructed with air-entrained concrete.

(b) D-cracking in highway and airfield pavements refers to a D-shaped pattern of closely spaced cracks which occur parallel to longitudinal transverse joints. This type of cracking is associated with coarse aggregates which contain a proportionately greater pore volume confined to a narrower pore size range (0.1 to 1 μm).

(c) Concrete scaling or flaking of the finished surface from freezing and thawing generally starts as localized small patches which later may merge and extend to expose large areas. Light scaling does not expose the coarse aggregate. Moderate scaling exposes the coarse aggregate and may involve loss of up to 3 to 9 mm of the surface mortar. In severe scaling more surface has been lost and the aggregate is clearly exposed and stands out. Most scaling is caused by (i) inadequate air entrainment, (ii) application of calcium and sodium chloride deicing salts, (iii) performing finishing operations while bleed water is still on the surface, and (iv) insufficient curing before exposure of concrete to frost action in the presence of moisture and deicing salts.

Deterioration by Frost Action

damage to concrete by frost action. The mechanisms by which frost damage occurs in the cement paste and air entrainment is able to prevent the damage are described below.

Frost Action on Hardened Cement Paste

Powers aptly described the mechanisms of frost action in cement paste, and also explained why air entrainment was effective in reducing expansions associated with the phenomenon:

> When water begins to freeze in a capillary cavity, the increase in volume accompanying the freezing of the water requires a dilation of the cavity equal to 9% of the volume of frozen water, or the forcing of the amount of excess water out through the boundaries of the specimen, or some of both effects. During this process, *hydraulic pressure* is generated and the magnitude of that pressure depends on the distance to an "escape boundary," the permeability of the intervening material, and the rate at which ice is formed. Experience shows that disruptive pressures will be developed in a saturated specimen of paste unless every capillary cavity in the paste is not farther than three or four thousandths of an inch from the nearest escape boundary. Such closely spaced boundaries are provided by the correct use of a suitable air-entraining agent.[8]

Powers's data and a diagrammatic representation of his hypothesis are shown in Fig. 5–6. During freezing to −24°C, the saturated cement paste specimen containing no entrained air elongated about 1600 millionths, and on thawing to the original temperature about 500 millionths permanent elongation (Fig. 5–6a) was observed. The specimen containing 2 percent entrained air showed about 800 millionths elongation on freezing, and a residual elongation of less than 50 millionths on thawing (Fig. 5–6b). The specimen containing 10 percent entrained air showed no appreciable dilation during freezing and no residual dilation at the end of the thawing cycle. Also, it may be noted that the air-entrained paste showed contraction during freezing (Fig. 5–6c). A diagrammatic illustration of Powers's hypothesis is shown in Fig. 5–6d.

Powers also proposed that, in addition to hydraulic pressure caused by water freezing in large cavities, the *osmotic pressure* resulting from partial freezing of solutions in capillaries can be another source of destructive expansions in cement paste. Water in the capillaries is not pure; it contains several soluble substances, such as alkalies, chlorides, and calcium hydroxide. Solutions freeze at lower temperatures than pure water; generally, the higher the concentration of a salt in a solution, the lower the freezing point. The existence of local salt concentration gradients between capillaries is envisaged as the source of osmotic pressure.

The hydraulic pressure due to an increase in the specific volume of water on

---

[8] T. C. Powers, *The Physical Structure and Engineering Properties of Concrete,* Bulletin 90, Portland Cement Association, Skokie, Ill., 1958.

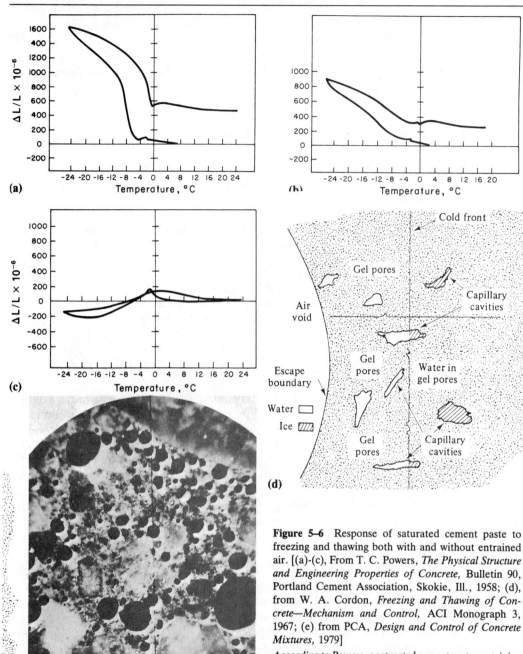

Figure 5–6 Response of saturated cement paste to freezing and thawing both with and without entrained air. [(a)-(c), From T. C. Powers, *The Physical Structure and Engineering Properties of Concrete,* Bulletin 90, Portland Cement Association, Skokie, Ill., 1958; (d), from W. A. Cordon, *Freezing and Thawing of Concrete—Mechanism and Control,* ACI Monograph 3, 1967; (e) from PCA, *Design and Control of Concrete Mixtures,* 1979]

According to Powers, a saturated cement paste containing no entrained air expands on freezing due to the generation of hydraulic pressure (a). With increasing air entrainment, the tendency to expand decreases because the entrained air voids provide escape boundaries for the hydraulic pressure [(b), (c), and (d)]. (e) Polished section of air-entrained concrete as seen through a microscope.

Deterioration by Frost Action

freezing in large cavities, and the osmotic pressure due to salt concentration differences in the pore fluid, do not appear to be the only causes of expansion of cement pastes exposed to frost action. Expansion of cement paste specimens was observed[9] even when benzene, which contracts on freezing, was used as a pore fluid instead of water.

Analogous to the formation of ice lenses in soil, a *capillary effect*,[10] involving large-scale migration of water from small pores to large cavities, is believed to be the primary cause of expansion in porous bodies. According to the theory advanced by Litvan,[11] the rigidly held water by the C-S-H (both interlayer and adsorbed in gel pores) in cement paste cannot rearrange itself to form ice at the normal freezing point of water because the mobility of water existing in an ordered state is rather limited. Generally, the more rigidly a water is held, the lower will be the freezing point. It may be recalled (p. 29) that three types of water are physically held in cement paste; in order of increasing rigidity these are the capillary water in small capillaries (10 to 50 nm), the adsorbed water in gel pores, and the interlayer water in the C-S-H structure.

It is estimated that water in gel pores does not freeze above $-78°C$. Therefore, when a saturated cement paste is subjected to freezing conditions, while the water in large cavities turns into ice the gel pore water continues to exist as liquid water in a supercooled state. This creates a thermodynamic disequilibrium between the frozen water in capillaries, which acquires a low-energy state, and the supercooled water in gel pores, which is in a high-energy state. The difference in entropy of ice and supercooled water forces the latter to migrate to the lower-energy sites (large cavities), where it can freeze. This fresh supply of water from the gel pores to the capillary pores increases the volume of ice in the capillary pores steadily until there is no room to accommodate more ice. Any subsequent tendency for the supercooled water to flow toward the ice-bearing regions would obviously cause internal pressures and expansion of the system. Further, according to Litvan, the moisture transport associated with cooling of saturated porous bodies may not necessarily lead to mechanical damage. Mechanical damage occurs when the rate of moisture transport is considerably less than demanded by the conditions (e.g., a large temperature gradient, low permeability, and high degree of saturation).

It may be noted that during frost action on cement paste, the tendency for certain regions to expand is balanced by other regions that undergo contraction (e.g., loss of adsorbed water from C-S-H). The net effect on a specimen is, obviously, the result of the two opposite tendencies. This satisfactorily explains why the cement paste containing no entrained air showed a large elongation (Fig. 5–6a), while the cement paste containing 10 percent entrained air showed contraction during freezing (Fig. 5–6c).

---

[9] J. J. Beaudoin and C. McInnis, *Cem. Concr. Res.*, Vol. 4, pp. 139–48, 1974.

[10] U. Meier and A. B. Harnik, *Cem. Concr. Res.*, Vol. 8, pp. 545–51, 1978.

[11] G. G. Litvan, *Cem. Concr. Res.*, Vol. 6, pp. 351–56, 1976.

Frost Action on Aggregate

Depending on how the aggregate responds to frost action, a concrete containing entrained air in the cement paste matrix can still be damaged. The mechanism underlying the development of internal pressure on freezing a saturated cement paste is also applicable to other porous bodies; this includes aggregates produced from porous rocks, such as certain cherts, sandstones, limestones, and shales. Not all porous aggregates are susceptible to frost damage; the behavior of an aggregate particle when exposed to freeze-thaw cycles depends primarily on the size, number, and continuity of pores (i.e., on pore size distribution and permeability).

From the standpoint of lack of durability of concrete to frost action, which can be attributed to the aggregate, Verbeck and Landgren[12] proposed three classes of aggregate. In the first category are aggregates of *low permeability* and high strength, so that on freezing of water the elastic strain in the particle is accommodated without causing fracture. In the second category are aggregates of *intermediate permeability*, that is, those having a significant proportion of the total porosity represented by small pores of the order of 500 nm and smaller. Capillary forces in such small pores cause the aggregate to get easily saturated and to hold water. On freezing, the magnitude of pressure developed depends primarily on the rate of temperature drop and the distance that water under pressure must travel to find an escape boundary to relieve the pressure. Pressure relief may be available either in the form of any empty pore within the aggregate (analogous to entrained air in cement paste) or at the aggregate surface. The critical distance for pressure relief in a hardened cement paste is of the order of 0.2 mm; it is much greater for most rocks because of their higher permeability than cement paste.

These considerations have given rise to the concept of *critical aggregate size* with respect to frost damage. With a given pore size distribution, permeability, degree of saturation, and freezing rate, a large aggregate may cause damage but smaller particles of the same aggregate would not. For example, when 14-day-old concrete specimens containing a 50:50 mixture of varying sizes of quartz and chert used as coarse aggregate were exposed to freeze-thaw cycles, those containing 25- to 12-mm chert required 183 cycles to show a 50 percent reduction in the modulus of elasticity, compared to 448 cycles for similarly cured concretes containing 12- to 5-mm chert.[13]

There is no single critical size for an aggregate type because this will depend on freezing rate, degree of saturation, and permeability of the aggregate. The permeability plays a dual role: first, it determines the degree of saturation or the rate at which water will be absorbed in a given period of time; and second, it determines the rate at which water will be expelled from the aggregate on freezing (and thus the development of hydraulic pressure). Generally, when aggregates larger than the critical size are present in a concrete, freezing is accompanied by **pop-outs,** that is,

---

[12] G. J. Verbeck and R. Landgren, *Proc. ASTM,* No. 60, pp. 1063–79, 1960.

[13] D. L. Bloem, *Highway Res. Rec.,* No. 18, pp. 48–60, 1963.

failure of the aggregate in which a part of the aggregate piece remains in the concrete and the other part comes out with the mortar flake.

Aggregates of high permeability, which generally contain a large number of big pores, belong to the third category. Although they permit easy entry and egress of water, they are capable of causing durability problems. This is because the transition zone between the aggregate surface and the cement paste matrix may be damaged when water under pressure is expelled from an aggregate particle. In such cases, the aggregate particles themselves are not damaged as a result of frost action. Incidentally, this shows why the results from freeze-thaw and soundness tests on aggregate alone are not always reliable in predicting its behavior in concrete.

It is believed that in concrete pavements exposed to frost action, some sandstone or limestone aggregates are responsible for the D-cracking phenomenon. The aggregates that are likely to cause D-cracking seem to have a specific *pore-size distribution* that is characterized by a large volume of very fine pores (i.e., <1 μm in diameter).

Factors Controlling Frost Resistance of Concrete

By now it should be obvious that the ability of a concrete to resist damage due to frost action depends on the characteristics of both the cement paste and the aggregate. In each case, however, the outcome is controlled actually by the interaction of several factors, such as the location of escape boundaries (the distance by which water has to travel for pressure relief), the pore structure of the system (size, number, and continuity of pores), the degree of saturation (amount of freezable water present), the rate of cooling, and the tensile strength of the material that must be exceeded to cause rupture. As discussed below, provision of escape boundaries in the cement paste matrix and modification of its pore structure are the two parameters that are relatively easy to control; the former can be controlled by means of air entrainment in concrete, and the latter by the use of proper mix proportions and curing.

Air entrainment. It is not the total air, but void spacing of the order of 0.1 to 0.2 mm within every point in the hardened cement, which is necessary for protection of concrete against frost damage. By adding small amounts of certain air-entraining agents to the cement paste (e.g., 0.05 percent by weight of the cement) it is possible to incorporate 0.05- to 1-mm bubbles. Thus, for a given volume of air *depending on the size of air bubbles,* the number of voids, void spacings, and degree of protection against frost action can vary a great deal. In one experiment,[14] 5 to 6 percent air was incorporated into concrete by using one of five different air-entraining agents. Agents A, B, D, E, and F produced 24,000, 49,000, 55,000, 170,000, and 800,000 air voids per cubic centimeter of hardened cement paste, and the corre-

---

[14] H. Woods, *Durability of Concrete,* ACI Monograph 4, 1968, p. 20.

TABLE 5–4 TOTAL AIR CONTENT FOR FROST-RESISTANT CONCRETE

Nominal maximum aggregate size, in. (mm).[a]	Air content (%)	
	Severe exposure	Moderate exposure
⅜ (9)	7½	6
½ (12.5)	7	5½
¾ (19)	6	5
1 (25)	6	4½
1½ (37.5)	5½	4½
2[b] (50)	5	4
3[b] (76)	4½	3½

[a] See ASTM C 33 for tolerances on oversize for various nominal maximum size designations.

[b] These air contents apply to total mix, as for the preceding aggregate sizes. When testing these concretes, however, aggregate larger than 37.5 mm is removed by handpicking or sieving and air content is determined on the minus 37.5 mm fraction of mix. (Tolerance on air content as delivered applies to this value.) Air content of total mix is computed from value determined on the minus 37.5 mm fraction.
Source: ACI Building Code 318.

sponding concrete specimens required 29, 39, 82, 100, and 550 freeze-thaw cycles to show 0.1 percent expansion, respectively.

Although the volume of entrained air is not a sufficient measure for protection of concrete against frost action, assuming that mostly small air bubbles are present, it is the easiest criterion for the purpose of quality control of concrete mixtures. Since the cement paste content is generally related to the maximum aggregate size, lean concretes with large aggregates have less cement paste than rich concretes with small aggregates; therefore, the latter would need more air entrainment for an equivalent degree of frost resistance. Total air contents specified for frost resistance, according to the ACI Building Code 318, are shown in Table 5–4.

The aggregate grading also affects the volume of entrained air, which is decreased by an excess of very fine sand particles. Addition of mineral admixtures such as fly ash, or the use of very finely ground cements, has a similar effect. In general, a more cohesive concrete mixture is able to hold more air than either a very wet or a very stiff concrete. Also, insufficient mixing or overmixing, excessive time for handling or transportation of fresh concrete, and overvibration tend to reduce the air content. For these reasons, it is recommended that air content should be determined on concrete as placed, and the adequacy of void spacing be estimated by microscopical determination as described by the ASTM Standard Method C 457.

Deterioration by Frost Action

Water/cement ratio and curing. Earlier it was explained how the pore structure of a hardened cement paste is determined by the water/cement ratio and degree of hydration. In general, the higher the water/cement ratio for a given degree of hydration or the lower the degree of hydration for a given water/cement ratio, the higher will be the volume of large pores in the hydrated cement paste (Fig. 2–8). Since the readily freezable water resides in large pores, it can therefore be hypothesized that at a given temperature of freezing the amount of freezable water will be more with higher water/cement ratios and at earlier ages of curing. Experimental data of Verbeck and Klieger confirmed this hypothesis (Fig. 5–7a). The influence of water/cement ratio on frost resistance of concrete is shown in Fig. 5–7b.

The importance of the water/cement ratio on the frost resistance of concrete is recognized by building codes. For example, ACI 318-83 requires that normal-weight concrete subject to freezing and thawing in a moist condition should have a maximum 0.45 water/cement ratio in the case of curbs, gutters, guardrails, or their sections, and 0.50 for other elements. Obviously, these water/cement ratio limits assume adequate cement hydration; therefore, at least 7 days of moist curing at normal temperature is recommended prior to frost exposure.

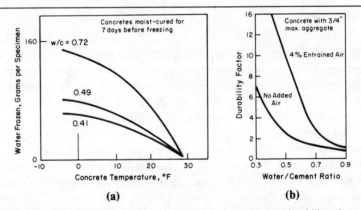

Figure 5–7 Influence of water/cement ratio and air content on durability of concrete to frost action. [(a), From G. Verbeck and P. Klieger, *Highway Research Board Bulletin 176*, Transportation Research Board, National Research Council, Washington, D.C., 1958, pp. 9–22; (b), from *Concrete Manual, 8th Edition*, U.S. Bureau of Reclamation, 1975, p. 35.]

The figure on the left shows that the amount of water that can be frozen in concrete with a given water/cement ratio increases with decreasing temperature. It also shows that the amount of water that will freeze at a given temperature increases with water/cement ratio. The observed effect of water/cement ratio is simply that higher ratios result in larger and a greater number of capillaries in which more freezable water can be present. The figure on the right shows that a combination of low water/cement ratio and entrained air ensures a high durability factor to frost action. With ASTM Method 666–80 it is required to continue freezing and thawing for 300 cycles or until the dynamic modulus of elasticity is reduced to 60 percent of the original value (whichever occurs first). The durability is then assessed by the formula: durability factor = percentage of original modulus × number of cycles at end of test ÷ 300.

Degree of saturation. It is well known that dry or partially dry substances do not suffer frost damage (see box). There is a critical degree of saturation above which concrete is likely to crack and spall when exposed to very low temperatures. In fact, it is the difference between the critical and the actual degree of saturation that determines the frost resistance of concrete, as explained in Fig. 5–8. A concrete may fall below the critical degree of saturation after adequate curing, but depending on the permeability, it may again reach or exceed the critical degree of saturation when exposed to a moist environment. The role of the permeability of concrete is therefore important in frost action because it controls not only the hydraulic pressure associated with internal water movement on freezing but also the critical degree of saturation prior to freezing. From the standpoint of frost damage the effect of increase in permeability, as a result of cracking due to any physical or chemical causes, should be apparent.

> Bugs do not freeze to death in winter. Some bugs are able to reduce the water content in their bodies so that they can hibernate without freezing; others contain a natural antifreeze in their blood.

Strength. Although there is generally a direct relationship between strength and durability, this does not hold in the case of frost damage. For example, when comparing non-air-entrained with air-entrained concrete, the former may be of higher strength, but the latter will have a better durability to frost action because of the protection against development of high hydraulic pressures. As a rule of thumb, in medium- and high-strength concretes, every 1 percent increase in the air content reduces the strength of concrete by about 5 percent. Without any change in water/cement ratio, a 5 percent air entrainment would, therefore, lower the concrete strength by 25 percent. Due to the improved workability as a result of entrained air, it is possible to make up a part of the strength loss by reducing the water/cement ratio a little while maintaining the desired level of workability. Nevertheless, air-entrained concrete is generally lower in strength than the corresponding non-air-entrained concrete.

Concrete Scaling

It is known that resistance of concrete against the combined influence of freezing and deicing salts,[15] which are commonly used to melt ice and snow from pavements, is generally lower than its resistance to frost alone. Many researchers have observed that the maximum damage to the concrete surface by scaling occurs at salt concentrations of about 4 to 5 percent.

[15] Typically, chlorides of ammonia, calcium, or sodium are used.

Deterioration by Frost Action

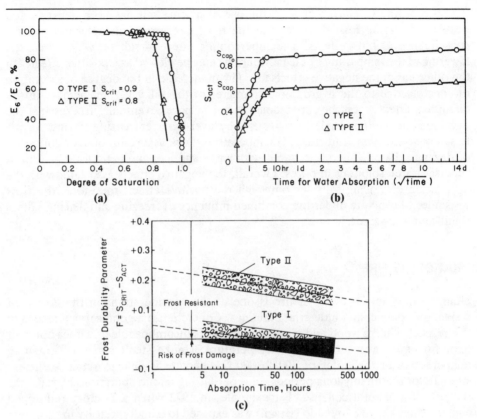

Figure 5–8 Method of predicting the frost resistance of concrete. (From *Betonghandboken*, Svensk Byggjanst, Stockholm, 1980, pp. 430–33.)

G. Fagerlund of the Swedish Cement and Concrete Research Institute proposed a method for predicting the frost resistance of concrete which emphasizes the importance of critical degree of saturation. The frost resistance (F) is evaluated as a difference between the critical degree of saturation (S_{crit}) and the actual degree of saturation (S_{act}). The degree of water saturation is defined as a ratio between the total volume of evaporable water at 105°C and the total open pore volume of available space before freezing. The micropoint on a plot of water saturation, S versus E_6/E_0 (i.e., the residual dynamic elastic modulus after six freeze-thaw cycles) gives S_{crit}. Examples of determining S_{crit} for a non-air-entrained concrete (Type I) and an air-entrained concrete containing 7.1 percent air are shown in part (a). An estimate of S_{act} can be obtained through a moisture absorption or a simple capillary suction test. As shown in part (b), the nick point on a plot of degree saturation versus square root of water uptake time corresponds to the capillary degree of saturation, S_{cap}. At the nick point all gel and capillary pores are filled with water; the larger air pores are the last to be filled and at a very slow rate. Frost resistance, expressed as a difference between S_{crit} and S_{act} for concrete exposed to water uptake for long periods, can be graphically determined. For example, as shown in part (c), concrete without entrained air (Type I) will be damaged by frost ($F \geq 0$) after about 200 hr of continuous water absorption, whereas the air-entrained concrete (Type II) will not be damaged even after very long exposure to water absorption.

According to Harnik et al.,[16] the use of deicing salt has both negative and positive effects on frost damage, and the most dangerous salt deterioration is a consequence of both effects. The supercooling effect of salt on water (i.e., the lowering of the temperature of ice formation) may be viewed as a positive effect. On the other hand, the negative effects are: (1) an increase in the degree of saturation of concrete due to the hygroscopic character of the salts; (2) an increase in the disruptive effect when the supercooled water in pores eventually freezes; (3) the development of differential stresses caused by layer-by-layer freezing of concrete due to salt concentration gradients; (4) temperature shock as a result of dry application of deicing salts on concrete covered with snow and ice; and (5) crystal growth in supersaturated solutions in pores. Overall, the negative effects associated with the application of deicing salts far outweigh the positive effect; therefore, the frost resistance of concrete under the combined influence of freezing and deicing salts is significantly lowered.

DETERIORATION BY FIRE

Human safety in the event of fire is one of the considerations in the design of residential, public, and industrial buildings. Concrete has a good service record in this respect. Unlike wood and plastics, concrete is incombustible and does not emit toxic fumes on exposure to high temperature. Unlike steel, when subjected to temperatures of the order of 700 to 800°C, concrete is able to retain sufficient strength for reasonably long periods, thus permitting rescue operations by reducing the risk of structural collapse. For example, in 1972 when a 31-story reinforced concrete building in São Paulo (Brazil) was exposed to a high-intensity fire for over 4 hr, more than 500 persons were rescued because the building maintained its structural integrity during the fire. It may be noted that from the standpoint of fire safety of steel structures, a 50-to-100 mm coating of concrete or any other fire-resisting material is routinely specified by building codes.

As is the case with other phenomena, many factors control the response of concrete to fire. Composition of concrete is important because both the cement paste and the aggregate consist of components that decompose on heating. The permeability of the concrete, the size of the element, and the rate of temperature rise are important because they govern the development of internal pressures from the gaseous decomposition products. Fire tests have shown that the degree of microcracking, and therefore the strength of concrete, is also influenced by test conditions (i.e., whether the specimens are tested hot and under load, or after cooling to the ambient humidity and temperature).

Again, the actual behavior of a concrete exposed to high temperature is the

---

[16] A. B. Harnik, U. Meier, and Alfred Rösli, ASTM STP 691, 1980, pp. 474–84.

Deterioration by Fire

result of many simultaneously interacting factors that are too complex for exacting analysis. However, for the purpose of understanding their significance, some of the factors are discussed below.

Effect of High Temperature on Cement Paste

The effect of increasing temperature on hydrated cement paste depends on the degree of hydration and moisture state. A well-hydrated portland cement paste, as described before, consists mainly of calcium silicate hydrate, calcium hydroxide, and calcium sulfoaluminate hydrates. A saturated paste contains large amounts of free water and capillary water, in addition to the adsorbed water. The various types of water are readily lost on raising the temperature of concrete. However, from the standpoint of fire protection, it may be noted that due to the considerable heat of vaporization needed for the conversion of water into steam, the temperature of concrete will not rise until all the evaporable water has been removed.

The presence of large quantities of evaporable water can cause one problem. If the rate of heating is high and the permeability of the cement paste is low, damage to concrete may take place in the form of surface spalling. Spalling occurs when the vapor pressure of steam inside the material increases at a faster rate than the pressure relief by the release of steam into the atmosphere.

By the time the temperature reaches about 300°C, the interlayer C-S-H water, and some of the chemically combined water from the C-S-H and sulfoaluminate hydrates, would also be lost. Further dehydration of the cement paste due to decomposition of calcium hydroxide begins at about 500°C, but temperatures on the order of 900°C, are required for complete decomposition of the C-S-H.

Effect of High Temperature on Aggregate

The porosity and mineralogy of the aggregate seem to exercise an important influence on the behavior of concrete exposed to fire. Porous aggregates, depending on the rate of heating and aggregate size, permeability, and moisture state, may themselves be susceptible to disruptive expansions leading to pop-outs of the type described in the case of frost attack. Low-porosity aggregates should, however, be free of problems related to moisture movements.

Siliceous aggregates containing quartz, such as granite and sandstone, can cause distress in concrete at about 573°C because the transformation of quartz from α to β form is associated with a sudden expansion of the order of 0.85 percent. In the case of carbonate rocks, a similar distress can begin above 700°C as a result of the decarbonation reaction. In addition to possible phase transformations and thermal decomposition of the aggregate, the response of concrete to fire is influenced in other ways by aggregate mineralogy. For instance, the aggregate mineralogy determines the differential thermal expansions between the aggregate and the cement paste and the ultimate strength of the transition zone.

Effect of High Temperature on Concrete

Abrams's data[17] shown in Fig. 5–9 illustrate the effect of short-duration exposure up to 1600°F (870°C) on compressive strength of concrete specimens with an average 3900 psi (27MPa) f'_c before the exposure. Variables included *aggregate type* (carbonate, siliceous, or lightweight expanded shale) and *testing conditions* (heated unstressed, i.e., without load and tested hot; heated with load at stress level that is 40 percent of the original strength and tested hot; and tested without load after cooling to ambient temperature).

When heated without load and tested hot (Fig. 5–9a), the specimens made with the carbonate aggregate or the sanded lightweight aggregate (60 percent of the fine aggregate replaced by natural sand) retained more than 75 percent of their original strengths at temperatures up to 1200°F (650°C). At this temperature, concrete specimens containing the siliceous aggregates retained only 25 percent of the original strength; they had retained 75 percent of the original strength up to about 800°F (427°C). The superior performance of the carbonate or the lightweight aggregate concretes at the higher temperature of exposure can be due to a strong transition zone and less difference in the coefficients of thermal expansion between the matrix and the aggregate.

Strengths of specimens tested hot but loaded in compression (Fig. 5–9b) were up to 25 percent higher than those of unloaded companion specimens, but the superior performance of carbonate and lightweight aggregate concretes was reaffirmed. However, the effect of aggregate mineralogy on concrete strength was significantly reduced when the specimens were tested after cooling to 70°F or 21°C (Fig. 5–9c). Microcracking in the transition zone associated with thermal shrinkage was probably responsible for this.

In the range 3300 to 6500 psi f'_c (23 to 45MPa), Abrams found that the original strength of the concrete had little effect on the percentage of compressive strength retained after high-temperature exposure. In a subsequent study[18] it was observed that compared to compressive strength of the heated specimens, the *elastic moduli* of concretes made with the three types of aggregate dropped more rapidly as the temperature was increased. For example, at 204 and 427°C, the moduli were 70 to 80 percent and 40 to 50 percent of the original value, respectively. This can be attributed to microcracking in the transition zone, which has a more damaging effect on flexural strength and elastic modulus than on compressive strength of concrete.

DETERIORATION BY CHEMICAL REACTIONS

The resistance of concrete to deterioration processes triggered by chemical reactions involves generally, but not necessarily, *chemical interactions between aggressive agents present in the external environment and the constituents of the cement paste.*

---

[17] M. S. Abrams, *Temperature and Concrete,* ACI SP-25, 1973, pp. 33–50.

[18] C. R. Cruz, *J. Res. & Dev.*, Portland Cement Association, Skokie, Ill., No. 1, pp. 37–45, 1966.

Deterioration by Chemical Reactions

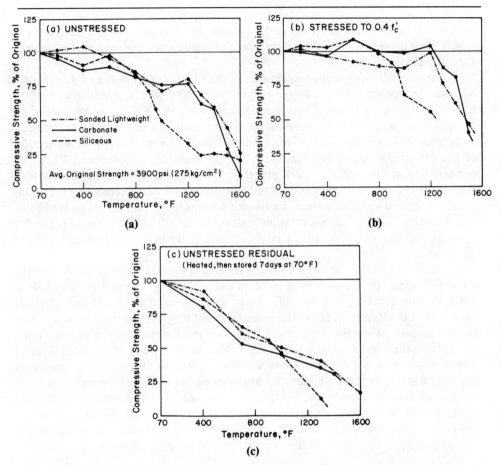

Figure 5-9 Effect of aggregate type and testing conditions on fire resistance. (From M. S. Abrams, *Temperature and Concrete*, ACI SP-25, 1973, pp. 33–58.)

Unloaded concrete specimens heated to 1200°F and tested hot [part (a)], showed that concretes containing limestone or lightweight aggregate retained 75 percent of the original strength, while concrete containing a siliceous aggregate retained only 25 percent of the original strength. When loaded to 40 percent of the original strength [part (b)], a similar trend was observed, although all strengths were higher by about 25 percent. However, irrespective of the aggregate type, all concretes showed considerable strength loss on cooling (c)].

Among the exceptions are alkali-aggregate reactions which occur between the alkalies in cement paste and certain reactive materials when present in aggregate, delayed hydration of crystalline CaO and MgO if present in excessive amounts in portland cement, and electrochemical corrosion of embedded steel in concrete.

In a well-hydrated portland cement paste, the solid phase, which is composed primarily of relatively insoluble hydrates of calcium (such as C-S-H, CH, and C-A-$\bar{\text{S}}$-H), exists in a state of stable equilibrium with a high-pH pore fluid. Large concentrations of Na^+, K^+, and OH^- ions account for the high pH value, 12.5 to 13.5, of the pore fluid in portland cement pastes. It is obvious that portland cement concrete would be in a state of chemical disequilibrium when it comes in contact with an acidic environment.

Theoretically, *any environment with less than 12.5 pH may be branded aggressive* because a reduction of the alkalinity of the pore fluid would, eventually, lead to destabilization of the cementitious products of hydration. Thus, from the standpoint of portland cement concrete, most industrial and natural waters can be categorized as aggressive. However, the rate of chemical attack on concrete will be a function of the pH of the aggressive fluid and the permeability of concrete. When the permeability of the concrete is low and the pH of the aggressive water is above 6, the rate of chemical attack is considered too slow to be taken seriously. Free CO_2 in soft water and stagnant waters, acidic ions such as SO_4^{2-} and Cl^- in groundwater and seawater, and H^+ in some industrial waters are frequently responsible for lowering the pH below 6, which is considered detrimental to portland cement concrete.

Again, it needs to be emphasized that chemical reactions manifest into detrimental physical effects, such as increase in porosity and permeability, decrease in strength, and cracking and spalling. In practice, several chemical and physical processes of deterioration act at the same time and may even reinforce each other. For the purpose of developing a clear understanding, the chemical processes can be divided into three subgroups shown in Fig. 5–10, and discussed one at a time. Special attention will be given to sulfate attack, alkali-aggregate attack, and corrosion of embedded steel, as these phenomena are responsible for deterioration of a large number of concrete structures. Finally, the last section of this chapter is devoted to durability of concrete in seawater, because coastal and offshore structures are exposed to a maze of interrelated chemical and physical processes of deterioration, which aptly demonstrate the complexities of concrete durability problems in practice.

Hydrolysis of Cement Paste Components

Water from ground, lakes, and rivers contains chlorides, sulfates, and bicarbonates of calcium and magnesium; it is generally **hard water** and it does not attack the constituents of the portland cement paste. **Pure water** from condensation of fog or water vapor, and *soft water* from rain or from melting of snow and ice, may contain little or no calcium ions. When these waters come in contact with portland cement paste, they tend to hydrolyze or dissolve the calcium-containing products. Once the contact solution attains chemical equilibrium, further hydrolysis of the cement paste would stop. However, in the case of flowing water or seepage under pressure, dilution of the contact solution will take place, thus providing the condition for continuous hydrolysis. In hydrated portland cement pastes, calcium hydroxide is the constituent that, because of its relatively high solubility in pure water (1230 mg/liter), is most susceptible to hydrolysis. Theoretically, the hydrolysis of the cement paste

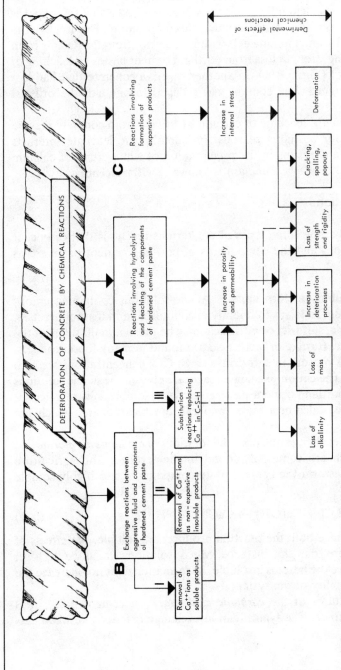

Figure 5-10 Types of chemical reactions responsible for concrete deterioration. A, Soft-water attack on calcium hydroxide and C-S-H present in hydrated portland cements; B(I), acidic solution forming soluble calcium compounds such as calcium chloride, calcium sulfate, calcium acetate, or calcium bicarbonate; B(II), solutions of oxalic acid and its salts, forming calcium oxalate; B(III), long-term seawater attack weakening the C-S-H by substitution of Mg^{2+} for Ca^{2+}; C, (1) sulfate attack forming ettringite and gypsum, (2) alkali-aggregate attack, (3) corrosion of steel in concrete, (4) hydration of crystalline MgO and CaO. (From P. K. Mehta and B. C. Gerwick, Jr., *Concr. Int.*, Vol. 4, pp. 45–51, 1982.)

continues until most of the calcium hydroxide has been leached away; this exposes the other cementitious constituents to chemical decomposition. Eventually, the process leaves behind silica and alumina gels with little or no strength. Results from two investigations showing strength loss from portland cement pastes by leaching of lime are cited by Biczok[19] (Fig. 5–18c). In another case,[20] a concrete that had lost about a fourth of its original lime content was reduced to one-half the original strength.

Besides loss of strength, leaching of calcium hydroxide from concrete may be considered undesirable for aesthetic reasons. Frequently, the leachate interacts with CO_2 present in air and results in the precipitation of white crusts of calcium carbonate on the surface. The phenomenon is known as **efflorescence.**

Cation-Exchange Reactions

Based on cation exchange, the three types of deleterious reactions that can occur between chemical solutions and the components of portland cement paste are as follows.

Formation of soluble calcium salts. Acidic solutions containing anions which form soluble calcium salts are frequently encountered in industrial practice. For example, hydrochloric, sulfuric, or nitric acid may be present in effluents of the chemical industry. Acetic, formic, or lactic acid are found in many food products. Carbonic acid, H_2CO_3, is present in soft drinks; high CO_2 concentrations are also found in natural waters. The cation-exchange reaction between the acidic solutions and the constituents of portland cement paste give rise to soluble salts of calcium, such as calcium chloride, calcium acetate, and calcium bicarbonate, which are removed by leaching.

Through the cation-exchange reaction, the solutions of *ammonium chloride* and *ammonium sulfate,* which are commonly found in the fertilizer and agriculture industry, are able to transform the cement paste components into highly soluble products, for example:

$$2NH_4Cl + Ca(OH)_2 \rightarrow CaCl_2 + 2NH_4OH \tag{5-1}$$

It should be noted that since both the reaction products are soluble, the effects of the attack are more severe than, for instance, $MgCl_2$ solution, which would form $CaCl_2$ and $Mg(OH)_2$. Since the latter is insoluble, its formation does not increase the porosity and the permeability of the system.

Due to certain features of the *carbonic acid attack* on cement paste, it is desirable to discuss it further. The typical cation-exchange reactions between car-

---

[19] I. Biczok, *Concrete Corrosion and Concrete Protection,* Chemical Publishing Company, Inc., New York, 1967, p. 291.

[20] R. D. Terzaghi, Inc., *J. ACI,* Proc., Vol. 44, p. 977, 1948.

Deterioration by Chemical Reactions

bonic acid and calcium hydroxide present in hydrated portland cement paste can be shown as follows:

$$Ca(OH)_2 + H_2CO_3 \rightarrow CaCO_3 + 2H_2O \qquad (5\text{--}2)$$

$$CaCO_3 + CO_2 + H_2O \rightleftharpoons Ca(HCO_3)_2 \qquad (5\text{--}3)$$

After the precipitation of calcium carbonate which is insoluble, the first reaction would stop unless some free CO_2 is present in the water. By transforming calcium carbonate into soluble bicarbonate in accordance with the second reaction, the presence of free CO_2 aids the hydrolysis of calcium hydroxide. Since the second reaction is reversible, a certain amount of free CO_2, referred to as the *balancing CO_2*, is needed to maintain the reaction equilibrium. Any *free CO_2 over and above the balancing CO_2* would be aggressive to the cement paste because by driving the second reaction to the right it would accelerate the process of transformation of calcium hydroxide present in the hydrated paste into the soluble bicarbonate of calcium. The balancing CO_2 content of a water depends on its hardness (i.e., the amount of calcium and magnesium present in solution).

It should be noted that the acidity of naturally occurring water is generally due to the dissolved CO_2 which is found in significant concentrations in mineral waters, seawater, and groundwater when decaying vegetable or animal wastes are in contact with the water. Normal groundwaters contain 15 to 40 mg/liter CO_2; however, concentrations of the order of 150 mg/liter are not uncommon; seawater contains 35 to 60 mg/liter CO_2. As a rule, when the pH of groundwater or seawater is 8 or above, the free CO_2 concentration is generally negligible; when the pH is below 7, harmful concentrations of free CO_2 may be present.

Formation of insoluble and nonexpansive calcium salts. Certain anions when present in aggressive water may react with cement paste to form insoluble salts of calcium; their formation may not cause damage to concrete unless the reaction product is either expansive (see below), or removed by erosion due to flowing solution, seepage, or vehicular traffic. The products of reaction between calcium hydroxide and oxalic, tartaric, tannic, humic, hydrofluoric, or phosphoric acid belong to the category of insoluble, nonexpansive, calcium salts. When concrete is exposed to decaying animal waste or vegetable matter, it is the presence of humic acid that causes chemical deterioration.

Chemical attack by solutions containing magnesium salts. Chloride, sulfate, or bicarbonate of magnesium are frequently found in groundwaters, seawater, and some industrial effluents. The magnesium solutions readily react with the calcium hydroxide present in portland cement pastes to form soluble salts of calcium. As discussed in the next section, $MgSO_4$ solution is the most aggressive because the sulfate ion can be deleterious to the alumina-bearing hydrates present in portland cement paste.

A characteristic feature of the *magnesium ion attack* on portland cement paste is that the attack is, eventually, extended to the calcium silicate hydrate, which is the principal cementitious constituent. It seems that on prolonged contact with magnesium solutions, the C-S-H in hydrated portland cement paste gradually loses calcium ions, which are replaced by magnesium ions. The ultimate product of the substitution reaction is a magnesium silicate hydrate, the formation of which is associated with loss of cementitious characteristics.

REACTIONS INVOLVING FORMATION OF EXPANSIVE PRODUCTS

Chemical reactions that involve formation of expansive products in hardened concrete can lead to certain harmful effects. Expansion may, at first, take place without any damage to concrete, but increasing buildup of internal stress eventually manifests itself by closure of expansion joints, deformation and displacements in different parts of the structure, cracking, spalling, and pop-outs. The four phenomena associated with expansive chemical reactions are: sulfate attack, alkali-aggregate attack, delayed hydration of free CaO and MgO, and corrosion of steel in concrete.

SULFATE ATTACK

A recent survey[21] of 42 concrete structures located along the Gulf coast of eastern Saudi Arabia showed that most structures suffered from an undesirably high degree of deterioration within a short period of 10 to 15 years; deterioration was attributed mainly to two causes: corrosion of reinforcement and sulfate attack. Sulfate attack on concrete has been reported from many other parts of the world, including the Canadian prairie provinces and the western United States. In fact, as early as 1936 a concrete construction manual published by the U.S. Bureau of Reclamation warned that concentrations of soluble sulfates greater than 0.1 percent in soil (150 mg/liter SO_4 in water) endanger concrete and more than 0.5 percent soluble sulfate in soil (over 2000 mg/liter SO_4 in water) may have a serious effect.

Most soils contain some sulfate in the form of gypsum (typically 0.01 to 0.05 percent expressed as SO_4); this amount is harmless to concrete. The solubility of gypsum in water at normal temperatures is rather limited (approximately 1400 mg/liter SO_4). Higher concentrations of sulfate in groundwaters are generally due to the presence of magnesium and alkali sulfates. Ammonium sulfate is frequently present in agricultural soil and waters. Effluent from furnaces that use high-sulfur fuels and from the chemical industry may contain sulfuric acid. Decay of organic matter in marshes, shallow lakes, mining pits, and sewer pipes often leads to the formation of H_2S, which can be transformed into sulfuric acid by bacterial action. According to ACI Committee 201, the water used in concrete cooling towers can also be a potential source of sulfate attack because of the gradual buildup of sulfates from evaporation of the water. Thus it is not uncommon to find deleterious concentrations of sulfate in natural and industrial environments.

[21] Rasheeduzzafar et al., *J. ACI*, Proc., Vol. 81, No. 1, pp. 13–20, 1984.

Sulfate Attack

Degradation of concrete as a result of chemical reactions between hydrated portland cement and sulfate ions from an outside source is known to take *two forms* that are distinctly different from each other. Which one of the deterioration processes is predominant in a given case depends on the concentration and source of sulfate ions (i.e., the associated cation) in the contact water and the composition of the cement paste in concrete. Sulfate attack can manifest in the form of *expansion* of concrete. When concrete cracks, its permeability increases and the aggressive water penetrates more easily into the interior, thus accelerating the process of deterioration. Sometimes, the expansion of concrete causes serious structural problems such as the displacement of building walls due to horizontal thrust by an expanding slab. Sulfate attack can also take the form of a *progressive loss of strength and loss of mass* due to deterioration in the cohesiveness of the cement hydration products. A brief review of some theoretical aspects of sulfate-generated failures, selected case histories, and control of sulfate attack follows.

Chemical Reactions Involved in Sulfate Attack

Calcium hydroxide and alumina-bearing phases of hydrated portland cement are more vulnerable to attack by sulfate ions. On hydration, portland cements with more than 5 percent potential C_3A[22] will contain most of the alumina in the form of monosulfate hydrate, $C_3A \cdot C\bar{S} \cdot H_{18}$. If the C_3A content of the cement is more than 8 percent, the hydration products will also contain $C_3A \cdot CH \cdot H_{18}$. In the presence of calcium hydroxide in portland cement pastes, when the cement paste comes in contact with sulfate ions, both the alumina-containing hydrates are converted to the high-sulfate form (ettringite, $C_3A \cdot 3C\bar{S} \cdot H_{32}$):

$$C_3A \cdot C\bar{S} \cdot H_{18} + 2CH + 2\bar{S} + 12H \rightarrow C_3A \cdot 3C\bar{S} \cdot H_{32} \quad (5\text{-}4)$$

$$C_3A \cdot CH \cdot H_{18} + 2CH + 3\bar{S} + 11H \rightarrow C_3A \cdot 3C\bar{S} \cdot H_{32} \quad (5\text{-}5)$$

There is general agreement that the sulfate-related expansions in concrete are associated with ettringite; however, the *mechanisms* by which ettringite formation causes expansion is still a subject of controversy.[23] Exertion of pressure by growing ettringite crystals, and swelling due to adsorption of water in alkaline environment by poorly crystalline ettringite, are two of the several hypotheses that are supported by most researchers.

Gypsum formation as a result of cation-exchange reactions is also capable of causing expansion. However, it has been observed[24] that deterioration of hardened portland cement paste by gypsum formation goes through a process leading to the reduction of stiffness and strength; this is followed by expansion and cracking, and eventual transformation of the material into a mushy or noncohesive mass.

Depending on the cation type present in the sulfate solution (i.e., Na^+ or

---

[22] For cement chemistry abbreviations see p. 23.

[23] M. D. Cohen and B. Mather, *ACI Materials Jour.*, Vol. 88, No. 1, pp. 62–69, 1991.

[24] P. K. Mehta, *Cem. Concr. Res.*, Vol. 13, No. 3, pp. 401–6, 1983.

Mg^{2+}), both calcium hydroxide and the C-S-H of portland cement paste may be converted to gypsum by sulfate attack:

$$Na_2SO_4 + Ca(OH)_2 + 2H_2O \rightarrow CaSO_4 \cdot 2H_2O + 2NaOH \qquad (5\text{--}6)$$

$$\left. \begin{array}{l} MgSO_4 + Ca(OH)_2 + 2H_2O \rightarrow CaSO_4 \cdot 2H_2O + Mg(OH)_2 \\ 3MgSO_4 + 3CaO \cdot 2SiO_2 \cdot 3H_2O + 8H_2O \rightarrow \\ \qquad 3(CaSO_4 \cdot 2H_2O) + 3Mg(OH)_2 + 2SiO_2 \cdot H_2O \end{array} \right\} \qquad (5\text{--}7)$$

In the first case (sodium sulfate attack), formation of sodium hydroxide as a by-product of the reaction ensures the continuation of high alkalinity in the system, which is essential for stability of the main cementitious phase (C-S-H). On the other hand, in the second case (magnesium sulfate attack) conversion of calcium hydroxide to gypsum is accompanied by formation of the relatively insoluble and poorly alkaline magnesium hydroxide; thus the stability of the C-S-H in the system is reduced and it is also attacked by the sulfate solution. The magnesium sulfate attack is, therefore, more severe on concrete.

Selected Case Histories

An interesting case history of sulfate attack by spring water on Elbe River bridge piers in Magdeburg, East Germany, is reported by Biczok.[25] The pier-sinking operation in a closed caisson opened up a spring. The spring water contained 2040 mg/liter SO_4. The expansion of the concrete lifted the piers by 8 cm in 4 years and caused extensive cracking, which made it necessary to demolish and rebuild the piers. Obviously, such occurrences of sulfate expansion can be avoided by a thorough survey of environmental conditions, and by providing suitable protection against sulfate attack when necessary.

Bellport[26] described the experience of the U.S. Bureau of Reclamation in regard to sulfate attack on hydraulic structures located in Wyoming, Montana, South Dakota, Colorado, and California. In some cases, the soluble sulfate content of soil was as high as 4.55 percent, and the sulfate concentration of water was up to 9900 mg/liter. Many cases of serious deterioration of concrete structures, 5 to 30 years old, were reported. Research studies showed that sulfate-resisting cements containing 1 to 3 percent potential C_3A performed better than 0 percent C_3A cements, which contained unusually large amounts of tricalcium silicate (58 to 76 percent).

As a result of sulfate exposure for 20 years, strength loss was reported from concrete structures of the Ft. Peck Dam in Montana (Fig. 5–11). The sulfate content of groundwaters, due entirely to alkali sulfates, was up to 10,000 mg/liter. An investigation of the deteriorated concrete specimens (Fig. 5–12) showed large amounts of gypsum formed at the expense of the cementitious constituents normally present in hydrated portland cement pastes.[27] Similar cases of sulfate deterioration

[25] Biczok, *Concrete Corrosion and Concrete Protection*.

[26] B. P. Bellport, in *Performance of Concrete*, ed. E. G. Swenson, University of Toronto Press, Toronto, 1968, pp. 77–92.

[27] T. E. Reading, ACI SP 47, 1975; pp. 343–66; and P. K. Mehta, *J. ACI*, Proc., Vol. 73, No. 4, pp. 237–38, 1976.

Sulfate Attack

Figure 5-11 Sulfate attack on concrete in Fort Peck Dam, 1971. (Photographs courtesy of T. J. Reading, formerly chief Materials Engineer of the Missouri River Division, U.S. Corps of Engineers.)

In the northern Great Plains states (the Dakotas and Montana), and extending up into the prairie provinces of Canada, groundwater may contain 1000 to 10,000 mg/liter SO_4 in areas of poor drainage. During 1935–1966, the U.S. Corps of Engineers constructed six earth-filled dams across the upper Missouri River; however, there are large auxiliary concrete structures such as tunnels, a stilling basin, a powerhouse, and a spillway. Four of the six projects, including the Ft. Peck (Montana) Dam contain more than 1 million cubic yards concrete each. Judged from the compressive strength (48 to 60 MPa) of cores on 20-year-old specimens, the Ft. Peck concrete made with Type I portland cement (7 to 9 percent C_3A), an 0.49 water/cement ratio, and 335 kg/m^3 cement content, is of good quality (low permeability).

Inspections of concrete structures in 1957–1958 after 20 years of use showed that the overall condition of the concrete at Ft. Peck was very good; however, appreciable sulfate attack was found in two areas: slabs in the penstock floor at the downstream end of Tunnel 1, and a tailrace training wall (shown in the photograph). The deteriorated concrete was mushy and disintegrated easily. The sulfate concentration of the groundwater, due almost entirely to sodium sulfate, was found to be about 10,000 mg/liter. Between 1958 and 1971, the deteriorated area in the tailrace training wall enlarged and increased in depth to about 200 mm. Mineralogical analysis of the cement paste from deteriorated concrete specimens showed that large amounts of gypsum had formed at the expense of C-S-H and calcium hydroxide.

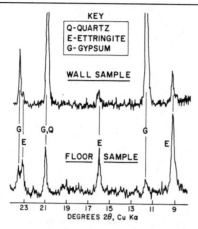

Figure 5–12 X-ray diffraction analysis of deteriorated concrete from Fort Peck Dam.

The X-ray diffraction (XRD) technique offers a convenient way to determine the mineralogical analysis of crystalline solids. If a crystalline mineral is exposed to X-rays of a particular wavelength, the layers of atoms diffract the rays and produce a pattern of peaks which is characteristic of the mineral. The horizontal scale (diffraction angle) of a typical XRD pattern gives the crystal lattice spacing, and the vertical scale (peak height) gives the intensity of the diffracted ray. When the specimen being X-rayed contains more than one mineral, the intensity of characteristic peaks from the individual minerals are proportional to their amount.

Using copper $K\alpha$ radiation, this XRD pattern was obtained from cement paste specimens taken from the deteriorated concrete of the Ft. Peck Dam. Large amounts of ettringite and gypsum are found to be present in the specimens instead of C-S-H, $Ca(OH)_2$, and monosulfate hydrate, which are normally present in mature portland cement concretes. This is an unmistakable evidence of strong sulfate attack on concrete. Contamination of the cement paste by aggregate is responsible for the presence of quartz peak in the XRD pattern.

---

are reported from the prairie soils in western Canada, which contain as high as 1½ percent alkali sulfates (groundwaters frequently contain 4000 to 9000 mg/liter sulfate). Typically, as a consequence of the sulfate attack, concrete was rendered relatively porous or weak and, eventually, reduced to a sandy (noncohesive) mass.

Verbeck[28] reported the results of a long-time investigation on concrete performance in sulfate soils located at Sacramento, California. Concrete specimens made with different types of portland cement at three cement contents were used. The soil in the basin contained approximately 10 percent sodium sulfate. The deterioration of the concrete specimens was evaluated by visual inspection, and by measurement of strength and dynamic modulus of elasticity after various periods of exposure. Verbeck's data regarding the effect of C_3A content of portland cement and the cement content of concrete on the average rate of deterioration are shown in Fig. 5–13. The results clearly demonstrate that the cement content (in other words, the permeability of concrete) had more influence on the sulfate resistance than the composition of cement. For example, the performance of concrete containing 390kg/m³ of the 10 percent C_3A cement was two to three times better than the concrete

---

[28] G. J. Verbeck, in *Performance of Concrete*, ed. E. G. Swenson, University of Toronto Press, Toronto, 1968.

Sulfate Attack

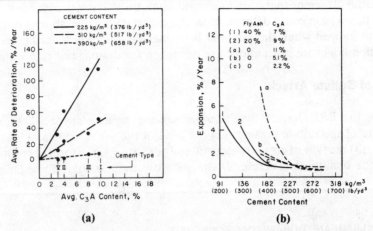

Figure 5–13 Effects of cement type and content and fly ash addition on sulfate attack in concrete. [(a), From G. J. Verbeck, in *Performance of Concrete*, ed. E. G. Swenson, University of Toronto Press, Toronto, 1968, pp. 113–24; (b), from G. E. Brown and D. B. Oates, *Concr. Int.*, Vol. 5, pp. 36–39, 1983.]

The deterioration of concrete due to sulfate attack can be affected by the cement content, cement type, and mineral admixtures. The results from a long-time study of concrete specimens exposed to a sulfate soil (containing 10% Na_2SO_4) at Sacramento, California, showed (figure on the left) that the low permeability of concrete (high cement content) was more important in reducing the rate of deterioration than the C_3A content of the cement. The figure on the right shows that in the case of a high-C_3A portland cement, addition of mineral admixtures (fly ash) offers another way of controlling sulfate attack, by reducing the effective C_3A content in the total cementitious material.

containing 310 kg/m³ of the 4 percent C_3A cement (Fig. 5–13a). With a high-C_3A cement (11 percent C_3A), the effective C_3A content in the cementitious mixture can be reduced by the addition of a pozzolanic admixture such as fly ash (Fig. 5–13b), thus causing a beneficial effect on the sulfate resistance.

An interesting case of sulfate attack was brought to the attention of the author, which showed that soil, groundwaters, seawater, and industrial waters are not the only sources of sulfate. Recently, deterioration of the drypack grout between the cantilevered precast concrete girders and the cast-in-place concrete bleacher girders was reported from Candlestick Park Stadium (Fig. 1–7) in San Francisco, California.[29] Apparently, the grout was not compacted properly during construction; therefore, leaching of the cementitious material resulted in a high strength loss[30] and caused the formation of calcium carbonate stalactites in the vicinity. X-ray diffraction analysis of the deteriorated material showed the presence of considerable amounts of ettringite and gypsum formed as a result of sulfate attack. It may be noted that the joint containing the grout is located 18 to 30 m above ground level. Due to inadequate drainage, it was found that rain water had accumulated in the vicinity

---

[29] *Engineering News Record*, p. 32, January 5, 1984.

[30] Grout cores showed about 600 to 1000 psi against the normal 4000 psi compressive strength.

of the mortar. It seems that, due to air pollution, the sulfates present in rain water (see box below) can cause deterioration of mortar or concrete above ground. This is likely to happen when the material is permeable, and when during design and construction adequate precautions are not taken for drainage of rain water.

Control of Sulfate Attack

According to a *BRE Digest*,[31] the *factors influencing sulfate attack* are (1) the amount and nature of the sulfate present, (2) the level of the water table and its seasonal variation, (3) the flow of groundwater and soil porosity, (4) the form of construction, and (5) the quality of concrete. If the sulfate water cannot be prevented from

---

ACID RAIN AND DURABILITY OF CONCRETE

Samples collected for the Air Resources Board have shown that the average pH value of rain in Northern California ranged from pH 4.4 at San Jose... [to] pH 5.2 at Davis.... Nor is the occurrence of **acid rain** confined to the state's urban centers.... In Sequoia National Park and in the Mammoth Lakes region, the average pH value of rainfall during 1980 and 1981 was 4.9, and one week averaged 3.5.

Yet those readings pale in comparison with the *disturbing levels of acidity found in fog*. In December 1982 the fog blanketing Orange County reached an all-time low reading... of pH 1.7 at Corona del Mar. [Even the coastal fog that rolls through at San Francisco's Golden Gate has registered as low as pH 3.5.] According to Dr. Michael Hoffman at California Institute of Technology at Pasadena, fog near urban areas routinely registers between 2.5 and 3 on the pH scale and is laden with such pollutants as *sulfate*, nitrate, ammonium ions, lead, copper, nickel, vanadium, and aldehydes.

Source: Report by K. Patrick Conner,
published in the *San Francisco Chronicle*, June 3, 1984

Acid rain is a manmade, not a natural phenomenon, with 90 percent of such pollution in the [Northeastern part of the United States] coming from industrial and automotive combustion of fossil fuels. These pollutants [the chief component is sulfur dioxide, with nitrogen oxide also playing an important role] are transported through the atmosphere long distances from their sources.... Several thousand lakes and streams... have been acidified, with life in them killed or reduced. [Among other factors,] acid rain may be a contributor to the decline of the forests. *Buildings, monuments and other man-made structures are being eroded by air pollution and could be the "sleeper" issue of the acid rain problem* said Dr. J. Christopher Bernabo, executive director of the national assessment program.

Source: Report by Philip Shabecoff,
published in the *San Francisco Chronicle*, February 24, 1985
Copyright © 1985 by the New York Times Company. Reprinted by permission.

---

[31] *Building Research Establishment Digest 250*, Her Majesty's Stationery Office, London, 1981. Hereafter, *BRE Digest 250*.

Sulfate Attack

reaching the concrete, the only defense against attack lies in the control of factor 5, as discussed below. It is observed that the rate of attack on a concrete structure with all faces exposed to sulfate water is less than if moisture can be lost by evaporation from one or more surfaces. Therefore, basements, culverts, retaining walls, and slabs on ground are more vulnerable than foundations and piles.

The *quality of concrete,* specifically a low permeability, is the best protection against sulfate attack. Adequate concrete thickness, high cement content, low water/cement ratio and proper compaction and curing of fresh concrete are among the important factors that contribute to low permeability. In the event of cracking due to drying shrinkage, frost action, corrosion of reinforcement, or other causes, additional safety can be provided by the *use of sulfate-resisting cements.*

Portland cement containing less than 5 percent C_3A (ASTM Type V) is sufficiently sulfate resisting under moderate conditions of sulfate attack (i.e., when ettringite forming reactions are the only consideration). However, when high sulfate concentrations of the order of 1500 mg/liter or more are involved (which are normally associated with the presence of magnesium and alkali cations), the Type V portland cement may not be effective against the cation-exchange reactions involving gypsum formation, especially if the C_3S content of the cement is high.[32] Under these conditions, experience shows that cements potentially containing a little or no calcium hydroxide on hydration perform much better: for instance, high-alumina cements, portland blast-furnace slag cements with more than 70 percent slag, and portland pozzolan cements with at least 25 percent pozzolan (natural pozzolan, calcined clay, or low-calcium fly ash).

Based on standards originally developed by the U.S. Bureau of Reclamation, *sulfate exposure is classified into four degrees of severity* in the ACI Building Code 318-83, which contains the following requirements:

- **Negligible attack:** When the sulfate content is under 0.1 percent in soil, or under 150 ppm (mg/liter) in water, there shall be no restriction on the cement type and water/cement ratio.
- **Moderate attack:** When the sulfate content is 0.1 to 0.2 percent in soil, or 150 to 1500 ppm in water, ASTM Type II portland cement or portland pozzolan or portland slag cement shall be used, with less than an 0.5 water/cement ratio for normal-weight concrete.
- **Severe attack:** When the sulfate content is 0.2 to 2.00 percent in soil, or 1500 to 10,000 ppm in water, ASTM Type V portland cement, with less than an 0.45 water/cement ratio, shall be used.
- **Very severe attack:** When the sulfate content is over 2 percent in soil, or over 10,000 ppm in water, ASTM Type V cement plus a pozzolanic admixture shall be used, with less than an 0.5 water/cement ratio. For lightweight-aggregate concrete, the ACI Building Code specifies a minimum 28-day compressive strength of 4250 psi for severe or very severe sulfate-attack conditions.

---

[32] Bellport, in *Performance of Concrete,* p. 33.

It is suggested that with normal-weight concrete a lower water/cement ratio (or higher strength in the case of lightweight concrete) may be required for watertightness or for protection against corrosion of embedded items and freezing and thawing. For very severe attack conditions *BRE Digest 250* requires the use of sulfate-resisting portland cement, a maximum 0.45 water/cement ratio, a minimum 370 kg/m^3 cement content, and a protective coating on concrete. It may be noted that concrete coatings are not a substitute for high-quality or low-permeability concrete because it is difficult to ensure that thin coatings will remain unpunctured and that thick ones will not crack. ACI Committee 515 recommendations should be considered for barrier coatings to protect concrete from various chemicals.

ALKALI-AGGREGATE REACTION

Expansion and cracking, leading to loss of strength, elasticity, and durability of concrete can also result from chemical reactions involving alkali ions from portland cement (or from other sources), hydroxyl ions, and certain siliceous constituents that may be present in the aggregate. In recent literature, the phenomenon is referred to as ***alkali-silica reaction.*** Pop-outs and exudation of a viscous alkali-silicate fluid are other manifestations of the phenomenon, a description of which was first published in 1940 by Stanton[33] from investigations of cracked concrete structures in California. Since then, numerous examples of concrete deterioration from other parts of the world have been reported to show that the alkali-silica reaction is at least one of the causes of distress in structures located in humid environments, such as dams, bridge piers, and sea walls. Characteristics of cements and aggregates that contribute to the reaction, mechanisms associated with expansion, selected case histories, and methods of controlling the phenomenon, are discussed below.

Cements and Aggregate Types Contributing to the Reaction

Raw materials used in portland cement manufacture account for the presence of alkalies in cement, typically in the range 0.2 to 1.5 percent equivalent[34] Na_2O. Depending on the alkali content of a cement, the pH of the pore fluid in normal concretes is generally 12.5 to 13.5. This pH represents a caustic or strongly alkaline liquid in which some acidic rocks (i.e., aggregates composed of silica and siliceous minerals) do not remain stable on long exposure.

Both laboratory and field data from several studies in the United States showed that portland cements containing more than 0.6 percent equivalent Na_2O, when used

---

[33] T. E. Stanton, *Proc. ASCE,* Vol. 66, p. 1781–812, 1940.

[34] Both sodium and potassium compounds are usually present in portland cements. However, it is customary to express the alkali content of cement as acid-soluble sodium oxide equivalent, which is equal to $Na_2O + 0.658K_2O$.

Alkali-Aggregate Reaction

in combination with an alkali-reactive aggregate, caused large expansions due to the alkali-aggregate reaction (Fig. 5–14a). ASTM C150 therefore designated the cements with less than 0.6 percent equivalent Na_2O as *low-alkali,* and with more than 0.6 percent equivalent Na_2O as *high-alkali.* In practice, cement alkali contents of 0.6 percent or less are usually found sufficient to prevent damage due to the alkali-aggregate reaction irrespective of the type of reactive aggregate. With concretes containing very high cement content, even less than 0.6 percent alkali in cement may prove harmful. Investigations in Germany and England have shown that if the total alkali content of the concrete from all sources is below 3 kg/m^3, damage will probably not occur.

As discussed below, the presence of both hydroxyl ions and alkali-metal ions appear to be necessary for the expansive phenomenon. Due to the large amount of calcium hydroxide in hydrated portland cements, the concentration of hydroxyl ions in the pore fluid remains high even in low-alkali cements; in this case the expansive phenomenon will therefore be limited by the short supply of alkali-metal ions unless these ions are furnished by any other source, such as alkali-containing admixtures, salt-contaminated aggregates, and penetration of seawater or deicing solutions containing sodium chloride into concrete.

In regard to *alkali-reactive aggregates,* depending on time, temperature, and particle size, all silicate or silica minerals, as well as silica in hydrous (opal) or amorphous form (obsidian, silica glass), can react with alkaline solutions, although a large number of minerals react only to an insignificant degree. Feldspars, pyroxenes, amphiboles, micas, and quartz, which are the constituent minerals of granites, gneisses, schists, sandstones, and basalts, are classified as innocuous minerals. Opal, obsidian, cristobalite, tridymite, chalcedony, cherts, cryptocrystalline volcanic rocks (andesites and rhyolites), and strained or metamorphic quartz have been found to be alkali reactive, generally in the decreasing order of reactivity. A comprehensive list of substances responsible for deterioration of concrete by alkali-aggregate reaction is given in Fig. 5–14b. A few cases of reaction between alkali and carbonate rocks are also reported in the literature, but they will not be discussed here.

Mechanisms of Expansion

Depending on the degree of disorder in the aggregate structure and its porosity and particle size, alkali-silicate gels of variable chemical composition are formed in the presence of hydroxyl and alkali-metal ions. The mode of attack in concrete involves depolymerization or breakdown of silica structure[35] of the aggregate by hydroxyl ions followed by adsorption of the alkali-metal ions on newly created surface of the reaction products. Like marine soils containing surface-adsorbed sodium or potassium, when the alkali silicate gels come in contact with water, they swell by imbibing

---

[35] In the case of sedimentary rocks composed of clay minerals such as phyllites, graywackes, and argillites, *exfoliation* of the sheet structure due to hydroxyl ion attack and water adsorption is the principal cause of expansion. In the case of dense particles of glass and flint, reaction rims form around the particles with the *onion-ring type of progressive cracking and peeling.*

DELETERIOUSLY REACTIVE ROCKS, MINERALS, AND SYNTHETIC SUBSTANCES

Reactive substance	Chemical composition	Physical character
Opal	$SiO_2 \cdot nH_2O$	Amorphous
Chalcedony	SiO_2	Microcrystalline to cryptocrystalline; commonly fibrous
Certain forms of quartz	SiO_2	(a) Microcrystalline to cryptocrystalline; (b) Crystalline, but intensely fractured, strained, and/or inclusion-filled
Cristobalite	SiO_2	Crystalline
Tridymite	SiO_2	Crystalline
Rhyolitic, dacitic, latitic, or andesitic glass or cryptocrystalline devitrification products	Siliceous, with lesser proportions of Al_2O_3, Fe_2O_3, alkaline earths, and alkalies	Glass or cryptocrystalline material as the matrix of volcanic rocks or fragments in tuffs
Synthetic siliceous glasses	Siliceous, with less proportions of alkalies, alumina, and/or other substances	Glass

The most important deleteriously alkali-reactive rocks (that is, rocks containing excessive amounts of one or more of the substances listed above) are as follows:

Opaline cherts	Andesites and tuffs
Chalcedonic cherts	Siliceous shales
Quartzose cherts	Phyllites
Siliceous limestones	Opaline concretions
Siliceous dolomites	Fractured, strained, and
Rhyolites and tuffs	inclusion-filled quartz
Dacites and tuffs	and quartzites

Note: A rock may be classified as, for example, a "siliceous limestone" and be innocuous if its siliceous constituents are other than those indicated above.

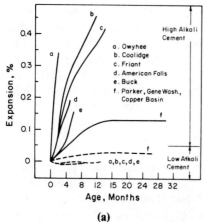

(a)

(b)

Figure 5–14 Alkali-reactive rocks in portland cement concrete. [(a), Based on R. F. Blanks and H. L. Kennedy, *The Technology of Cement and Concrete*, Vol. 1, John Wiley & Sons, Inc., New York, 1955; (b), from ACI Committee 201, *ACI Mater, J.*, Vol. 88, No. 5, p. 565, 1991.]

Combinations of high-alkali portland cement (>0.6 percent equivalent Na_2O) and certain siliceous aggregates used for making concrete for several U.S. dams showed undesirably large expansions in mortar prisms [part (a)]. The same aggregates showed only small expansions when a low-alkali cement was used in the test. A comprehensive list of alkali-reactive aggregates is given in part (b).

a large amount of water through osmosis. The hydraulic pressure so developed may lead to expansion and cracking of the affected aggregate particles, the cement paste matrix surrounding the aggregates, and the concrete.

Solubility of the alkali silicate gels in water accounts for their mobility from the interior of aggregate to the microcracked regions both within the aggregate and the concrete. Continued availability of water to the concrete causes enlargement and extension of the microcracks, which eventually reach the outer surface of the concrete. The crack pattern is irregular and is therefore referred to as *map cracking*.

It should be noted that the evidence of alkali-aggregate reaction in a cracked concrete does not necessarily prove that the reaction is the principal cause of cracking. Among other factors, development of internal stress depends on the amount, size, and type of the reactive aggregate present and the chemical composition of the alkali-silicate gel formed. It is observed that when a large amount of a reactive material is present in a finely divided form (i.e., under 75 μm), there may be considerable evidence of the reaction, but expansions to any significant extent do not occur. On the other hand, many case histories of expansion and cracking of concrete attributable to the alkali-aggregate reaction are associated with the presence in aggregate of sand-size alkali-reactive particles, especially in the size range 1 to 5 mm. Satisfactory explanations for these observations are not available due to the simultaneous interplay of many complex factors; however, a lower water adsorption tendency of alkali-silica gels with a higher silica/alkali ratio, and relief of hydraulic pressure at the surface of the reactive particle when its size is very small, may partially explain these observations.

Selected Case Histories

From published reports of concrete deterioration due to alkali-aggregate reaction, it seems that deposits of alkali-reactive aggregates are widespread in the United States, eastern Canada, Australia, New Zealand, South Africa, Denmark, Germany, England, and Iceland. Blanks and Kennedy[36] described some of the earlier cases in the United States. According to the authors, deterioration was first observed in 1922 at the Buck hydroelectric plant on the New River (Virginia), 10 years after construction. As early as 1935, R. J. Holden had concluded from petrographic studies of concrete that the expansion and cracking were caused by chemical reaction between the cement and the phyllite rock used as aggregate. Linear expansion in excess of 0.5 percent caused by the alkali-aggregate reaction had been detected in concrete. The crown of an arch dam in California deflected upstream by about 127 mm in 9 years following construction in 1941. Measurements at Parker Dam (California-Arizona) showed that expansion of the concrete increased from the surface to a depth of 3 m, and linear expansions in excess of 0.1 percent were detected.

---

[36] R. F. Blanks and H. L. Kennedy, *The Technology of Cement and Concrete,* Vol. 1, John Wiley & Sons, Inc., New York, 1955, pp. 316–341.

Figure 5–15 Alkali-aggregate expansion in concrete. (Photograph courtesy of J. Figg, Ove Arup Partnership, U.K.)

Parapet of the Val-de-la-Mare dam (Jersey Island, U.K.), showing misalignment caused by differential movement of adjacent blocks resulting from expansion due to alkali-aggregate reactivity.

Since chemical reactions are a function of temperature, in colder countries, such as Denmark, Germany, and England, it was first thought that the alkali-silica reaction may not be a problem. Subsequent experience with certain alkali-reactive rocks has shown that the assumption was incorrect. For example, in 1971[37] it was discovered that concrete of the Val-de-la-Mare dam in the United Kingdom (Fig. 5–15) was suffering from alkali-silica reaction, possibly as a result of the use of a crushed diorite rock containing veins of amorphous silica. Extensive remedial measures were needed to ensure the safety of the dam. By 1981,[38] evidence of concrete deterioration attributable to alkali-silica reaction was found in 23 structures, 6 to 17 years old, located in Scotland, the Midlands, Wales, and other parts of southwestern England. Many of the structures contained concrete made with inadequately washed sea-dredged aggregates.

Control of Expansion

From the foregoing description of case histories and mechanisms underlying expansion associated with the alkali-aggregate reaction it may be concluded that the most important factors influencing the phenomenon are: (1) the alkali content of the

[37] J. W. Figg, *Concrete,* Cement and Concrete Association, Grosvenor Crescent, London, Vol. 15, No. 7, 1981, pp. 18–22.

[38] D. Palmer, *Concrete,* Cement and Concrete Association, Vol. 15, No. 3, 1981, pp. 24–27.

cement and the cement content of the concrete; (2) the alkali-ion contribution from sources other than portland cement, such as admixtures, salt-contaminated aggregates, and penetration of seawater or deicing salt solution into concrete; (3) the amount, size, and, reactivity of the alkali-reactive constituent present in the aggregate; (4) the availability of moisture to the concrete structure; and (5) the ambient temperature.

When cement is the only source of alkali ions in concrete and alkali-reactive constituents are suspected to be present in the aggregate, experience shows that the use of low-alkali portland cement (less than 0.6 percent equivalent Na_2O) offers the best protection against alkali attack. If beach sand or sea-dredged sand and gravel are to be used, they should be washed with sweet water to ensure that the total alkali content[39] from the cement and aggregates in concrete does not exceed 3 kg/m^3. If a low-alkali portland cement is not available, the total alkali content in concrete can be reduced by replacing a part of the high-alkali cement with cementitious or pozzolanic admixtures such as granulated blast-furnace slag, volcanic glass (ground pumice), calcined clay, fly ash, or condensed silica fume.[40] It should be noted that, similar to the well-bound alkalies in most feldspar minerals, the alkalies present in slags and pozzolans are acid-insoluble and probably are not available for reaction with aggregate.

In addition to reduction in the effective alkali content, the use of pozzolanic admixtures results in the formation of less expansive alkali-silicate products with a high silica/alkali ratio. In countries such as Iceland where only alkali-reactive volcanic rocks are available and the cement raw materials are such that only high-alkali portland cement is produced, all portland cement is now blended with condensed silica fume.

With mildly reactive aggregates, another approach for reducing concrete expansion is to "sweeten" the reactive aggregate with 25 to 30 percent limestone or any other nonreactive aggregate if it is economically feasible. Finally, it should be remembered that subsequent to or simultaneously with the progress of the reaction, moisture availability to the structure is essential for expansion to occur. Control of the access of water to concrete by prompt repair of any leaking joints is therefore highly desirable to prevent excessive concrete expansions.

HYDRATION OF CRYSTALLINE MgO AND CaO

Numerous reports, including a review by Mehta,[41] indicate that hydration of crystalline MgO or CaO, when present in substantial amounts in cement, can cause expansion and cracking in concrete. The expansive effect of high MgO in cement

---

[39] The total alkali content is the acid-soluble alkali determined by a standard chemical test method.

[40] Condensed silica fumes is an industrial by-product which consists of highly reactive silica particles on the order of 0.1 μm in size.

[41] P. K. Mehta, ASTM STP 663, 1978, pp. 35–60.

was first recognized in 1884, when a number of concrete bridges and viaducts in France failed two years after construction. About the same time the town hall of Kassel in Germany had to be rebuilt as a result of expansion and cracking attributed to MgO in cement. The French and the German cements contained 16 to 30 percent and 27 percent MgO, respectively. This led to restrictions in the maximum permissible MgO. For example, the current ASTM Standard Specification for Portland Cement (ASTM C 150-83) requires that the MgO content in cement shall not exceed 6 percent.

Although expansion due to hydration of crystalline CaO has been known for a long time, in the United States the deleterious effect associated with the phenomenon was recognized in the 1930s when certain 2- to 5-year-old concrete pavements cracked. Initially suspected to be due to MgO, later the expansion and cracking were attributed to the presence of hard-burnt CaO in the cements used for construction of the pavements.[42] Laboratory tests showed that cement pastes made with a low-MgO portland cement, which contained 2.8 percent hard-burnt CaO, showed considerable expansion. However, in concrete, due to the restraining effect of aggregate, relatively large amounts of hard-burnt CaO are needed to obtain a significant expansion. The phenomenon is virtually unknown in modern concretes because better manufacturing controls ensure that the content of uncombined or crystalline CaO in portland cements seldom exceeds 1 percent.

The crystalline MgO (periclase) in a portland cement clinker which has been exposed to 1400 to 1500°C is essentially inert to moisture at room temperature because reactivity of periclase drops sharply when heated above 900°C. No cases of structural distress due to the presence of periclase in cement are reported from countries such as Brazil, where raw material limitations compel some cement producers to manufacture portland cements containing more than 6 percent MgO. Recently, several cases of expansion and cracking of concrete structures were reported from Oakland (California) when the aggregate used for making concrete was found to have been accidentally contaminated with crushed dolomite bricks, which contained large amounts of MgO and CaO calcined at temperatures far below 1400°C.

CORROSION OF EMBEDDED STEEL IN CONCRETE

Deterioration of concrete containing embedded metals, such as conduits, pipes, and reinforcing and prestressing steel, is attributable to the combined effect of more than one cause; however, corrosion of the embedded metal is invariably one of the principal causes. A survey[43] of collapsed buildings in England showed that from 1974

---

[42] Conversion of $CaCO_3$ to CaO can occur at 900 to 1000°C. The CaO thus formed can hydrate rapidly and is called *soft-burnt lime*. Since portland cement clinker is heat-treated to 1400 to 1500°C, any uncombined CaO present is called *hard-burnt*, and it hydrates slowly. It is the slow hydration of hard-burnt CaO in a hardened cement paste that causes expansion.

[43] *Building Research Establishment News,* Her Majesty's Stationery Office, London, Winter 1979.

to 1978, the immediate cause of failure of at least eight concrete structures was corrosion of reinforcing or prestressing steel. These structures were 12 to 40 years old at the time of collapse, except one which was only 2 years old.

It is to be expected that when the embedded steel is protected from air by an adequately thick cover of low-permeability concrete, the corrosion of steel and other problems associated with it would not arise. That this expectation is not fully met in practice is evident from the unusually high frequency with which even properly built reinforced and prestressed concrete structures continue to suffer damage due to steel corrosion. The magnitude of damage is especially large in structures exposed to marine environments and to deicing chemicals. For example, in 1975 it was reported[44] that the U.S. Interstate Highway System alone needed $6 billion for repair and replacement of reinforced concrete bridges and bridge decks. Approximately 4800 of the 25,000 bridges in the state of Pennsylvania were found to be in need of repair.

The damage to concrete resulting from corrosion of embedded steel manifests in the form of expansion, cracking, and eventually spalling of the cover (Fig. 5–16a). In addition to loss of cover, a reinforced-concrete member may suffer structural damage due to loss of bond between steel and concrete and loss of rebar cross-sectional area—sometimes to the extent that structural failure becomes inevitable.[45] A review of the mechanisms involved in concrete deterioration due to corrosion of embedded steel, selected case histories, and measures for control of the phenomenon are given here.

Mechanisms Involved in Concrete Deterioration by Corrosion of Embedded Steel

Corrosion of steel in concrete is an *electrochemical process*. The electrochemical potentials to form the *corrosion cells* may be generated in two ways:

1. Composition cells may be formed when two dissimilar metals are embedded in concrete, such as steel rebars and aluminum conduit pipes, or when significant variations exist in surface characteristics of the steel.
2. Concentration cells may be formed due to differences in concentration of dissolved ions in the vicinity of steel, such as alkalies, chlorides, and oxygen.

As a result, one of the two metals (or some parts of the metal when only one metal is present) becomes anodic and the other cathodic. The fundamental chemical changes occurring at the anodic and cathodic areas[46] are as follows (see also Fig. 5–16b).

---

[44] R. E. Carrier and P. D. Cady, ACI SP-47, 1975, pp. 121–68.

[45] P. D. Cady, ASTM STP 169B, 1978, pp. 275–99.

[46] B. Erlin and G. J. Verbeck, ACI SP-49, 1978, pp. 39–46.

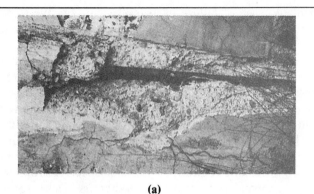

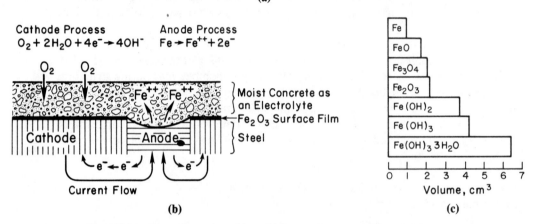

Figure 5-16 Expansion and cracking of concrete due to corrosion of the embedded steel. [(b), (c), *Beton-Bogen,* Aalborg Cement Company, Aalborg, Denmark, 1981.]

Part (a) shows that deterioration of concrete due to corrosion of embedded steel is manifested in the form of expansion, cracking, and loss of cover. Loss of steel-concrete bond and reduction of rebar cross section may lead to structural failure. Part (b) illustrates the electrochemical process of steel corrosion in moist and permeable concrete. The galvanic cell constitutes an anode process and a cathode process. The anode process cannot occur until the protective or the passive iron oxide film is either removed in an acidic environment (e.g., carbonation of concrete) or made permeable by the action of Cl^- ions. The cathode process cannot occur until a sufficient supply of oxygen and water is available at the steel surface. The electrical resistivity of concrete is also reduced in the presence of moisture and salts. Part (c) shows that, depending on the oxidation state, metallic iron can increase more than six times in volume.

$$\text{Anode:} \quad \underset{\text{(metallic iron)}}{Fe} \longrightarrow 2e^- + Fe^{2+}$$

$$\text{Cathode:} \quad \underset{\text{(air)}}{\tfrac{1}{2}O_2} + \underset{\text{(water)}}{H_2O} + 2e^- \longrightarrow 2(OH)^- \qquad \underset{\text{(rust)}}{FeO \cdot (H_2O)_x} \quad (5\text{-}8)$$

The transformation of metallic iron to rust is accompanied by an increase in volume which, depending on the state of oxidation, may be as large as 600 percent of the original metal (Fig. 5–16c). This volume increase is believed to be the principal cause of concrete expansion and cracking. It should be noted that the anodic reaction involving ionization of metallic iron will not progress far unless the electron flow to the cathode is maintained by consumption of the electrons at the cathode; for this *the presence of both air and water at the surface of the cathode is absolutely necessary.* Also, ordinary iron and steel products are covered by a thin iron-oxide film which becomes impermeable and strongly adherent to the steel surface in alkaline environments, thus *making the steel passive to corrosion*; that is, metallic iron will not be available for the anodic reaction until the passivity of steel is destroyed.

In the absence of chloride ions in the solution, the protective film on steel is reported to be stable as long as the pH of the solution stays above 11.5. Since hydrated portland cements contain alkalies in the pore fluid and about 20 weight percent solid calcium hydroxide, normally there is sufficient alkalinity in the system to maintain the pH above 12. In exceptional conditions (e.g., when concrete has high permeability and alkalies and most of the calcium hydroxide are either *carbonated or neutralized by an acidic solution*), the pH of concrete in the vicinity of steel may be reduced to less than 11.5, thus destroying the passivity of steel and setting the stage for the corrosion process.

In the *presence of chloride ions,* depending on the Cl^-/OH^- ratio, it is reported that the protective film may be destroyed even at pH values considerably above 11.5. When Cl^-/OH^- molar ratios are higher than 0.6, steel seems to be no longer protected against corrosion, probably because the iron-oxide film becomes either permeable or unstable under these conditions. For the typical concrete mixtures normally used in practice, the threshold chloride content to initiate corrosion is reported to be in the range 0.6 to 0.9 kg of Cl^- per cubic meter of concrete. Furthermore, when large amounts of chloride are present, concrete tends to hold more moisture, which also increases the risk of steel corrosion by lowering the electrical resistivity of concrete. Once the passivity of the embedded steel is destroyed, it is the electrical resistivity and the availability of oxygen that control the rate of corrosion; in fact, significant corrosion is not observed as long as *the electrical resistivity* of concrete is above 50 to $70 \times 10^3 \, \Omega \cdot cm$. It should be noted that the common sources of chloride in concrete are admixtures, salt-contaminated aggregate, and penetration of deicing salt solutions or seawater.

Selected Case Histories

A survey of the recent collapsed buildings and their immediate causes by the British Building Research Establishment[47] showed that in 1974 a sudden collapse of one main beam of a 12-year-old roof with post-tensioned prestressed concrete beams was due to corrosion of tendons. Poor grouting of ducts and the use of 2 to 4 percent calcium chloride by weight of cement as an accelerating admixture in the concrete were diagnosed as the factors responsible for the corrosion of steel. A number of similar mishaps in Britain provided support for the 1979 amendment to the British

---

[47] *Building Research Establishment News,* see reference 43.

Code of Practice 110 that *calcium chloride should never be added to prestressed concrete, reinforced concrete, and concrete containing embedded metal.*

A survey by the Kansas State Transportation Department showed that in bridge decks exposed to deicing salt treatment there was a strong relation between depth of cover and concrete deterioration in the form of delaminations or horizontal cracking. Generally, good protection to steel was provided when the cover thickness was 50 mm or more (at least thrice the nominal diameter of the rebar, which was 15mm); however, the normal distribution of variation in cover depth was such that about 8 percent of the steel was 37.5 mm or less deep. It was corrosion of this shallower steel that was responsible for the horizontal cracks or delaminations. On one bridge deck, a combination of the freeze-thaw cracking and corrosion of steel in concrete extended the area of delamination about eightfold in 5 years, so that 45 percent of the deck surface had spalled by the time the bridge was only 16 years old. Similar case histories of bridge deck damage on numerous highways including those in Pennsylvania are reported (Fig. 5–17a).

Furthermore, the Kansas survey showed that corrosion of the reinforcing steel

(a)   (b)

Figure 5–17 Damage to reinforced concrete structures due to corrosion of steel. [(a) Photograph courtesy of P. D. Cady, The Pennsylvania State University, University Park, Pennsylvania; (b), photograph from P. K. Mehta and B. C. Gerwick, Jr., *Concr. Int.*, Vol. 4, pp. 45–51, 1982.]

When the $Cl^-/(OH)^-$ ratio of the moist environment in contact with the reinforcing steel in concrete exceeds a certain threshold value, the passivity of steel is broken. This is the first step necessary for the onset of the anodic and cathodic reactions in a corrosion cell. In cold climates, reinforced concrete bridge decks are frequently exposed to the application of deicing chemicals containing chlorides. Progressive penetration of chlorides in permeable concrete leads to scaling, potholes, and delaminations at the concrete surface, finally rendering it unfit for use. Part (a) shows a typical concrete failure (scaling and potholes in the surface of a concrete pavement in Pennsylvania) due to a combination of frost action, corrosion of the embedded reinforcement, and other causes. Part (b) shows deterioration of concrete due to corrosion of the reinforcing steel in the spandrel beams of the San Mateo-Hayward Bridge after 17 years of service. In this case, seawater was the source of chloride ions.

also produced vertical cracks in the deck, which contributed to corrosion of the steel girders supporting the deck. Drawing attention to bridge deck corrosion problems due to deicing salt applications, Carl Crumpton of the Kansas State Transportation Department said:

> The wedding of concrete and steel was an ideal union and we used lots of reinforced concrete for bridge decks. Unfortunately, we began tossing salt to melt snow and ice instead of rice for good fertility. That brought irritation, tensions, and erosion of previously good marital relations. No longer could the two exist in blissful union; the seeds of destruction had been planted and the stage had been set for today's bridge deck cracking and corrosion problems.[48]

It was recently reported[49] that many heavily reinforced, 26- by 12- by 6-ft (8- by 3.7- by 1.8 m) spandrel beams of the San Mateo-Hayward bridge at the San Francisco Bay (California) had to undergo expensive repairs due to serious cracking of concrete associated with corrosion of embedded steel (Fig. 5–17b). The beams were made in 1963 with a high-quality concrete (370 kg/m^3 cement, 0.45 water/cement ratio). The damage was confined to the underside and to the windward faces which were exposed to seawater spray, and occurred only in the precast steam-cured beams. No cracking and corrosion were in evidence in the naturally cured cast-in-place beams made at the same time with a similar concrete mixture. It was suggested that a combination of heavy reinforcement and differential cooling rates in the massive steam-cured beams might have resulted in the formation of microcracks in concrete, which were later enlarged due to severe weathering conditions on the windward side of the beams. Thereafter, penetration of the salt water promoted the corrosion-cracking-more corrosion type of chain reaction which led to the serious damage. More discussion of cracking-corrosion interaction and case histories of seawater attack on concrete are presented later.

Control of Corrosion

Since water, oxygen, and chloride ions play important roles in the corrosion of embedded steel and cracking of concrete, it is obvious that *permeability of concrete is the key* to control the various processes involved in the phenomena. Concrete-mixture parameters to ensure low permeability, e.g., low water/cement ratio, adequate cement content, control of aggregate size and grading, and use of mineral admixtures have been discussed earlier. Accordingly, ACI Building Code 318 specifies a maximum 0.4 water/cement ratio for reinforced normal-weight concrete exposed to deicing chemicals and seawater. Proper consolidation and curing of concrete are equally essential. Design of concrete mixtures should also take into account the possibility of a permeability increase in service due to various physical-chemical causes, such as frost action, sulfate attack, and alkali-aggregate expansion.

---

[48] C. F. Crumpton, ACI Convention Paper, Dallas, 1981.
[49] P. K. Mehta and B. C. Gerwick, Jr., *Concr. Int.,* Vol. 4, No. 10, pp. 45–51, 1982.

For corrosion protection, **maximum permissible chloride contents** of concrete mixtures are also specified by ACI Building Code 318. For instance, maximum water-soluble Cl^- ion concentration in hardened concrete, at an age of 28 days, from all ingredients (including aggregates, cementitious materials, and admixtures) should not exceed 0.06, 0.15, and 0.30 percent by weight of cement, for prestressed concrete, reinforced concrete exposed to chloride in service, and other reinforced concretes, respectively. Reinforced concretes that will remain dry or protected from moisture in service are permitted to contain up to 1.00 percent Cl^- by weight of cement.

Certain design parameters also influence permeability. That is why for concrete exposed to corrosive environment, Section 7.7 of the ACI Building Code 318-83 specifies **minimum concrete cover** requirements. A minimum concrete cover of 50 mm for walls and slabs and 63 mm for other members is recommended. Current practice for coastal structures in the North Sea is to provide minimum 50 mm of cover on conventional reinforcement and 70 mm on prestressing steel. Also, ACI 224R-80 specifies 0.15 mm as the **maximum permissible crack width** at the tensile face of reinforced concrete structures subject to wetting-drying or seawater spray. The CEB Model Code recommends limiting crack widths to 0.1 mm at the steel surface for concrete members exposed to frequent flexural loads, and 0.2 mm to others. The FIP (International Prestressing Federation) recommendations specify that crack widths at points near the main reinforcement should not exceed 0.004 times the nominal cover (i.e., maximum permissible crack width for a 50-mm cover is 0.2 mm and for a 75-mm cover is 0.3 mm). Many researchers find no direct relation between the crack width and corrosion; however, it appears that by increasing the permeability of concrete and exposing it to numerous physical-chemical processes of deterioration, the presence of cracks would eventually have a deleterious effect.

The repair and replacement costs associated with concrete bridge decks damaged by corrosion of reinforcing steel have become a major maintenance expense. Many highway agencies now prefer the extra initial cost of providing a waterproof membrane, or a thick overlay of an impervious concrete mixture on newly constructed or thoroughly repaired surfaces of reinforced and prestressed concrete elements that are large and have flat configuration. **Waterproof membranes,** usually preformed and of the sheet-type variety, are used when they are protected from physical damage by asphaltic concrete wearing surfaces; therefore, their surface life is limited to the life of the asphaltic concrete, which is about 15 years. **Overlay of watertight concrete** 37.5 to 63 mm thick provides a more durable protection to penetration of aggressive fluids into reinforced or prestressed concrete members. Typically, concrete mixtures used for overlay are of low slump, very low water/cement ratio (made possible by adding a superplasticizing admixture), and high cement content. Portland cement mortars containing polymer emulsions (latexes) also show excellent impermeability and have been used for overlay purposes; however, vinylidene chloride-type latexes are suspected to be the cause of corrosion problems in some cases, and it is now preferred that styrene-butadiene-type products be used.

Reinforcing bar coatings and cathodic protection provide other approaches to prevent corrosion and are costlier than producing a low-permeability concrete through quality, design, and construction controls. ***Protective coatings for reinforcing steel*** are of two types: anodic coatings (e.g., zinc-coated steel) and barrier coatings (e.g., epoxy-coated steel). Due to concern regarding the long-term durability of zinc-coated rebars in concrete, the U.S. Federal Highway Administration in 1976 placed a temporary moratorium on its use in bridge decks. Long-time performance of epoxy-coated rebars is under serious investigation in many countries. ***Cathodic protection*** techniques involve suppression of current flow in the corrosion cell, either by supplying externally a current flow in the opposite direction or by using sacrificial anodes. According to Carrier and Cady,[50] both systems have been used extensively with mixed results.

CONCRETE IN SEAWATER

For several reasons, effect of seawater on concrete deserves special attention. First, coastal and offshore sea structures are exposed to the simultaneous action of a number of physical and chemical deterioration processes, which provide an excellent opportunity to understand the complexity of concrete durability problems in practice. Second, oceans make up 80 percent of the surface of the earth; therefore, a large number of structures are exposed to seawater either directly or indirectly (e.g., winds can carry seawater spray up to a few miles inland from the coast). Concrete piers, decks, break-waters, and retaining walls are widely used in the construction of harbors and docks. To relieve land from pressures of urban congestion and pollution, floating offshore platforms made of concrete are being considered for location of new airports, power plants, and waste disposal facilities. The use of concrete offshore drilling platforms and oil storage tanks is already on the increase.

Most seawaters are fairly uniform in chemical composition, which is characterized by the presence of about 3.5 percent soluble salts by weight. The ionic concentrations of Na^+ and Cl^- are the highest, typically 11,000 and 20,000 mg/liter, respectively. However, from the standpoint of aggressive action to cement hydration products, sufficient amounts of Mg^{2+} and SO_4^{2-} are present, typically 1400 and 2700 mg/liter, respectively. The pH of seawater varies between 7.5 and 8.4, the average value in equilibrium with the atmospheric CO_2 being 8.2. Under exceptional conditions (i.e., in sheltered bays and estuaries) pH values lower than 7.5 may be encountered; these are usually due to a higher concentration of dissolved CO_2, which would make the seawater more aggressive to portland cement concrete.

Concrete exposed to marine environment may deteriorate as a result of combined effects of chemical action of seawater constituents on cement hydration products, alkali-aggregate expansion (when reactive aggregates are present), crystallization pressure of salts within concrete if one face of the structure is subject to

---

[50] R. E. Carrier and P. D. Cady, in ACI SP-47.

wetting and others to drying conditions, frost action in cold climates, corrosion of embedded steel in reinforced or prestressed members, and physical erosion due to wave action and floating objects. Attack on concrete due to any one of these causes tends to increase the permeability; not only would this make the material progressively more susceptible to further action by the same destructive agent but also to other types of attack. Thus a maze of interwoven chemical as well as physical causes of deterioration are found at work when a concrete structure exposed to seawater is in an advanced stage of degradation. The theoretical aspects, selected case histories of concrete deteriorated by seawater, and recommendations for construction of concrete structures in marine environment are discussed next.

Theoretical Aspects

From the standpoint of chemical attack on hydrated portland cement in unreinforced concrete, when alkali-reactive aggregates are not present, one might anticipate that sulfate and magnesium are the harmful constituents in seawater. It may be recalled that with groundwaters, sulfate attack is classified as *severe* when the sulfate ion concentration is higher than 1500 mg/liter; similarly, portland cement paste can deteriorate by cation-exchange reactions when magnesium ion concentration exceeds, for instance, 500 mg/liter.

Interestingly, in spite of the undesirably high sulfate content of seawater, it is a common observation that even when a high-C_3A portland cement has been used and large amounts of ettringite are present as a result of sulfate attack on the cement paste, the deterioration of concrete is not characterized by expansion; instead, it mostly takes the form of erosion or loss of the solid constituents from the mass. It is proposed that ettringite expansion is suppressed in environments where $(OH)^-$ ions have essentially been replaced by Cl^- ions. Incidentally, this view is consistent with the hypothesis that alkaline environment is necessary for swelling of ettringite by water adsorption. Irrespective of the mechanism by which the sulfate expansion associated with ettringite is suppressed in high-C_3A portland cement concretes exposed to seawater, the influence of chloride on the system demonstrates the error too often made in modeling the behavior of materials when, for the sake of simplicity, the effect of an individual factor on a phenomenon is predicted without sufficient regard to the other factors present, which may modify the effect significantly.

It may be noted that according to ACI Building Code 318-83, sulfate exposure to seawater is classified as *moderate,* for which the use of ASTM Type II portland cement (maximum 8 percent C_3A) with a 0.50 maximum water/cement ratio in normal-weight concrete is permitted. In fact, it is stated in the ACI 318R-21 *Building Code Commentary* that cements with C_3A up to 10 percent may be used if the maximum water/cement ratio is further reduced to 0.40.

The fact that the presence of uncombined calcium hydroxide in concrete can cause deterioration by an exchange reaction involving magnesium ions was known as early as 1818 from investigations on disintegration of lime-pozzolan concretes by

Vicat, who undoubtedly is regarded as one of the founders of the technology of modern cement and concrete. Vicat made the profound observation:

> On being submitted to examination, the deteriorated parts exhibit much less lime than the others; what is deficient then, has been dissolved and carried off; it was in excess in the compound. Nature, we see, labors to arrive at exact proportions, and to attain them, corrects the errors of the hand which has adjusted the doses. Thus the effects which we have just described, and in the case alluded to, become the more marked, the further we deviate from these exact proportions.[51]

Several state-of-the-art reviews[52] on the performance of structures in marine environments confirm that Vicat's observation is equally valid for portland cement concrete. From long-term studies of portland cement mortars and concretes exposed to seawater, the evidence of magnesium ion attack is well established by the presence of white deposits of $Mg(OH)_2$, also called *brucite* (Fig. 5–18b), and magnesium silicate hydrate. In seawater, well-cured concretes containing large amounts of slag or pozzolan in cement usually outperform reference concrete containing only portland cement,[53] partly because the former contain less uncombined calcium hydroxide after curing. The implication of loss of lime by cement paste, whether by magnesium ion attack or by CO_2 attack, is obvious from Fig. 5–18c.

Since seawater analyses seldom include the dissolved CO_2 content, the potential for loss of concrete mass by leaching away of calcium from hydrated cement paste due to carbonic acid attack is often overlooked. According to Feld,[54] in 1955, after 21 years of use, the concrete piles and caps of the trestle bends of the James River Bridge at Newport News, Virginia, required a $1.4 million repair and replacement job which involved 70 percent of the 2500 piles. Similarly, 750 precast concrete piles driven in 1932 near Ocean City, New Jersey had to be repaired in 1957 after 25 years of service; some of the piles had been reduced from the original 550 mm diameter to 300 mm. In both cases, the loss of material was associated with higher than normal concentrations of dissolved CO_2 present in the seawater.

It should be noted that in *permeable concrete* the normal amount of CO_2 present in seawater is sufficient to decompose the cementitious products eventually. The presence of thaumasite (calcium silicocarbonate), hydrocalumite (calcium carboaluminate hydrate), and aragonite (calcium carbonate) have been reported in cement pastes derived from deteriorated concretes exposed to seawater for long periods.

---

[51] L. J. Vicat, *A Practical and Scientific Treatise on Calcareous Mortars and Cements*, 1837 (translated by J. T. Smith, London).

[52] W. G. Atwood and A. A. Johnson, *Trans. ASCE*, Vol. 87, Paper 1533, pp. 204–75, 1924; F. M. Lea, *The Chemistry of Cement and Concrete*, Chemical Publishing Company, Inc., New York, 1971, pp. 623–38; P. K. Mehta, *Performance of Concrete in Marine Environment*, ACI SP-65, 1980, pp. 1–20.

[53] O. E. Gjorv, *J. ACI*, Proc., Vol. 68, pp. 67–70, 1971.

[54] J. Feld, *Construction Failures*, John Wiley & Sons, Inc., New York, 1968, pp. 251–55.

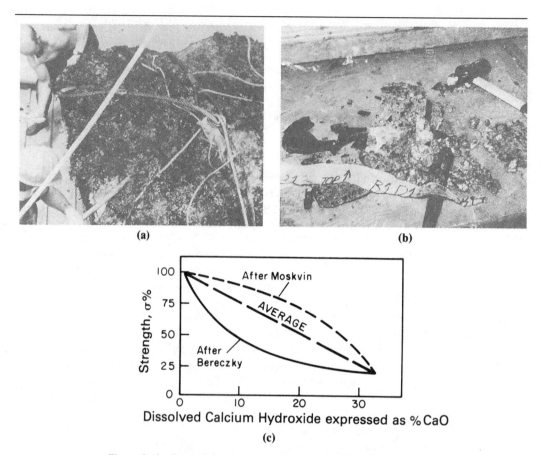

Figure 5–18 Strength loss in permeable concrete due to lime leaching. [(a), (b), Photographs from P. K. Mehta and H. Haynes, *J. ASCE, Struct Div.*, Vol. 101, No. ST-8, pp. 1679–86, 1975; (c), adapted from I. Biczok, *Concrete Corrosion and Concrete Protection*, Chemical Publishing Company, Inc., New York, 1967, p. 291.]

Unreinforced concrete test blocks (1.75 by 1.75 by 1.07 m) partially submerged in seawater at San Pedro harbor in Los Angeles, California, were examined after 67 years of continuous exposure. Low-permeability concretes, irrespective of the portland cement composition were found to be in excellent condition. Concretes containing a low cement content (higher permeability) showed so much reduction in surface hardness that deep grooves were made by a wire rope on the test blocks when they were being hauled up with the help of an amphibious crane [part (a)]. Test cores showed that concrete was very porous and weak. The pores contained large deposits of a white precipitate [part (b)], which was identified to be $Mg(OH)_2$ by X-ray diffraction analysis; the products of portland cement hydration, C-S-H and $Ca(OH)_2$, were no longer present.

Numerous researchers have found that portland cement pastes, mortars, and concretes undergo strength loss when the cementitious products are decomposed and leached out as a result of attack by acidic or magnesium-containing solutions. The severity of leaching can be evaluated from the content of dissolved CaO. On the average, the compressive strength drops by about 2 percent when 1 percent CaO is removed from the portland cement paste [part (c)].

Concrete in Seawater

Case Histories of Deteriorated Concrete

Compared to other structural materials, generally, concrete has a satisfactory record of performance in seawater. However, published literature contains descriptions of a large number of both plain and reinforced concretes which showed serious deterioration in marine environment. For the purpose of drawing useful lessons for construction of concrete sea structures, six case histories of deterioration of concrete as a result of long-term exposure to seawater are summarized in Table 5-5 and discussed below.

In the *mild climates* of southern France and southern California, *unreinforced mortar and concrete specimens* remained in excellent condition after more than 60 years of seawater exposure, except when permeability was high. Permeable specimens showed considerable loss of mass associated with magnesium ion attack, CO_2 attack, and calcium leaching. In spite of the use of high-C_3A portland cements, expansion and cracking of concrete due to ettringite was not observed in low-permeability concretes. Therefore, the effect of cement composition on durability to seawater appears to be less significant than the permeability of concrete.

Reinforced concrete members in a mild climate (Piers 26 and 28 of the San Francisco Ferry Building in California), in spite of a low-permeability concrete mixture 390 kg/m^3 cement content), showed cracking due to corrosion of the reinforcing steel after 46 years of service. Since corrosion requires permeation of seawater and air to the embedded steel, poor consolidation of concrete and structural microcracking were diagnosed to be the causes of the increase in permeability (which made the corrosion of steel possible).

In the *cold climates* of Denmark and Norway, concrete that was not protected by entrained air was subject to expansion and cracking by frost action. (It may be noted that air entrainment was not prevalent before the 1950s.) Therefore, cracking due to freeze-thaw cycles was probably responsible for the increase in permeability, which promoted other destructive processes such as alkali-aggregate attack and corrosion of the reinforcing steel.

Investigations of reinforced concrete structures have shown that, generally, concrete fully immersed in seawater suffered only a little or no deterioration; concrete exposed to salts in air or water spray suffered some deterioration, especially when permeable; and concrete subject to tidal action suffered the most.

Lessons from Case Histories

For the construction of concrete sea structures, important lessons from case histories of concrete deteriorated by seawater can be summed up as follows:

1. *Permeability is the key to durability.* Deleterious interactions of serious consequence between constituents of hydrated portland cement and seawater take place when seawater is not prevented from penetrating into the interior of concrete. Typical causes of insufficient watertightness are poorly proportioned

TABLE 5-5 PERFORMANCE OF CONCRETE EXPOSED TO SEAWATER

History of structures	Results of examination
MILD CLIMATE	
Forty-centimeter mortar cubes made with different cements and three different cement contents, 300, 450, and 600 kg/m^3, were exposed to seawater at La Rochell, southern France, in 1904–1908.[a]	After 66 years of exposure to seawater, the cubes made with 600 kg/m^3 cement were in good condition even when they contained a high-C_3A (14.9%) portland cement. Those containing 300 kg/m^3 were destroyed; therefore, chemical resistance of the cement was of major importance for low-cement-content cubes. In general, pozzolan and slag cements showed better resistance to seawater than portland cements. Electron micrographic studies of deteriorated specimens showed the presence of aragonite, brucite, ettringite, magnesium silicate hydrate, and thaumasite.
Eighteen 1.75 × 1.75 × 1.07 m unreinforced concrete blocks, made with six different portland cements and three different concrete mixtures, partially submerged in seawater in Los Angeles in 1905.[b]	After 67 years of exposure, the dense concrete (1:2:4) blocks, some made with 14% C_3A portland cement, were still in excellent condition. Lean concrete (1:3:6) blocks lost some material and were much softer (Fig. 5–18a). X-ray diffraction analyses of the weakened concrete showed the presence of brucite, gypsum, ettringite, and hydrocalumite. The cementing constituents, CSH gel and $Ca(OH)_2$, were not detected.
Concrete structures of the San Francisco Ferry Building, built in 1912. Type I portland cement with 14–17% C_3A was used. 1:5 concrete mixture contained 658 lb/yd^3 (390 kg/m^3) cement. (a) Precast concrete cylinders jacket for Pier 17. (b) Cast-in-place concrete cylinders for Piers 30 and 39. (c) Cast-in-place concrete cylinders and transverse girders for Piers 26 and 28.[c]	After 46 years of service (a) was found in excellent condition, and 90% of piles in (b) were in good condition. In (c), 20–30% of piles were attacked in tidal zone, and about 35% of the deep transverse girders had their underside and part of the vertical face cracked or spalled due to corrosion of reinforcement. Presence of microcracks due to deflection under load might have exposed the reinforcing steel to corrosion by seawater. Poor workmanship was held responsible for differences in behavior of concrete, which was of the same quality in all the structures.
COLD CLIMATE	
Many 20- to 50-year coastal structures were included in a 1953–55 survey of 431 concrete structures in Denmark.[d] Among the severely deteriorated structures were the following in Jutland.	Of the coastal structures, about 40% showed overall deterioration, and about 35% showed from severe surface damage to slight deterioration.
Oddesund Bridge, Pier 7. History of structure indicated initial cracking of caissons due to	Examination of deteriorated concrete from the Oddesund Bridge indicated decomposi-

Concrete in Seawater

TABLE 5–5 Continued

History of structures	Results of examination
thermal stresses. This permitted considerable percolation of water through the caisson walls and the interior mass concrete filling. General repairs commenced after 8 years of service.	tion of cement and loss of strength due to sulfate attack below low-tide level and cracking due to freezing and thawing as well as alkali-aggregate reaction above high-tide level. Reaction products from cement decomposition were aragonite, ettringite, gypsum, brucite, and alkali-silica gel.
Highway Bridge, North Jutland. Severe cracking and spalling of concrete at the mean water level provided a typical hourglass shape to the piers. Concrete in this area was very weak. Corrosion of reinforcement was everywhere and pronounced in longitudinal girders.	Examination of concrete piers of the highway bridge showed evidence of poor concrete quality (high w/c). Symptoms of general decomposition of cement and severe corrosion of the reinforcement were superimposed on the evidences for the primary deleterious agents, such as freezing-thawing and alkali-aggregate reaction.
Groin 71, north barrier, Lim Fiord. Lean concrete blocks (220 kg/m^3 cement) exposed to windy weather, repeated wetting and drying, high salinity, freezing and thawing, and severe impact of gravel and sand in the surf. Some blocks disappeared in the sea in the course of 20 years.	Examination of the severely deteriorated concrete blocks from Groin 71 showed very weak, soapy matrix with loose aggregate pebbles. In addition to the alkali-silica gel, the presence of gypsum and brucite was confirmed.
Along the Norwegian seaboard, 716 concrete structures were surveyed in 1962–64. About 60% of the structures were reinforced concrete wharves of the slender-pillar type containing tremie-poured underwater concrete. Most wharves had decks of the beam and slab type. At the time of survey, about two-thirds of the structures were 20–50 years old.[e]	Below the low-tide level and above the high-tide level, concrete pillars were generally in good condition. In the splashing zone, about 50% of the surveyed pillars were in good condition; 14% had their cross-sectional area reduced by 30% or more, and 24% had 10–30% reduction in area of cross section. Deck slabs were generally in good condition but 20% deck beams needed repair work because of major damage due to corrosion of reinforcement. Deterioration of pillars in the tidal zone was ascribed mainly due to frost action on poor-quality concrete.

[a] M. Regourd, *Annales de l'Institute Technique du Bâtiment et des Travaux Publics,* No. 329, June 1975, and No. 358, February 1978.

[b] P. K. Mehta and H. Haynes, *J. Struct. Div. ASCE,* Vol. 101, No. ST-8, August 1975.

[c] P. J. Fluss and S. S. Gorman, *J. ACI,* Proc., Vol. 54, 1958.

[d] G. M. Idorn, "Durability of Concrete Structures in Denmark," D.Sc. dissertation, Tech. Univ., Copenhagen, Denmark, 1967.

[e] O. E. Gjorv, Durability of Reinforced Concrete Wharves in Norwegian Harbors, The Norwegian Committee on Concrete in Sea Water, 1968.

concrete mixtures, absence of properly entrained air if the structure is located in a cold climate, inadequate consolidation and curing, insufficient concrete cover on embedded steel, badly designed or constructed joints, and microcracking in hardened concrete attributable to lack of control of loading conditions and other factors, such as thermal shrinkage, drying shrinkage, and alkali aggregate expansion.

It is interesting to point out that engineers on the forefront of concrete technology are becoming increasingly conscious of the significance of permeability to durability of concrete exposed to aggressive waters. For example, concrete specifications for offshore structures in Norway now specify the maximum permissible permeability directly ($k \leq 10^{-13}$ kg/Pa·m·sec).

2. *Type and severity of deterioration may not be uniform throughout the structure* (Fig. 5–19). For example, with a concrete cylinder the section that always remains above the high-tide line will be more susceptible to frost action and corrosion of embedded steel. The section that is between high- and low-tide lines will be vulnerable to cracking and spalling, not only from frost action and steel corrosion but also from wet-dry cycles. Chemical attacks due to alkali-aggregate reaction and seawater-cement paste interaction will also be at work

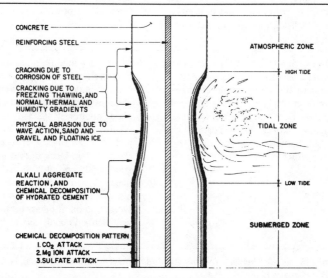

Figure 5–19 Diagrammatic representation of deterioration of a concrete cylinder exposed to seawater. (From P. K. Mehta, *Performance of Concrete in Marine Environment,* ACI SP-65, 1980, pp. 1–20.)

The type and severity of attack on a concrete sea structure depend on the conditions of exposure. The section of the structure that remains fully submerged is rarely subjected to frost action or corrosion of the embedded steel. Concrete at this exposure condition will be susceptible to chemical attacks. The general pattern of chemical attack from the concrete surface to the interior is shown. The section above the high-tide mark will be vulnerable to both frost action and corrosion of the embedded steel. The most severe deterioration is likely to take place in the tidal zone because here the structure is exposed to all kinds of physical and chemical attacks.

Concrete in Seawater

here. Concrete weakened by microcracking and chemical attacks will eventually disintegrate by wave action and the impact of sand, gravel, and ice; thus maximum deterioration occurs in the tidal zone. On the other hand, the fully submerged part of the structure will only be subject to chemical attack by seawater; since it is not exposed to subfreezing temperatures there will be no risk of frost damage, and due to lack of oxygen there will be little corrosion.

It appears that progressive chemical deterioration of cement paste by seawater from the surface to the interior of the concrete follows a general pattern.[55] The formation of aragonite and bicarbonate by CO_2 attack is usually confined to the surface of concrete, the formation of brucite by magnesium ion attack is found below the surface of concrete, and evidence of ettringite formation in the interior shows that sulfate ions are able to penetrate even deeper. Unless concrete is very permeable, no damage results from chemical action of seawater on cement paste because the reaction products (aragonite, brucite, and ettringite), being insoluble, tend to reduce the permeability and stop further ingress of seawater into the interior of the concrete. This kind of protective action would not be available under conditions of dynamic loading and in the tidal zone, where the reaction products are washed away by wave action as soon as they are formed.

3. *Corrosion of embedded steel is, generally, the major cause of concrete deterioration in reinforced and prestressed concrete structures exposed to seawater, but in low-permeability concrete this does not appear to be the first cause of cracking.* Based on numerous case histories, it appears that cracking-corrosion interactions probably follow the route diagrammatically illustrated in Fig. 5–20. Since the corrosion rate depends on the cathode/anode area, significant corrosion and expansion accompanying the corrosion should not occur until there is sufficient supply of oxygen at the surface of the reinforcing steel (i.e., an

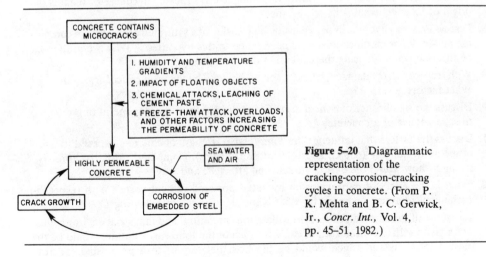

Figure 5–20 Diagrammatic representation of the cracking-corrosion-cracking cycles in concrete. (From P. K. Mehta and B. C. Gerwick, Jr., *Concr. Int.*, Vol. 4, pp. 45–51, 1982.)

---

[55] Biczok, *Concrete Corrosion and Concrete Protection*, pp. 357–58.

increase in the cathode area). This does not happen as long as the permeability of steel-cement paste interfacial zone remains low.

Pores and microcracks already exist in the interfacial zone, but their enlargement through a variety of phenomena other than corrosion seems to be necessary before conditions exist for significant corrosion of the embedded steel in concrete. Once the conditions for significant corrosions are established, a progressively escalating cycle of cracking-corrosion-more cracking begins, eventually leading to complete deterioration of concrete.

TEST YOUR KNOWLEDGE

1. What do you understand by the term *durability*? Compared to other considerations, how much importance should be given to durability in the design and construction of concrete structures?
2. Write a short note on the structure and properties of water, with special reference to its destructive effect on materials.
3. How would you define the coefficient of permeability? Give typical values of the coefficient for (a) fresh cement pastes; (b) hardened cement pastes; (c) commonly used aggregates; (d) high-strength concretes; and (e) mass concrete for dams.
4. How does aggregate size influence the coefficient of permeability of concrete? List other factors that determine the permeability of concrete in a structure.
5. What is the difference between erosion and cavitation? From the standpoint of durability to severe abrasion, what recommendations would you make in the design of concrete and construction of an industrial floor?
6. Under what conditions may salt solutions damage concrete without involving chemical attack on the portland cement paste? Which salt solutions commonly occur in natural environments?
7. Briefly explain the causes and control of scaling and D-cracking in concrete. What is the origin of laitance; what is its significance?
8. Discuss Powers's hypothesis of expansion on freezing of a saturated cement paste containing no air. What modifications have been made to this hypothesis? Why is entrainment of air effective in reducing the expansion due to freezing?
9. With respect to frost damage, what do you understand by the term *critical aggregate size*? What factors govern it?
10. Discuss the significance of *critical degree of saturation* from the standpoint of predicting frost resistance of a concrete.
11. Discuss the factors that influence the compressive strength of concrete exposed to a fire of medium intensity (650°C, short-duration exposure). Compared to the compressive strength, how would the elastic modulus be affected, and why?
12. What is the effect of pure water on hydrated portland cement paste? With respect to carbonic acid attack on concrete, what is the significance of *balancing CO_2*?
13. List some of the common sources of sulfate ions in natural and industrial environments. For a given sulfate concentration, explain which of the following solutions would be the most deleterious and which would be the least deleterious to a permeable concrete containing a high-C_3A portland cement: Na_2SO_4, $MgSO_4$, $CaSO_4$.

14. What chemical reactions are generally involved in sulfate attack on concrete? What are the physical manifestations of these reactions?
15. Critically review the *BRE Digest 250* and the ACI Building Code 318 requirements for control of sulfate attack on concrete.
16. What is the alkali-aggregate reaction? List some of the rock types that are vulnerable to attack by alkaline solutions. Discuss the effect of aggregate size on the phenomenon.
17. With respect to the corrosion of steel in concrete, explain the significance of the following terms: carbonation of concrete, passivity of steel, Cl^-/OH^- ratio of the contact solution, electrical resistivity of concrete, state of oxidation of iron.
18. Briefly describe the measures that should be considered for the control of corrosion of embedded steel in concrete.
19. With coastal and offshore concrete structures directly exposed to seawater, why does most of the deterioration occur in the tidal zone? From the surface to the interior of concrete, what is the typical pattern of chemical attack in sea structures?
20. A heavily reinforced and massive concrete structure is to be designed for a coastal location in Alaska. As a consultant to the primary contractor, write a report explaining the state of the art on the choice of cement type, aggregate size, admixtures, mix proportions, concrete placement, and concrete curing procedures.

SUGGESTIONS FOR FURTHER STUDY

General

ACI Committee 201, "Guide to Durable Concrete," *ACI Mat. Jour.*, Vol. 88, No. 5, pp. 544–582, 1991.

MALHOTRA, V. M., ed., *Durability of Concrete*, ACI SP 126, Vol. 1&2, 1991.

SCANLON, J. M., ed., *Concrete Durability*. ACI SP 100, Vol. 1&2, 1987.

MEHTA, P. K., SCHIESSL P., and RAUPACH, M., "Performance and Durability of Concrete Systems," Proc. 9th International Congress on the Chemistry of Cements, Vol. 1, New Delhi, 1992.

Chemical Aspects of Durability

BICZOK, I., *Concrete Corrosion and Concrete Protection*, Chemical Publishing Company, Inc., New York, 1967.

LEA, F. M., *The Chemistry of Cement and Concrete*, Chemical Publishing Company, Inc., New York, 1971, pp. 338–59, 623–76.

Alkali-Aggregate Expansion

BLANK, R. F., and H. L. KENNEDY, *The Technology of Cement and Concrete*, Vol. 1, John Wiley & Sons, Inc., New York, 1955, pp. 318–42.

DIAMOND, S., *Cem. Concr. Res.*, Vol. 5, pp 329–46, 1975; Vol. 6, pp. 549–60, 1976.

GRATTEN-BELLEUE, P. E., ed., Proc. 7th Int'l. Conf. on Alkali-Aggregate Reactions, National Research Council, Ottawa, Canada, 1987.

HOBBS, D.W., *Alkali-silica Reaction in Concrete*, Thomas Telford, London, 1988.

Corrosion of Embedded Steel

CRANE, A. P., ed., *Corrosion of Reinforcement in Concrete Construction,* Ellis Horwood Ltd., Chichester, West Sussex, U.K., 1983.

TONINI, D. E., and S. W. DEAN, JR., *Chloride Corrosion of Steel in Concrete,* ASTM STP 629, 1977.

Seawater Attack

MALHOTRA, V. M., ed., *Performance of Concrete in Marine Environment,* ACI SP-65, 1980.

MALHOTRA, V. M., ed., *Performance of Concrete in Marine Environment,* ACI SP109, 1988.

MEHTA, P. KUMAR, *Concrete in the Marine Environment,* Elsevier Publishing, 1991.

Frost Action and Fire

ACI, *Behavior of Concrete under Temperature Extremes,* SP-39, 1973.

Betonghandboken (in Swedish), Svensk Byggtjanst, Stockholm, 1980; and Report of RILEM Committee 4 CDC, *Materials and Structures,* Vol. 10, No. 58, 1977.

LITVAN, G. G., and P. J. SEREDA, eds., *Durability of Building Materials and Components,* ASTM STP 691, 1980.

A SIMPLE CODE FOR BUILDERS

Hammurabi, a king of Babylon, who lived four thousand years ago, had the following rule about the responsibility of builders enforced:

"If a building falls down causing the death of the owner or his son, whichever may be the case, the builder or his son will be put to death. If the slave of the home owner dies, he shall be given a slave of the same value. If other possessions are destroyed, these shall be restored, and the damaged parts of the home shall be reconstructed at builder's cost."

To those engaged in the concrete construction industry, Hammurabi's code should be a reminder of the individual's responsibility toward durability of structures.

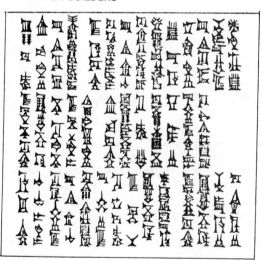

CHAPTER 6

Hydraulic Cements

PREVIEW

Hydraulic or water-resisting cements consist essentially of portland cement and its several modifications. To understand the properties of portland cement, it is helpful to acquire some familiarity with its manufacturing process, chemical and mineralogical composition, crystal structure, and the reactivity of constituent compounds such as calcium silicates and calcium aluminates. Furthermore, the properties of concrete containing portland cement develop as a result of chemical reactions between the compounds in portland cement and water since these hydration reactions are accompanied by changes in matter and energy.

In this chapter the composition and characteristics of the principal compounds in portland cement are described. Hydration reactions of aluminates with their influence on the various types of setting behavior, and of silicates with their influence on strength development, are fully discussed. The relationships between the chemistry of reactions and the physical aspects of setting and hardening of portland cements are covered in detail. Classification of portland cements and specifications according to ASTM C 150 are also reviewed.

Portland cements do not fulfill all the needs of the concrete construction industry; to fill certain unmet needs special cements have been developed. The compositions, hydration characteristics, and important properties of pozzolan cements, blast-furnace slag cements, expansive cements, rapid setting and hardening cements, white or colored cements, oil-well cements, and calcium aluminate cements are also described in this chapter.

HYDRAULIC AND NONHYDRAULIC CEMENTS

Definitions, and the Chemistry of Gypsum and Lime Cements

Hydraulic cements are defined as cements that not only harden by reacting with water but also form a water-resistant product. Cements derived from calcination of gypsum or carbonates such as limestone are **nonhydraulic** because their products of hydration are not resistant to water. Lime mortars that were used in ancient structures built by Greeks and Romans were rendered hydraulic by the addition of pozzolanic materials which reacted with lime to produce a water-resistant cementitious product. The chemistry underlying gypsum and lime cements is illustrated in Fig. 6–1.

Compared to gypsum and lime cements, portland cement and its various modifications are the principal cements used today for making structural concrete. This is because portland cement is truly hydraulic; it does not require the addition of a pozzolanic material to develop water-resisting properties.

PORTLAND CEMENT

Definition

ASTM C 150 defines **portland cement** as a hydraulic cement produced by pulverizing clinkers consisting essentially of hydraulic calcium silicates, usually containing one or more of the forms of calcium sulfate as an interground addition. Clinkers are 5- to 25-mm-diameter nodules of a sintered material which is produced when a raw mixture of predetermined composition is heated to high temperatures.

Manufacturing Process

Since calcium silicates are the primary constituents of portland cement, the raw materials for the production of cement must provide calcium and silica in suitable forms and proportions. Naturally occurring calcium carbonate materials such as limestone, chalk, marl, and seashells are the common industrial sources of calcium, but clay and dolomite ($CaCO_3 \cdot MgCO_3$) are present as principal impurities. Clays and shales, rather than quartz or sandstone, are the preferred sources of additional silica in the raw mix for making calcium silicates because quartzitic silica does not react easily.

Clays also contain alumina (Al_2O_3), and frequently iron oxide (Fe_2O_3) and alkalies. The presence of Al_2O_3, Fe_2O_3, MgO, and alkalies in the raw mix has a mineralizing effect on the formation of calcium silicates; that is, it helps the formation of calcium silicates at considerably lower temperatures than would otherwise be possible. Therefore, when sufficient amounts of Al_2O_3 and Fe_2O_3 are not present in

Portland Cement

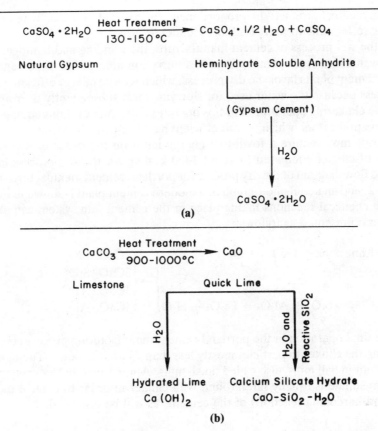

Figure 6–1 Chemistry of gypsum and lime cements: (a) production of gypsum cement, and hydration reaction; (b) production of lime cements, and hydration reactions both with and without pozzolans.

Crystallization of gypsum needles from a hydrated gypsum cement is the cause of setting and hardening. However, gypsum is not stable in water; therefore, the gypsum cement is nonhydraulic.

Hydrated lime, $Ca(OH)_2$, is also not stable in water. However, it can slowly carbonate in air to form a stable product ($CaCO_3$). When a pozzolan (reactive silica) is present in the system, the calcium silicate hydrates formed as a result of the reaction between lime and pozzolan are stable in water.

the principal raw materials, these are purposefully incorporated into the raw mix through addition of secondary materials such as bauxite and iron ore. As a result, besides the calcium silicates the final product also contains aluminates and aluminoferrites of calcium.

To facilitate the formation of desired compounds in portland cement clinker it is necessary that the raw mix be well homogenized before the heat treatment. This explains why the quarried materials have to be subjected to a series of crushing, grinding, and blending operations. From chemical analyses of the stockpiled materials, their individual proportions are determined by the compound composition desired in the final product; the proportioned raw materials are usually interground in ball or roller mills to particles mostly under 75 μm.

In the wet process of cement manufacture, the grinding and homogenization of the raw mix is carried out in the form of a slurry containing 30 to 40 percent water. Modern cement plants favor the dry process, which is more energy efficient than the wet process because the water used for slurrying must subsequently be evaporated before the clinkering operation. For this operation, dry-process kilns equipped with suspension preheaters, which permit efficient heat exchange between the hot gases and the raw mix, require a fossil-fuel energy input on the order of 800 kcal per kilogram of clinker compared to about 1400 kcal/kg for the wet-process kilns. A simplified flow diagram of the dry process for portland cement manufacture is shown in Fig. 6–2, and an aerial photograph of a modern cement plant is shown in Fig. 6–3.

The chemical reactions taking place in the cement kiln system can approximately be represented as follows:

$$\left.\begin{array}{l} \text{Limestone} \rightarrow \text{CaO} + \text{CO}_2 \\ \\ \text{Clay} \rightarrow \text{SiO}_2 + \text{Al}_2\text{O}_3 + \text{Fe}_2\text{O}_3 + \text{H}_2\text{O} \end{array}\right\} \rightarrow \begin{cases} 3\text{CaO} \cdot \text{SiO}_2 \\ 2\text{CaO} \cdot \text{SiO}_2 \\ 3\text{CaO} \cdot \text{Al}_2\text{O}_3 \\ 4\text{CaO} \cdot \text{Al}_2\text{O}_3 \cdot \text{Fe}_2\text{O}_3 \end{cases} \quad (6\text{–}1)$$

The final operation in the portland cement manufacturing process consists of pulverizing the clinker to particles mostly less than 75 μm diameter. The operation is carried out in ball mills, also called finish mills. Approximately 5 percent gypsum or calcium sulfate is usually interground with clinker in order to control the early setting and hardening reactions of the cement, as will be discussed.

Chemical Composition

Although portland cement consists essentially of various compounds of calcium, the results of routine chemical analyses are reported in terms of oxides of the elements present. This is because direct determination of the compound composition requires special equipment and techniques. Also, it is customary to express the individual oxides and clinker compounds by using the following *abbreviations:*

Oxide	Abbreviation	Compound	Abbreviation
CaO	C	$3\text{CaO} \cdot \text{SiO}_2$	C_3S
SiO_2	S	$2\text{CaO} \cdot \text{SiO}_2$	C_2S
Al_2O_3	A	$3\text{CaO} \cdot \text{Al}_2\text{O}_3$	C_3A
Fe_2O_3	F	$4\text{CaO} \cdot \text{Al}_2\text{O}_3 \cdot \text{Fe}_2\text{O}_3$	C_4AF
MgO	M	$4\text{CaO} \cdot 3\text{Al}_2\text{O}_3 \cdot \text{SO}_3$	$C_4A_3\bar{S}$
SO_3	$\bar{S}$	$3\text{CaO} \cdot 2\text{SiO}_2 \cdot 3\text{H}_2\text{O}$	$C_3S_2H_3$
H_2O	H	$\text{CaSO}_4 \cdot 2\text{H}_2\text{O}$	$C\bar{S}H_2$

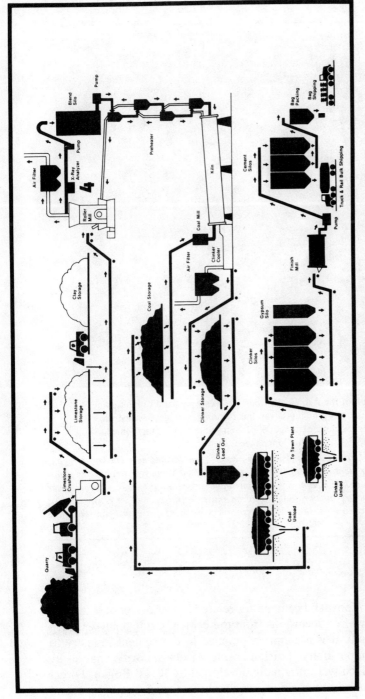

Figure 6-2 Flow diagram of the dry process for portland cement manufacture. (Courtesy of Southwestern Portland Cement Company, Fairborn, Ohio.)

A simplified flow diagram of the dry process of portland cement manufacture is shown. A major step in the process is the clinkering operation carried out in a rotary kiln, which consists of an inclined steel cylinder lined with refactory bricks. The preheated and partially calcined raw mix enters at the higher end of the continuously rotating kiln and is transported to the lower end at a rate controlled by the slope and the speed of the kiln rotation. Pulverized coal, oil, or a fuel gas is injected at the lower end of the burning zone, where temperatures on the order of 1450 to 1550°C may be reached and the chemical reactions involving the formation of portland cement compounds are completed.

Figure 6–3 Aerial view of the Ash Grove Cement (West) portland cement plant at Durkee, Oregon: 1, raw mix grinding; 2, raw mix blending, storage; 3, suspension preheater; 4, rotary kiln; 5, clinker storage; 6, cement grinding. (Photograph courtesy of Vagn Johansen, F. L. Smidth, Copenhagen, Denmark.)

An aerial photograph of the Ash Grove Cement (West) dry-process plant located near Durkee, Oregon, is shown. This 500,000 tonne/year plant, which in 1979 replaced a 200,000 tonne/year wet-process plant, contains a 4.35- by 66-m-long rotary kiln equipped with a four-stage suspension preheater. The preheater exhaust gases go to an electrostatic precipitator designed for an emission efficiency of 99.93 percent. All process loops are monitored and controlled with a 2000 microprocessor-based distributed control system utilizing fuzzy logic.

Since the properties of portland cement are related to the compound composition, it is difficult to draw any conclusions from the cement oxide analyses, such as those shown in Table 6–1. It is a common practice in the cement industry to compute the compound composition of portland cement from the oxide analysis by using a set of equations which were originally developed by R. H. Bogue. Direct determination of the compound composition, which requires special equipment and skill (Fig. 6–4), is not necessary for routine quality control.

Portland Cement

TABLE 6–1 OXIDE ANALYSES OF PORTLAND CEMENTS (%)

Oxide	Cement no. 1	Cement no. 2	Cement no. 3	Cement no. 4	Cement no. 5
S	21.1	21.1	21.1	20.1	21.1
A	6.2	5.2	4.2	7.2	7.2
F	2.9	3.9	4.9	2.9	2.9
C	65.0	65.0	65.0	65.0	64.0
$\bar{S}$	2.0	2.0	2.0	2.0	2.0
Rest	2.8	2.8	2.8	2.8	2.8

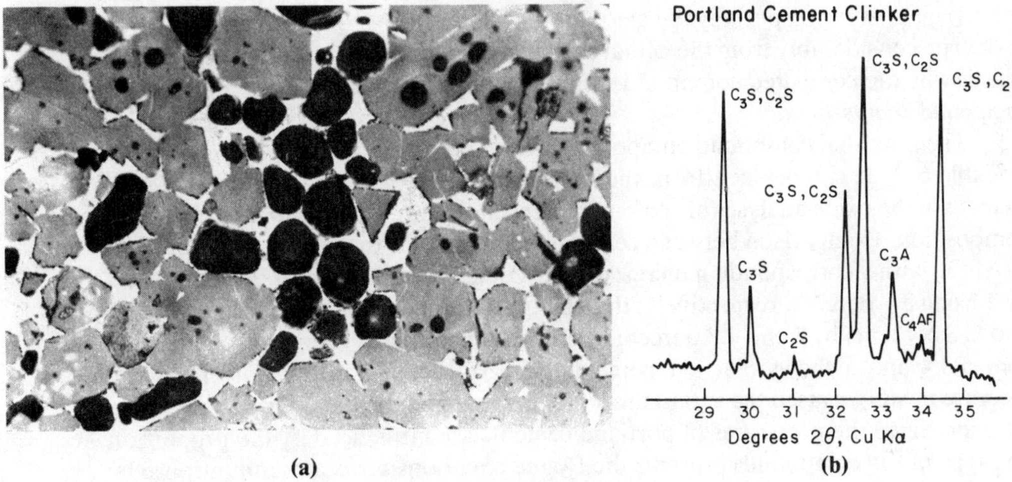

Figure 6–4 (a) Photomicrograph of a polished clinker specimen by reflected light microscopy; (b) X-ray diffraction pattern of a powdered clinker specimen.

Two methods are commonly used for direct quantitative analyses of portland cement clinker. The first method involves reflected-light optical microscopy of polished and etched sections, followed by a point count of areas occupied by the various compounds. Typically, C_3S appears as hexagonal-plate crystals, C_2S as rounded grains with twinning bands, and C_3A as well as C_4AF as interstitial phases. The second method, which is also applicable to pulverized cements, involves X-ray diffraction of powder specimens. Calibration curves based on known mixtures of pure compounds and an internal standard are required; an estimate of the compound is made using these curves and the intensity ratios between a selected diffraction peak of the compound and the internal standard.

Determination of Compound Composition from Chemical Analysis

The **Bogue equations** for estimating the theoretical or potential compound composition of portland cement are as follows:

$$\% \ C_3S = 4.071C - 7.600S - 6.718A - 1.430F - 2.850\bar{S} \qquad (6\text{–}2a)$$

$$\% \text{ C}_2\text{S} = 2.867\text{S} - 0.7544\text{C}_3\text{S} \tag{6-2b}$$

$$\% \text{ C}_3\text{A} = 2.650\text{A} - 1.692\text{F} \tag{6-2c}$$

$$\% \text{ C}_4\text{AF} = 3.043\text{F} \tag{6-2d}$$

The equations are applicable to portland cements with an A/F ratio 0.64 or higher; should the ratio be less than 0.64 another set of equations apply, which are included in ASTM C 150. Also, it should be noted that the Bogue equations assume the chemical reactions of formation of clinker compounds to have proceeded to completion, and that the presence of impurities such as MgO and alkalies can be ignored. Both assumptions are not valid; hence in some cases the computed compound composition, especially the amounts of C_3A and C_4AF in cements, are known to deviate considerably from the actual compound composition determined directly. This is why the computed compound composition is also referred to as the *potential compound composition*.

The potential compound compositions of the cements in Table 6–1 are shown in Table 6–2. It can be seen from the data in Tables 6–1 and 6–2 that even small changes in the oxide analyses of cements can result in large changes in the compound composition. Comparison between cements 1 and 2 shows that a 1 percent decrease in Al_2O_3, with a corresponding increase in Fe_2O_3, lowered the C_3A and C_2S contents by 4.3 and 3.7 percent, respectively; this change also caused an increase in the C_4AF and C_3S contents by 3 and 4.3 percent, respectively. Similarly, comparison between cements 4 and 5 shows that a 1 percent decrease in CaO, with a corresponding increase in SiO_2, caused the C_3S to drop 11.6 percent and the C_2S to rise by the same amount. Since the properties of portland cements are influenced by the proportion and type of the compounds present, the Bogue equations serve a useful purpose by offering a simple way of determining the compound composition of a portland cement from its chemical analysis.

Crystal Structures and Reactivity of Compounds

The chemical composition of compounds present in the industrial portland cements is not exactly what is expressed by the commonly used formulas, C_3S, C_2S, C_3A, and C_4AF. This is because at the high temperatures prevalent during clinker formation the elements present in the system, including the impurities such as magnesium,

TABLE 6–2 COMPOUND COMPOSITION OF PORTLAND CEMENTS (%)

Compound composition	Cement no. 1	Cement no. 2	Cement no. 3	Cement no. 4	Cement no. 5
C_3S	53.7	58.0	62.3	53.6	42.0
C_2S	19.9	16.2	12.5	17.2	28.8
C_3A	11.4	7.1	2.8	14.0	14.0
C_4AF	8.8	11.9	14.9	8.8	8.8

sodium, potassium, and sulfur, possess the capability of entering into solid solutions with each of the major compounds in clinker. Small amounts of impurities in solid solution may not significantly alter the crystallographic nature and the reactivity of a compound with water, but larger amounts can do so.

Besides factors such as the particle size and the temperature of hydration, the reactivity of the portland cement compounds with water is influenced by their crystal structures. Under the high-temperature and nonequilibrium conditions of the cement kiln and with a variety of metallic ions present, the crystal structures formed are far from perfect. The structural imperfections account for the instability of the cement compounds in aqueous environments. In fact, the differences between the reactivity of two compounds having a similar chemical composition can only be explained from the degree of their structural instability. It is beyond the scope of this book to discuss in detail the highly complex crystal structures of cement compounds; however, essential features that account for differences in the reactivity are described next.

Calcium silicates. Tricalcium silicate (C_3S) and beta-dicalcium silicate (βC_2S) are the two hydraulic silicates commonly found in industrial portland cement clinkers. Both invariably contain small amounts of magnesium, aluminum, iron, potassium, sodium, and sulfur ions; the impure forms of C_3S and βC_2S are known as *alite* and *belite,* respectively.

Although three main crystalline forms of alite—triclinic, monoclinic, and trigonal—have been detected in industrial cements, these forms are a slight distortion of an ideal C_3S pseudostructure built from SiO_4 tetrahedra, calcium ions, and oxygen ions (Fig. 6–5a). According to Lea,[1] a notable feature of the ionic packing is that the coordination of oxygen ions around the calcium is irregular, so that the oxygens are concentrated to one side of each calcium ion. This arrangement leaves large structural holes which account for the high lattice energy and reactivity.

Similarly, the structure of belite in industrial cements is irregular, but the interstitial holes thus formed are much smaller than in C_3S and this makes belite far less reactive than alite. By way of contrast, another crystallographic form of dicalcium silicate, namely γC_2S, has a regularly coordinated structure (Fig. 6–5b), thus making the compound nonreactive.

Calcium aluminate and ferroaluminate. Several hydraulic calcium aluminates can occur in the $CaO\text{-}Al_2O_3$ system; however, the tricalcium aluminate (C_3A) is the principal aluminate compound in portland cement clinker. Calcium ferrites are not found in normal portland cement clinker; instead, calcium ferroaluminates which belong to the $C_2A\text{-}C_2F$ ferrite solid solution (F_{ss}) series are formed, and the most common compound corresponds approximately to the equimolecular composition, C_4AF.

---

[1] F. M. Lea, *The Chemistry of Cement and Concrete,* Chemical Publishing Company, Inc., New York, 1971, pp. 317–37.

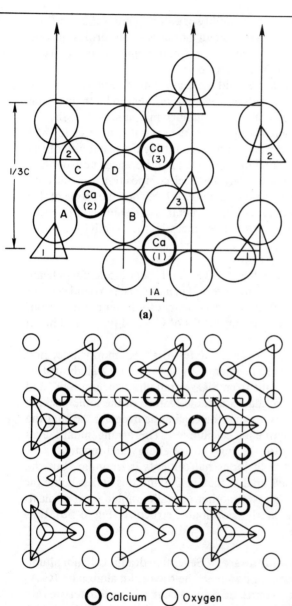

Figure 6-5 Crystal structures of (a) $3CaO \cdot SiO_2$: vertical section of the bottom layer of the pseudostructure of $3CaO \cdot SiO_2$ through the long diagonal of the cell. Only the oxygen atoms in the symmetry plane are shown as plain circles. 1, 2, and 3 are sections of SiO_4 tetrahedron. Calcium atoms are labeled; (b) $'\gamma - 2CaO \cdot SiO_2$: silicon atoms are not shown; they occur at the center of the tetrahedra. [From F. M. Lea, *The Chemistry of Cement and Concrete*, Chemical Publishing Company, Inc., New York, 1971, by permission of Edward Arnold (Publishers) Ltd.]

The irregular coordination of the oxygen ions around calcium leaves large voids, which account for the high reactivity of C_3S. On the other hand, γC_2S has a regularly coordinated structure and is, therefore, nonreactive.

Similar to the calcium silicates, both C_3A and C_4AF in industrial clinkers contain in their crystal structures significant amounts of such impurities as magnesium, sodium, potassium, and silica. The crystal structure of pure C_3A is cubic; however, C_4AF and C_3A containing large amounts of alkalies are both orthorhombic. The crystal structures are very complex but are characterized by large structural holes which account for high reactivity.

Magnesium oxide and calcium oxide. The source of magnesium oxide in cement is usually dolomite, which is present as an impurity in most limestones. A part of the total magnesium oxide in portland cement clinker (i.e., up to 2 percent) may enter into solid solution with the various compounds described above; however, the rest occurs as crystalline MgO, also called *periclase*. Hydration of periclase to magnesium hydroxide is a slow and expansive reaction which under certain conditions can cause *unsoundness* (i.e., cracking and pop-outs in hardened concrete).

Uncombined or *free calcium oxide* is rarely present in significant amounts in modern portland cements. Improper proportioning of raw materials, inadequate grinding and homogenization, and insufficient temperature or hold time in the kiln burning zone are among the principal factors that account for the presence of free or crystalline calcium oxide in portland cement clinker. Like MgO, crystalline CaO exposed to high temperature in the cement kiln hydrates slowly, and the hydration reaction is capable of causing unsoundness in hardened concretes.

Both MgO and CaO form cubic structure, with each magnesium or calcium ion surrounded by six oxygens in a regular octahedron. The size of the Mg^{2+} ion is such that in the MgO structure the oxygen ions are in close contact with the Mg^{2+} well packed in the interstices. However, in the case of the CaO structure, due to the much larger size of the Ca^{2+} ion, the oxygen ions are forced apart, so that the Ca^{2+} ions are not well packed. Consequently, the crystalline MgO formed from a high-temperature ($>1400°C$) melt in a portland cement kiln is much less reactive with water than the crystalline CaO, which has been exposed to the same temperature condition. This is the reason why **under ordinary curing temperatures** the presence of significant amounts of crystalline CaO in portland cement may cause unsoundness in concrete, whereas a similar amount of crystalline MgO generally proves harmless.

Alkali and sulfate compounds. The alkalies, sodium and potassium, in portland cement clinker are derived mainly from the clay components present in the raw mix and coal; their total amount, expressed as Na_2O equivalent ($Na_2O + 0.64K_2O$), may range from 0.3 to 1.5 percent. The sulfates in a cement kiln generally originate from fuel. Depending on the amount of sulfate available, soluble double-sulfates of alkalies such as langbeinite ($2C\bar{S} \cdot N\bar{S}$) and aphthitalite ($3N\bar{S} \cdot K\bar{S}$) are known to be present in portland cement clinker. Their presence is known to have a significant influence on the early hydration reactions of the cement.

When sufficient sulfate is not present in the kiln system, the alkalies are preferentially taken up by C_3A and C_2S, which may then be modified to compositions of the type NC_8A_3 and $KC_{23}S_{12}$, respectively. Sometimes large amounts of sulfate in the form of gypsum are purposefully added to the raw mix either for lowering the burning temperature or for modification of the C_3A phase to $C_4A_3\bar{S}$, which is an important constituent of certain types of expansive as well as rapid-hardening cements (as will be described later).

In ordinary portland cement the source of most of the sulfate (expressed as SO_3) is gypsum, or calcium sulfate in one of its several possible forms, added to the clinker. The main purpose of this additive is to retard the quick-setting tendency of the ground portland cement clinker due to the very high reactivity of C_3A present. Calcium sulfate can occur as gypsum ($CaSO_4 \cdot 2H_2O$), hemihydrate or plaster of paris ($CaSO_4 \cdot \frac{1}{2}H_2O$), and anhydrite ($CaSO_4$). Compared to clinker compounds gypsum dissolves quickly in water; however, the hemihydrate is much more soluble and is invariably present in cements due to decomposition of the gypsum during the finish grinding operation.

Fineness

In addition to the compound composition, the fineness of a cement affects its reactivity with water. Generally, the finer a cement, the more rapidly it will react. For a given compound composition the rate of reactivity and hence the strength development can be enhanced by finer grinding of cements; however, the cost of grinding and the heat evolved on hydration set some limits on fineness.

For quality control purposes in the cement industry, the fineness is easily determined as residue on standard sieves such as No. 200 mesh (75 μm) and No. 325 mesh (45 μm). It is generally agreed that cement particles larger than 45 μm are difficult to hydrate and those larger than 75 μm may never hydrate completely. However, an estimate of the relative rates of reactivity of cements with similar compound composition cannot be made without knowing the complete particle size distribution by sedimentation methods. Since the determination of particle size distribution by sedimentation is either cumbersome or requires expensive equipment, it is a common practice in the industry to obtain a relative measure of the particle size distribution from surface area analysis of the cement by the Blaine Air Permeability Method (ASTM C 204). Typical data on particle size distribution and Blaine surface area for two samples of industrially produced portland cements are shown in Fig. 6–6.

HYDRATION OF PORTLAND CEMENT

Significance

Anhydrous portland cement does not bind sand and gravel; it acquires the adhesive property only when mixed with water. This is because the chemical reaction of cement with water, commonly referred to as the **hydration of cement,** yields products

Hydration of Portland Cement

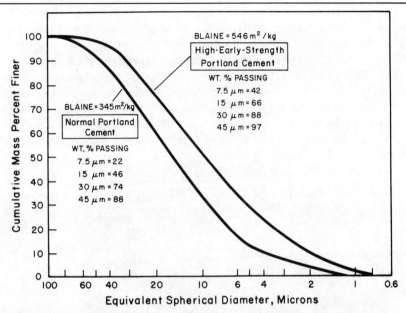

Figure 6–6 Plots of typical particle size distribution analyses data of ASTM Types I and III portland cement samples. (Data courtesy of Southwestern Portland Cement Company, Victorville, California.)

that possess setting and hardening characteristics. Brunauer and Copeland aptly described the significance of portland cement hydration to concrete technology:

> The chemistry of concrete is essentially the chemistry of the reaction between portland cement and water.... In any chemical reaction the main features of interest are the changes in matter, the changes in energy, and the speed of the reaction. These three aspects of a reaction have great practical importance for the user of portland cement. Knowledge of the substances formed when portland cement reacts is important because the cement itself is not a cementing material; its hydration products have the cementing action. Knowledge of the amount of heat released is important because the heat is sometimes a help and sometimes a hindrance.... Knowledge of reaction speed is important because it determines the time of setting and hardening. The initial reaction must be slow enough to enable the concrete to be poured into place. On the other hand, after the concrete has been placed rapid hardening is often desirable.[2]

Mechanism of Hydration

Two mechanisms of hydration of portland cement have been proposed. The ***through-solution hydration*** involves dissolution of anhydrous compounds to their ionic constituents, formation of hydrates in the solution, and due to their low solubility,

---

[2] S. Brunauer and L. E. Copeland, "The Chemistry of Concrete," *Sci. Am.*, April 1964.

eventual precipitation of the hydrates from the supersaturated solution. Thus the through-solution mechanism envisages complete reorganization of the constituents of the original compounds during the hydration of cement. According to the other mechanism, called the *topochemical* or *solid-state hydration* of cement, the reactions take place directly at the surface of the anhydrous cement compounds without their going into solution. From electron microscopic studies of hydrating cement pastes (Fig. 6–7), it appears that the through-solution mechanism is dominant in the early stages of cement hydration. At later ages when the ionic mobility in solution becomes restricted, the hydration of the residual cement particle may occur by solid-state reactions.

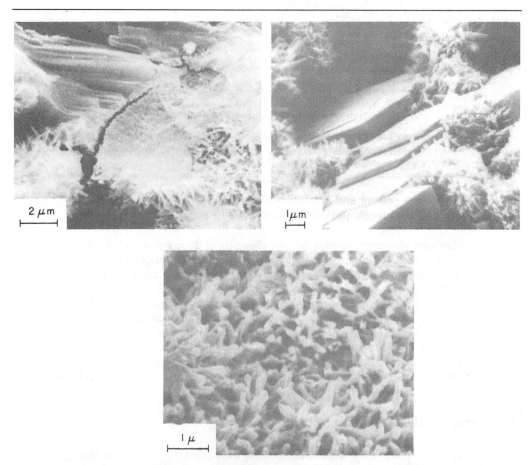

Figure 6–7 Scanning electron micrograph of a fractured specimen of a 3-day-old portland cement paste.

Massive crystals are of calcium hydroxide, whereas C-S-H crystals are poorly crystalline and show a fibrous morphology.

Since portland cement is composed of a heterogeneous mixture of several compounds, the hydration process consists of simultaneously occurring reactions of the anhydrous compounds with water. All the compounds, however, do not hydrate at the same rate. The aluminates are known to hydrate at a much faster rate than the silicates. In fact, the *stiffening* (loss of consistency) and *setting* (solidification), characteristics of a portland cement paste, are largely determined by the hydration reactions involving aluminates.

The silicates, which make up about 75 percent of ordinary portland cement, play a dominant role in determining the *hardening* (rate of strength development) characteristics. For the purpose of obtaining a clear understanding of the chemical and physical changes during the hydration of portland cement, it is desirable to discuss separately the hydration reactions of the aluminates and the silicates.

Hydration of the Aluminates

The reaction of C_3A with water is immediate. Crystalline hydrates, such as C_3AH_6, C_4AH_{19}, and C_2AH_8, are formed quickly, with liberation of a large amount of heat of hydration. Unless the rapid reaction of C_3A hydration is slowed down by some means, the portland cement will be useless for most construction purposes. This is generally accomplished by the addition of gypsum. Therefore, for practical purposes it is not the hydration reactions of C_3A alone but the hydration of C_3A in the presence of gypsum which is important.

From the standpoint of hydration reactions of portland cement, it is convenient to discuss C_3A and ferroaluminate together because the products formed when the latter reacts with water in the presence of sulfate are structurally similar to those formed from C_3A. For instance, depending on the sulfate concentration the hydration of C_4AF may produce either $C_6A(F)\bar{S}_3H_{32}$ or $C_4A(F)\bar{S}H_{18}$,[3] which have variable chemical compositions but structures similar to ettringite and low sulfate, respectively. However, the part played by the ferroaluminate in portland cement in early setting and hardening reactions of the cement paste depends mainly on its chemical composition and temperature of formation. Generally, the reactivity of the ferrite phase is somewhat slower than C_3A, but it increases with increasing alumina content and decreasing temperature of formation during the cement manufacturing process. In any case, it may be noted that the hydration reaction of the aluminates described below are applicable to both the C_3A and the F_{ss} in portland cement, although for the sake of simplicity only C_3A is discussed.

Several theories have been postulated to explain the *mechanism of retardation of C_3A by gypsum.* According to one theory, since gypsum and alkalies go into solution quickly, the solubility of C_3A is depressed in the presence of hydroxyl, alkali, and sulfate ions. Depending on the concentration of aluminate and sulfate ions in solution, the precipitating crystalline product is either the calcium aluminate

---

[3] In recent literature the terms AF_t and AF_m are employed to designate the products which may have variable chemical compositions but are structurally similar to ettringite and monosulfate hydrate, respectively.

trisulfate hydrate or the calcium aluminate monosulfate hydrate. In solutions saturated with calcium and hydroxyl ions, the former crystallizes as short prismatic needles and is also referred to as *high-sulfate* or by the mineralogical name *ettringite*. The monosulfate is also called *low-sulfate* and crystallizes as thin hexagonal plates. The relevant chemical reactions may be expressed as

$$[AlO_4]^- + 3[SO_4]^{2-} + 6[Ca]^{2+} + aq. \rightarrow C_6A\bar{S}_3H_{32} \text{ (ettringite)} \quad (6\text{-}3a)$$

$$[AlO_4]^- + [SO_4]^{2-} + 4[Ca]^{2+} + aq. \rightarrow C_4A\bar{S}H_{18} \quad \text{(monosulfate)} \quad (6\text{-}3b)$$

Ettringite is usually the first hydrate to crystallize because of the high sulfate/aluminate ratio in the solution phase during the first hour of hydration. In normally retarded portland cements which contain 5 to 6 percent gypsum, the precipitation of ettringite contributes to stiffening (loss of consistency), setting (solidification of the paste), and early strength development. Later, after the depletion of sulfate in the solution when the aluminate concentration goes up again due to renewed hydration of C_3A and C_4AF, ettringite becomes unstable and is gradually converted into monosulfate, which is the final product of hydration of portland cements containing more than 5 percent C_3A:

$$C_6A\bar{S}_3H_{32} + 2C_3A \rightarrow C_4A\bar{S}H_{18} \quad (6\text{-}4)$$

Since the aluminate-to-sulfate balance in the solution phase of a hydrated portland cement paste primarily determines whether the setting behavior is normal or not, various setting phenomena affected by an imbalance in the $A/\bar{S}$ ratio, which have a practical significance in the concrete technology, are illustrated by Fig. 6–8 and discussed below:

- **Case I:** When the rates of availability of the aluminate ions and the sulfate ions to the solution phase are low, the cement paste will remain workable for about 45 min; thereafter it will start stiffening as the water-filled space begins to get filled with ettringite crystals. Most so-called *normal-setting* portland cements belong to this category. The paste becomes less workable between 1 and 2 hr after the addition of water, and may begin to solidify within 2 to 3 hr.

- **Case II:** When the rates of availability of the aluminate ions and the sulfate ions to the solution phase are high, large amounts of ettringite form rapidly and cause a considerable loss of consistency in 10 to 45 min, with solidification of the paste between 1 and 2 hr. Freshly produced high-C_3A cements containing more than normal amounts of alkali sulfates or calcium sulfate hemihydrate are generally characterized by this type of behavior.

- **Case III:** When the amount of reactive C_3A is high but the soluble sulfate present is less than required for normal retardation, hexagonal-plate crystals of monosulfate and calcium aluminate hydrates form quickly and in large amounts with the cement paste setting in less than 45 min after the addition of water. This phenomenon is known as *quick set*.

- **Case IV:** When little or no gypsum has been added to a ground portland cement clinker, the hydration of C_3A is rapid and the hexagonal-plate calcium alumi-

Hydration of Portland Cement

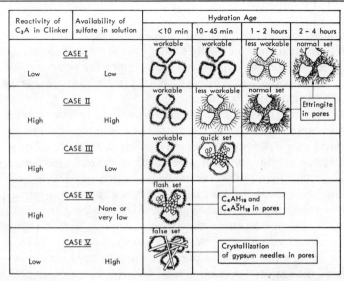

Figure 6–8 Influence of the aluminate/sulfate ratio in the solution phase on the setting characteristics of portland cement pastes. (From F. W. Locher, W. Richartz, and S. Sprung, *Zement-Kalk Gips,* No. 6, pp. 271–77, 1980.)

nate hydrates start forming in large amounts soon after the addition of water, causing almost an instantaneous set. This phenomenon, known as *flash set,* is associated with large heat evolution and poor ultimate strengths.

- **Case V:** When the C_3A in cement is of low reactivity, as is the case in partially hydrated or carbonated cements which have been improperly stored, and at the same time a large amount of calcium hemihydrate is present in the cement, the solution phase will contain a low concentration of aluminate ions but will quickly become supersaturated with respect to calcium and sulfate ions. This situation will lead to the rapid formation of large crystals of gypsum with a corresponding loss of consistency. The phenomenon, called *false set,* is not associated with large heat evolution and can be remedied by vigorous mixing of the cement paste with or without additional water.

Although gypsum is added to cement to serve as a retarder, what is known as the ***optimum gypsum content of cement*** is generally determined from standard tests which show maximum cement strength and minimum shrinkage at given ages of hydration. Sulfate ions contributed to the solution by the dissolution of gypsum have a retarding effect on the aluminates but an accelerating effect on the hydration of the silicates (see Chapter 8), which are the principal compounds in portland cement. Therefore, depending on the composition of a cement, a specific gypsum content is indicated for optimum performance of the cement.

Hydration of the Silicates

The hydration of C_3S and βC_2S in portland cement produces a family of calcium silicate hydrates which are structurally similar but vary widely in calcium/silica ratio and the content of chemically combined water. Since the structure determines the properties, the compositional differences among the calcium silicate hydrates have little effect on their physical characteristics.

The structure and properties of the calcium silicate hydrates formed in portland cement pastes were described in Chapter 2. In general, the material is poorly crystalline and forms a porous solid which exhibits characteristics of a rigid gel. In the literature, this gel has sometimes been referred to as **tobermorite gel,** after a naturally occurring mineral of seemingly similar structure. The use of this name is no longer favored because the similarity in crystal structures is rather poor. Also, since the chemical composition of the calcium silicate hydrates in hydrating portland cement pastes varies with the water/cement ratio, temperature, and age of hydration, it has become rather customary to refer to these hydrates simply as C-S-H, a notation that does not imply a fixed composition. On complete hydration the approximate composition of the material corresponds to $C_3S_2H_3$; this composition is therefore used for stoichiometric calculations.

The stoichiometric reactions for fully hydrated C_3S and C_2S pastes may be expressed as

$$2C_3S + 6H \rightarrow C_3S_2H_3 + 3CH \quad (6\text{–}5a)$$

$$2C_2S + 4H \rightarrow C_3S_2H_3 + CH \quad (6\text{–}5b)$$

In addition to the fact that similar reaction products are formed on hydration of both the calcium silicates present in portland cement, there are several points that need to be noted.

First, stoichiometric calculations show that hydration of C_3S would produce 61 percent $C_3S_2H_3$ and 39 percent calcium hydroxide, whereas the hydration of C_2S would produce 82 percent $C_3S_2H_3$ and 18 percent calcium hydroxide. If the surface area and, consequently, the adhesive property of hydrated cement paste are mainly due to the formation of the calcium silicate hydrate, it is expected that the ultimate strength of a high-C_3S portland cement would be lower than a high-C_2S cement. This, indeed, is confirmed by the data from many investigations.

Second, if the durability of a hardened cement paste to acidic and sulfate waters is reduced due to the presence of calcium hydroxide, it may be expected that the cement containing a higher proportion of C_2S will be more durable in acidic and sulfate environments than the cement containing a higher proportion of C_3S. This observation is also generally confirmed by laboratory and field experiences. From the standpoint of durability to chemical attacks, many standard specifications attempt to limit the maximum permissible C_3S in cements; some recommend the use of pozzolans in order to remove the excess calcium hydroxide from the hydrated cement paste. Third, it can be calculated from the equations above that for complete hydration, C_3S and C_2S require 24 percent and 21 percent water, respectively.

Hydration of Portland Cement

The stoichiometric equations of C_3S and C_2S hydration do not tell anything about the reaction rates. From the standpoint of structural instability described earlier and the heat of hydration data given below, it will be apparent that C_3S hydrates at a faster rate than C_2S. In the presence of gypsum, C_3S in the fine particles begins to hydrate within an hour of the addition of water to cement, and probably contributes to the final time of set and early strength of the cement paste. In fact, the relatively quick rate of C_3S hydration is an important factor in the design of high-early-strength portland cements, as will be discussed later.

Hydration reactions of alite and belite are accelerated in the presence of sulfate ions in solution. Numerous researchers have found that, unlike the depression of solubility shown by the aluminate compounds, the solubility of the calcium silicate compounds, both C_3S and C_2S, is actually increased in sulfate solutions, which explains the acceleration of hydration. Typical data on the effect of gypsum addition on the hydration rate of alite are shown in Table 6–3. In conclusion, although the primary purpose of gypsum in portland cement is to retard the hydration of aluminates, a side effect is the acceleration of alite hydration without which the industrial cements would harden at a slower rate.

TABLE 6–3 ACCELERATING EFFECT OF GYPSUM ON SETTING TIME, HEAT OF HYDRATION, AND STRENGTH OF ALITE

	Type I/II[a] portland cement	Alite cement[a]	
		No gypsum	3% gypsum
Setting time[b] (hr)			
Initial	3.0	8.5	4.5
Final	6.0	11.5	7.5
Heat of hydration[b] (cal/g)			
3 days	61	59	63
7 days	75	61	66
28 days	83	85	81
Compressive strength[b] [psi (MPa)]			
3 days	1940	1250	1610
	(13.4)	(8.62)	(11.0)
7 days	3100	2060	2440
	(21.4)	(14.2)	(16.8)
28 days	5070	3650	4010
	(34.9)	(25.2)	(27.6)
90 days	5740	5360	5375
	(36.9)	(36.9)	(37.0)

[a] Alite cement made by grinding a laboratory preparation of high-purity monoclinic alite to fineness 330 m²/kg Blaine. An industrial portland cement meeting the requirements of both ASTM Types I and II, and a fineness of 330 m²/g Blaine, was included for reference purposes.

[b] ASTM Methods C 266, C 186, and C 109 were used for determination of setting time, heat of hydration, and compressive strength, respectively.

Source: Data from P. K. Mehta, D. Pirtz, and M. Polivka, *Cem. Concr. Res.*, Vol. 9, pp. 439–50, 1979.

HEAT OF HYDRATION

The compounds of portland cement are nonequilibrium products of high-temperature reactions and are therefore in a high-energy state. When a cement is hydrated, the compounds react with water to acquire stable low-energy states, and the process is accompanied by the release of energy in the form of heat. In other words, the hydration reactions of portland cement compounds are exothermic.

The significance of heat of cement hydration in concrete technology is manifold. The heat of hydration can sometimes be a hindrance (e.g., mass concrete structures), and at other times a help (e.g., winter concreting when ambient temperatures may be too low to provide the activation energy for hydration reactions). The total amount of heat liberated and the rates of heat liberation from hydration of the individual compounds can be used as indices of their reactivity. As discussed below, the data from heat of hydration studies can be used for characterizing the setting and hardening behavior of cements, and for predicting the temperature rise.

By using a conduction calorimeter, Lerch[4] recorded the rate of heat evolution from cement pastes during the setting and early hardening period. A typical plot of the data is shown in Fig. 6–9. In general, on mixing the cement with water, a rapid heat evolution (ascending peak A) lasting a few minutes occurs. This probably represents the heat of solution of aluminates and sulfates. This initial heat evolution ceases quickly (descending peak A) when the solubility of aluminates is depressed in the presence of sulfate in the solution. The next heat evolution cycle, culminating in the second peak after about 4 to 8 hours of hydration for most portland cements, represents the heat of formation of ettringite (ascending peak B). Many researchers believe that the heat evolution period includes some heat of solution due to C_3S and heat of formation of C-S-H. The paste of a properly retarded cement will retain much of its plasticity before the commencement of this heat cycle and will stiffen and show the *initial set* (beginning of solidification) before reaching the apex at B, which corresponds to the *final set* (complete solidification and beginning of hardening).

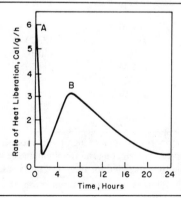

Figure 6–9 Heat liberation rate of a portland cement paste during the setting and early hardening period.

[4] W. Lerch, *Proceedings Am. Soc. Test. Mat.*, Vol. 46, p. 1252, 1946.

TABLE 6-4 HEATS OF HYDRATION OF PORTLAND CEMENT COMPOUNDS

Compound	Heats of hydration at the given age (cal/g)		
	3 days	90 days	13 years
C_3S	58	104	122
C_2S	12	42	59
C_3A	212	311	324
C_4AF	69	98	102

From analysis of heat of hydration data on a large number of cements, Verbeck and Foster[5] computed the individual rates of heat evolution due to the four principal compounds in portland cement (Table 6-4). Since the heat of hydration of cement is an additive property, it can be predicted from an expression of the type

$$H = aA + bB + cC + dD \qquad (6\text{-}6)$$

where H represents the heat of hydration at a given age and under given conditions; A, B, C, and D are the percentage contents of C_3S, C_2S, C_3A, and C_4AF present in the cement; and a, b, c, and d are coefficients representing the contribution of 1 percent of the corresponding compound to the heat of hydration. The values of the coefficients will be different for the various ages of hydration.

For a typical portland cement, it appears that approximately 50 percent of the potential heat is liberated within the first 3 days, and 90 percent within the first 3 months of hydration. For low-heat portland cements (ASTM Type IV), ASTM C 150 requires the 7- and 28-day heats of hydration to be limited to 60 and 70 cal/g, respectively. Normal portland cements, ASTM Type I, generally produce 80 to 90 cal/g in 7 days, and 90 to 100 cal/g in 28 days.

PHYSICAL ASPECTS OF THE SETTING AND HARDENING PROCESS

The chemical aspects of the hydration reactions of portland cement compounds have already been discussed. For application to concrete technology it is desirable to review the physical aspects, such as stiffening, setting, and hardening, which are different manifestations of the ongoing chemical processes.

Stiffening is the loss of consistency by the plastic cement paste, and is associated with the slump loss phenomenon in concrete. It is the free water in a cement paste that is responsible for its plasticity. The gradual loss of free water from the system due to early hydration reactions, physical adsorption at the surface of poorly crystalline hydration products such as ettringite and the C-S-H, and evaporation causes the paste to stiffen and, finally, to set and harden.

[5] G. J. Verbeck and C. W. Foster, *Proceedings Am. Soc. Test. Mat.*, Vol. 50, p. 1235, 1950.

The term *setting* implies solidification of the plastic cement paste. The beginning of solidification, called *the initial set,* marks the point in time when the paste has become unworkable. Accordingly, placement, compaction, and finishing of concrete beyond this stage will be very difficult. The paste does not solidify suddenly; it requires considerable time to become fully rigid. The time taken to solidify completely marks *the final set,* which should not be too long in order to resume construction activity within a reasonable time after placement of concrete. Almost universally, the initial and the final setting times are determined by the Vicat apparatus, which measures the resistance of a cement paste of a standard consistency to the penetration of a needle under a total load of 300 g. The initial set is an arbitrary time in the setting process which is said to be reached when the needle is no longer able to pierce the 40-mm-deep pat of the cement paste to within about 5 to 7 mm from the bottom. The final set is said to be reached when the needle makes an impression on the surface of the paste but does not penetrate. ASTM C 150, *Standard Specification for Portland Cement,* requires the initial setting time to be not less than 45 min, and the final setting time to be not more than 375 min as determined by the Vicat Needle (ASTM C 191).

A freshly set portland cement paste has little or no strength because it represents only the beginning of the hydration of C_3S, the principal compound present. Once the C_3S hydration starts, the reaction continues rapidly for several weeks. The process of progressive filling of the void spaces in the paste with the reaction products results in a decrease in porosity and permeability, and an increase in strength. In concrete technology the phenomenon of strength gain with time is called *hardening*. Fig. 6–10 shows a graphic representation of the relation between the chemistry of the hydration process of a normal portland cement paste and the physical phenomena of gradual stiffening, setting, and hardening with a corresponding decrease in porosity and permeability.

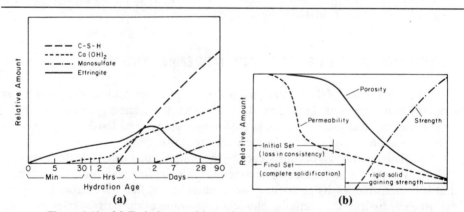

Figure 6–10 (a) Typical rates of formation of hydration products in an ordinary portland cement paste; (b) influence of formation of hydration products on setting time, porosity, permeability, and strength of cement paste. [(a) Adapted from J. Soroka, *Portland Cement Paste and Concrete,* The Macmillan Press, p. 35, 1979.]

EFFECT OF CEMENT CHARACTERISTICS ON STRENGTH AND HEAT OF HYDRATION

Since the rates of reactivity of the individual portland cement compounds with water vary considerably, it is possible to change the strength development characteristics of cements simply by altering the compound composition. For instance, the early strengths at 3, 7, and 28 days would be high if the cement contains relatively large amounts of C_3S and C_3A; and the early strength would be low if the cement contains a larger proportion of C_2S. Also, from theoretical considerations already given (p. 187), the ultimate strength of the cement high in C_2S should be greater than that of a low-C_2S cement. Laboratory studies confirm these expectations (Fig. 6–11a).

Also, since the compound composition of the cement affects the heat of hydration, it is to be expected that cements containing high C_2S will not only be slow hardening but also less heat producing (Fig. 6–11b).

In addition to compound composition, the rates of strength development and heat evolution can be readily controlled by adjusting the fineness of cement. For instance, depending on the specific compound composition, by making a change in the surface area of the cement from 320 to 450 m^2/kg Blaine, it is possible to increase the 1-, 3-, and 7-day compressive strengths of the cement mortar by about 50 to 100, 30 to 60, and 15 to 40 percent, respectively. Typical data on the influence of fineness on strength are shown in Fig. 6–11c. Additional data on the influence of compound composition, fineness, and hydration temperature on heat development are shown in Fig. 6–12.

TYPES OF PORTLAND CEMENT

A summary of the main characteristics of the principal compounds in portland cement is shown in Table 6–5. From the knowledge of relative rates of reactivity and products of hydration of the individual compounds it is possible to design cements with special characteristics such as high early strength, low heat of hydration, high sulfate resistance, and moderate heat of hydration or moderate sulfate resistance. Accordingly, ASTM C 150, *Standard Specification for Portland Cement,* covers the following 8 types of portland cement:

- **Type I:** For use when the special properties specified for any other type are not required. No limits are imposed on any of the four principal compounds.
- **Type IA:** Air-entraining Type I cement, where air entrainment is desired (e.g., for making frost-resisting concrete).
- **Type II:** For general use, more especially when moderate sulfate resistance or moderate heat of hydration is desired. Since C_3A and C_3S produce high heats of hydration, the Specification limits the C_3A content of the cement to maximum 8 percent, and has an optional limit of maximum 58 percent on the sum of C_3S and C_3A (this limit applies when a moderate heat of hydration is required and test data for heat of hydration are not available).

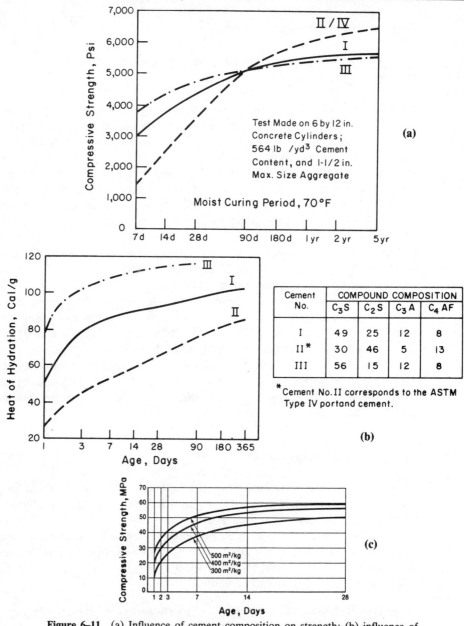

Figure 6–11 (a) Influence of cement composition on strength; (b) influence of cement composition on heat of hydration; (c) influence of cement fineness on strength. [Data for (a) and (b) are taken from *Concrete Manual,* U.S. Bureau of Reclamation, 1975, pp. 45–46; data for (c) is taken from *Beton-Bogen,* Aalborg Cement Company, Aalborg, Denmark, 1979.]

In industrial practice, the modification of compound composition and fineness of portland cement provides effective methods of controlling both the rates of early strength development and the heat of hydration.

Types of Portland Cement

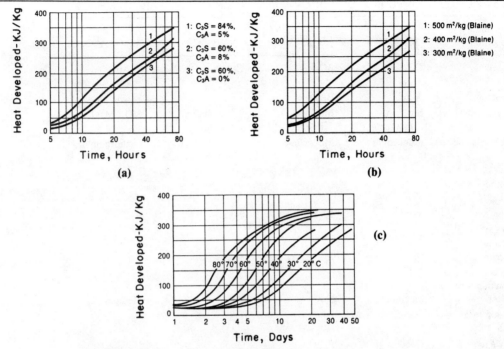

Figure 6-12 Influence of compound composition, fineness, and hydration temperature on heat development in cement pastes (0.4 water/cement ratio). (From *Beton-Bogen*, Aalborg Cement Company, Aalborg, Denmark.)

Both the rate and the total amount of heat developed on hydration of cement are influenced by the compound composition, fineness, and temperature of hydration.

- **Type IIA:** Air-entraining Type II cement, where air entrainment is desired.
- **Type III:** For use when high early strength is desired. To ensure that the high strength is not due mainly to the hydration products of C_3A, the Specification limits the C_3A content of the cement to maximum 15 percent.

  It may be noted from Fig. 6-11 that generally the high early strength of the Type III portland cement is in part due to the higher specific surface approximately 500 m^2/kg Blaine, instead of 330 to 400 m^2/kg for Type I portland cement.

- **Type IIIA:** Air-entraining Type III cement, where air entrainment is desired.
- **Type IV:** For use when a low heat of hydration is desired. Since C_3S and C_3A produce high heats of hydration, but C_2S produces much less heat, the Specification calls for maximum limits of 35 and 7 percent on C_3S and C_3A, respectively, and requires a minimum of 40 percent C_2S in the cement.
- **Type V:** For use when high sulfate resistance is desired. The Specification calls for a maximum limit of 5 percent on C_3A which applies when the sulfate expansion test is not required.

TABLE 6-5 PRINCIPAL COMPOUNDS OF PORTLAND CEMENT AND THEIR CHARACTERISTICS

Approximate composition	$3CaO \cdot SiO_2$	$\beta 2CaO \cdot SiO_2$	$3CaO \cdot Al_2O_3$	$4CaO \cdot Al_2O_3 \cdot Fe_2O_3$
Abbreviated formula	C_3S	βC_2S	C_3A	C_4AF
Common name	Alite	Belite	–	Ferrite phase, Fss
Principal impurities	MgO, Al_2O_3, Fe_2O_3	MgO, Al_2O_3, Fe_2O_3	SiO_2, MgO, alkalies	SiO_2, MgO
Common crystalline form	Monoclinic	Monoclinic	Cubic, orthorhombic	Orthorhombic
Proportion of compounds present (%)				
Range	35–65	10–40	0–15	5–15
Average in ordinary cement	50	25	8	8
Rate of reaction with water	Medium	Slow	Fast	Medium
Contribution to strength				
Early age	Good	Poor	Good	Good
Ultimate	Good	Excellent	Medium	Medium
Heat of hydration	Medium	Low	High	Medium
Typical (cal/g)	120	60	320	100

It should be noted that the ultimate product of hydration is cements containing more than 5 percent potential C_3A, as calculated by Bogue equations, is monosulfate hydrate which is unstable when exposed to a sulfate solution; ettringite is the stable product in sulfate environments, and conversion of the monosulfate to ettringite is generally associated with expansion and cracking.

Although ASTM C 150 covers the production and use of air-entraining portland cements, concrete producers prefer cements without entrained air because the application of air-entraining admixtures during concrete manufacture offers better control for obtaining the desired amount and distribution of air in the product. Consequently, there is little demand for the air-entrained cements. Similarly, low-heat cement is no longer made in the United States because the use of mineral admixtures in concrete offers, in general, a less expensive way to control the temperature rise. It may be noted that more than 90 percent of the hydraulic cements produced in the country belong to the ASTM Types I and II portland cements, approximately 3 percent to the ASTM Type III, and the rest to special cements such

TABLE 6-6 TYPICAL COMPOUND COMPOSITION OF VARIOUS TYPES OF PORTLAND CEMENT AVAILABLE IN THE UNITED STATES

ASTM type	General description	Compound composition range (%)			
		C_3S	C_2S	C_3A	C_4AF
I	General purpose	45–55	20–30	8–12	6–10
II	General purpose with moderate sulfate resistance and moderate heat of hydration	40–50	25–35	5–7	6–10
III	High early strength	50–65	15–25	8–14	6–10
V	Sulfate resistant	40–50	25–35	0–4	10–20

as oil-well cement, which comprises approximately 5 percent of the total cement production.

Typical compound compositions of the commonly available portland cements in the United States are shown in Table 6–6. Important features of the *physical requirements according to ASTM C 150,* with reference to methods of testing, are summarized in Table 6–7. The test methods and specifications are useful mainly for the purposes of quality control of cement; they must not be used to predict properties of concrete which, among other factors, are greatly influenced by the water/cement ratio, curing temperature, and cement-admixture interaction when applicable. For instance, compared to the standard test conditions, the time of set of a cement will increase with increasing water/cement ratios and decrease with increasing curing temperatures.

The *cement standards in the world* are generally similar in principle but vary from one another in minor details. However, some exceptions may be noted. For example, cement standards do not differentiate between the ASTM Types I and II portland cements. Also, most cement standards do not favor the ASTM C 151 autoclave expansion test and the C 150 soundness specification, because the hydration behavior of portland cement is considerably distorted under the autoclaving conditions, and because a correlation has never been demonstrated between the maximum permissible expansion according to the test and the soundness of cement in service.[6] Instead, the Le Châtelier's test, which involves exposure of a cement paste to boiling water (not autoclaving conditions), is preferred.

SPECIAL HYDRAULIC CEMENTS

Classification and Nomenclature

Portland cements do not satisfy all the needs of the concrete industry; therefore, special cements have been developed to meet certain needs. Compared to portland cement, their volume is small and prices generally higher, but due to unique characteristics the special cements deserve to be better known to the structural engineer.

[6] P. K. Mehta, ASTM STP 663, 1978, pp. 35–60.

TABLE 6-7 PRINCIPAL PHYSICAL REQUIREMENTS AND ESSENTIAL FEATURES OF ASTM TEST METHODS FOR PORTLAND CEMENTS

Requirement specified by ASTM C 150	Type I	Type II	Type III	Type V	Method of test
Fineness: minimum (m^2/kg)	280	280	None	280	ASTM Method C 204 covers determination of fineness of cements using Blaine Air Permeability Apparatus. Fineness is expressed in terms of specific surface of the cement.
Soundness: maximum, autoclave expansion (%)	0.8	0.8	0.8	0.8	ASTM Method C 151 covers determination of soundness of cements by measuring expansion of neat cement paste prisms cured normally for 24 hr and subsequently at 2 MPa (295 psi) steam pressure in an autoclave for 3 hr.
Time of setting Initial set minimum (min) Final set maximum (min)	45 375	45 375	45 375	45 375	ASTM Method C 191 covers determination of setting time of cement pastes by Vicat apparatus. Initial setting time is obtained when the 1-mm needle is able to penetrate the 35-mm depth of a 40-mm-thick pat of the cement paste. Final setting time is obtained when a hollowed-out 5-mm needle does not sink visibly into the paste.
Compressive strength: minimum [MPa (psi)]					ASTM Method C 109 covers determination of compressive strength of mortar cubes composed of 1 part cement, 0.485 part water, and 2.75 parts graded standard sand by weight.
1 day in moist air	None	None	12.4 (1800)	None	
1 day moist air + 2 days water	12.4 (1800)	10.3[a] (1500)	24.1 (3500)	8.3 (1200)	
1 day moist air + 6 days water	19.3 (2800)	17.2[a] (2500)	None	15.2 (2200)	
1 day moist air + 27 days water	None[b]	None[b]	None	20.7 (3000)	

[a] The 3-day and the 7-day minimum compressive strength shall be 6.0 MPa (1000 psi) and 11.7 MPa (1700 psi), respectively, when the optional heat of hydration or the chemical limits on the sum of C_3S and C_3A are specified.

[b] When specifically requested, minimum 28-day strength values for Types I and II cements shall be 27.6 MPa (4000 psi).

Special Hydraulic Cements

With one notable exception, the special hydraulic cements may be considered as modified portland cements in the sense that they are made either by altering the compound composition of portland cement clinker, or by blending certain additives with portland cement, or by doing both. A clear classification of the special cements is difficult; however, in the American practice the use of the term **blended portland cements** is confined to blends of portland cements with either rapidly cooled blast-furnace slag or pozzolanic materials such as fly ash. Other special cements are generally classified under the term **modified portland cements** because they are made by modification of the compound composition of portland cement clinker. The hydraulic calcium silicates, C_3S and βC_2S, continue to be the primary cementitious constituents; only the aluminate and the ferrite phases are suitably altered to obtain the desired properties.

The exception to the blended and modified portland cements is the calcium aluminate cement, which does not derive its cementing property from the presence of hydraulic calcium silicates. Noteworthy special cements, their compositions, and major applications are summarized in Table 6–8. Important features, hydration characteristic, and properties of the cements listed in the table are discussed next.

Blended Portland Cements

Cost saving was probably the original reason for the development of blended portland cements. However, the impetus to rapid growth in the production of blended cements in many countries of Europe and Asia came as a result of energy-saving potential. Also, in certain respects, the blended cements perform better than portland cement. Presently, the production of slag cements represents nearly one-fourth of the total cement production of Germany, and the production of pozzolan cements represents about one-third of the total cement produced in Italy. In the United States the production of blended cements is still in infancy; however, there is a growing interest to use pozzolanic (e.g., fly ash) and cementitious materials (e.g., ground blast-furnace slag) as mineral admixtures in concrete. The composition and properties of pozzolanic and cementitious materials are described in Chapter 8.

ASTM C 595, Standard Specification for Blended Hydraulic Cements, covers five classes of cements, but commercial production is limited to portland blast-furnace slag cement (Type IS), and portland pozzolan cement (Type IP). According to the Specification, **Type IS cement** shall consist of an intimate and uniform blend of portland cement and fine granulated blast-furnace slag in which the slag constituent is between 25 and 70% of the weight of portland blast-furnace slag cement. *Blast-furnace slag* is a nonmetallic product consisting essentially of silicates and aluminosilicates of calcium and other bases; *granulated slag* is the glassy or noncrystalline product which is formed when molten blast-furnace slag is rapidly chilled, as by immersion in water. **Type IP cement** shall consist of an intimate and uniform blend of portland cement (or portland blast-furnace slag cement) and fine pozzolan in which the pozzolan content is between 15 and 40 percent of the weight of the total cement. A **pozzolan** is defined as a siliceous or siliceous and aluminous material which in itself possesses little or no cementing property but will in a finely divided

TABLE 6–8 SPECIAL HYDRAULIC CEMENTS: THEIR COMPOSITION AND USES

Classification and types	Composition	Major uses
Blended portland cements Portland blast-furnace slag cement (ASTM Type IS) Portland pozzolan cement (ASTM Type IP)	Consist essentially of an intimate and uniform blend of granulated blast-furnace slag or a pozzolan or both with portland cement, and often containing calcium sulfate. Industrial Type IS cements contain typically 30–40% slag, while Type IP cements contain 20–25% pozzolan. Compared to portland cement, both types are ground to more fine particle size to partly compensate for the loss of early strength.	1. Low heat of hydration 2. Excellent durability when properly designed and cured 3. Energy-saving and resource-conserving, and generally less expensive than portland cement
Expansive cements Type K Type M Type S Type O	Consist essentially of portland cement containing an expansive additive. Types K, M, and S cements, covered by ASTM C 845, derive their expansion by ettringite formation from $C_4A_3\bar{S}$, CA, and C_3A, respectively. Hard-burnt CaO is the expansive agent in Type O cements.	1. Production of crack-resistant concrete by offsetting the tensile stress due to drying shrinkage 2. Production of chemically prestressed concrete elements 3. Demolition of old concrete without shattering

Rapid setting and hardening cements
  Regulated set cement (RSC) (or jet cement)
  Very high early strength cement (VHE)
  High iron cement (HIC)
  Ultra high early strength cement (UHE)
	Most cements derive their rapid setting and hardening properties from compounds capable of forming a large amount of ettringite rapidly; C-S-H subsequently. For ettringite formation the main source of aluminate ions is a calcium fluoroaluminate in RSC, while it is $C_4A_3\bar{S}$ in VHE and HIC. UHE is a high-C_3S portland cement containing extra fine particles.	1. Emergency repairs, shotcreting
		2. Fabrication of precast-prestressed concrete products without steam curing
		3. Agglomeration of particulate matter in mining and metallurgical industries

Oil-well cements	Consist of portland cements with a little or no C_3A, relatively coarser particles, and with or without a retarder present:	To allow time for placement of cement slurry, the thickening time at service temperature is retarded:
API Class A-C	Low-C_3A cements without any retarder; Class C is sulfate-resistant	For well depths up to 6000 ft or 1830 m (80–170 °F or 27–77 °C)
API Class D, E	Low-C_3A cements with retarder	For well depths 6000–14,000 ft or 1830 to 4260 m (170–290 °F or 77–143 °C)

Special Hydraulic Cements

TABLE 6-8 (Cont.)

Classification and types	Composition	Major uses
API Class F	Low-C_3A cement with retarder	For well depths 10,000–16,000 ft or 3048–4877 m (230–320 °F or 110–160 °C)
API Class G, H	Essentially coarse-ground ASTM Types II and V portland cements, without retarder	For well temperatures 80–200 °F (27–93 °C)
API Class J	Essentially βC_2S and pulverized silica sand	For well depths below 20,000 ft or 6100 m (> 350 °F or 177 °C)
White or colored cements	Consist of portland cements with a little or no iron present (Fss < 1%). Colored cements are produced by adding suitable pigments to white cement.	Production of architectural concrete
Calcium aluminate cements	Consists essentially of pulverized clinker containing hydraulic calcium aluminates, such as $C_{12}A_7$, CA, and CA_2.	1. High-temperature concrete 2. Emergency repairs, especially in cold weather

form and in the presence of moisture chemically react with calcium hydroxide *at ordinary temperatures* to form compounds possessing cementitious properties.

Compared to pozzolans, finely ground granulated blast-furnace slag is self-cementing; that is, it does not require calcium hydroxide to form cementitious products such as C-S-H. However, when granulated blast-furnace slag hydrates by itself, the amount of cementitious products formed and the rates of formation are insufficient for application of the material to structural purposes. When used in combination with portland cement, the hydration of slag is accelerated in the presence of calcium hydroxide and gypsum. During the hydration of Type IS cement, some calcium hydroxide produced by the portland cement is consumed by the slag constituent of the cement. In this respect, and also due to the general similarity of the microstructure between Type IS and Type IP hydrated cement pastes, it is desirable to discuss the hydration characteristics and properties of the two types of cement together.

The pozzolanic reaction and its significance. With respect to the main C-S-H-forming reaction, a comparison between portland cement and portland pozzolan cement is useful for the purpose of understanding the reasons for differences in their behavior:

Portland Cement

$$\boxed{C_3S + H \xrightarrow{\text{fast}} C\text{-}S\text{-}H + CH}$$

Portland Pozzolan Cement

$$\boxed{\text{Pozzolan} + CH + H \xrightarrow{\text{slow}} C\text{-}S\text{-}H}$$

(6–6)

The reaction between a pozzolan and calcium hydroxide is called the *pozzolanic reaction*. The technical significance of pozzolan cements (and also slag cements) is derived mainly from three features of the pozzolanic reaction. First, the reaction is slow; therefore, the rates of heat liberation and strength development will be accordingly slow. Second, the reaction is lime consuming instead of lime producing, which has an important bearing on the durability of the hydrated paste to acidic environments. Third, pore size distribution studies of hydrated IP and IS cements have shown that the reaction products are very efficient in filling up large capillary space, thus improving the strength and impermeability of the system. Pore size distribution data of portland pozzolan cements containing a Greek pozzolan (Santorin earth) are shown in Fig. 6–13, and a graphic representation of the *pore refinement process* associated with the pozzolanic reaction is shown in Fig. 6–14.

In addition to reactive silica, slags and pozzolans contribute reactive alumina, which in the presence of calcium hydroxide and sulfate ions in the system, also forms cementitious products such as C_4AH_{13}, AF_t, and AF_m. Properties of Types IP and IS vary widely depending on the curing conditions and the proportions as well as the physical-chemical characteristics of the constituent materials present. The properties given below may therefore be considered as indicative of general trends.

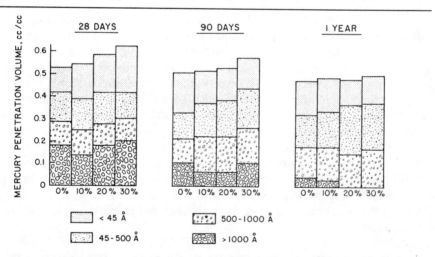

Figure 6–13 Changes in pore size distribution of cement pastes with varying pozzolan content. (Reprinted with permission from *Cem. Concr. Res.*, Vol. 11, No. 4, P. K. Mehta, copyright 1981, Pergamon Press, Ltd.)

In a laboratory investigation portland pozzolan cements containing 10, 20, or 30 weight percent of a Greek natural mineral pozzolan were hydrated at a given water/cement ratio, and the pore size distributions were determined at 28, 90, and 365 days by mercury penetration porosimetry. With 20 or 30 percent pozzolan content, no large pores (> 0.1 μm) were found in the pastes cured for 1 year. Water permeability tests showed that these cement pastes were much more impermeable than the reference portland cement paste.

Special Hydraulic Cements

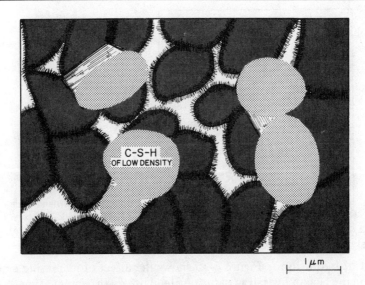

Figure 6–14 Diagrammatic representation of well-hydrated cement pastes made with a portland pozzolan cement. Compared to a portland cement paste (see Fig. 2–6 for identification of the phases present) it is shown here that, as a result of the pozzolanic reaction, the capillary voids are either eliminated or reduced in size, and dense crystals of calcium hydroxide are replaced with additional C-S-H of a lower density.

On the basis of scanning electron microscopic and pore-size distribution studies of hydrated cement pastes both with and without a pozzolan, it is possible to conclude that there are two physical effects of the chemical reaction between the pozzolanic particles and calcium hydroxide: (i) pore-size refinement, and (ii) grain-size refinement. The formation of secondary hydration products (mainly calcium silicate hydrates) around the pozzolan particles tends to fill the large capillary voids with a microporous and, therefore, a low-density material. The process of transformation of a system containing large capillary voids into a microporous product containing numerous fine pores is referred to as "pore-size refinement." Also, nucleation of calcium hydroxide around the fine and well distributed particles of the pozzolan will have the effect of replacing the large and oriented crystals of calcium hydroxide with numerous, small, and less oriented crystals plus poorly crystalline reaction products. The process of transformation of a system containing large grains of a component into a product containing smaller grains is referred to as "grain-size refinement." Both the pore-size and the grain-size refinement processes strengthen the cement paste.

From the standpoint of impermeability and durability the effects of the pozzolanic reaction are probably more important in concrete than in the hydrated cement paste. As discussed in Chapter 5, the permeability of concrete is generally much higher than the permeability of cement paste because of microcracks in the transition zone. It is suggested that the processes of pore-size and grain-size refinement strengthen the transition zone, thus reducing the microcracking and increasing impermeability of concrete.

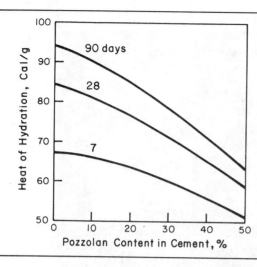

Figure 6-15 Effect of substituting an Italian natural pozzolan on the heat of hydration of a portland cement. (From F. Massazza and U. Costa, *Il Cemento*, Vol. 76, p. 13, 1979).

Heat of hydration. Figure 6-15 shows the effect of increasing amounts of pozzolan on the heat of hydration of the portland pozzolan cement. Type IS cements containing 50 percent slag show comparable results (i.e., 45 to 50 cal/g heat of hydration at 7 days).

Strength development. Figure 6-16a shows strength development rates up to 1 year in cements containing 10, 20, or 30 percent pozzolan, and Fig. 6-16b shows similar data for cements containing 40, 50, or 60 percent granulated slag. In general, pozzolan cements are somewhat slower than slag cements in developing strength; whereas Type IS cements usually make a significant contribution to the 7-day strength, a Type IP cement containing an ordinary pozzolan shows considerable strength gain between the 7- and the 28-day test period. When adequately reactive materials are used in moderate proportion (e.g., 15 to 30 percent pozzolan or 25 to 50 percent slag), and moist curing is available for long periods, the ultimate strengths of Types IP and IS cements are higher than the strength of the portland cement from which these cements are made. This is because of the pore refinement associated with the pozzolanic reactions and increase in the C-S-H and other hydration products at the expense of calcium hydroxide.

Durability. Compared to portland cement, the superior durability of Type IP cement to sulfate and acidic environments is due to the combined effect of better impermeability at given water/cement ratio and degree of hydration, and reduced calcium hydroxide content in the hydrated cement paste (Fig. 6-17a). In one investigation it was found that compared to portland cement the depth of penetration of water was reduced by about 50 percent in 1-year-old pastes of cements containing 30 weight percent of a Greek volcanic ash. Also, compared to 20 percent calcium hydroxide in the 1-year-old paste of the reference portland cement, there was only 8.4 percent calcium hydroxide in a similarly hydrated paste of the cement containing

Special Hydraulic Cements

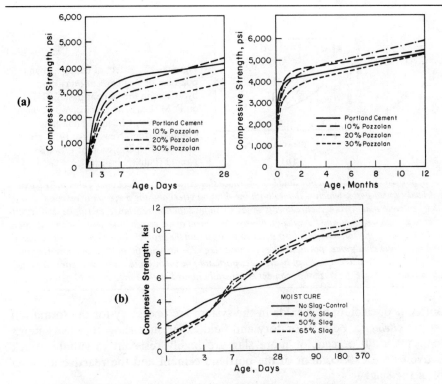

Figure 6–16 Strength of blended portland cement containing a pozzolan or a blast-furnace slag. [(a), Reprinted with permission from *Cem. Concr. Res.*, Vol. 11, No. 4, P. K. Mehta, Copyright 1981, Pergamon Press, Ltd.; (b), reprinted with permission from F. J. Hogan and J. W. Meusel, *Cem. Concr. Aggregates*, Vol. 3, No. 1, 1981, Copyright, ASTM, 1916 Race Street, Philadelphia PA 19103.]

The upper figures show the compressive strength of portland cements (<400 m^2/kg Blaine) made with a Greek natural mineral pozzolan. The lower figure shows the compressive strengths of portland blast-furnace slag cements (>500 m^2/kg) made with an American granulated blast-furnace slag.

30 weight percent of the Greek pozzolan. It may be noted that due to the dilution effect in the latter case, without the pozzolanic reaction the amount of calcium hydroxide should have been about 14 percent.

Type IS cements behave in a similar manner. Figure 6–17b shows the effect of increasing slag content on the amount of calcium hydroxide in portland blast-furnace slag cements at 3 and 28 days after hydration. At about 60 percent slag content, the amount of calcium hydroxide becomes so low that even slags containing large amounts of reactive alumina can be used to make sulfate-resisting cements. It may be recalled (see Chapter 5) that the rate of sulfate attack depends on the permeability and the amount of calcium hydroxide and reactive alumina phases present. Some high-alumina slags and fly ashes tend to increase in the hydrated cement paste the amounts of C-A-H and monosulfate, which are vulnerable to sulfate attack. Since

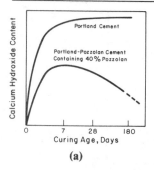
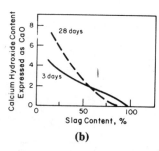

Figure 6–17 Effect of curing age and pozzolan or slag content on calcium hydroxide in the cement paste. [Based on F. M. Lea, *The Chemistry of Cement and Concrete,* Chemical Publishing Company, Inc., New York, 1971, pp. 442, 481, by permission of Edward Arnold (Publishers) Ltd.].

In the case of portland pozzolan and portland blast-furnace slag cements the reduction of calcium hydroxide in the hydrated cement paste, which is due to both the dilution effect and the pozzolanic reaction, is one reason that concrete made from such cements tends to show superior resistance to sulfate and acidic environments. Initially, with curing the calcium hydroxide content of the cement increases due to hydration of the portland cement present; however, later it begins to drop with the progress of the pozzolanic reaction. Depending on curing conditions, portland blast-furnace slag cements with 60 percent or more slag may contain as little as 2 to 3 percent calcium hydroxide; portland pozzolan cement products contain higher calcium hydroxide because the pozzolan content is generally limited to the range 20 to 40 percent.

large amounts of calcium hydroxide in the system are necessary for the formation of *expansive ettringite,* both laboratory and field experience show that IS cements containing 60 to 70 percent or more slag are highly resistant to sulfate attack irrespective of the C_3A content of the portland cement and the reactive alumina content of the slag.

In regard to the deleterious expansion associated with the alkali-aggregate reaction, combinations of high-alkali portland cement and pozzolans or slags are generally known to produce durable products (Fig. 6–18). Sometimes the alkali content of pozzolans and slags are high, but if the alkali-containing mineral is not soluble in the high-pH environment of portland cement concrete, the high-alkali content of the blended cement should not cause any problem.

Expansive Cements

Expansive cements are hydraulic cements which, unlike portland cement, expand during the early hydration period after setting. Large expansion occurring in an unrestrained cement paste can cause cracking; however, if the expansion is properly restrained, its magnitude will be reduced and a prestress or self-stress will develop. When the magnitude of expansion is small such that the prestress developed in concrete is on the order of 25 to 100 psi (0.2 to 0.7 MPa), which is usually adequate to offset the tensile stress due to drying shrinkage, the cement is known as *shrinkage compensating.* Cements of this type have proved very useful for making crack-free pavements and slabs. When the magnitude of expansion is large enough to produce prestress levels on the order of 1000 psi (6.9 MPa), the cement is called *self-stressing* and can be used for the production of chemically prestressed concrete elements.

Formation of ettringite and hydration of hard-burnt CaO are the two phenomena known to cement chemists that can cause disruptive expansion in concrete

Special Hydraulic Cements

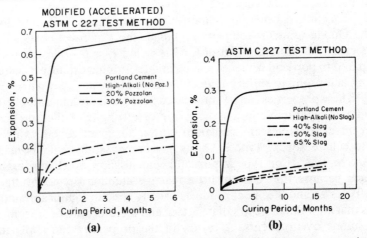

Figure 6-18 Influence of pozzolan or slag addition on alkali-aggregate expansion. [(a), From *Cem. Concr. Res.*, Vol. 11, No. 4, P. K. Mehta, Copyright 1981, Pergamon Press, Ltd.; (b), reprinted with permission from F. J. Hogan and J. M. Meusel, *Cem. Concr. Aggregates*, Vol. 3, No. 1, 1981. Copyright, ASTM, 1916 Race Street, Philadelphia PA 19103.]

Pozzolans and slag are generally very effective in reducing the expansion associated with the alkali-aggregate reaction. Santorin Earth from Greece was used for the test data shown in part (a); a granulated blast-furnace slag from the United States was used for the test data shown in part (b). Since different test methods were used, the data in the two figures are not directly comparable; however, the trend is similar in both cases.

(Chapter 5). Both phenomena have been harnessed to produce expansive cements. Developed originally by Alexander Klein of the University of California in the 1960s, the sulfoaluminate-type clinker (which is commonly used in the U.S. practice) is a modified portland cement clinker containing significant amounts of $C_4A_3\bar{S}$ and $C\bar{S}$ in addition to the cementitious compounds (C_3S, C_2S, and C_4AF). The cement produced by grinding this clinker is called *Type K expansive cement.* To achieve a better control of the potential expansion in industrial expansive cements, it is customary to blend a suitable proportion of the sulfoaluminate clinker with normal portland cement clinker.

ASTM C 845 covers two other expansive hydraulic cements which also derive their expansion characteristic from ettringite but are no longer commercially produced in the United States. The cements differ from the Type K cement and from each other with respect to the source of aluminate ions for ettringite formation. *Type M expansive cement* is a mixture of portland cement, calcium aluminate cement (with CA as the principal compound), and calcium sulfate. *Type S expansive cement* is composed of a very high C_3A portland cement (approximately 20 percent C_3A) and large amounts of calcium sulfate. The stoichiometry of the expansive reactions in the three cements can be expressed as

$$C_4A_3\bar{S} + 8C\bar{S} + 6CH + H \rightarrow 3C_6A\bar{S}_3H_{32} \quad \text{(Type K)} \quad (6\text{-}7a)$$

$$CA + 3C\bar{S} + 2CH + H \rightarrow C_6A\bar{S}_3H_{32} \quad \text{(Type M)} \quad (6\text{-}7b)$$

$$C_3A + 3C\bar{S} + H \rightarrow C_6A\bar{S}_3H_{32} \quad \text{(Type S)} \quad (6\text{-}7c)$$

The CH in the reaction shown above is provided by portland cement hydration, although Type K clinkers generally contain some uncombined CaO. Initially developed by the Onoda Cement Company of Japan, the expansive portland cement deriving its expansion from hard-burnt CaO is called *Type O expansive cement*.

Compared to portland cements, the ettringite-forming expansive cements are quick setting and prone to suffer rapid slump loss. However, they show excellent workability. These properties can be anticipated from the large amounts of ettringite formed and the water-imbibing characteristic of ettringite. Other properties of expansive cement concretes are similar to portland cement concretes except for durability to sulfate attack. Type K shrinkage-compensating cements made with blending ASTM Type II or Type V portland cement show excellent durability to sulfate attack because they contain little reactive alumina or monosulfate after hydration. Types M and S cement products usually contain significant amounts of compounds that are vulnerable to sulfate attack and therefore are not recommended for use in sulfate environments. A review of the properties and applications of expansive cement concrete is included in Chapter 11.

Rapid Setting and Hardening Cements

It may be noted that ASTM Type III cement is rapid hardening (high early strength) but not rapid setting because the initial and final setting times of the cement are generally similar to Type I portland cement. For applications such as emergency repair of leaking joints and shotcreting, hydraulic cements are needed that not only are rapid hardening but rapid setting. This need is frequently fulfilled by using mixtures of portland cement and plaster of paris ($CaSO_4 \cdot \frac{1}{2}H_2O$) or portland cement and calcium aluminate cement, which give setting times as low as 10 min. The durability and ultimate strengths of the hardened products are rather poor.

During the 1970s a new generation of cements were developed which derive rapid setting and hardening characteristics from ettringite formation. After the initial rapid hardening period, these cements continue to harden subsequently at a normal rate due to the formation of C-S-H from hydraulic calcium silicates.

Regulated-set cement, also called *jet cement* in Japan, is manufactured under patents issued to the U.S. Portland Cement Association. A modified portland cement clinker containing mainly alite and a calcium fluoroaluminate ($11CaO \cdot 7Al_2O_3 \cdot CaF_2$) is made. A suitable proportion of the fluoroaluminate clinker is blended with normal portland cement clinker and calcium sulfate so that the final cement contains 20 to 25 percent of the fluoroaluminate compound and about 10 to 15 percent calcium sulfate. The cement is generally very fast setting (3 to 5 min setting time) but can be retarded to the desired time of set by using citric acid, sodium sulfate, calcium hydroxide, and other retarders.

The high reactivity of the cement is confirmed by the high heat of hydration (100 to 110 cal/g at 3 days), and over 1000 psi (6.9 MPa) and 4000 psi (28 MPa) compressive strengths (ASTM C 109 mortar) at 1 hour and 3 days after hydration, respectively. The ultimate strength and other physical properties of the cement are

Special Hydraulic Cements

comparable to those of portland cement except that due to the high content of the reactive aluminate, the sulfate resistance is poor. Studies at the concrete laboratory of the U.S. Army Engineer Waterways Experiment Station[7] have shown that the high heat of hydration of the regulated-set cement can help produce concretes with adequate strengths even when the concrete is placed and cured at temperatures as low as 15°F (−9.5°C).

Very high early strength and high iron cements. In addition to regulated-set cements, two other modified portland cements, *very high early strength cement* (VHE) and *high iron cement* (HIC), derive their rapid setting and hardening characteristics from the formation of large amounts of ettringite during the early hydration period. With the VHE cement $C_4A_3\bar{S}$ is the main source of aluminate for ettringite formation, whereas with HIC cement both $C_4A_3\bar{S}$ and C_4AF provide the aluminate ions. Although there are certain basic differences in their composition, both cements exhibit strength development rates that are suitable for application to precast and prestressed concrete products. In the precast and prestressed concrete industry, quick turnover of forms or molds is an economic necessity. The VHE and HIC cements are still under development, but such cements should have a considerable appeal to the construction industry because under normal curing temperatures (i.e., without steam curing) they are capable of developing compressive strengths of 15 MPa and 25 MPa within 8 and 24 hr, respectively, with about 50 MPa ultimate strength.

Oil-Well Cements

As discussed below, oil-well cements are not used for making structural concrete. Since approximately 5 percent of the total portland cement produced in the United States is consumed by the petroleum industry, it may be desirable to know the purpose for which they are used and to have an idea of the composition and properties required.

Once an oil well (or gas well) has been drilled to the desired depth, cementing a steel casing to the rock formation offers the most economic way to achieve the following purposes:

- To prevent unwanted migration of fluids from one formation to another
- To prevent pollution of valuable oil zones
- To protect the casing from external pressures that may be able to collapse it
- To protect the casing from possible damage due to corrosive gases and waters

For the purposes of cementing a casing, a high water/cement ratio mortar or cement slurry is pumped to depths which in some instances may be below 6100 m, and where

---

[7] G. C. Hoff, B. J. Houston, and F. H. Sayles, *U.S. Army Engineer Waterway Experiment Station*, Vicksburg, Miss., Miscellaneous Paper C-75-5, 1975.

the slurry may be exposed to temperatures above 204°C and pressures above 20,000 psi (140 MPa). In the Gulf coast region the static bottom hole temperature increases by 1.5°F (0.8°C) for every 100 ft (30 m) of well depth. It is desired that the slurry must remain sufficiently fluid under the service conditions for the several hours needed to pump it into position, and then harden quickly. Oil-well cements are modified portland cements that are designed to serve this need.

Nine classes of oil-well cements (Classes A to J in Table 6–8) that are applicable at different well depths are covered by the API (American Petroleum Institute) Standard 10A. The discovery that the *thickening time of cement slurries at high temperatures* can be increased by reducing the C_3A content and fineness of ordinary portland cement (i.e., by using coarsely ground cement) led to the initial development of oil-well cements. Later it was found that for applications above 82°C, the cement must be further retarded by addition of lignosulfonates, cellulose products, or salts of acids containing one or more hydroxyl groups (p. 265). Subsequently, it was also discovered that in the case of oil-well temperatures above 110°C the CaO/SiO_2 ratio of the cement hydration product must be lowered to below 1.3 by the addition of silica flour in order to achieve high strength after hardening. These findings became the basis for the development of numerous cement additives for application to the oil-well cement industry.

The petroleum industry generally prefers the basic low-C_3A, coarse-ground portland cements (API Classes G and H), to which one or more admixtures of the type listed below are added at the site:

1. **Cement retarders:** to increase the setting time of cement and allow time for placement of the slurry
2. **Cement accelerators:** to reduce the setting time of cement for early strength development when needed (i.e., in permafrost zones)
3. **Lightweight or heavyweight additives:** to reduce or increase the weight of the column of cement slurry as needed
4. **Friction reducers:** to allow placement of slurry with less frictional pressure (2 to 3 percent bentonite clay is commonly used for this purpose)
5. **Low water-loss additives:** to retain water in the slurry when passing permeable zones downhole (i.e., latex additives)
6. **Strength-retrogression reducers:** to reduce the CaO/SiO_2 ratio of the hydration product at temperatures above 110°C (i.e., silica flour or pozzolans)

Since organic retarders are unstable at high temperatures, API Class J cement represents a relatively recent development in the field of modified portland cements that can be used for case-cementing *at temperatures above 300°F (150 °C) without the addition of a retarder.* The cement consists mainly of βC_2S, is ground to about 200 m²/kg Blaine, and contains 40 weight percent silica flour. It may be noted that slurry thickening times and strength values for oil-well cements are determined with special procedures set forth in API RP-10B, *Recommended Practice for Testing Oil-Well Cements and Cement Additives.*

White or Colored Cements

The uniformly gray color of portland cement products limits an architect's opportunity for creating surfaces with aesthetic appeal. A white cement, with exposed-aggregate finish, can be used to create desired aesthetic effects. Furthermore, by adding appropriate pigments, white cements are used as a basis for producing cements with varying colors.

White cement is produced by pulverizing a white portland cement clinker. The gray color of ordinary portland cement clinker is generally due to the presence of iron. Thus by lowering the iron content of clinker, light-colored cements can be produced. When the total iron in clinker corresponds to less than 0.5% Fe_2O_3, and the iron is held in the reduced Fe^{2+} state, the clinker is usually white (see box on this page). These conditions are achieved in cement manufacturing by using iron-free clay and carbonate rock as raw materials, special ball mills with ceramic liners and balls for grinding the raw mix, and clean fuel such as oil or gas for production of clinker under a reducing environment in the high-temperature zone of the cement rotary kiln. Consequently, white cements are approximately three times as expensive as normal portland cement.

The importance of the reducing environment in making white clinker is underscored by an experience which the author had during a consulting trip to a South American cement plant. The raw mix contained more iron than normally acceptable, and the clinker from the kiln was persistently off-white. In order to prolong the reducing environment around the clinker particles by increasing the amount of oil sprayed on hot clinkers leaving the burning zone, I requested a heat-resisting steel pipe of large diameter. Since there was none in stock and the cement plants are generally located far away from urban areas, I was getting nowhere, while the low-iron raw mix specially made for the experiment was running out fast.

The communication problem added to the difficulty; I could not speak Spanish and the foreman did not understand English. To emphasize my point about one pipe with a large diameter I raised one finger. Suddenly, he waved two fingers in my face. Somehow, this brought to mind the story of princess Vidyotama in the Sanskrit literature. Once a king in ancient India had a very beautiful daughter who refused to marry until she found someone wiser than herself. When many scholarly princes failed to win her in debates on philosophical and religious issues, they decided to play a practical joke. A dumb and stupid man was dressed in scholarly robes and presented for debate with the princess. When the princess raised one finger the fool, assuming that the princess was threatening to poke one of his eyes, raised two fingers. The judges interpreted the one finger to mean that God is the only important thing in the universe, and the two to mean that nature reveals the glory and splendor of God and is important, too, thus giving the victory to the fool.

The foreman really meant that since he did not have a pipe with a large diameter, he would like to install two pipes of a smaller diameter. When the thought of Tilotama's fool trying to blind me in both eyes came to me, I yielded without further argument. The foreman installed the two small pipes for spraying oil on hot clinkers; subsequently, the whitest clinker I have ever seen came out of the kiln.

Colored cements fall into two groups; most are derived from pigment addition to white cement, but some are produced from clinkers having the corresponding colors. A buff-colored cement marketed in the United States under the name *warm tone cement* is produced from the clinker made from a portland cement raw mix containing a higher iron content than normal (approximately 5 percent Fe_2O_3), and processed under reducing conditions.

For producing colored cements by adding pigments to white cements, it should be noted that not all the pigments that are used in the paint industry are suitable for making colored cements. To be suitable, a pigment should not be detrimental to the setting, hardening, and durability characteristics of the cement, and should produce durable color when exposed to light and weather. Red, yellow, brown, or black cements can be produced by intergrinding 5 to 10 weight percent iron oxide pigments of the corresponding color with a white clinker. Green and blue colors in cement can be achieved by using chromium oxide and cobalt blue, respectively.

Calcium Aluminate Cement

Compared to portland cement, calcium aluminate cement (CAC) possesses many *unique properties,* such as high early strength, hardening even under low-temperature conditions, and superior durability to sulfate attack. However, several structural failures due to gradual loss in strength associated with concrete containing CAC have been instrumental in limiting the use of this cement for structural applications. In most countries, now CAC is used mainly for making castable refractory lining for high-temperature furnaces.

According to ASTM C 219 definitions, **calcium aluminate cement** is the product obtained by pulverizing calcium aluminate cement clinker; the clinker is a partially fused or a completely fused product consisting of hydraulic calcium aluminates. Thus unlike portland and modified portland cements, in which C_3S and C_2S are the principal cementing compounds, in CAC the monocalcium aluminate (abbreviated as CA) is the principal cementing compound, with $C_{12}A_7$, CA, C_2AS, βC_2S, and Fss as minor compounds. Typically, the chemical analysis of ordinary CAC corresponds to approximately 40 percent Al_2O_3, and some cements contain even higher alumina content (50 to 80 percent); therefore, the cement is also called *high-alumina cement* (HAC).

Bauxite, a hydrated alumina mineral, is the commonly used source of alumina in raw materials used for the manufacture of CAC. Most bauxite ores contain considerable amounts of iron as an impurity, which accounts for the 10 to 17 percent iron (expressed as Fe_2O_3) usually present in ordinary CAC. This is why, unlike portland cement clinker, the CAC clinker containing high iron is in the form of completely fused melts which are made in specially designed furnaces. This is also the reason why in France and Germany the cement is called *ciment fondu* and *tonerdeschmelz zement,* respectively. Products meant to be used for making very high-temperature concretes contain very low iron and silica, and can be made by sintering in rotary kilns.

Like portland cement, the properties of CAC are dependent on the hydration characteristics of the cement and the microstructure of the hydrated cement paste.

Special Hydraulic Cements

The principal compound in cement is CA, which usually amounts to 50 to 60 weight percent. Although CAC products have setting times comparable to ordinary portland cement, the rate of strength gain at early ages is quite high mainly due to the high reactivity of CA. Within 24 hours of hydration, the strength of normally cured CAC concretes can attain values equal to or exceeding the 7-day strength of ordinary portland cement (Fig. 6–19a). Also, the strength gain characteristic under subzero curing conditions (Fig. 6–19b) is much better than for portland cements; hence the material is quite attractive for cold weather applications. It may be noted that the rate of heat liberation from a freshly hydrated CAC can be as high as 9 cal/g per hour, which is about three times as high as the rate for high-early-strength portland cement.

The composition of the hydration products shows a time-temperature dependency; the low-temperature hydration product (CAH_{10}) is thermodynamically unstable, especially in warm and humid storage conditions, under which a more stable compound, C_3AH_6, is formed (see the left-hand side of the boxed area below). Laboratory and field experience with CAC concretes show that on prolonged storage the hexagonal CAH_{10} and C_2AH_8 phases tend to convert to the cubic C_3AH_6. As a *consequence of the CAH_{10}-to-C_3AH_6 conversion,* a hardened CAC paste would show more than 50 percent reduction in the volume of solids (see the right-hand side of the box), which causes an increase in porosity (Fig. 6–20a) and a loss in strength associated with the phenomenon (Fig. 6–20b).

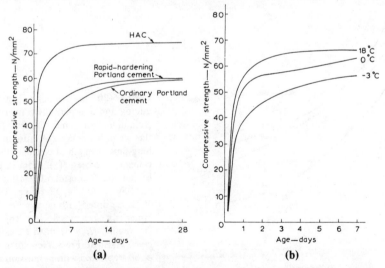

Figure 6–19 (a) Strength development rates for various cements at normal temperature; (b) effect of low curing temperatures on the strength of high-alumina cement concretes. [From A. M. Neville, in *Progress in Concrete Technology,* ed. V. M. Malhotra, CANMET, Ottawa, 1980, pp. 293–331.]

Calcium aluminate or high-alumina cements are able to develop very high strengths in relatively short periods of time. Unlike portland cements, they can develop high strengths even at lower than normal temperatures.

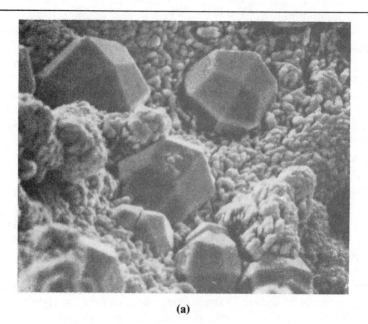

(a)

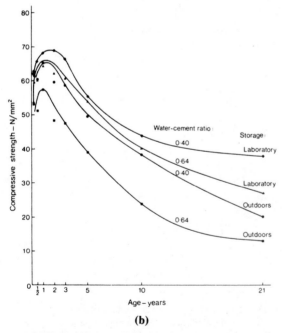

(b)

Figure 6–20 (a) Scanning electron micrograph of a partially converted calcium aluminate cement system; (b) influence of water/cement ratio on the long-time strength of calcium aluminate cement concretes. [(a), From P. K. Mehta and G. Lesnikoff, *J. Am. Ceram. Soc.*, Vol. 54, No. 4, pp. 210–212, 1971, reprinted with permission of American Ceramics Society; (b), from A. Neville, *High Alumina Cement Concrete*, Halstead Press, New York, 1975, p. 58, reprinted with permission from Construction Press (Longman Group Ltd.)]

Calcium aluminate cement concretes are generally not recommended for structural use. This is because the principal hydration product, CAH_{10} is unstable under ordinary conditions. It gradually transforms into a stable phase, C_3AH_6, which has a cubic structure and is denser. The CAH_{10}-to-C_3AH_6 conversion is associated with a large increase in porosity and therefore a corresponding decrease in strength.

Special Hydraulic Cements

$$CA + H \begin{cases} \xrightarrow{<10°C} CAH_{10} \\ \xrightarrow{10\text{-}20°C} C_2AH_8 + AH_3 \\ \xrightarrow{>30°C} C_3AH_6 + 2AH_3 \end{cases}$$

	$3CAH_{10}$	$=$	C_3AH_6	$+$	$2AH_3$	$+ 18H \uparrow$
g	1014		378		312	
g/cm^3	1.72		2.52		2.4	
cm^3	590→		150 +		136	

Formerly, it was assumed that the strength-loss problem in concrete could be ignored when low water/cement ratios were used, and the height of a casting was limited to reduce the temperature rise due to heat of hydration. The data in Fig. 6–20a show that this may not be the case. The real concern is not that the residual strength will be inadequate for structural purposes but that, as a result of increase in porosity, the resistance to atmospheric carbonation and to corrosion of the embedded steel in concrete would be reduced.

From the hydration reaction of CAC, it may be noted that there is no calcium hydroxide in the hydration product; this feature also distinguishes CAC from portland cements and is the main reason why CAC concretes show *excellent resistance to acidic environments* (dilute acids, 4 to 6 pH), seawater, and sulfate waters. As discussed below, the absence of calcium hydroxide in hydrated CAC is also helpful in utilizing the material for making high-temperature concrete.

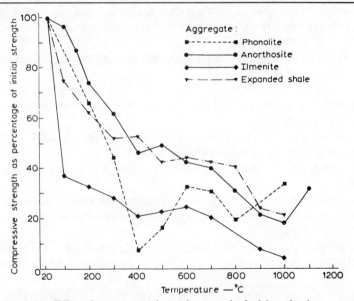

Figure 6–21 Effect of temperature rise on the strength of calcium aluminate cement concretes. [From A. M. Neville, in *Progress in Concrete Technology*, ed. V. M. Malhotra, CANMET, Ottawa, 1980, pp. 293–331.]

Calcium aluminate cement concrete mostly finds application in monolithic refractory lining for high-temperature furnaces. With increasing temperature, the cement hydration products decompose and this causes a loss in strength. However, at high temperatures the strength increases due to the formation of a stable sintered material (ceramic bond).

In practice, the use of portland cement for concrete exposed to high temperature is rather limited to about 500°C, because at higher temperatures the free CaO formed on decomposition of calcium hydroxide would cause the heated concrete to become unsound on exposure to moist air or water. Not only does CAC not produce any calcium hydroxide on hydration, but also it is rapid hardening under normal curing temperatures. Also, at temperatures above 1000°C, CAC is capable of developing a *ceramic bond* which is as strong as the original hydraulic bond. The green or the unfired strength of the CAC concrete drops considerably during the first-heating cycle due to the CAH_{10}-to-C_3AH_6 conversion phenomenon. With a high cement content of the concrete, however, the green strength may be adequate to prevent damage until the strength increases again due to the development of the ceramic bond (Fig. 6–21).

TEST YOUR KNOWLEDGE

1. When producing a certain type of portland cement it is important that the oxide composition remains uniform. Why?
2. In regard to sulfate resistance and rate of strength development, evaluate the properties of the portland cement which has the following chemical analysis: SiO_2 = 20.9 percent; Al_2O_3 = 5.4 percent; Fe_2O_3 = 3.6 percent; CaO = 65.1 percent; MgO = 1.8 percent; and SO_3 = 2.1 percent.
3. What do you understand by the following terms: alite, belite, periclase, langbeinite, plaster of paris, tobermorite gel?
4. Why is C_3S more reactive, and γC_2S nonreactive with water at normal temperatures? MgO and CaO have similar crystal structures, but their reactivities are very different from each other. Explain why.
5. What is the significance of fineness in cement? How is it determined? Can you give some idea of the fineness range in industrial portland cements?
6. Why is gypsum added to the cement clinker? Typically, how much is the amount of added gypsum?
7. The presence of high free-lime in portland cement can lead to unsoundness. What is meant by the term, "unsoundness"? Which other compound can cause unsoundness in portland cement products?
8. Approximately, what is the combined percentage of calcium silicates in portland cement? What are the typical amounts of C_3A and C_4AF in ordinary (ASTM Type I) portland cement?
9. Which one of the four major compounds of portland cement contributes most to the strength development during the first few weeks of hydration? Which compound or compounds are responsible for rapid stiffening and early setting problems of the cement paste?
10. Discuss the major differences in the physical and chemical composition between an ordinary (ASTM Type I) and a high early strength (ASTM Type III) portland cement.
11. Why do the ASTM Specifications for Type IV cement limit the minimum C_2S content to 40 percent and the maximum C_3A content to 7 percent?

12. Explain which ASTM type cement would you use for:
    (a) Cold-weather construction
    (b) Construction of a dam
    (c) Making reinforced concrete sewer pipes
13. The aluminate-sulfate balance in solution is at the heart of several abnormal setting problems in concrete technology. Justify this statement by discussing how the phenomena of quick-set, flash set, and false set occur in freshly hydrated portland cements.
14. Assuming that the chemical composition of the calcium silicate hydrate formed on hydration of C_3S or C_2S corresponds to $C_3S_2H_3$, make calculations to show the proportion of calcium hydroxide in the final products and the amount of water needed for full hydration.
15. Define the terms *initial set* and *final set*. For a normal portland cement draw a typical heat evolution curve for the setting and early hardening period, label the ascending and descending portions of the curve with the underlying chemical processes at work, and show the points where the initial set and final set are likely to take place.
16. Discuss the two methods that the cement industry employs to produce cements having different rates of strength development or heat of hydration. Explain the principle behind the maximum limit on the C_3A content in the ASTM C 150 Standard Specification for Type V portland cements.
17. With the help of the "pozzolanic reaction," explain why under given conditions, compared to portland cement, portland pozzolan and portland blast-furnace slag cements are likely to produce concrete with higher ultimate strengths and superior durability to sulfate attack.
18. What is the distinction between shrinkage-compensating and self-stressing cements? What are Types K, M, S, and O expansive cements? Explain how the expansive cements function to make concrete crack-free.
19. Write short notes on the compositions and special characteristics of the following cements: regulated-set cement, very high early strength cement, API Class J cement, white cement, and calcium aluminate cement.
20. Discuss the physical-chemical factors involved in explaining the development of strength in products containing the following cementitious materials, and explain why portland cement has come to stay as the most commonly used cements for structural purposes:
    (a) lime
    (b) plaster of paris
    (c) calcium aluminate cement

SUGGESTIONS FOR FURTHER STUDY

LEA, F. M., *The Chemistry of Cement and Concrete,* Chemical Publishing Company, Inc., New York, 1971.

MALHOTRA, V. M., ed., *Progress in Concrete Technology,* CANMET, Ottawa, 1980 Chap. 7, Expansive Cements and Their Applications, by P. K. Mehta, and M. Polivka; and Chap. 8, High Alumina Cement—Its Properties, Application, and Limitations, by A. M. Neville.

SKALNY, J. P. ed., *Material Science of Concrete,* The American Ceramic Society Inc., 1989: Cement Production and Cement Quality by V. Johansen; Hydration Mechanisms by E. M. Gartner and J. M. Gaidis; The Microtextures of Concrete by K. L. Scrivner.

TAYLOR, H. W. F., *Cement Chemistry,* Academic Press, Inc., San Diego, CA, 1990.

CHAPTER 7
Aggregates

PREVIEW

Aggregate is relatively inexpensive and does not enter into complex chemical reactions with water; it has been customary, therefore, to treat it as an inert filler in concrete. However, due to increasing awareness of the role played by aggregates in determining many important properties of concrete, the traditional view of the aggregate as an inert filler is being seriously questioned.

Aggregate characteristics that are significant to concrete technology include porosity, grading or size distribution, moisture absorption, shape and surface texture, crushing strength, elastic modulus, and the type of deleterious substances present. These characteristics are derived from mineralogical composition of the parent rock (which is affected by geological rock-formation processes), exposure conditions to which the rock has been subjected before making the aggregate, and the type of operation and equipment used for producing the aggregate. Therefore, fundamentals of rock formation, classification and description of rocks and minerals, and industrial processing factors that influence aggregate characteristics are briefly described in this chapter.

Natural mineral aggregates, which comprise over 90 percent of the total aggregates used for making concrete, are described in more detail. Due to their greater potential use, the aggregates from industrial by-products such as blast-furnace slag, fly ash, municipal waste, and recycled concrete are also described. Finally, the principal aggregate characteristics that are important to concrete technology are covered in detail.

SIGNIFICANCE

From Chapter 6 we know that cements consist of chemical compounds that enter into chemical reactions with water to produce complex hydration products with adhesive property. Unlike cement, although the aggregate in concrete occupies 60 to 80 percent of the volume, it is frequently looked upon as an inert filler and therefore not worthy of much attention in regard to its possible influence on the properties of concrete. The considerable influence that the aggregate can exercise on strength, dimensional stability, and durability of concrete has been described in Chapters 3, 4, and 5, respectively. In addition to these important properties of hardened concrete, aggregates also play a major role in determining the cost and workability of concrete mixtures (Chapter 9); therefore, *it is inappropriate to treat them with less respect than cements.*

CLASSIFICATION AND NOMENCLATURE

Classifications of aggregates according to particle size, bulk density, or source have given rise to a special nomenclature which should be clearly understood. For instance, the term *coarse aggregate* is used to describe particles larger than 4.75 mm (retained on No. 4 sieve), and the term *fine aggregate* is used for particles smaller than 4.75 mm; typically, fine aggregates contain particles in the size range 75 μm (No. 200 sieve) to 4.75 mm, and coarse aggregates from 4.75 to about 50 mm, except for mass concrete, which may contain up to 150-mm coarse aggregate.

Most natural mineral aggregates such as sand and gravel have a bulk density of 95 to 105 lb/ft^3 (1520 to 1680 kg/m^3) and produce *normal-weight* concrete with approximately 150 lb/ft^3 (2400 kg/m^3) unit weight. For special needs, aggregates with lighter or heavier density can be used to make correspondingly lightweight and heavyweight concretes. Generally, the aggregates with bulk densities less than 70 lb/ft^3 (1120 kg/m^3) are called *lightweight,* and those weighing more than 130 lb/ft^3 (2080 kg/m^3) are called *heavyweight.*

For the most part, concrete aggregates are comprised of sand, gravel, and crushed rock derived from natural sources, and are therefore referred to as *natural mineral aggregates*. On the other hand, thermally processed materials such as expanded clay and shale, which are used for making lightweight concrete, are called *synthetic aggregates*. Aggregates made from industrial by-products, for instance, blast-furnace slag and fly ash, also belong to this category. Municipal wastes and recycled concrete from demolished buildings and pavements have also been investigated for use as aggregates, as described below.

NATURAL MINERAL AGGREGATES

Natural mineral aggregates form the most important class of aggregates for making portland cement concrete. Approximately half of the total coarse aggregate consumed by the concrete industry in the United States consists of gravels; most of the remainder is crushed rock. Carbonate rocks comprise about two-thirds of the crushed aggregate; sandstone, granite, diorite, gabbro, and basalt make up the rest. Natural silica sand is predominantly used as fine aggregate, even with most lightweight concretes. Natural mineral aggregates are derived from rocks of several types; most rocks are themselves composed of several minerals. A **mineral** is defined as a naturally occurring inorganic substance of more or less definite chemical composition and usually of a specific crystalline structure. An elementary review of aspects of rock formation and the classification of rocks and minerals is essential for understanding not only why some materials are more abundantly used as aggregates than others, but also the microstructure-property relations in aggregate.

DESCRIPTION OF ROCKS

Rocks are classified according to origin into three major groups: igneous, sedimentary, and metamorphic; these groups are further subdivided according to mineralogical and chemical composition, texture or grain size, and crystal structure.

Igneous rocks are formed on cooling of the magma (molten rock matter) above, below, or near the earth's surface. The degree of crystallinity and the grain size of igneous rocks, therefore, vary with the rate at which magma was cooled at the time of rock formation. It may be noted that grain size has a significant effect on the rock characteristics; rocks having the same chemical composition but different grain size may behave differently under the same condition of exposure.

Magma intruded at great depths cools at a slow rate and forms completely crystalline minerals with coarse grains (>5 mm grain size); rocks of this type are called *intrusive* or *plutonic*. However, due to a quicker cooling rate, the rocks formed near the surface of the earth contain minerals with smaller crystals, are fine-grained (1 to 5 mm grain size), and may contain some glass; they are called *shallow-intrusive* or *hypabyssal*. Rapidly cooled magma, as in the case of volcanic eruptions, contains mostly noncrystalline or glassy matter; the glass may be dense (quenched lava) or cellular (pumice), and the rock type is called *extrusive* or *volcanic*.

Also, a magma may be supersaturated, saturated, or undersaturated with respect to the amount of silica present for mineral formation. From an oversaturated magma, the free or uncombined silica crystallizes out as quartz after the formation of minerals such as feldspars, mica, and hornblende. In saturated or unsaturated magma, the silica content is insufficient to form quartz. This leads to a classification of igneous rocks based on the total SiO_2 present; rocks containing more than 65 percent SiO_2, 55 to 65 percent SiO_2, and less than 55 percent SiO_2 are called *acid*, *intermediate*, and *basic*, respectively. Again, the classifications of igneous rocks on the basis of crystal structure and silica content are useful because it appears that it is the combination of the acidic character and the fine-grained or glassy texture of the rock that determines whether an aggregate would be vulnerable to alkali attack in portland cement concrete.

Sedimentary rocks are stratified rocks that are usually laid down under water but are, at times, accumulated through wind and glacial action. The siliceous sedimentary rocks are derived from existing igneous rocks. Depending on the method of deposition and consolidation, it is convenient to subdivide them into three groups: (1) mechanically deposited either in an unconsolidated or physically consolidated state, (2) mechanically deposited and consolidated usually with chemical cements, and (3) chemically deposited and consolidated.

Gravel, sand, silt, and *clay* are the important members of the group of unconsolidated sediments. Although the distinction between these four members is made on the basis of particle size, a trend in the mineral composition is generally seen. Gravel and coarse sands usually consist of rock fragments; fine sands and silt consist predominately of mineral grains, and clays consist exclusively of mineral grains.

Sandstone, quartzite, and *graywacke* belong to the second category. Sandstones and quartzite consist of rock particles in the sand-size range; if the rock breaks around the sand grains, it is called *sandstone*; if the grains are largely quartz and the rock breaks through the grains, it is called *quartzite.* Quartzite may be sedimentary or metamorphic. The cementing or interstitial materials of sandstone may be opal (silica gel), calcite, dolomite, clay, or iron hydroxide. *Graywackes* are a special class of sandstones which contain angular and sand-size rock fragments in an abundant matrix of clay, shale, or slate.

Chert and *flint* belong to the third group of siliceous sedimentary rocks. Chert is usually fine-grained and can vary from porous to dense. Dense black or gray cherts, which are quite hard, are called flint. In regard to mineral composition, chert consists of poorly crystalline quartz, chalcedony, and opal; often all three are present.

Limestones are the most widespread of carbonate rocks. They range from pure limestone, consisting of the mineral calcite, to pure *dolomite,* consisting of the mineral dolomite. Usually, they contain both the carbonate minerals in various proportions and significant amounts of noncarbonate impurities, such as clay and sand.

It should be noted that compared to igneous rocks, the aggregates produced from stratified sediments can vary widely in characteristics, such as shape, texture, porosity, strength, and soundness. This is because the conditions under which they are consolidated vary widely. The rocks tend to be porous and weak when formed under relatively low pressures. They are dense and strong if formed under high pressure. Some limestones and sandstones may have less than 100 MPa crushing strength and are therefore unsuitable for use in high-strength concrete. Also, compared to igneous rocks, sedimentary rocks frequently contain impurities which at times jeopardize their use as aggregate. For instance, limestone, dolomite, and sandstone may contain opal or clay minerals, which adversely affect the behavior of aggregate under certain conditions of exposure.

Metamorphic rocks are igneous or sedimentary rocks that have changed their original texture, crystal structure, or mineralogical composition in response to physical and chemical conditions below the earth's surface. Common rock types belonging to this group are marble, schist, phyllite, and gneiss. The rocks are dense but frequently foliated. Some phyllites are reactive with the alkalies of portland cement.

Earth's crust consists of 95 percent igneous and 5 percent sedimentary rocks. Sedimentary rocks are composed of approximately 4 percent shale, 0.75 percent sandstone, and 0.25 percent limestone. Whereas igneous rocks crop out in only 25 percent of the earth's land area, sedimentary rocks cover 75 percent of the area. This is why most of the natural mineral aggregates used in concrete—sand, gravel, and crushed carbonate rocks—are derived from sedimentary rocks. Although some sedimentary deposits are up to 13 kilometers thick, over the continental areas the average is about 2300 m.

Description of Minerals

ASTM Standard C 294 contains the descriptive nomenclature which provides a basis for understanding the terms used to designate aggregate constituents. Based on this standard, a brief description of the constituent minerals that commonly occur in natural rocks is given below.

Silica minerals. *Quartz* is a very common hard mineral composed of crystalline SiO_2. The hardness of quartz as well as that of feldspar is due to the framework Si-O structure, which is very strong. Quartz is present in acidic-type igneous rocks (>65 percent SiO_2), such as granite and rhyolite. Due to its resistance to weathering, it is an important constituent of many sand and gravel deposits, and of sandstones. Tridymite and cristobalite are also crystalline silica minerals, but they are metastable at ordinary temperatures and pressures and are rarely found in nature except in volcanic rocks. Noncrystalline minerals are referred to as *glass*.

Opal is a hydrous (3 to 9 percent water) silica mineral which appears noncrystalline by optical microscopy but may show short-order crystalline arrangement by X-ray diffraction analysis. It is usually found in sedimentary rocks, especially in cherts, and is the principal constituent of diatomite. *Chalcedony* is a porous silica mineral, generally containing microscopic fibers of quartz. The properties of chalcedony are intermediate between those of opal and quartz.

Silicate minerals. Feldspars, ferromagnesium, micaceous, and clay minerals belong to this category. The minerals of the *feldspar group* are the most abundant rock-forming minerals in the earth's crust and are important constituents of igneous, sedimentary, and metamorphic rocks. Almost as hard as quartz, the various members of the group are differentiated by chemical composition and crystallographic properties.

Orthoclase, sanidine, and microcline are potassium aluminum silicates, which are frequently referred to as the *potash feldspars*. The *plagioclase* or soda-lime feldspars include sodium aluminum silicates (albite), calcium aluminum silicates (anorthite), or both. The alkali feldspars containing potassium or sodium occur typically in igneous rocks of high silica content, such as granites and rhyolites, whereas those of higher calcium content are found in igneous rocks of lower silica content, such as diorite, gabbro, and basalt.

Ferromagnesium minerals, which occur in many igneous and metamorphic rocks, consist of silicates of iron or magnesium or both. Minerals with the amphibole and pyroxene arrangements of crystal structure are referred to as hornblende and augite, respectively. Olivine is a common mineral of this class which occurs in igneous rocks of relatively low silica content.

Muscovite, biotite, chlorite, and vermiculite, which form the group of *micaceous minerals,* also consist of silicates of iron and magnesium, but their internal sheet structure arrangement is responsible for the tendency to split into thin flakes. The micas are abundant and occur in all three major rock groups.

The *clay mineral* group covers sheet-structure silicates less than 2 μm (0.002 mm) in size. The clay minerals, which consist mainly of hydrous aluminum, magnesium, and iron silicates, are major constituents of clays and shales. They are soft and disintegrate on wetting; some clays (know as montmorillonites in the United States and smectites in the United Kingdom) undergo large expansions on wetting. Clays and shales are therefore not directly used as concrete aggregates. However, clay minerals may be present as contaminants in a natural mineral aggregate.

Carbonate minerals. The most common carbonate mineral is *calcite* or calcium carbonate, $CaCO_3$. The other common mineral *dolomite* consists of equimolecular proportions of calcium carbonate and magnesium carbonate (corresponding to 54.27 and 45.73 weight percent $CaCO_3$ and $MgCO_3$, respectively). Both carbonate minerals are softer than quartz and feldspars.

Sulfide and sulfate minerals. The sulfides of iron (e.g., *pyrite, marcasite,* and *pyrrohotite*), are frequently found in natural aggregates. Marcasite, which is found mainly in sedimentary rocks, readily oxidizes to form sulfuric acid and hydroxides of iron. The formation of acid is undesirable, especially from the standpoint of the potential corrosion of steel in prestressed and reinforced concretes. Marcasite and certain forms of pyrite and pyrrohotite are suspected of being responsible for expansive volume changes in concrete, causing cracks and pop-outs.

Gypsum (hydrous calcium sulfate) and *anhydrite* (anhydrous calcium sulfate) are the most abundant sulfate minerals which may be present as impurities in carbonate rocks and shales. Sometimes found as coatings on sand and gravel, both gypsum and anhydrite, when present in aggregate, increase the chances of sulfate attack in concrete.

Since the largest amounts of concrete aggregates are derived from sedimentary and igneous rocks, descriptions of rock types in each class, principal minerals present, and characteristics of the aggregates are summarized in Tables 7–1 and 7–2, respectively.

LIGHTWEIGHT AGGREGATES

Aggregates that weigh less than 70 lb/ft^3 (1120 kg/m^3) are generally considered lightweight, and find application in the production of various types of lightweight concretes. The light weight is due to the cellular or highly porous microstructure.

TABLE 7-1 CHARACTERISTICS OF AGGREGATES FROM SEDIMENTARY ROCKS

Rock type	Common name	Principal minerals present	Aggregate characteristics
Siliceous rocks			
Mechanically deposited either in an unconsolidated or physically consolidated state.	Cobbles (>75 mm) Gravel (4.75–75 mm) Sand (0.075–4.75 mm) Silt (0.002–0.075 mm) Clay (>0.002 mm) Shale (consolidated clay)	All types of rocks and minerals may be present in cobbles, gravel, and sand. Silt consists predominately of grains of silica and silicate minerals. Clays are composed largely of a group of clay minerals.	Since natural cobbles, gravel, and sand are derived from geological weathering processes, they consist of hard rocks and minerals that have a rounded shape and a smooth surface. When uncontaminated with clay and silt, they make strong and durable aggregates for concrete. Shales may appear to be hard, but they give platy fragments and disintegrate in water.
Mechanically deposited and consolidated usually with chemical cements	Sandstone	Sand-size fragments consisting mainly of quartz and feldspar, usually cemented with opal, calcite, dolomite, clay, or iron hydroxide.	Generally, sandstones produce aggregates of satisfactory quality. Like carbonate rocks, the porosity, moisture absorption capacity, strength, and durability of sandstones can vary widely and will therefore affect properties of the aggregate.
	Graywacke	Graywackes are gray sandstones containing angular and sand-sized rock fragments in an abundant matrix of clay, shale, or slate.	
Chemically deposited and consolidated	Chert, flint	Chert consists of poorly crystalline quartz, chalcedony, or opal; often, all three are present. Flint is the name given to dense varieties of chert.	Dense cherts make good aggregates. Predominately opaline or chalcedonic cherts are capable of reacting with the alkalies in portland cement paste.

Lightweight Aggregates

TABLE 7-1 Continued

Rock type	Common name	Principal minerals present	Aggregate characteristics
Carbonate rocks	Limestone	Predominately calcite	Carbonate rocks are softer than siliceous sedimentary rocks. However, they generally produce aggregates of satisfactory quality. Like sandstone, the porosity, moisture absorption capacity, strength, and durability of carbonate rocks can vary widely and will therefore affect the properties of the aggregate. Being stratified rocks, they tend to produce flat or elongated fragments.
	Dolomite	Predominately dolomite	
	Dolomitic calcite	50–90% calcite; rest is dolomite	
	Calcitic dolomite	50–90% dolomite; rest is calcite	
	Arenaceous limestone (or dolomite)	Carbonate rocks containing 10–50% sand	
	Argillaceous limestone (or dolomite)	Carbonate rocks containing 10–50% clay	

TABLE 7-2 CHARACTERISTICS OF AGGREGATES FROM IGNEOUS ROCKS

Rock type	Common name	Principal minerals present	Aggregate characteristics
Intrusive or plutonic	Granite	Quartz, feldspar (O, P),[a] mica	The rocks of this group generally make excellent aggregate because: (1) They are medium- to coarse-grained, strong, and produce equidimensional fragments on crushing. (2) They have very low porosity and moisture absorption. (3) They do not react with alkalies in portland cement concrete.
	Syenite	Feldspar (O, P), hornblende, biotite	
	Diorite	Feldspar (P), hornblende, biotite	
	Gabbro	Hornblende, augite, feldspar (P)	
	Diabase or dolerite	Same minerals as in gabbro, but medium- to fine-grained	
	Trap rock	Gabbro, diabase, and basalt	
Shallow-intrusive or hypabyssal	Felsite group: rhyolite, trachyte, andesite	The mineral composition of the rocks of the felsite group— rhyolite, trachyte,	The rocks of this group are fine-grained and hard and make good aggregate except

TABLE 7–2 Continued

Rock type	Common name	Principal minerals present	Aggregate characteristics
	Basalt	and andesite—is equivalent of granite, syenite, and diorite, respectively. With respect to mineral composition, basalt is the shallow-intrusive or extrusive equivalent of gabbro and diabase.	that felsite rocks, when microcrystalline or containing natural glass, are reactive with the alkalies in portland cement concrete. However, in the case of basalt, even when it contains natural glass, the glass is generally basic and therefore nonreactive with alkalies in portland cement concrete.
Extrusive or volcanic	Obsidian	A dense, dark, natural glass of high silica content	Obsidian and pitchstone are dense and hard, but do not occur commonly.
	Pitchstone	Natural glass containing up to 10% water	
	Perlite	High-silica glass with onionlike texture and pearly luster; contains 2 to 5% water	Perlite is generally used for making insulating concretes after its structure is altered to a pumicelike vesicular structure by heat treatment.
	Pumice	Porous glass with elongated voids	Pumice, scoria, and tuffs are porous and weak, and are useful for making lightweight and insulating concretes.
	Scori	Porous glass with spherical voids	
	Tuff	Porous glass formed by consolidation of volcanic ash	

[a] The abbreviations O and P stand for orthoclase and plagioclase feldspar, respectively.

It may be noted that cellular organic materials such as wood chips should not be used as aggregate because of lack of durability in the moist alkaline environment in portland cement concrete.

Natural lightweight aggregates are made by processing igneous volcanic rocks such as pumice, scoria, or tuff. Synthetic lightweight aggregates can be manufactured by thermal treatment from a variety of materials, for instance, clays, shale, slate, diatomite, perlite, vermiculite, blast-furnace slag, and fly ash.

Heavyweight Aggregates

Actually, there is a whole spectrum of lightweight aggregates (Fig. 7–1) weighing from 5 to 55 lb/ft^3 (80 to 900 kg/m^3). Very porous aggregates, which are at the lighter end of the spectrum, are generally weak and are therefore more suitable for making nonstructural insulating concretes. At the other end of the spectrum are those lightweight aggregates that are relatively less porous; when the pore structure consists of uniformly distributed fine pores, the aggregate is usually strong and capable of producing structural concrete. ASTM has separate specifications covering lightweight aggregates for use in structural concrete (ASTM C 330), insulating concrete (ASTM C 332), and concrete for production of masonry units (ASTM C 331). These specifications contain requirements for grading, undesirable substances, and unit weight of aggregate, as well as for unit weight, strength, and drying shrinkage of concrete containing the aggregate. Properties of lightweight aggregate concrete are described in Chapter 11.

HEAVYWEIGHT AGGREGATES

Compared to normal-weight aggregate concrete, which typically has a unit weight of 150 lb/ft^3 (2400 kg/m^3), heavyweight concretes weigh from 180 to 380 lb/ft^3, and find application for nuclear radiation shields. Heavyweight aggregates (i.e., those which have higher density than normal-weight aggregates) are used for the production of heavy-weight concrete. Natural rocks suitable for making heavyweight aggregates consist predominately of two barium minerals, several iron ores, and a titanium ore (Table 7–3).

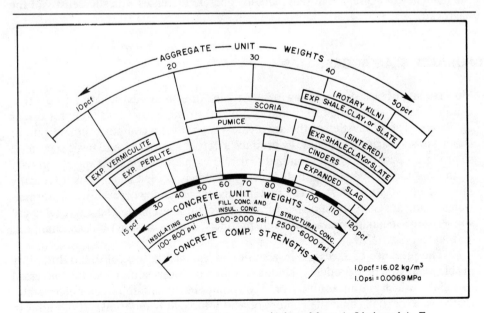

Figure 7–1 Lightweight aggregate spectrum. (Adapted from A. Litvin and A. E. Fiorato, *Concr. Int.*, Vol. 3, No. 3, p. 49, 1981.)

TABLE 7-3 COMPOSITION AND DENSITY OF HEAVYWEIGHT AGGREGATES

Type of aggregate	Chemical composition of principal mineral	Specific gravity of pure mineral	Typical bulk density [lb/ft^3 (kg/m^3)]
Witherite	$BaCO_3$	4.29	145 (2320)
Barite	$BaSO_4$	4.50	160 (2560)
Magnetite	Fe_3O_4	5.17	170 (2720)
Hematite	Fe_2O_3	4.9–5.3	190 (3040)
Lepidocrocite	Hydrous iron ores containing 8–12% water	3.4–4.0	140 (2240)
Geothite			
Limonite			
Ilmenite	$FeTiO_3$	4.72	160 (2560)
Ferrophosphorus	Fe_3P, Fe_2P, FeP	5.7–6.5	230 (3680)
Steel aggregate	Fe	7.8	280 (4480)

A synthetic product called ferrophosphorus can also be used as heavyweight aggregate. ASTM C 632 and C 637, which cover Standard Specifications and Descriptive Nomenclature, respectively, of aggregates for radiation-shielding concrete warn that ferrophosphorus aggregate when used in portland cement concrete will generate flammable and possibly toxic gases which can develop high pressures if confined. Hydrous iron ores and boron minerals and frits are at times included in the aggregates for making heavyweight concretes because boron and hydrogen are very effective in neutrons attenuation (capture). Steel punchings, sheared iron bars, and iron shots have also been investigated as heavyweight aggregates, but generally the tendency of aggregate to segregate in concrete increases with the density of the aggregate.

BLAST-FURNACE SLAG AGGREGATES

Slow cooling of blast-furnace slag in ladles, pits, or iron molds yields a product that can be crushed and graded to obtain dense and strong particles suitable for use as aggregate. The properties of the aggregate vary with the composition and rate of cooling of slag; acid slags generally produce a denser aggregate, and basic slags tend to produce vesicular or honeycombed structure with lower apparent specific gravity (2 to 2.8). On the whole the bulk density of slowly cooled slags, which typically ranges from 70 to 85 lb/ft^3 (1120 to 1360 kg/m^3), is somewhere between normal-weight natural aggregate and structural lightweight aggregate. The aggregates are widely used for making precast concrete products, such as masonry blocks, channels, and fence posts.

The presence of excessive iron sulfide in slag may cause color and durability problems in concrete products. Under certain conditions sulfide can be converted to sulfate, which is undesirable from the standpoint of sulfate attack on concrete. British specifications limit the content of acid soluble SO_3 and total sulfide sulfur in slag to 0.7 and 2 percent, respectively. It should be noted that blast-furnace slags have also been used for the production of lightweight aggregates meeting ASTM C 330 or C 331 requirements. For this purpose, molten slag is treated with limited amounts of water or steam, and the product is called *expanded* or *foamed slag*.

AGGREGATE FROM FLY ASH

Fly ash consists essentially of small spherical particles of aluminosilicate glass which is produced on combustion of pulverized coal in thermal power plants. Since large quantities of the ash remain unutilized in many industrialized parts of the world, attempts have been made to use the ash for making lightweight aggregates. In a typical manufacturing process fly ash is pelletized and then sintered in a rotary kiln, shaft kiln, or a traveling grate at temperatures in the range 1000 to 1200°C. Variations in fineness and carbon content of the fly ash are a major problem in controlling the quality of sintered fly-ash aggregate. Aggregate from fly ash is being commercially produced in the United Kingdom.

AGGREGATES FROM RECYCLED CONCRETE AND MUNICIPAL WASTES

Rubble from demolished concrete buildings yields fragments in which the aggregate is contaminated with hydrated cement paste, gypsum, and minor quantities of other substances. The size fraction that corresponds to fine aggregate contains mostly hydrated cement paste and gypsum, and it is unsuitable for making fresh concrete mixtures. However, the size fraction that corresponds to coarse aggregate, although coated with cement paste, has been used successfully in several laboratory and field studies. A review of several studies [1] indicates that compared with concrete containing a natural aggregate, the recycled-aggregate concrete would have at least two-thirds of the compressive strength and the modulus of elasticity, and satisfactory workability and durability (Table 7-4).

A major obstacle in the way of using rubble as aggregate for concrete is the cost of crushing, grading, dust control, and separation of undesirable constituents. Recycled concrete or waste concrete that has been crushed can be an economically feasible source of aggregate where good aggregates are scarce and when the cost of waste disposal is included in the economic analysis. From the largest concrete pavement recycling job ever undertaken, it was reported by the Michigan State Department of Transportation that recycling rubble from crushing an existing pavement was cheaper than using all new material (Fig. 7-2).

Investigations have also been made to evaluate municipal wastes and incinerator residues as possible sources for concrete aggregates. Glass, paper, metals, and

---

[1] S. A. Frondistou-Yannas, in *Progress in Concrete Technology,* ed. V. M. Malhotra, CANMET, Ottawa, 1980, pp. 639–84.

TABLE 7-4 COMPARSION OF PROPERTIES OF UNCONTAMINATED RECYCLED AGGREGATE CONCRETE AND NATURAL AGGREGATE CONCRETE OF SIMILAR COMPOSITION

Aggregate-mortar bond strength	
Aggregate primarily gravel from the old concrete	Comparable to that of control
Aggregate primarily mortar from the old concrete	55% that of control
Compressive strength	64–100% that of control
Static modulus of elasticity in compression	60–100% that of control
Flexural strength	80–100% that of control
Freeze-thaw resistance	Comparable to that of control
Linear coefficient of thermal expansion	Comparable to that of control
Length changes of concrete specimens stored for 28 days at 90% RH and 23°C	Comparable to that of control
Slump	Comparable to that of control

Source: Based on S. A. Frondistou-Yannas, in *Progress in Concrete Technology,* ed. V. M. Malhotra, CANMET, 1980, p. 672.

organic materials are major constituents of municipal wastes. The presence of crushed glass in aggregate tends to produce unworkable concrete mixtures and, due to the high alkali content, affects the long-term durability and strength. Metals such as aluminum react with alkaline solutions and cause excessive expansion. Paper and organic wastes, with or without incineration, cause setting and hardening problems in portland cement concrete. In general, therefore, municipal wastes are not suitable for making aggregates for use in structural concrete.

PRODUCTION OF AGGREGATES

Deposits of coarse-grained soil are a good source of **natural sand and gravel.** Since soil deposits usually contain varying quantities of silt and clay, which adversely affect the properties of both fresh and hardened concrete, these must be removed by washing or dry screening. The choice of the process between washing and dry screening of silt and clay will clearly influence the amount of deleterious substances in the aggregate; for instance, clay coatings may not be as efficiently removed by dry screening as by washing.

Generally, crushing equipment is a part of the aggregate production facilities because oversize gravel may be crushed and blended suitably with the uncrushed material of similar size. Again, the choice of crushing equipment may determine the shape of particles. With laminated sedimentary rocks, jaw and impact-type crushers

Figure 7-2 (a) Broken concrete ready for hauling to the crusher; (b) crushed concrete; (c) finishing the pavement made with concrete containing the recycled-concrete aggregate. (Photographs courtesy of Michigan State Department of Transportation.)

In 1983, Interstate 94, one of the oldest and most heavily traveled freeways in Michigan, became the first major freeway in the United States to recycle concrete. A 5.7-mile-long (9 km) deteriorated section of the concrete pavement was crushed, and then the crushed concrete was used as aggregate in the construction of the new pavement. The project, involving about 125,000 m^2 of 250 mm-thick pavement, was completed in 4 months at a cost of about $4.5 million. In 1984, about 22 miles (35 km) of 7.3 to 11 m-wide pavements on I-75 and I-94 were similarly recycled.

tend to produce flat particles. The importance of proper aggregate grading on cost of concrete is so well established now that modern aggregate plants, whether producing sand and gravel or crushed rock, have the necessary equipment to carry out operations involving crushing, cleaning, size separation, and combining two or more fractions to meet customer specifications. A photograph of a modern aggregate processing plant is shown in Fig. 7-3.

Synthetic lightweight aggregates such as expanded clays, shales, and slate are produced by heat treatment of suitable materials. Crushed and sized, or ground and pelletized raw material is exposed to temperatures generally of the order of 1000 to 1100°C, such that a portion of the material fuses to provide a viscous melt. Gases evolved as a result of chemical decomposition of some of the constituents present in raw materials are entrapped by the viscous melt, thus expanding the sintered mass. Generally, carbonaceous matter or carbonate minerals are the sources of these gases; alkalies and other impurities in clay or shale are responsible for the formation of melt at lower temperature. The heat treatment is generally carried out in a gas- or oil-fired rotary kiln similar to those used for the manufacture of portland cement. Many plants vacuum-saturate the product with moisture before delivery to the customer to facilitate better control on the consistency of fresh concrete.

Figure 7-3 Aerial view of a sand and gravel processing plant: a, scrubber; b, 1½-to 6-in. surge pile; c, clay and silt settling pond; d, crusher for oversize; e, sand pile; f, gravel piles; g, batching various gravel fractions for shipment; h, recreation area at the site of old quarry. (Photograph courtesy of Lone Star Industries, Inc., Pleasanton, California.)

Aggregate Characteristics and Their Significance

The properties of aggregate are greatly influenced by the external coating on the aggregate particles and the internal void distribution. Modern lightweight aggregate plants crush, grind, blend, and pelletize materials to obtain a uniform distribution of fine pores, which is necessary for producing materials possessing high crushing strength. Stable and impervious glassy coatings tend to reduce the moisture absorption capacity of the aggregate, which affects water demand and soundness.

AGGREGATE CHARACTERISTICS AND THEIR SIGNIFICANCE

A knowledge of certain aggregate characteristics (i.e., density, grading, and moisture state) is required for *proportioning concrete mixtures* (Chapter 9). Porosity or density, grading, shape, and surface texture determine the *properties of plastic concrete mixtures*. In addition to porosity, the mineralogical composition of aggregate affects its crushing strength, hardness, elastic modulus, and soundness, which in turn influence various properties of hardened concrete containing the aggregate. From a diagram illustrating the various interrelations (Fig. 7-4) it is evident that the

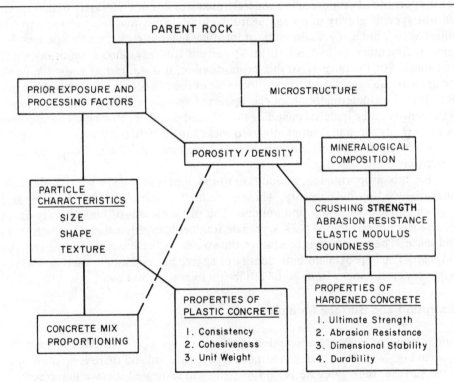

Figure 7-4 Diagram illustrating how microstructure, prior exposure, and processing factors determine aggregate characteristics that affect mix proportions and properties of fresh as well as hardened concrete.

aggregate characteristics significant to concrete technology are derived from microstructure of the material, prior exposure conditions, and processing factors.

Generally, aggregate properties are discussed in two parts on the basis of properties affecting (1) mix proportions and (2) the behavior of fresh and hardened concrete. Due to a considerable overlap between the two, it is more appropriate to divide the properties into the following groups, which are based on microstructural and processing factors:

1. **Characteristics dependent on porosity:** density, moisture absorption, strength, hardness, elastic modulus, and soundness
2. **Characteristics dependent on prior exposure and processing factors:** particle size, shape, and texture
3. **Characteristics dependent on chemical and mineralogical composition:** strength, hardness, elastic modulus, and deleterious substances present

Density and Apparent Specific Gravity

For the purpose of proportioning concrete mixtures it is not necessary to determine the true specific gravity of an aggregate. Natural aggregates are porous; porosity values up to 2 percent are common for intrusive igneous rocks, up to 5 percent for dense sedimentary rocks, and 10 to 40 percent for very porous sandstones and limestones. For the purpose of mix proportioning, it is desired to know the space occupied by the aggregate particles, inclusive of the pores existing within the particles. Therefore determination of the **apparent specific gravity,** which is defined as the density of the material including the internal pores, is sufficient. The apparent specific gravity for many commonly used rocks ranges between 2.6 and 2.7; typical values for granite, sandstone, and dense limestone are 2.69, 2.65, and 2.60, respectively.

For mix proportioning, in addition to the apparent specific gravity, data are usually needed on **bulk density,** which is defined as the weight of the aggregate fragments that would fill a unit volume. The phenomenon of bulk density arises because it is not possible to pack aggregate fragments together such that there is no void space. The term *bulk* is used since the volume is occupied by both *aggregates* and *voids*. The approximate bulk density of aggregates commonly used in normal-weight concrete ranges from 80 to 110 lb/ft^3 (1300 to 1750 kg/m^3).

Absorption and Surface Moisture

Various states of moisture absorption in which an aggregate particle can exist are shown in Fig. 7-5a. When all the permeable pores are full and there is no water film on the surface, the aggregate is said to be in the **saturated-surface dry condition (SSD);** when the aggregate is saturated and there is also free moisture on the surface, the aggregate is in the **wet** or **damp condition.** In the **oven-dry condition,** all the evaporable water has been driven off by heating to 100°C. **Absorption capacity** is

Aggregate Characteristics and Their Significance

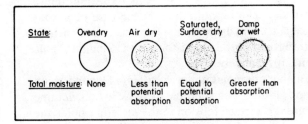

Moisture conditions of aggregates.

(a)

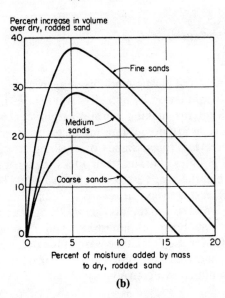

(b)

Figure 7–5 (a) Aggregate in various moisture states; (b) bulking due to moisture in fine aggregate. (From *Design and Control of Concrete Mixtures, 13th Edition,* Portland Cement Association, Skokie, Ill., 1988, pp. 36–37.)

The figure at the left illustrates how the concept of saturated-surface dry (SSD) condition is useful in determining potential absorption or free moisture in aggregate. The figure at the right shows that surface moisture on fine aggregate can cause considerable bulking, which varies with the amount of moisture present and the aggregate grading.

defined as the total amount of moisture required to bring an aggregate from the oven-dry to the SSD condition; **effective absorption** is defined as the amount of moisture required to bring an aggregate from the air-dry to the SSD condition. The amount of water in excess of the water required for the SSD condition is referred to as the **surface moisture.** The absorption capacity, effective absorption, and surface moisture data are invariably needed for correcting the batch water and aggregate proportions in concrete mixtures made from stock materials (Chapter 9). As a first approximation, the absorption capacity of an aggregate, which is easily determined, can be used as a measure of porosity and strength.

Normally, moisture correction values for intrusive igneous rocks and dense sedimentary rocks are very low, but they can be quite high in the case of porous sedimentary rocks, lightweight aggregates, and damp sands. For example, typically, the effective absorption values of trap rock, porous sandstone, and expanded shale aggregates are ½, 5, and 10 precent, respectively.

Damp sands may suffer from a phenomenon known as *bulking*. Depending on the amount of moisture and aggregate grading, considerable increase in bulk volume of sand can occur (Fig. 7–5b) because the surface tension in the moisture holds the particles apart. Since most sands are delivered at the job site in a damp condition, wide variations can occur in batch quantities if the batching is done by volume. For this reason, proportioning concrete mixture by weight has become the standard practice in most countries.

Crushing Strength, Abrasion Resistance, and Elastic Modulus

Crushing strength, abrasion resistance, and elastic modulus of aggregate are interrelated properties, which are greatly influenced by porosity. Aggregates from natural sources that are commonly used for making normal-weight concretes are generally dense and strong, therefore are seldom a limiting factor to strength and elastic properties of hardened concrete. Typical values of crushing strength and dynamic elastic modulus for most granites, basalts, trap rocks, flints, quartzitic sandstone, and dense limestones are in the range 30 to 45 $\times$ 10^3 psi (210–310 MPa) and 10 to 13 $\times$ 10^6 psi (70–90 GPa), respectively. In regard to sedimentary rocks, the porosity varies over a wide range, and so will their crushing strength and related characteristics. In one investigation involving 241 limestones and 79 sandstones, while the maximum crushing strengths for each rock type were of the order of 35,000 psi (240 MPa), some limestones and sandstones showed as low as 14,000 psi (96 MPa) and 7000 psi (48 MPa) crushing strengths, respectively.

Soundness

Aggregate is considered **unsound** when volume changes in the aggregate induced by weather, such as alternate cycles of wetting and drying, or freezing and thawing, result in deterioration of concrete. Unsoundness is shown generally by rocks having a certain characteristic pore structure. Concretes containing some cherts, shales, limestones, and sandstones have been found susceptible to damage by frost and salt crystallization within the aggregate. Although high moisture absorption is often used as an index for unsoundness, many aggregates, such as pumice and expanded clays, can absorb large amounts of water but remain sound. Unsoundness is therefore related to pore size distribution rather than to the total porosity of aggregate. Pore size distributions that allow aggregate particles to get saturated on wetting (or thawing in the case of frost attack), but prevent easy drainage on drying (or freezing), are capable of causing high hydraulic pressures within the aggregate. Soundness of aggregate to weathering action is determined by ASTM Method C 88, which describes a standard procedure for directly determining the resistance of aggregate to disintegration on exposure to five wet-dry cycles; saturated sodium or magnesium sulfate solution is used for the wetting cycle.

In the case of frost attack, in addition to pore size distribution and degree of saturation, as described in Chapter 5, there is a critical aggregate size below which high internal stresses capable of cracking the particle will not occur. For most aggregate, this critical aggregate size is greater than the normal size of coarse aggregates used in practice; however, for some poorly consolidated rocks (sandstones, limestones, cherts, shales), this size is reported to be in the range 12 to 25 mm.

Size and Grading

Grading is the distribution of particles of granular materials among various sizes, usually expressed in terms of cumulative percentages larger or smaller than each of a series of sizes of sieve openings, or the percentages between certain ranges of sieve openings. ASTM C 33 (Standard Specification for Concrete Aggregates) grading requirements for coarse and fine aggregates are shown in Tables 7–5 and 7–6, respectively.

There are several *reasons for specifying grading limits and maximum aggregate size,* the most important being their influence on workability and cost. For example, very coarse sands produce harsh and unworkable concrete mixtures, and very fine sands increase the water requirement (therefore, the cement requirement for a given water/cement ratio) and are uneconomical; aggregates that do not have a large deficiency or excess of any particular size produce the most workable and economical concrete mixtures.

The maximum size of aggregate is conventionally designated by the sieve size on which 15 percent or more particles are retained. In general, the larger the maximum aggregate size, the smaller will be the surface area per unit volume which has to be covered by the cement paste of a given water/cement ratio. Since the price of cement is usually about 10 times (in some cases even 20 times) as much as the price of aggregate, any action that saves cement without reducing the strength and workability of concrete can result in significant economic benefit. In addition to cost economy, there are other factors that govern the choice of maximum aggregate size for a concrete mixture. According to one rule of thumb used in the construction industry, the maximum aggregate size should not be larger than one-fifth of the narrowest dimension of the form in which the concrete is to be placed; also, it should not be larger than three-fourths of the maximum clear distance between the reinforcing bars. Since large particles tend to produce more microcracks in the transition zone between the coarse aggregate and the cement paste, in high-strength concretes the maximum aggregate size is generally limited to 19 mm.

The effect of a variety of sizes in reducing the total volume of voids between aggregates can be demonstrated by the method shown in Fig. 7–6a. One beaker is

TABLE 7-5 GRADING REQUIREMENTS FOR COARSE AGGREGATES

Size number	Nominal size (sieves with square openings)	Amounts finer than each laboratory sieve (square openings) (wt %)											
		100 mm (4 in.)	90 mm (3½ in.)	75 mm (3 in.)	63 mm (2½ in.)	50 mm (2 in.)	37.5 mm (1½ in.)	25.0 mm (1 in.)	19.0 mm (¾ in.)	12.5 mm (½ in.)	9.5 mm (⅜ in.)	4.75 mm (No. 4)	2.36 mm (No. 8)	1.18 mm (No. 16)
1	90–37.5 mm (3½–1½ in.)	100	90–100	—	25–60	—	0–15	—	0–5	—	—	—	—	—
2	63–37.5 mm (2½–1½ in.)	—	—	100	90–100	35–70	0–15	—	0–5	—	—	—	—	—
3	50–25.0 mm (2–1 in.)	—	—	—	100	90–100	35–70	0–15	—	0–5	—	—	—	—
357	50–4.75 mm (2 in.–No. 4)	—	—	—	100	95–100	—	35–70	—	10–30	—	0–5	—	—
4	37.5–19.0 mm (1½–¾ in.)	—	—	—	—	100	90–100	20–55	0–15	—	0–5	—	—	—
467	37.5–4.75 mm (1½ in.–No. 4)	—	—	—	—	100	95–100	—	35–70	—	10–30	0–5	—	—
5	25.0–12.5 mm (1–½ in.)	—	—	—	—	—	100	90–100	20–55	0–10	0–5	—	—	—
56	25.0–9.5 mm (1–⅜ in.)	—	—	—	—	—	100	90–100	40–85	10–40	0–15	0–5	—	—
57	25.0–4.75 mm (1 in.–No. 4)	—	—	—	—	—	100	95–100	—	25–60	—	0–10	0–5	—
6	19.0–9.5 mm (¾–⅜ in.)	—	—	—	—	—	—	100	90–100	20–55	0–15	0–5	—	—
67	19.0–4.75 mm (¾ in.–No. 4)	—	—	—	—	—	—	100	90–100	—	20–55	0–10	0–5	—
7	12.5–4.75 mm (½ in.–No. 4)	—	—	—	—	—	—	—	100	90–100	40–70	0–15	0–5	—
8	9.5–2.36 mm (⅜ in.–No. 8)	—	—	—	—	—	—	—	—	100	85–100	10–30	0–10	0–5

Source: Reprinted, with permission, from the 1991 Annual Book of ASTM Standards, Section 4, Vol. 04.02. Copyright, ASTM, 1916 Race Street, Philadelphia, PA 19103.

TABLE 7-6 GRADING REQUIREMENTS FOR FINE AGGREGATES

Sieve (Specification E 11)	Percent passing
9.5-mm (3/8-in.)	100
4.75-mm (No. 4)	95–100
2.36-mm (No. 8)	80–100
1.18-mm (No. 16)	50–85
600-μm (No. 30)	25–60
300-μm (No. 50)	10–30
150-μm (No. 100)	2–10

Source: Reprinted, with permission, from the 1991 Annual Book of ASTM Standards, Section 4, Vol. 04.02. Copyright, ASTM, 1916 Race Street, Philadelphia, PA 19103.

filled with 25-mm particles of relatively uniform size and shape; a second beaker is filled with a mixture of 25- and 9-mm particles. Below each beaker is a graduate cylinder holding the amount of water required to fill the voids in that beaker. It is evident that when two aggregate sizes are combined in one beaker, the void content is decreased. If particles of several more sizes smaller than 9 mm are added to the combination of 25 mm and 9 mm aggregate, a further reduction in voids will result. In practice, low void contents can be achieved by using smoothly graded coarse aggregates with suitable proportions of graded sand (Fig. 7–6b). The data show that as low as 21 percent void content was obtained when 40 percent sand was mixed with the 3/8- to 1½ in. (9- to 37-mm) gravel. From the standpoint of workability of concrete mixtures, it is now known that the smallest percentage of voids (greatest dry-rodded density) with given materials is not the most satisfactory; the optimum void content is somewhat more than the smallest possible.

In practice, an empirical factor called fineness modulus is often used as an index of the fineness of an aggregate. The **fineness modulus** is computed from screen analysis data by adding the cumulative percentages of aggregate retained on each of a specified series of sieves, and dividing the sum by 100. The sieves used for determining the fineness modulus are: No. 100 (150 μm), No. 50 (300 μm), No. 30 (600 μm), No. 16 (1.18 mm), No. 8 (2.36 mm), No. 4 (4.75 mm), 3/8 in. (9.5 mm), ¾ in. (19 mm), 1½ in. (37.5 mm), and larger—increasing in the ratio of 2 to 1. Examples of the method for determining the fineness modulus of fine aggregates from three different sources are shown by the tabulated data in Fig. 7–7, together with a typical grading curve. It may be noted that the higher the fineness modulus, the coarser the aggregate.

Shape and Surface Texture

The shape and surface texture of aggregate particles influence the properties of fresh concrete mixtures more than hardened concrete; compared to smooth and rounded particles, rough-textured, angular, and elongated particles require more cement paste to produce workable concrete mixtures, thus increasing the cost.

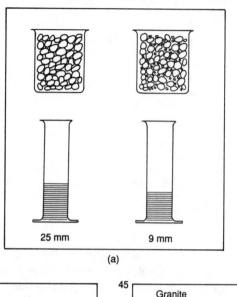

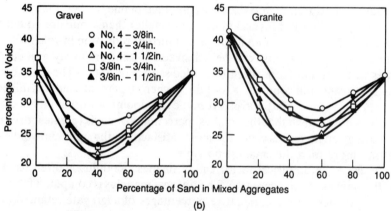

Figure 7–6 Reduction in the volume of voids on mixing fine and coarse aggregates. [(a), From *Design and Control of Concrete Mixtures, 13th Edition,* Portland Cement Association, Skokie, Ill., 1988, p. 33; (b), from S. Walker, Circular 8, National Sand and Gravel Association, 1930.]

Shape refers to geometrical characteristics such as round, angular, elongated, or flaky. Particles shaped by attrition tend to become **rounded** by losing edges and corners. Wind-blown sands, as well as sand and gravel from seashore or riverbeds, have a generally well-rounded shape. Crushed intrusive rocks possess well-defined edges and corners and are called **angular.** They generally produce equidimensional particles. Laminated limestones, sandstones, and shale tend to produce elongated and flaky fragments, especially when jaw crushers are used for processing. Those particles in which thickness is small relative to two other dimensions are referred to

UNIVERSITY OF CALIFORNIA
DEPARTMENT OF CIVIL ENGINEERING

ENGINEERING MATERIALS LABORATORY

AGGREGATE GRADING CALCULATIONS

Date	June 1, 1984			June 1, 1984			June 1, 1984		
Source	A (fine sand for blending)			B (Concrete Sand)			C (Concrete Sand)		
Sample wt.	455g			450g			456g		
Sieve Size	Weight Retained	Retained, %		Weight Retained	Retained, %		Weight Retained	Retained, %	
		Individual	Cumulative		Individual	Cumulative		Individual	Cumulative
No. 4	0	0	0	0	0	0	0	0	0
8	0	0	0	40.5	9.1	9	42.1	9.2	9
16	2.8	0.6	1	86.0	19.1	28	137.0	30.2	39
30	10.1	2.2	3	94.5	21.0	49	112.1	24.7	64
50	259.2	56.9	60	135.9	30.2	79	84.9	18.7	83
100	173.1	38.0	98	77.0	17.1	96	48.8	10.8	94
200	5.6	1.2	99	13.5	3.0	99	29.1	6.4	100
Pan	3.3	0.7	100	2.1	0.5	100	1.0	0.2	100
Total	454.1	F.M.	1.62	449.5	F.M.	2.61	455.0	F.M.	2.89

(a)

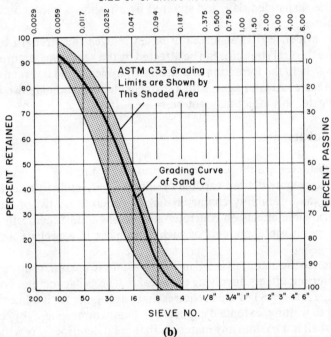

(b)

Figure 7–7 (a) Determination of fineness modulus from the sieve analysis data; (b) typical grading curve for sand with ASTM C 33 grading limits.

as **flat** or **flaky,** while those in which length is considerably larger than the other two dimensions are called **elongated.** Another term sometimes used to describe the shape of coarse aggregate is **sphericity,** which is defined as a ratio of surface area to volume. Spherical or well-rounded particles have low sphericity, but elongated and flaky particles possess high sphericity.

Photographs of particles of various shapes are shown in Fig. 2–3. Aggregates should be relatively free of flat and elongated particles. Elongated, blade-shaped aggregate particles should be avoided or limited to a maximum of 15 percent by weight of the total aggregate. This requirement is important not only for coarse aggregate but also for manufactured sands (made by crushing stone), which contain elongated grains and produce very harsh concrete.

The classification of **surface texture,** which is defined as the degree to which the aggregate surface is smooth or rough, is based on visual judgment. Surface texture of aggregate depends on the hardness, grain size, and porosity of the parent rock and its subsequent exposure to forces of attrition. Obsidian, flint, and dense slags show a smooth, glassy texture. Sand, gravels, and chert are smooth in their natural state. Crushed rocks such as granite, basalt, and limestone show a rough texture. Pumice, expanded slag, and sintered fly ash show a honeycombed texture with visible pores.

There is evidence that at least during early ages the strength of concrete, particularly the flexural strength, can be affected by the aggregate texture; a rougher texture seems to help the formation of a stronger physical bond between the cement paste and aggregate. At later ages, with a stronger chemical bond between the paste and the aggregate, this effect may not be so important.

Deleterious Substances

Deleterious substances are those which are present as minor constituents of either fine or coarse aggregate but are capable of adversely affecting workability, setting and hardening, and durability characteristics of concrete. A list of harmful substances, their possible effects on concrete, and ASTM C 33 Specification limits on the maximum permissible amounts of such substances in aggregate are shown in Table 7–7.

In addition to the materials listed in Table 7–7, there are other substances which may have deleterious effects, involving chemical reactions in concrete. For both fine and coarse aggregates, ASTM C 33 requires that "aggregate for use in concrete that will be subject to wetting, extended exposure to humid atmosphere, or contact with moist ground shall not contain any materials that are deleteriously reactive with the alkalies in the cement in an amount sufficient to cause excessive expansion except that if such materials are present in injurious amounts, the aggregate may be used with a cement containing less than 0.6 percent alkalies or with the addition of a material that has been shown to prevent harmful expansion due to the alkali-aggregate reaction." A description of th alkali-aggregate reaction, including a list of the reactive aggregates, is given in Chapter 5.

TABLE 7-7 LIMITS FOR DELETERIOUS SUBSTANCES IN CONCRETE AGGREGATES

Substance	Possible harmful effects on concrete	Maximum permitted (wt %)	
		Fine aggregate	Coarse aggregate[a]
Material finer than 75-μm (No. 200) sieve			
Concrete subject to abrasion	Affect workability; increase water requirement	3[b]	1
All other concrete		5[b]	
Clay lumps and friable particles	Affect workability and abrasion resistance	3	5
Coal and lignite			
Where surface appearance of concrete is important	Affect durability; cause staining	0.5	0.5
All other concrete		1.0	
Chert (less than 2.4 specific gravity)	Affect durability	—	5

[a] ASTM C 33 limits for deleterious substances in coarse aggregate vary with the conditions of exposure and type of concrete structure. The values shown here are for outdoor structures exposed to moderate weather conditions.

[b] In the case of manufactured sand, if the material finer than 75-μm sieve consists of the dust of fracture, essentially free of clay or shale, these limits may be increased to 5 and 7%, respectively.

Source: Reprinted, with permission, from the 1991 Annual Book of ASTM Standards, Section 4, Vol. 04.02. Copyright, ASTM, 1916 Race Street, Philadelphia, PA 19103.

Iron sulfides, especially marcasite, present as inclusions in certain aggregates have been found to cause an expansive reaction. In the lime-saturated environment of portland cement concrete, reactive iron sulfides can oxidize to form ferrous sulfate, which causes sulfate attack on concrete and corrosion of the embedded steel. Aggregates contaminated with gypsum or other soluble sulfates, such as magnesium, sodium, or potassium sulfate, also promote sulfate attack.

Recently,[2] cases of failure of concrete to set were reported from two block-making plants in southern Ireland. The problem was attributed to the presence of significant amounts of lead and zinc (mostly in the form of sulfides) in calcite aggregate. Those blocks had failed to set which contained 0.11 percent or more lead compound or 0.15 percent or more zinc compound by weight of concrete. Soluble lead or zinc salts are such powerful retarders of portland cement hydration that experimental concretes made with samples of contaminated aggregate failed to develop any strength after 3 days of curing. It should be noted that concrete setting and hardening problems can also be caused by organic impurities in aggregate, such as decayed vegetable matter that may be present in the form of organic loam or humus.

---

[2] *Concrete Construction*, Concrete Construction Publications, Inc., Vol. 22, No. 4, 1977, p. 237.

METHODS OF TESTING AGGREGATE CHARACTERISTICS

It is beyond the scope of this text to describe in detail the test methods used for evaluation of aggregate characteristics. For reference purposes, however, a list of ASTM test methods for determining various characteristics of aggregates, including the significance of tests, is given in Table 7–8.

TABLE 7–8 ASTM STANDARD TESTS FOR AGGREGATE CHARACTERISTICS

Characteristic	Significance	Test designation*	Requirement or item reported
Resistance to abrasion and degradation	Index of aggregate quality; wear resistance of floors, pavements	ASTM C 131 ASTM C 535 ASTM C 779	Maximum percentage of weight loss. Depth of wear and time
Resistance to freezing and thawing	Surface scaling, roughness, loss of section, and unsightliness	ASTM C 666 ASTM C 682	Maximum number of cycles or period of frost immunity; durability factor
Resistance to disintegration by sulfates	Soundness against weathering action	ASTM C 88	Weight loss, particles exhibiting distress
Particle shape and surface texture	Workability of fresh concrete	ASTM C 295 ASTM D 3398	Maximum percentage of flat and elongated pieces
Grading	Workability of fresh concrete; economy	ASTM C 117 ASTM C 136	Minimum and maximum percentage passing standard sieves
Bulk unit weight or bulk density	Mix design calculations; classification	ASTM C 29	Compact weight and loose weight
Specific gravity	Mix design calculations	ASTM C 127, fine aggregate ASTM C 128, coarse aggregate	—
Absorption and surface moisture	Control of concrete quality	ASTM C 70 ASTM C 127 ASTM C 128 ASTM C 566	—
Compressive and flexural strength	Acceptability of fine aggregate failing other tests	ASTM C 39 ASTM C 78	Strength to exceed 95% of strength achieved with purified sand

TABLE 7-8 Continued

Characteristic	Significance	Test designation*	Requirement or item reported
Definitions of constituents	Clear understanding and communication	ASTM C 125 ASTM C 294	—
Aggregate constituents	Determine amount of deleterious and organic materials	ASTM C 40 ASTM C 87 ASTM C 117 ASTM C 123 ASTM C 142 ASTM C 295	Maximum percentage of individual constituents
Resistance to alkali reactivity and volume change	Soundness against volume change	ASTM C 227 ASTM C 289 ASTM C 295 ASTM C 342 ASTM C 586	Maximum length change, constituents and amount of silica, and alkalinity

*The majority of the tests and characteristics listed are referenced in ASTM C 33.

TEST YOUR KNOWLEDGE

1. The aggregate in concrete is looked down upon as an "inert filler." Explain why this viewpoint is erroneous.
2. What is the distinction between the terms *rocks* and *minerals*? Write a short report on the influence of rock-forming process on characteristics of aggregates derived from the rock. In particular, explain why:
   (a) Basalts, which are generally fine-grained or glassy, are not alkali-reactive.
   (b) Limestones tend to form flat aggregate particles.
   (c) Pumice is useful for the production of lightweight aggregate.
3. What do you know about the following rocks and minerals: dolomite, graywacke, flint, opal, plagioclase, smectites, and marcasite?
4. Give typical ranges of aggregate unit weights for making structural lightweight, normal-weight, or heavyweight concretes. What types of natural and synthetic aggregates are used for making lightweight masonry blocks and insulating concrete; what types of natural minerals are useful for producing heavyweight aggregates?
5. You are the civil engineer in charge of rehabilitating some old concrete pavements in your area. In a brief note to your superiors, discuss the equipment needed, deleterious constituents to be avoided, and the cost economy of using the crushed concrete from old pavements as a source of aggregates for the construction of new pavements.
6. How are aggregates made from expandable clays, fly ash, and blast-furnace slag? What are some of the interesting characteristics of the products?
7. In concrete technology, what distinction is made between the terms *sp. gr.* and *bulk density*? With the help of suitable sketches, explain the following terms and discuss their significance: absorption capacity, saturated-surface-dry condition, damp condition.

8. What is the cause of the *bulking* phenomenon and what role does it play in concrete manufacturing practice?
9. List any three characteristics of concrete aggregate and discuss their influence on both the properties of fresh concrete and hardened concrete.
10. Briefly discuss the following propositions:
    (a) Specific gravity and absorption capacity are good indicators of aggregate quality.
    (b) The crushing strength of an aggregate can have a large influence on the compressive strength of concrete.
    (c) Using natural mineral aggregates, the unit weight of structural quality concrete can be varied between 1600 and 3200 kg/m^3 (2700 − 5400 lb/yd^3).
    (d) Pores in aggregate that are smaller than 4 microns can become **critically saturated.**
11. (a) If the dry-rodded unit weight of an aggregate is 105 lb/ft^3 (1680 kg/m^3) and its specific gravity is 2.65, determine the void content.
    (b) A sample of sand weighs 500 g and 480 g in "as received" and oven-dried condition, respectively. If the absorption capacity of the sand is 1%, calculate the percentage of free moisture.
    (c) From the following data on a sample of coarse aggregate, compute the bulk specific gravity on S.S.D. basis:
    Weight in S.S.D. condition = 5.00 lb
    Weight in oven-dry condition = 4.96 lb
    Weight under water = 3.15 lb
12. Define the terms *grading* and *maximum aggregate size,* as used in concrete technology. What considerations control the choice of the maximum aggregate size of aggregate in a concrete? Discuss the reasons why grading limits are specified.
13. Assuming that the workability of concrete is of no consequence, would sand be necessary in concrete mixtures? What is the significance of *fineness modulus?* Calculate the fineness modulus of sand with the following sieve analysis:
    Weight retained on No.   8 sieve, g = 30
    Weight retained on No. 16 sieve, g = 70
    Weight retained on No. 30 sieve, g = 125
    Weight retained on No. 50 sieve, g = 135
    Weight retained on No. 100 sieve, g = 120
    Weight retained on No. 200 sieve, g = 20
    Is this sand suitable for making concrete?
14. With the help of appropriate sketches, explain the terms *sphericity, flat particles,* and *angular particles.* Explain how the surface texture of fine aggregate may influence the properties of concrete.
15. When present as deleterious substances in concrete aggregates, how can the following materials affect properties of concrete: clay lumps, silt, zinc sulfide, gypsum, humus?

SUGGESTIONS FOR FURTHER STUDY

Report of ACI Committee 221, "Guide for Use of Normalweight Aggregates in Concrete," *ACI Manual of Concrete Practice, Part 1,* 1989.

ASTM, *Significance of Tests and Properties of Concrete and Concrete-Making Materials,* STP 169-A, 1966, pp. 381–512.

Suggestions for Further Study

ASTM, *Significance of Tests and Properties of Concrete and Concrete-Making Materials,* STP 169-B, 1978, pp. 539–761.

DOLAR-MANTUANI, L., *Handbook of Concrete Aggregates,* Noyes Publications, 1984.

ORCHARD, D. F., *Properties and Testing of Aggregates, Concrete Technology,* Vol. 3, John Wiley & Sons, Inc. (Halstead Press), New York, 1976.

POPOVICS, S., *Concrete-Making Materials,* McGraw-Hill Book Company, 1979, pp. 157–354.

CHAPTER 8

Admixtures

PREVIEW

The realization that properties of concrete, in both the fresh and hardened states, can be modified by adding certain materials to concrete mixtures is responsible for the large growth of the concrete admixtures industry during the last 40 years. Hundreds of products are being marketed today, and in some countries it is not uncommon that 70 to 80 percent of all the concrete produced contains one or more admixtures; therefore, it is quite important that civil engineers be familiar with the commonly used admixtures, together with their typical applications and limitations.

Admixtures vary in composition from surfactants and soluble salts and polymers to insoluble minerals. The purposes for which they are generally used in concrete include improvement of workability, acceleration or retardation of setting time, control of strength development, and enhancement of resistance to frost action, thermal cracking, alkali-aggregate expansion, and acidic and sulfate solutions. Important classes of concrete admixtures, their physical-chemical characteristics, mechanism of action, applications, and side effects are described in this chapter.

SIGNIFICANCE

ASTM C 125 defines an **admixture** as a material other than water, aggregates, hydraulic cements, and fiber reinforcement, used as an ingredient of concrete or mortar and added to the batch immediately before or during mixing. ACI Committee 212[1] lists 20 important purposes for which admixtures are used, for example, to increase the plasticity of concrete without increasing the water content, to reduce bleeding and segregation, to retard or accelerate the time of set, to accelerate the rates of strength development at early ages, to reduce the rate of heat evolution, and to increase the durability of concrete to specific exposure conditions. The realization that important properties of concrete, in both the freshly made and hardened states, can be modified to advantage by application of admixtures gave such an impetus to the admixture industry that within 20 years after the beginning of development of the industry in the 1940s, nearly 275 different products were marketed in England and 340 in Germany.[2] Today, most of the concrete produced in some countries contains one or more admixtures; it is reported that chemical admixtures are added to 88 percent concrete placed in Canada, 85 percent in Australia, and 71 percent in the United States.

NOMENCLATURE, SPECIFICATIONS, AND CLASSIFICATIONS

Admixtures vary widely in chemical composition and many perform more than one function; therefore, it is difficult to classify them according to their functions. The chemicals used as admixtures can broadly be divided into two types. Some chemicals begin to act on the cement-water system instantaneously by influencing the surface tension of water and by adsorbing on the surface of cement particles; others break up into their ionic constituents and affect the chemical reactions between cement compounds and water from several minutes to several hours after the addition. Finely ground insoluble materials, either from natural sources or by-products from some industries, are also used as admixtures. The physical effect of the presence of these admixtures on rheological behavior of fresh concrete becomes immediately apparent, but it takes several days to several months for the chemical effects to manifest.

The soluble salts and polymers, both surface-active agents and others, are added in concrete in very small amounts mainly for the purposes of entrainment of air, plasticization of fresh concrete mixtures, or control of setting time. By plasticizing concrete it is possible either to increase the consistency without increasing the water content, or to reduce the water content while maintaining a given consistency. Therefore, in the United States the plasticizing chemicals are called **water-reducing admixtures.**

The ASTM has separate specifications for air-entraining chemicals and for water-reducing and/or set-controlling chemicals. ASTM C 260, *Standard Specification for Air-Entraining Admixtures for Concrete,* sets limits on the effect that a given admixture under test may exert on bleeding, time of set, compressive and flexural strengths, drying shrinkage, and freeze-thaw resistance of concrete, compared to a reference air-entraining admixture. ASTM C 494, *Standard Specification for Chem-*

---

[1] ACI Committee 212, "Admixtures for Concrete," *Concr. Int.,* Vol. 3, No. 5, pp. 24–52, 1981.
[2] R. C. Mielenz, *Concr. Int.,* Vol. 6, No. 4, pp. 40–53, 1984.

ical Admixtures for Concrete, divides the water-reducing and/or set-controlling chemicals into the following seven types: **Type A**, water-reducing; **Type B**, retarding; **Type C**, accelerating; **Type D**, water-reducing and retarding; **Type E**, water-reducing and accelerating; **Type F**, high-range water-reducing; and **Type G**, high-range water-reducing and retarding. The distinction between the normal (Types A, D, and E) and the high-range water-reducing agents is that, compared to the reference concrete mixture of a given consistency, the former should be able to reduce at least 5 percent water and the latter at least 12 percent water. The specification also sets limits on time of set, compressive and flexural strengths, and drying shrinkage.

The *mineral admixtures* are usually added to concrete in large amounts. Besides cost reduction and enhancement of workability of fresh concrete, they can successfully be employed to improve the resistance of concrete to thermal cracking, alkali-aggregate expansion, and sulfate attack. Natural pozzolanic materials and industrial by-products such as fly ash and slag are the commonly used mineral admixtures. Again, the ASTM has separate classifications covering natural pozzolans and fly ashes on the one hand, and iron blast-furnace slag on the other. ASTM C 618, *Standard Specification for Fly Ash and Raw or Calcined Natural Pozzolan for Use as a Mineral Admixture in Portland Cement Concrete,* covers the following three classes of mineral admixtures: **Class N**, raw or calcined pozzolans such as diatomaceous earths, opaline cherts and shales, tuffs, and volcanic ashes or pumicite, and calcined materials such as clays and shales; **Class F**, fly ash normally produced from burning anthracite or bituminous coal; **Class C**, fly ash normally produced from burning lignite or subbituminous coal (in addition to being pozzolanic, this fly ash is also cementitious). The specification sets limits on fineness, water requirement, pozzolanic activity, soundness, and chemical constituents. Among others, a serious objection to ASTM C 618 is the arbitrary chemical requirements, which have no proven relation to the performance of the mineral admixtures in concrete.

ASTM C 989, *Standard Specification for Ground Iron Blast-Furnace Slag for Use in Concrete and Mortars,* is a new standard that is refreshingly free from cumbersome chemical and physical requirements. Unlike ASTM C 618, this standard takes the approach that when a material meets the performance specification, there should be no further need for prescriptive specifications. At the heart of ASTM C 989 is a slag activity test with portland cement which is intended to grade slags based on their contribution to the strength of a slag-portland cement mixture. According to the test method, slag activity is evaluated by determining the compressive strength of both portland cement mortars and corresponding mortars made with the same amount of a 50-50 weight combination of slag and portland cement. A strength index is calculated as the percentage strength between the slag-portland cement mortar and the portland cement mortar. The specification covers three strength grades of finely ground granulated iron blast-furnace slag for use as cementitious material: Grades 80, 100, and 120 correspond to a minimum strength index of 75, 95, and 115 percent in the 28-day cured mortars. Grades 100 and 120 also have a 7-day strength index requirement.

Since there is no single classification covering all the concrete admixtures, for the purpose of presenting a detailed description of their composition, mechanism

Surface-Active Chemicals

of action, and applications, they are grouped here into the following three categories: (1) surface-active chemicals, (2) set-controlling chemicals, and (3) mineral admixtures.

SURFACE-ACTIVE CHEMICALS

Nomenclature and Composition

Surface-active chemicals, also known as *surfactants*, cover admixtures that are generally used for air entrainment or reduction of water in concrete mixtures. An *air-entraining admixture* is defined as a material that is used as an ingredient of concrete for the purpose of entraining air; a *water-reducing admixture* is an admixture that reduces the quantity of mixing water required to produce concrete of a given consistency.

Surface-active admixtures consist essentially of long-chain organic molecules, one end of which is *hydrophilic* (water-attracting) and the other *hydrophobic* (water-repelling). The hydrophilic end contains one or more polar groups, such as —COO^-, —SO_3^-, or —NH_3^+. In concrete technology, mostly anionic admixtures are used either with a nonpolar chain or with a chain containing some polar groups. The former serve as air-entraining and the latter as water-reducing admixtures. As explained below, the surfactants become adsorbed at the air-water and the cement-water interfaces with an orientation of the molecule that determines whether the predominant effect is the entrainment of air or plasticization of the cement-water system.

Surfactants used as air-entraining admixtures generally consist of salts of wood resins, proteinaceous materials and petroleum acids, and some synthetic detergents. Surfactants used as plasticizing admixtures usually are salts, modifications, and derivatives of lignosulfonic acids, hydroxylated carboxylic acids, and polysaccharides, or any combinations of the foregoing three, with or without other subsidiary constituents. Superplasticizers or high range water-reducing admixtures, which are also discussed below, consist of sulfonated salts of melamine or naphthalene formaldehyde condensates.

Mechanism of Action

Air-entraining surfactants. The chemical formula of a typical air-entraining surfactant which consists of a nonpolar hydrocarbon chain with an anionic polar group is shown in Fig. 8–1a, and the mechanism of action is illustrated in Fig. 8–1b. Lea describes the mechanisms by which air is entrained and stabilized when a surfactant is added to the cement-water system:

> At the air-water interface the polar groups are oriented towards the water phase lowering the surface tension, promoting bubble formation and counteracting the tendency for the dispersed bubbles to coalesce. At the solid-water interface where directive

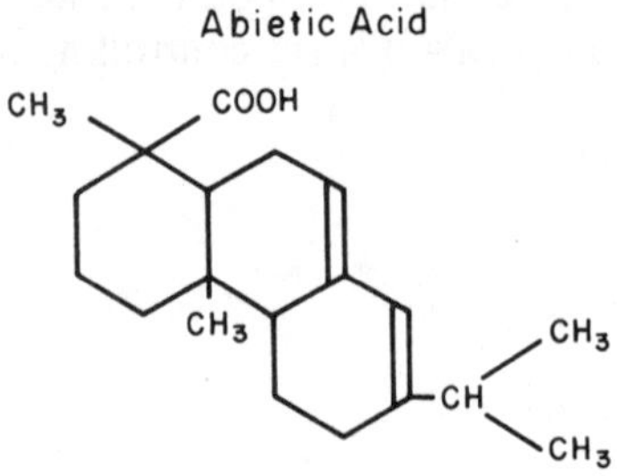

Figure 8–1 (a) Formula of a typical air-entraining surfactant derived from pine oil or tall oil processing; (b) mechanism of air entrainment when an anionic surfactant with a nonpolar hydro-carbon chain is added to the cement paste. (Adapted from P.C. Kreijger, in *Admixtures,* The Construction Press, 1980.)

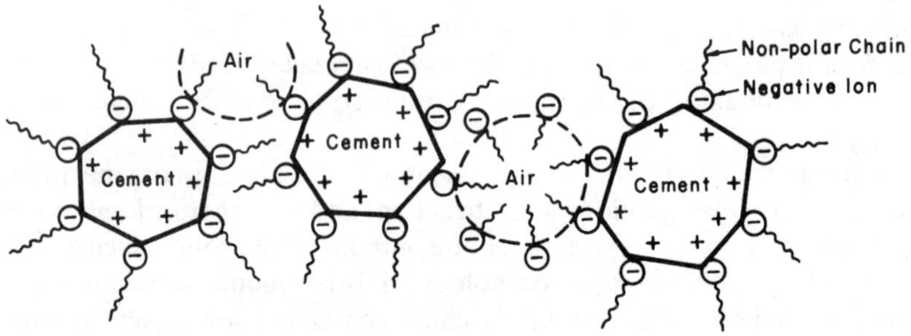

forces exist in the cement surface, the polar groups become bound to the solid with the nonpolar groups oriented towards the water, making the cement surface hydrophobic so that air can displace water and remain attached to the solid particles as bubbles.[3]

Water-reducing surfactants. The formulas of three typical plasticizing surfactants are shown in Fig. 8–2a. It may be noted that, in contrast to air-entraining surfactants, in the case of plasticizers the anionic polar group is joined to a hydrocarbon chain which itself is polar or hydrophilic (i.e., several OH groups are present in the chain). When a small quantity of water is added to the cement, without the presence of the surfactant a well-dispersed system is not attained because, first, the water possesses high surface tension (hydrogen-bonded molecular structure), and second, the cement particles tend to cluster together or form flocs (attractive force exists between positively and negatively charged edges, corners, and surfaces when crystalline minerals or compounds are finely ground). A diagram representing such a flocculated system is shown in Fig. 8–2c.

[3] F. M. Lea, *The Chemistry of Cement and Concrete,* Chemical Publishing Company, Inc., New York, 1971, p. 596.

Surface-Active Chemicals

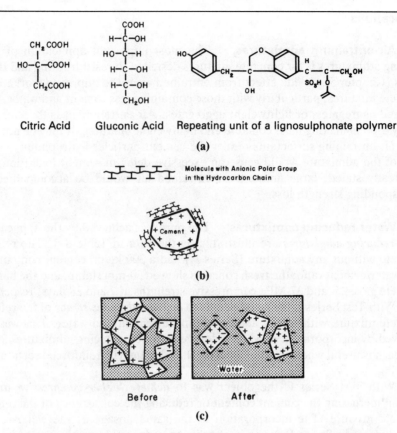

Figure 8–2 (a) Formulas of typical surfactants used as water-reducing admixtures. These are hydrocarbons containing anionic polar groups; (b) when a surfactant with several anionic polar groups in the hydrocarbon chain is added to the cement-water system, the polar chain gets absorbed on the surface of cement particle. Thus, not only the surface tension of water is lowered but also the cement particles are made hydrophilic; (c) diagrammatic representation of floc formation by cement particles before the addition of a water-reducing surfactant, and dispersion of flocs after the addition. (Adapted from P.C. Kreijger, in *Admixtures,* The Construction Press, 1980).

When a surfactant with a hydrophilic chain is added to the cement-water system, the polar chain is adsorbed alongside the cement particle; instead of directing a nonpolar end toward water, in this case the surfactant directs a polar end, thus lowering the surface tension of water and making the cement particle hydrophilic (Fig. 8–2b). As a result of layers of water dipoles surrounding the hydrophilic cement particles, their flocculation is prevented and a well-dispersed system is obtained (Fig. 8–2c).

Applications

Air-entraining admixtures. The most important application of air-entraining admixtures is for concrete mixtures designed to resist freezing and thawing cycles (Chapter 5). A side effect from entrained air is the improved workability of concrete mixtures, particularly with those containing less cement and water, rough-textured aggregates, or lightweight aggregates. Air entrainment is therefore commonly used in making mass concrete and lightweight mixtures. It may be noted that since air-entraining surfactants render the cement particles hydrophobic, an overdose of the admixture would cause an excessive delay in cement hydration. Also, as already stated, large amounts of entrained air would be accompanied by a corresponding strength loss.

Water-reducing admixtures. The purposes achieved by the application of *water-reducing admixtures* are illustrated by the data in Table 8–1. The reference concrete without any admixture (Series A) had a 300 kg/m^3 cement content and a 0.62 water/cement ratio; the fresh concrete showed 50-mm slump, and the hardened concrete gave 25- and 37-MPa compressive strengths at 7 and 28 days, respectively.

With Test Series B the purpose was to *increase the consistency* of the reference concrete mixture without the addition of more cement and water. This was easily achieved by incorporation of a given dosage of water-reducing admixture. Such an approach is useful when concrete is to be placed in heavily reinforced sections or by pumping.

With Test Series C the object was to *achieve higher compressive strengths* without increasing the cement content or reducing the consistency of the reference concrete mixture. The incorporation of the same dosage of the water-reducing admixture as in Series B made it possible to reduce the water content by about 10 percent (from 186 to 168 kg/m^3) while maintaining the 50-mm slump. As a result

TABLE 8–1 BENEFITS ACHIEVED BY USING WATER-REDUCING ADMIXTURES

Test series	Cement content (kg/m^3)	Water/ cement ratio	Slump (mm)	Compressive strength (MPa)	
				7 days	28 days
Ⓐ Reference concrete (no admixture)	300	0.62	50	25	37
A given dosage of a water-reducing admixture is added with the purpose of:					
Ⓑ Consistency increase	300	0.62	100	26	38
Ⓒ Strength increase	300	0.56	50	34	46
Ⓓ Cement saving	270	0.62	50	25.5	37.5

Source: Based on P. C. Hewlett, in *Concrete Admixtures: Use and Applications,* ed. M. R. Rixom, The Construction Press, London, 1978, p. 16. By permission of Longman.

of reduction in the water/cement ratio, the 7-day compressive strength increased from 25 to 34 MPa and the 28-day strength from 37- to 46-MPa. This approach may be needed when job specifications limit the maximum water/cement ratio but require high early strength to develop.

Test Series D demonstrates how the addition of the water-reducing admixture made it possible to effect a 10 percent *cement saving* without compromising either the consistency or the strength of the reference concrete mixture. Besides cost economy, such cement savings may be important when reduction of temperature rise in mass concrete is the primary goal.

It should be noted from the foregoing description of benefits resulting from the application of water-reducing admixtures that *all three benefits were not available at the same time*.

The *period of effectiveness of surfactants* is rather limited because soon after the onset of hydration reactions between portland cement compounds and water, large amounts of products such as ettringite begin to form. The cement hydration products engulf the small quantity of the surfactant present in the system. Thus the ambient temperature and cement fineness and composition, especially the C_3A, SO_3, and alkali contents, which control the rate of ettringite formation, may have an effect on cement-admixture interactions. Obviously, the *amount or concentration of the admixture* in the system will also determine the effect. Larger amounts of an admixture than normally needed for the plasticizing or water-reducing effect may retard the time of set by preventing the hydration products to flocculate (form bonds). Thus, depending on the dosage, most surfactants can serve simultaneously as water reducers and set retarders. Industrial water reducers may contain accelerating agents to offset the retarding tendency when this is unwanted. It seems that except for possible retardation of the time of set, other mechanical properties of concrete are not affected by the presence of water-reducing agents; in fact, early strengths may be somewhat accelerated due to better dispersion of the cement particles in water.

Also, some water-reducing admixtures, especially those derived from lignin products, are known to entrain considerable air; to overcome this problem commercial lignin-based admixtures usually contain air-detraining agents. Since commercial products may contain many unknown ingredients, it is always desirable to make a laboratory investigation before using a new admixture or a combination of two or more admixtures.

Superplasticizers

Superplasticizers, also called high range water-reducing admixtures because they are able to reduce three to four times water in a given concrete mixture compared to normal water-reducing admixtures, were developed in the 1970s and have already found a wide acceptance in the concrete construction industry. They consist of long-chain, high-molecular-weight (20,000 to 30,000) anionic surfactants with a large number of polar groups in the hydrocarbon chain. When adsorbed on cement particles, the surfactant imparts a strong negative charge, which helps to lower the surface tension of the surrounding water considerably and greatly enhances the fluidity of the system.

Compared to normal water-reducing admixtures, relatively large amounts of superplasticizers, up to 1% by weight of cement, can be incorporated into concrete mixtures without causing excessive bleeding and set retardation, in spite of a consistency on the order of 200- to 250-mm slump. It is probably the colloidal size of the long-chain particles of the admixture which obstructs the bleed-water flow channels in concrete, so that segregation is generally not encountered in superplasticized concretes. Excellent dispersion of the cement particles in water (Fig. 8–3) seems to accelerate the rate of hydration; therefore, retardation is rarely observed; instead, acceleration of setting and hardening is a common occurrence. In fact, the first generation of superplasticizers acquired a bad reputation for rapid loss of consistency or slump. Currently available products often contain lignosulfonate or other retarding materials in order to offset the rapid loss of high consistency achieved immediately after the addition of the admixture.

Compared to 5 to 10 percent water reduction made possible by the application of ordinary plasticizing admixtures, water reductions in the range 20 to 25 percent can often be achieved in reference concrete without reducing the consistency. The increase in mechanical properties (i.e., compressive and flexural strength) is generally commensurate with reduction in the water/cement ratio. Frequently, due to a greater rate of cement hydration in the well-dispersed system, concretes containing superplasticizers show even higher compressive strengths at 1, 3, and 7 days than reference concretes having the same water/cement ratio (Table 8–2). This is of special importance in the precast concrete industry, where high early strengths are required for faster turnover of the formwork. By using higher cement contents and water/

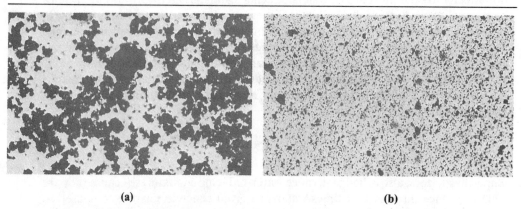

(a)   (b)

Figure 8–3 (a) Photomicrograph of flocculated cement particles in a portland cement-water suspension with no admixture present; (b) photomicrograph of the system after it is dispersed with the addition of a superplasticizing admixture. (Photographs courtesy of M. Collepardi, University of Anacona, Italy.)

TABLE 8–2 EXAMPLES OF HIGH EARLY STRENGTHS MADE POSSIBLE BY THE USE OF SUPERPLASTICIZING ADMIXTURES

Test	Cement content (kg/m^3)	Water/cement ratio	Slump (mm)	Compressive strength (MPa)			
				1-day	3-day	7-day	28-day
Ⓐ Reference concrete (no admixture)	360	0.60	225	10	21	32	45
Ⓑ Concrete of same consistency as Ⓐ but containing less water and 2% superplasticizer by weight of cement	360	0.45	225	20	35	43	55
Ⓒ Concrete of same water/cement ratio as Ⓑ but containing no superplasticizer, and having lower slump	360	0.45	30	16	28	37	52

cement ratios much lower than 0.45, it is possible to achieve even more rapid strength development rates. A description of properties and some recent applications of superplasticized concrete are given in Chapter 11.

SET-CONTROLLING CHEMICALS

Nomenclature and Composition

Besides certain types of surfactants already described, there are a large number of chemicals that can be used as retarding admixtures; on the other hand, there are chemicals that can accelerate the setting time and rate of strength development at early ages. Interestingly, some chemicals act as retarders when used in small amounts (e.g., 0.3 percent by weight of cement), but in large dosage (e.g., 1 percent by weight of cement) they behave as accelerators.

Forsen[4] was first in presenting a comprehensive analysis of the action of chemical admixtures on the setting of portland cement. He divided retarders into several groups according to the type of curve obtained when initial setting time was plotted against the concentration of the retarder in the system. A modified version of Forsen's classification which covers both retarders and accelerators is shown in Fig. 8–4. The compositions of commonly used chemicals under each class are also shown in the figure.

[4] L. Forsen, *Proc. Int. Symp. on Chemistry of Cements,* Stockholm, 1983, p. 298.

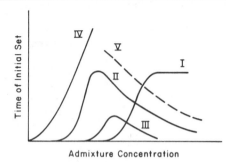

Figure 8–4 Classification and composition of set-controlling chemicals: Class I: $CaSO_4 \cdot 2H_2O$; Class II: $CaCl_2$, $Ca(NO_3)_2$; Class III: K_2CO_3, $NaCO_3$, $NaSiO_3$; Class IV: (1) surfactants with polar groups in the hydrocarbon chain (i.e., gluconates, lignosulfonates, and sugars), (2) sodium salts of phosphoric, boric, oxalic, or hydrofluoric acid, (3) zinc or lead salts; Class V: salts of formic acid and triethanolamine.

Mechanism of Action

It is generally accepted now that at least the early reactions of portland cement compounds with water are through-solution; that is, the compounds first ionize and then the hydrates form in solution. Due to their limited solubility the hydration products crystallize out; the stiffening, setting, and hardening phenomena with portland cement pastes are derived from the progressive crystallization of the hydration products. It is therefore reasonable to assume that by adding certain soluble chemicals to the portland cement-water system, one may be able to influence either the rate of ionization of cement compounds or the rate of crystallization of the hydration products, consequently affecting the setting and hardening characteristics of the cement paste.

According to Joisel[5] the action of set-controlling chemicals on portland cement can be attributed mainly to dissolving of the anhydrous constituents, rather than the crystallization of the hydrates. To understand the mechanism of acceleration or retardation it is helpful to consider a hydrating portland cement paste as being composed of certain anions (silicate and aluminate) and cations (calcium), the solubility of each being dependent on the type and concentration of the acid and base ions present in the solution. Since most chemical admixtures will ionize in water, by adding them to the cement–water system it is possible to alter the type and concentration of the ionic constituents in the solution phase, thus influencing the dissolution of the cement compounds according to the following guidelines proposed by Joisel:

1. An accelerating admixture must promote the dissolution of the cations (calcium ions) and anions from the cement. Since there are several anions to dissolve, the accelerator should promote the dissolving of that constituent

---

[5] A. Joisel, *Admixtures for Cement,* published by the author, 3 Avenue André, 95230 Soisy, France, 1973.

which has the lowest dissolving rate during the early hydration period (e.g., silicate ions).

2. A retarding admixture must impede the dissolution of the cement cations (calcium ions) and anions, preferably that anion which has the highest dissolving rate during the early hydration period (e.g., aluminate ions).
3. The presence of monovalent cations in solution (i.e., K^+ or Na^+) reduces the solubility of Ca^{2+} ions but tends to promote the solubility of silicate and aluminate ions. In small concentrations, the former effect is dominant; in large concentrations, the latter effect becomes dominant.
4. The presence of certain monovalent anions in solution (i.e., Cl^-, NO_3^-, or SO_4^{2-}) reduces the solubility of silicates and aluminates but tends to promote the solubility of calcium ions. In small concentrations, the former effect is dominant; in large concentrations, the latter effect becomes dominant.

From the foregoing it can be concluded that the overall outcome when a chemical admixture is added to a portland cement-water system will be determined by a number of both complementary and opposing effects, which are dependent on the type and concentration of ions contributed by the admixture to the system. With small concentrations (e.g., 0.1 to 0.3 percent by weight of cement) of the salts of weak bases and strong acids (e.g., $CaCl_2$), or strong bases and weak acids (e.g., K_2CO_3), the retardation of solubility of the calcium and the aluminate ions from the cement is the dominant effect rather than the acceleration of solubility of the silicate ions; hence the overall effect is that of retardation. With larger concentrations (e.g., 1 percent or more) of these salts, the accelerating effects of the ions in solution on the silicate and the aluminate ions of the cement become more dominant than the retarding effects; thus it is possible for the same salt to change its role and become an accelerator instead of a retarder (Fig. 8–4). It should be noted that $CaCl_2$, 1 to 3 percent by weight of cement, is the most commonly used accelerator for plain concrete.

Gypsum ($CaSO_4 \cdot 2H_2O$) is a salt of a weak base and a strong acid, but it does not show the retarder-to-accelerator role-reversal phenomenon on the setting time of cement when the amount of gypsum added to a portland cement paste is gradually increased. This is because the solubility of gypsum in water is low (2 g/liter at 20°C). Until the sulfate ions from gypsum go into solution, they will not be able to accelerate the solubility of calcium from the cement compounds. With the gradual removal of sulfate ions from the solution due to the crystallization of calcium sulfoaluminate hydrates (mostly ettringite), more gypsum goes into solution; this has a beneficial effect on the C_3S hydration and therefore on the rate of strength development. However, instead of gypsum, if a large amount of sulfate is introduced in a highly soluble form (e.g., plaster of paris or hemihydrate), both the setting time and early strengths will be accelerated.

It is expected that low molecular-weight organic acids and their soluble salts, which are weakly acidic, would serve as accelerators because of their ability to promote the solution of calcium ions from the cement compounds. Actually, calcium formate and formic acid, HCOOH, are accelerators, but other acids with long hydrocarbon chains generally act as retarders by counteracting bond formations among the hydration products. Triethanolamine, $N(CH_2 - CH_2OH)_3$, is another organic chemical, which in small amounts (0.1 to 0.5 percent) is used as an accelerating ingredient of some water-reducing admixtures because of its ability to accelerate the hydration of C_3A and the formation of ettringite. However, triethanolamine tends to retard the C_3S hydration and therefore reduces the rate of strength development. Both organic accelerators play an important part in prestressed and reinforced concrete applications, where the use of accelerating admixtures containing chloride is considered to be undesirable.

Chemical admixtures listed under Class IV (Fig. 8-4) act as powerful retarders by mechanisms other than those discussed above. Surfactants, such as gluconates and lignosulfonates, act as retarders by delaying the bond formation among the hydration products; others reduce the solubility of the anhydrous constituents from cement by forming insoluble and impermeable products around the particles. Sodium salts of phosphoric, boric, oxalic, and hydrofluoric acid are soluble, but the calcium salts are highly insoluble and therefore readily form in the vicinity of hydrating cement particles. Once insoluble and dense coatings are formed around the cement grains, further hydration slows down considerably. Phosphates are commonly found to be present as ingredients of commercial set-retarding admixtures.

Applications

Accelerating admixtures. According to the report by ACI Committee 212:

> Accelerating admixtures are useful for modifying the properties of portland cement concrete, particularly in cold weather, to: (a) expedite the start of finishing operations and, when necessary, the application of insulation for protection; (b) reduce the time required for proper curing and protection; (c) increase the rate of early strength development so as to permit earlier removal of forms and earlier opening of the construction for service; and (d) permit more efficient plugging of leaks against hydraulic pressures.[6]

Since calcium chloride is by far the best known and most widely used accelerator, the effects of $CaCl_2 \cdot 2H_2O$ additions in amounts from 0.5 to 2.0 percent by weight of cement on setting times and relative compressive strengths are shown in Fig. 8-5. Properties of concrete as affected by the use of calcium chloride are summarized in Table 8-3.

Retarding admixtures. According to ACI Committee 212, the following applications of retardation of setting are important in the construction practice.

---

[6] ACI Committee 212, *Concr. Int.*, Vol. 3, No. 5, pp. 24-25, 1981.

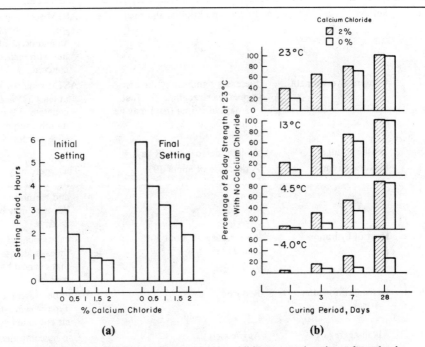

Figure 8–5 (a) Effect of calcium chloride addition on setting time of portland cement; (b) effect of calcium chloride addition on strength at various curing temperatures. [From V.S. Ramachandran, in *Progress in Concrete Technology,* ed. V.M. Malhotra, CANMET, Ottawa, 1980, pp. 421–50.]

1. Compensation for adverse ambient temperature conditions particularly in hot weather. Extensive use is made of retarding admixtures to permit proper placement and finishing and to overcome damaging and accelerating effects of high temperatures.
2. Control of setting of large structural units to keep concrete workable throughout the entire placing period. This is particularly important for the elimination of cold joints and discontinuities in large structural units. Also control of setting may prevent cracking of concrete beams, bridge decks, and composite construction due to form deflection of movement associated with placing of adjacent units. Adjustments of the dosage as placement proceeds can permit various portions of a unit, a large post-tensioned beam for example, to attain a given level of early strength at approximately the same time.

Effect of addition of a retarding and water-reducing admixture (Type D, ASTM C 494) on setting time and strength of concrete, at normal temperature, is shown by the data in Table 8–4. The effectiveness of the admixture in maintaining concrete in a workable condition for longer periods, even at a higher than normal ambient temperature, is illustrated by the data in Table 8–5.

TABLE 8-3 SOME OF THE PROPERTIES INFLUENCED BY THE USE OF CALCIUM CHLORIDE ADMIXTURE IN CONCRETE

No.	Property	General effect	Remarks
1	Setting	Reduces both initial and final setting	ASTM standard requires that initial and final setting times should occur at least 1 hr earlier with respect to reference concrete.
2	Compressive strength	Increases significantly the compressive strength in the first 3 days of curing (gain may be about 30–100%)	ASTM requires an increase of at least 125% over control concrete at 3 days. At 6–12 months, requirement is only 90% of control specimen.
3	Tensile strength	A slight decrease at 28 days	
4	Flexural strength	A decrease of about 10% at 7 days	This figure may vary depending on the starting materials and method of curing. The decrease may be more at 28 days.
5	Heat of hydration	An increase of about 30% in 24 hr	Total amount of heat at longer times is almost the same as that evolved by reference concrete.
6	Resistance to sulfate attack	Reduced	Can be overcome by use of Type V cement with adequate air entrainment.
7	Alkali-aggregate reaction	Aggravated	Can be controlled by use of low-alkali cement or pozzolan.
8	Corrosion	Causes no problems in normal reinforced concrete if adequate precautions taken. Dosage should not exceed 1.5% $CaCl_2$ and adequate cover to be given. Should not be used in concrete containing a combination of dissimilar metals or where there is a possibility of stray currents	Calcium chloride admixture should not be used in prestressed concrete or in a concrete containing a combination of dissimilar metals. Some specifications do not allow use of $CaCl_2$ in reinforced concretes.
9	Shrinkage and creep	Increased	
10	Volume change	Increase of 0–15% reported	
11	Resistance to damage by freezing and thawing	Early resistance improved	At later ages may be less resistant to frost attack
12	Watertightness	Improved at early ages	
13	Modulus of elasticity	Increased at early ages	At longer periods almost same with respect to reference concrete.
14	Bleeding	Reduced	

Source: V. S. Ramachandran, in *Progress in Concrete Technology,* ed. V. M. Malhotra, CANMET, Ottawa, 1980, pp. 421–50.

Mineral Admixtures

TABLE 8-4 EFFECT OF APPLICATION OF ASTM TYPE D ADMIXTURE[a] ON SETTING TIME AND STRENGTH

Admixture dosage by weight of cement (liters/kg)	Setting time (hr) (ASTM C 403)		Water/cement ratio	Compressive strength (MPa)		
	Initial	Final		3-day	7-day	28-day
0	4.5	9	0.68	20.3	28.0	37.0
0.14	8	13	0.61	28.0	36.5	46.8
0.21	11.5	16	0.58	29.6	40.1	49.7

[a] According to ASTM C 494, Type D admixtures are both retarding and water reducing.

Source: Based on P. C. Hewlett, in *Concrete Admixtures: Use and Applications,* ed. M. R. Rixom, The Construction Press, London, 1978, p. 18. By permission of Longman.

TABLE 8-5 EFFECT OF APPLICATION OF ASTM TYPE D ADMIXTURE ON CONSISTENCY AT HIGH AMBIENT TEMPERATURES

Test	Ambient temperature (°C)	Concrete slump (mm)					
		0	1 hr	2 hr	3 hr	4 hr	5 hr
Control concrete (no admixture)	20	127	89	76	57	38	32
Concrete with admixture	20	127	127	114	102	70	57
Control concrete (no admixture)	43	114	57	7	0	0	0
Concrete with admixture	43	127	70	25	19	13	0

Source: Based on P. C. Hewlett, in *Concrete Admixtures: Use and Applications,* ed. M. R. Rixom, The Construction Press, London, 1978, p. 18. By permission of Longman.

MINERAL ADMIXTURES

Significance

Mineral admixtures are finely divided siliceous materials which are added to concrete in relatively large amounts, generally in the range 20 to 100 percent by weight of portland cement. Although pozzolans in the raw state or after thermal activation are being used in some parts of the world, for economic reasons many industrial by-products are fast becoming the primary source of mineral admixtures in concrete.

Power generation units[7] using coal as fuel, and metallurgical furnaces producing cast iron, silicon metal, and ferrosilicon alloys, are the major source of by-products, which are being produced at the rate of millions of tons every year in many

---
[7] Power generation furnaces using rice husks as fuel have been developed; under controlled combustion conditions they produce a highly pozzolanic ash.

industrial countries. Dumping away these by-products represents a waste of the material and causes serious environmental pollution problems. Disposal as aggregates for concrete and roadbase construction is a low-value use which does not utilize the potential of these pozzolanic and cementitious materials. With proper quality control, large amounts of many industrial by-products can be incorporated into concrete, either in the form of blended portland cement or as mineral admixtures. When the pozzolanic and/or cementitious properties of a material are such that it can be used as a partial replacement for portland cement in concrete, this results in significant energy and cost savings.

The mechanism by which the pozzolanic reaction exercises a beneficial effect on the properties of concrete remains the same, irrespective of whether a pozzolanic material has been added to concrete in the form of a mineral admixture or as a component of blended portland cements. From the description of the pozzolanic reaction and properties of blended cements in Chapter 6 (p. 201) it is clear that the engineering benefits likely to be derived from the use of mineral admixtures in concrete include improved resistance to thermal cracking because of lower heat of hydration, enhancement of ultimate strength and impermeability due to pore refinement, and (as a result of reduced alkalinity) a better durability to chemical attacks such as by sulfate water and alkali-aggregate expansion.

Classification

Some mineral admixtures are pozzolanic (e.g., low-calcium fly ash), some are cementitious (e.g., granulated iron blast-furnace slag), whereas others are both cementitious and pozzolanic (e.g., high-calcium fly ash). A classification of mineral admixtures according to their pozzolanic and/or cementitious characteristics is shown in Table 8–6. The table also contains a description of mineralogical compositions and particle characteristics since these two properties, rather than the chemical composition or the source of the material, determine the effect of a mineral admixture on the behavior of concrete containing the admixture.

For the purposes of a detailed description of the important mineral admixtures given below, the materials are divided into two groups:

1. **Natural materials:** those materials which have been processed for the sole purpose of producing a pozzolan. Processing usually involves crushing, grinding, and size separation; in some cases it may also involve thermal activation.
2. **By-product materials:** those materials which are not the primary products of the industry producing them. Industrial by-products may or may not require any processing (e.g., drying and pulverization) before use as mineral admixtures.

Natural Materials

Except for diatomaceous earths, all natural pozzolans are derived from volcanic rocks and minerals. During explosive volcanic eruptions the quick cooling of the magma, composed mainly of aluminosilicates, results in the formation of glass or

Mineral Admixtures

TABLE 8-6 CLASSIFICATION, COMPOSITION, AND PARTICLE CHARACTERISTICS OF MINERAL ADMIXTURES FOR CONCRETE

Classification	Chemical and mineralogical composition	Particle characteristics
Cementitious and pozzolanic		
Granulated blast-furnace slag (cementitious)	Mostly silicate glass containing mainly calcium, magnesium, aluminum, and silica. Crystalline compounds of melilite group may be present in small quantity.	Unprocessed material is of sand size and contains 10–15% moisture. Before use it is dried and ground to particles less than 45 μm (usually about 500 m^2/kg Blaine). Particles have rough texture.
High-calcium fly ash (cementitious and pozzolanic)	Mostly silicate glass containing mainly calcium, magnesium, aluminum, and alkalies. The small quantity of crystalline matter present generally consists of quartz and C_3A; free lime and periclase may be present; $C\bar{S}$ and $C_4A_3\bar{S}$ may be present in the case of high-sulfur coals. Unburnt carbon is usually less than 2%.	Powder corresponding to 10–15% particles larger than 45 μm (usually 300–400 m^2/kg Blaine). Most particles are solid spheres less than 20 μm in diameter. Particle surface is generally smooth but not as clean as in low-calcium fly ashes.
Highly active pozzolans		
Condensed silica fume	Consists essentially of pure silica in noncrystalline form.	Extremely fine powder consisting of solid spheres of 0.1 μm average diameter (about 20 m^2/g surface area by nitrogen adsorption).
Rice husk ash	Consists essentially of pure silica in noncrystalline form.	Particles are generally less than 45 μm but they are highly cellular (about 60 m^2/g surface area by nitrogen adsorption).
Normal pozzolans		
Low-calcium fly ash	Mostly silicate glass containing aluminum, iron, and alkalies. The small quantity of crystalline matter present generally consists of quartz, mullite, sillimanite, hematite, and magnetite.	Powder corresponding to 15–30% particles larger than 45 μm (usually 200–300 m^2/kg Blaine). Most particles are solid spheres with average diameter 20 μm. Cenospheres and plerospheres may be present.

TABLE 8-6 Continued

Classification	Chemical and mineralogical composition	Particle characteristics
Natural materials	Besides aluminosilicate glass, natural pozzolans contain quartz, feldspar, and mica.	Particles are ground to mostly under 45 μm and have rough texture.
Weak pozzolans		
Slowly cooled blast-furnace slag, bottom ash, boiler slag, field burnt rice husk ash	Consists essentially of crystalline silicate materials, and only a small amount of noncrystalline matter.	The materials must be pulverized to very fine particle size in order to develop some pozzolanic activity. Ground particles are rough in texture.

vitreous phases with disordered structure. Due to the simultaneous evolution of dissolved gases, the solidified matter frequently acquires a porous texture with a high surface area which facilitates subsequent chemical attack. Since the aluminosilicates with a disordered structure will not remain stable on exposure to a lime solution, this becomes the basis for the pozzolanic properties of the volcanic glasses.

The alteration of the volcanic glass under hydrothermal conditions can lead to formation of zeolite minerals, which are compounds of the type $(Na_2Ca)O \cdot Al_2O_3 \cdot 4SiO_2 \cdot xH_2O$. This product, called volcanic tuff, is characterized by a compact texture. Zeolite minerals in finely ground tuffs are able to react with lime by a base-exchange process.

Progressive alteration of the aluminosilicates of a volcanic glass is believed to be responsible for the formation of clay minerals. Clays are not pozzolanic unless the crystalline structure of the aluminosilicate minerals in the clay is converted into an amorphous or disordered structure by heat treatment.

Diatomaceous earths consist of opaline or amorphous hydrated silica derived from the skeletons of diatoms, which are tiny water plants with cell walls composed of silica shells. The material is pozzolanic when pure, but is generally found contaminated with clay minerals and therefore must be thermally activated to enhance the pozzolanic reactivity.

It is difficult to classify natural pozzolans because the materials seldom contain only one reactive constituent. However, based on the principal reactive constituent present, a classification can be made into volcanic glasses, volcanic tuffs, calcined clays or shales, and diatomaceous earths. A description of their formation processes and relevent characteristics is given below.

Volcanic glasses. Santorin Earth of Greece, Bacoli Pozzolan of Italy, and Shirasu Pozzolan of Japan are examples of pozzolanic materials which derive their lime-reactivity characteristic mainly from unaltered aluminosilicate glass. A photograph of the pozzolan quarry on Santorini Island is shown in Fig. 8-6a, and the

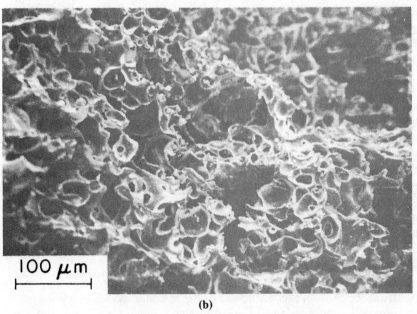

Figure 8–6 (a) Quarry for mining the pozzolan on Santorini Island, Greece; (b) scanning electron micrograph of the porous structure of the pozzolan.

pumiceous or porous texture that accounts for the high surface area and reactivity is evident from the scanning electron micrograph shown in Fig. 8–6b. Generally, small amounts of nonreactive crystalline minerals, such as quartz, feldspar, and mica, are found embedded in the glassy matrix.

Volcanic tuffs. Pozzolans of Segni-Latium (Italy), and trass of Rheinland and Bavaria (Germany), represent typical volcanic tuffs. The zeolite tuffs with their compact texture are fairly strong, possessing compressive strengths on the order of 100 to 300 kg/cm^2. The principal zeolite minerals are reported to be phillipsite and herschelite. After the compact mass is ground to a fine particle size, the zeolite minerals show considerable reactivity with lime and develop cementitious characteristics similar to those of pozzolans containing volcanic glass.

Calcined clays or shales. Volcanic glasses and tuffs do not require heat treatment to enhance their pozzolanic property. However, clay and shales will not show appreciable reactivity with lime unless the crystal structures of the clay minerals present are destroyed by heat treatment. Temperatures on the order of 600 to 900°C in oil-, gas-, or coal-fired rotary kilns are considered adequate for this purpose. The pozzolanic activity of the product is due mainly to the formation of an amorphous or disordered aluminosilicate structure as a result of the thermal treatment. *Surkhi*, a pozzolan made in India by pulverization of fired-clay bricks, belongs to this category. It should be obvious why heat treatment of clays and shales that contain large amounts of quartz and feldspar does not produce good pozzolans. In other words, pulverization of fired-clay bricks made from any clay may not yield a suitable mineral admixture for concrete.

Diatomaceous earth. This group of pozzolans is characterized by materials of organogen origin. Diatomite is a hydrated amorphous silica which is composed of skeletal shells from the cell walls of many varieties of microscopic aquatic algae. The largest known deposit is in California. Other large deposits are reported in Algeria, Canada, Germany, and Denmark. Diatomites are highly reactive to lime, but their skeletal microstructure accounts for a high water requirement, which is harmful for the strength and durability of concrete containing this pozzolan. Furthermore, diatomite deposits, such as Moler of Denmark, generally contain large amounts of clay and therefore must be thermally activated before use in order to enhance the pozzolanic reactivity.

By-Product Materials

Ashes from combustion of coal and some crop residues, volatilized silica from certain metallurgical operations, and granulated slag from both ferrous and nonferrous metal industries are the major industrial by-products that are suitable for use as mineral admixtures in portland cement concrete. Industrialized countries such as the United States, Russia, France, Germany, Japan, and the United Kingdom, are among the largest producers of fly ash, volatilized silica, and granulated blast-furnace slag. In addition to these materials, China and India have the potential for making large amounts of rice husk ash. The production and properties of the important by-product materials are described below.

Fly ash. During combustion of powdered coal in modern power plants, as coal passes through the high-temperature zone in the furnace, the volatile matter and carbon are burned off, whereas most of the mineral impurities, such as clays, quartz, and feldspar, will melt at the high temperature. The fused matter is quickly transported to lower-temperature zones, where it solidifies as spherical particles of glass. Some of the mineral matter agglomerates forming bottom ash, but most of it flies out with the flue gas stream and is called *fly ash* (pulverized fuel ash in U.K.). This ash is subsequently removed from the gas by electrostatic precipitators.

From the standpoint of significant differences in mineralogical composition and properties, fly ashes can be divided into two categories which differ from each other mainly in calcium content. The ash in the first category, containing less than 10 percent analytical CaO, is generally a product of combustion of anthracite and bituminous coals. The ash in the second category, typically containing 15 to 35 percent analytical CaO, is generally a product of combustion of lignite and subbituminous coals.

The low-calcium fly ashes, due to the high proportions of silica and alumina, consist principally of aluminosilicate glasses. In the furnace, when large spheres of molten glass do not get cooled rapidly and uniformly, sillimanite ($Al_2O_3 \cdot SiO_2$) or mullite ($3Al_2O_3 \cdot 2SiO_2$) may crystallize as slender needles in the interior of the glassy spheres. This kind of partial devitrification of glass in low-lime fly ashes accounts for the presence of the crystalline aluminosilicates. Also, depending on the fineness to which coal has been ground before combustion, remnants of α quartz are likely to be present in all fly ashes. In fact, X-ray diffraction analyses have confirmed that the principal crystalline minerals in low-calcium fly ashes are quartz, mullite, and hematite or magnetite. Since these crystalline minerals are nonreactive at ordinary temperature, their presence in large proportions, at the cost of the noncrystalline component or glass in a fly ash, tends to reduce the reactivity of the fly ash.

Compared to low-calcium fly ashes, the high-calcium variety is in general more reactive because it contains most of the calcium in the form of reactive crystalline compounds, such as C_3A, $C\bar{S}$, and $C_4A_3\bar{S}$; also there is evidence that the principal constitutent (i.e., the noncrystalline phase) contains enough calcium ions to enhance the reactivity of the aluminosilicate glass. Most fly ashes, whether low-calcium or high-calcium, contain approximately 60 to 85 percent glass, 10 to 30 percent crystalline compounds, and up to about 10 percent unburnt carbon. Carbon is generally present in the form of cellular particles larger than 45 μm. Under normal operating conditions, modern furnaces do not produce fly ashes containing more than 5 percent carbon (in fact, it is usually less than 2 percent in high-calcium fly ashes); larger amounts of carbon in a fly ash that is meant for use as a mineral admixture in concrete are considered harmful because the cellular particles of carbon tend to increase both the water requirement for a given consistency and the admixture requirement for entrainment of a given volume of air.

Micrographic evidence presented in Fig. 8-7 shows that most of the particles in fly ash occur as *solid spheres of glass,* but sometimes a small number of hollow spheres, which are called *cenospheres* (completely empty) and *plerospheres* (packed with numerous small spheres) may be present. Typically, the spheres in low-calcium fly ashes look cleaner than the spheres in high-calcium fly ash. Since alkalies and sulfate tend to occur in a relatively large proportion in the latter, the deposition of alkali sulfates on the surface may account for the dirty appearance of the spheres in high-calcium fly ash. Particle size distribution studies show that the particles vary from <1 μm to 100 μm in diameter, with more than 50 percent under 20 μm (Fig. 8-8). The particle size distribution, morphology, and surface characteristics of the fly ash used as a mineral admixture exercise a considerable influence on the water requirement and workability of freshly made concrete, and the rate of strength development in hardened concrete.

Blast-furnace slag. In the production of cast iron, also called pig iron, if the slag is cooled slowly in air, the chemical components of slag are usually present in the form of crystalline melilite (C_2AS-C_2MS_2 solid solution), which does not react with water at ordinary temperature. If ground to very fine particles, the material will be weakly cementitious and pozzolanic. However, when the liquid slag is rapidly quenched from a high temperature by either water or a combination of air and water, most of the lime, magnesia, silica, and alumina are held in noncrystalline or glassy state. The water-quenched product is called *granulated slag* due to the sand-size particles, while the slag quenched by air and a limited amount of water which is in

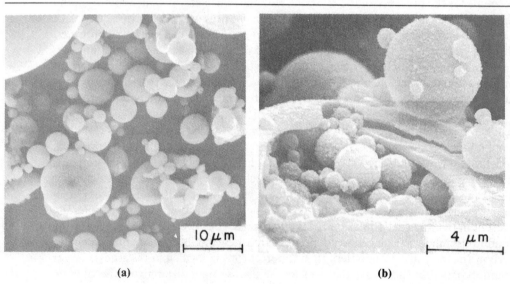

Figure 8-7 Scanning electron micrographs of a typical Class F fly ash: (a) spherical and glassy particles; (b) a plerosphere.

Mineral Admixtures

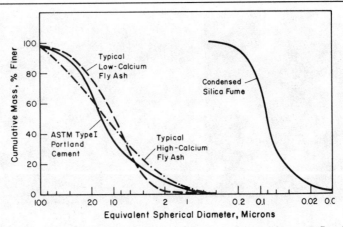

Figure 8-8 Comparison of particle size distributions of portland cement, fly ashes, and condensed silica fume.

the form of pellets is called *pelletized slag*. Normally, the former contains more glass; however, when ground to 400 to 500 m^2/kg Blaine, both products develop satisfactory cementitious properties.

Although high-calcium fly ashes are of relatively recent origin, and the production and use of granulated blast-furnace slag is more than 100 years old, there are similarities in the mineralogical character and reactivity of the two materials. Both are essentially noncrystalline, and the reactivity of the high-calcium glassy phase in both cases appears to be of a similar order. Compared to low-calcium fly ash, which usually does not make any significant contribution to the strength of portland cement concrete until after about 4 weeks of hydration, the strength contribution by high-calcium fly ash or granulated iron blast-furnace slag may become apparent as early as 7 days after hydration. It should be noted that although particle size characteristics, composition of glass, and the glass content are the primary factors determining the activity of fly ashes and slags, the reactivity of the glass itself varies with the thermal history of the material. The glass, chilled from a higher temperature and at a faster rate, will have a more disordered structure and will therefore be more reactive.

It is generally known that slag particles of less than 10 μm contribute to early strengths in concrete up to 28 days; particles of 10 to 45 μm contribute to later strengths, but particles coarser than 45 μm are difficult to hydrate. Since the slag obtained after granulation is very coarse and also moist, it is dried and pulverized to particle size mostly under 45 μm, which corresponds to approximately a 500-m^2/kg Blaine surface area.

Condensed silica fume. Condensed silica fume, also known by other names, such as volatilized silica, microsilica, or simply as silica fume, is a by-product

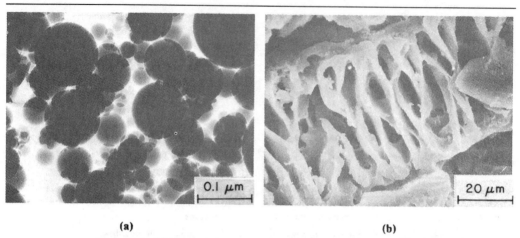

Figure 8–9 (a) Transmission electron micrograph of condensed silica fume; (b) scanning electron micrograph of rice husk ash. [(a), Courtesy of P. C. Aitcin, Univ. of Sherbrooke.]

of the induction arc furnaces in the silicon metal and ferrosilicon alloy industries. Reduction of quartz to silicon at temperatures up to 2000°C produces SiO vapors, which oxidize and condense in the low-temperature zone to tiny spherical particles (Fig. 8–9) consisting of noncrystalline silica. The material removed by filtering the outgoing gases in bag filters possesses an average diameter on the order of 0.1 μm and surface areas in the range 20 to 25 m^2/g. Compared to normal portland cement and typical fly ashes, condensed silica fume samples show particle size distributions that are two orders of magnitude finer (Fig. 8–8). This is why on the one hand the material is highly pozzolanic, but on the other it creates problems of handling and increases the water requirement in concrete appreciably unless water-reducing admixtures are used. The by-products from the silicon metal and the ferrosilicon alloy industries, producing alloys with 75 percent or higher silicon content, contain 85 to 95 percent noncrystalline silica; the by-product from the production of ferrosilicon alloy with 50 percent silicon contains a much lower silica content and is less pozzolanic.

Rice husk ash. Rice husks, also called rice hulls, are the shells produced during the dehusking operation of paddy rice. Since they are bulky, the husks present an enormous disposal problem for centralized rice mills. Each ton of paddy rice produces about 200 kg of husks, which on combustion yield approximately 40 kg of ash. The ash formed during open-field burning or uncontrolled combustion in industrial furnaces generally contains a large proportion of nonreactive silica minerals, such as cristobalite and tridymite, and must be ground to a very fine particle size in order to develop the pozzolanic property. On the other hand, a highly pozzolanic ash can be produced by controlled combustion when silica is retained in noncrystalline form and a cellular structure (Fig. 8–9). Industrially produced samples of this material showed 50 to 60 m²/g surface area by nitrogen adsorption.[8] Effect of processing conditions on the characteristics of rice husk ash and the beneficial effects of amorphous ash on concrete properties are described in a recent report.[9]

Applications

The significance of the pozzolanic reaction and the mechanisms by which properties of concrete are improved by the use of combinations of portland cement and pozzolanic materials were described in Chapter 6 (p. 209). Next, selected applications of mineral admixtures are described to illustrate the principles already discussed.

Improvement in workability. With fresh concrete mixtures that show a tendency to bleed or segregate, it is well known that incorporation of finely divided particles generally improves workability by reducing the size and volume of voids. The finer a mineral admixture, the less of it will be needed to enhance the cohesiveness and therefore the workability of a freshly mixed concrete.

The small size and the glassy texture of fly ashes and slags make it possible to reduce the amount of water required for a given consistency. In a review paper[10] on the use of fly ash in concrete it was reported that in one case a concrete made by substituting 30 percent of the cement with a Canadian fly ash required 7 percent less water than the control concrete of equal slump. In another case involving investigation of concrete materials for construction of the South Saskatchewan River Dam, it was found that with the addition of a fly ash the resulting concretes, with a lower ratio of water to total cementitious materials (portland cement + fly ash), showed improved cohesiveness and workability. Reduction of segregation and bleeding by the use of mineral admixture is of considerable importance when concrete is placed by pumping. The improvements in cohesiveness and finishability are particularly valuable in lean concretes or those made with aggregates deficient in fines.

It may be noted that although all mineral admixtures tend to improve the cohesiveness and workability of fresh concrete, many do not possess the water-reducing capability of fly ashes and slags. For a given consistency of concrete, the use of very high surface area materials, such as pumicite, rice husk ash, and condensed silica fume, tends to increase the water requirement.

Durability to thermal cracking. Assuming that due to heat of hydration the maximum temperature in a massive structure is reached within a week of concrete placement, the use of a mineral admixture (i.e., natural pozzolan, fly ash, or slag) offers the possibility of reducing the temperature rise almost in a direct proportion to the amount of portland cement replaced by the admixture. This is because, under normal conditions, these admixtures do not react to a significant degree for several

---

[8] P. K. Mehta and N. Pitt, *Resource Recovery and Conservation,* Elsevier Scientific Publishing Company, Amsterdam, Vol. 2, pp. 23–38, 1976.

[9] P. K. Mehta in *Advances in Concrete Technology,* ed. by V. M. Malhotra, CANMET, Ottawa, Canada, 1991.

[10] E. E. Berry and V. M. Malhotra, *J. ACI,* Proc., Vol. 77, No. 2, pp. 59–73, 1980.

days. As a rule of thumb, the total heat of hydration produced by the pozzolanic reactions involving mineral admixtures is considered to be half as much as the average heat produced by the hydration of portland cement.

Portland cement replacement by fly ash has been practiced in the United States since the 1930s. In mass concrete construction, where low cement contents and fly ash proportions as high as 60 to 100 percent by weight of portland cement are now commonly employed, the first successful application was in 1948 for building the Hungry Horse Dam, Montana. More than 3 million cubic yards (2.3 million m^3) of concrete was used; some of the concrete contained 32 percent portland cement replaced by fly ash shipped from the Chicago area. More recently, fly ash was used in concrete for the Dworshak Dam, Idaho, which is a 7-million cubic yard (5.4 million m^3) concrete structure.

Furthermore, construction engineers should be aware of another advantage from the use of mineral admixtures when concrete is to be exposed to considerably higher than normal temperatures, either because of heat of hydration or any other cause. Compared to specimens cured in the laboratory, field concretes without the presence of mineral admixtures are likely to undergo a strength loss due to microcracking on cooling, but concretes containing mineral admixtures frequently show a strength gain. Whereas high-temperature exposures can be harmful to portland cement concretes, concrete containing a mineral admixture may benefit from thermal activation (acceleration of the pozzolanic reaction). Recently, for the intake pressure tunnel of Kurobegowa Power Station in Japan, where the concrete is located in a hot base rock (100 to 160°C), the use of 25 percent fly ash as a cement replacement in the concrete mixture showed a favorable effect on strength.

Durability to chemical attacks. The permeability of concrete is fundamental to determining the rate of mass transfer related to destructive chemical actions such as the alkali-aggregate expansion and attack by acidic and sulfate solutions (Chapter 5). Since the pozzolanic reaction involving mineral admixtures is capable of pore refinement, which reduces the permeability of concrete, both field and laboratory studies have shown considerable improvement in the chemical durability of concretes containing mineral admixtures.

Investigations in the 1950s by R. E. Davis at the University of California on the permeability of concrete pipes with or without fly ash showed that the permeability of pipe containing 30 percent* of a low-calcium fly ash was higher at 28 days after casting, but considerably lower at 6 months. Recent work[11] has confirmed that in cement pastes containing 10 to 30 percent of a low-calcium fly ash, significant pore refinement occurred during the 28- to 90-day curing period; this caused a reduction in the permeability from 11 to 13 × 10^{-11} to 1 × 10^{-11} cm/sec. In the case of cement pastes containing 10 to 30 percent of a rice husk ash or a condensed silica fume, or

---

*Since mineral admixtures are used in large amounts, it is customary to show their proportion in concrete as a weight percent of the total cementitious material (i.e., cement + admixture).

[11] D. Manmohan and P. K. Mehta, *Cem. Concr. Aggregates*, Vol. 3, No. 1, pp. 63–67, 1981.

70 percent of a granulated blast-furnace slag, even at 28 days after hydration the system was found to be almost impermeable.

Depending on the individual characteristics of the mineral admixture used, usually the combinations of a high-alkali portland cement with 40 to 65 percent granulated blast-furnace slag, or 30 to 40 percent low-calcium fly ash, or 20 to 30 percent natural pozzolans, have been found to be quite effective in limiting the alkali-aggregate expansion to acceptable levels. In California, since aggregate deposits in many parts of the state contain alkali-reactive minerals, the Department of Water Resources has made it a standard practice to include pozzolanic admixtures in concrete for hydraulic structures. The use of large amounts of mineral admixtures to reduce the alkali-aggregate expansion is sometimes objectionable from the standpoint of low strengths at early ages. Highly active pozzolans, such as rice husk ash and condensed silica fume, are effective in amounts as low as 10 percent and tend to increase, rather than decrease, the strength at early ages. In Iceland, where only high-alkali portland cement is available and the aggregates are generally reactive, it is customary to blend 7 percent condensed silica fume with all portland cement.

The published literature contains sufficient evidence that, in general, the incorporation of mineral admixtures into concrete improves the resistance of the material to acidic water, sulfate water, and seawater. This is due mainly to the pozzolanic reaction, which is accompanied by a reduction in permeability as well as a reduction in the calcium hydroxide content of the hydrated product. However, as discussed below, not all portland cement-slag or fly-ash combinations are found satisfactory in combating the sulfate attack in concrete.

In the 1960s and 1970s extensive studies at the U.S. Bureau of Reclamation on concretes containing 30 percent low-calcium fly ashes showed greatly improved sulfate resistance to a standard sodium sulfate solution; however, the use of high-calcium fly ashes generally reduced the sulfate resistance. The explanation for this behavior probably lies in the mineralogical composition of the fly ash. In addition to the permeability and the calcium hydroxide content of the cement paste, the sulfate resistance is governed by the amount of reactive alumina present. Fly ashes that contain a large proportion of reactive alumina in glass or in crystalline constituents would not be expected to reduce the sulfate resistance of concrete. High-calcium fly ashes containing highly reactive alumina in the form of C_3A or $C_4A_3\bar{S}$ are therefore less suitable than low-calcium fly ashes for improving the sulfate resistance of concrete.

The sulfate resistance of portland cement concrete containing granulated blast-furnace slag depends on the amount of slag used and the alumina content of slag. The following excerpt from an appendix to the ASTM C 989 (*Standard Specification for Ground Iron Blast-Furnace Slag for Use in Concrete and Mortars*) explains why:

> *Effect of Ground Slag on Sulfate Resistance.* The use of ground slag will decrease the C_3A content of the cementing materials and decrease the permeability and calcium

hydroxide content of the mortar or concrete. Tests have shown that the alumina content of the slag also influences the sulfate resistance at low slag-replacement percentages. The data from these studies of laboratory exposures of mortars to sodium and magnesium sulfate solutions provide the following general conclusions.

The combinations of ground slag and portland cement, in which the slag content was greater than 60 to 65%, had high sulfate resistance, always better than the portland cement alone, irrespective of the alumina content of the slag. The low alumina (11%) slag tested increased the sulfate resistance independently of the C_3A content of the cement. To obtain adequate sulfate resistance, higher slag percentages were necessary with the higher C_3A cements. The high alumina (18%) slag tested, adversely affected the sulfate resistance of portland cements when blended in low percentages (50% or less). Some tests indicated rapid decreases in resistance for cements in the 8 and 11% C_3A ranges with slag percentages as low as 20% or less in the blends.

In regard to highly active pozzolans (i.e., rice husk ash and condensed silica fume), it seems that even when present in amounts as low as 30 percent these admixtures are able to consume the calcium hydroxide present in the hydrated cement paste almost completely and therefore are excellent not only for improving the resistance of concrete to acid attack but also to sulfate attack.

Production of high-strength concrete. Due to economic and durability considerations, mineral admixtures are generally used as a partial replacement for portland cement in concrete. In the amounts normally used, most low-calcium fly ashes and natural pozzolans tend to reduce the early strengths up to 28 days, but improve the ultimate strength. Compared to concrete without the admixture, concretes containing a granulated blast-furnace slag or a high-calcium fly ash usually show lower strengths at 1 and 3 days, but strength gains can be substantial after about 7 days of curing. Highly active pozzolans (i.e., rice husk ash and condensed silica fume) are capable of producing high strength in concrete at both early and late ages, especially when a water-reducing agent has been used to reduce the water requirement.

On the other hand, when used as a partial replacement for fine aggregates, all mineral admixtures are able to increase the strengths of concrete at both early and late ages. The strength gain at early ages is in part due to a slight acceleration in portland cement hydration; the strength gain at later ages, which can be substantial, is due mostly to the pozzolanic reaction, causing pore refinement and replacing the weaker component (calcium hydroxide) with the stronger one (calcium silicate hydrate).

If the elimination of large pores and reduction of calcium hydroxide are necessary elements for producing concretes with a high compressive strength, mineral admixtures seem well suited for playing a key role in the production of high-strength concrete, irrespective of whether they are used as a cement replacement, fine-aggregate replacement, or both. According to Malhortra,[12] *the development of*

---

[12] V. M. Malhotra, *Concr. Int.*, Vol. 6, No. 4, p. 21, 1984.

high-strength concrete for use in tall buildings in the Chicago area has shown that the use of fly ash is almost mandatory to achieve strengths greater than 8500 psi (59 MPa) at 56 days. Concrete mixtures containing 850 lb (386 kg) of portland cement and 100 lb (45 kg) of a low-calcium fly ash per cubic yard and a 0.33 ratio between water and the cementitious materials gave approximately 10,500 psi (72 MPa) compressive strength at 56 days. Cook[13] reported similar strength values for a high-strength concrete used for structural elements of a tall building in Houston, Texas; this concrete contained only 667 lb (303 kg) of portland cement and 167 lb (76 kg) of a high-calcium fly ash per cubic yard. Industrial applications of 70 to 80 MPa compressive strength concretes containing rice husk ash or condensed silica fumes and water-reducing admixtures have been reported. Using condensed silica fume, special aggregates of controlled particle size, and a superplasticizer, the Aalborg Cement Company in Denmark produced specimens which, with less than a 0.2 ratio between water and the cementitious materials, gave over 200 MPa compressive strength.

CONCLUDING REMARKS

For ready reference purposes a summary of the commonly used concrete admixtures, their primary function, principal active ingredients, applicable ASTM Standard Specification, and possible side effects is presented in Table 8–7.

In the 1940s and 1950s, efforts to promote the introduction of admixtures in concrete on a large scale met with considerable resistance because there was little understanding of their mode of action, and this led to many unsatisfactory experiences. Today, the situation is different; admixtures have become so much an integral part of concrete that in the near future the definition for the composition of concrete may have to be revised to include admixtures as one of the components.

Problems associated with the misuse of admixtures, however, continue to arise. The genesis of most of the problems appears to lie in the incompatibility between a particular admixture and a cement composition or between two or more admixtures that may be present in the system. Surfactants such as air-entraining chemicals, lignosulfonates, and superplasticizers are especially sensitive to interaction effects among aluminate, sulfate, and alkali ions present in the solution phase at the beginning of the hydration of cement. Loss of air or proper void spacing in concrete containing a superplasticizer or an exceedingly fine mineral admixture is a matter of great concern in the concrete industry. Therefore it is highly recommended to carry out laboratory tests involving field materials and conditions before the actual use of admixtures in concrete construction, particularly when large projects are undertaken or when the concrete-making materials are subject to significant variations in quality.

Finally, admixtures can certainly enhance the properties of a concrete but

---

[13] J. E. Cook, *Concr. Int.*, Vol. 7, pp. 72–80, 1982.

TABLE 8-7 COMMONLY USED CONCRETE ADMIXTURES

Primary function	Principal active ingredients/ ASTM specification	Side effects
Water-reducing		
Normal	Salts, modifications and derivatives of lignosulfonic acid, hydroxylated carboxylic acids, and polyhydroxy compounds. ASTM C 494 (Type A).	Lignosulfonates may cause air entrainment and strength loss; Type A admixtures tend to be set retarding when used in high dosage.
High range	Sulfonated naphthalene or melamine formaldehyde condensates. ASTM C 494 (Type F).	Early slump loss; difficulty in controlling void spacing when air entrainment is also required.
Set-controlling		
Accelerating	Calcium chloride, calcium formate, and triethanolamine. ASTM C 494 (Type C).	Accelerators containing chloride increase the risk of corrosion of the embedded metals.
Retarding	Same as in ASTM Type A; compounds such as phosphates may be present. ASTM C 494 (Type B).	
Water-reducing and set-controlling		
Water-reducing and retarding	Same as used for normal water reduction. ASTM C 494 (Type D).	See Type A above.
Water-reducing and accelerating	Mixtures of Types A and C. ASTM C 494 (Type E).	See Type C above.
High-range water-reducing and retarding	Same as used for Type F with lignosulfonates added. ASTM C 494 (Type G).	See Type F above.
Workability-improving		
Increasing consistency	Water-reducing agents, [e.g., ASTM C 494 (Type A)].	See Type A above.
Reducing segregation	(a) Finely divided minerals (e.g., ASTM C 618).	Loss of early strength when used as cement replacement.
	(b) Air-entrainment surfactants (ASTM C 260).	Loss of strength.

TABLE 8–7 Continued

Primary function	Principal active ingredients/ ASTM specification	Side effects
Strength-increasing		
By water-reducing admixtures	Same as listed under ASTM C 494 (Types A, D, F, and G).	See Types A and F above.
By Pozzolanic and cementitious admixtures	Same as listed under ASTM C 618 and C 989.	Workability and durability may be improved.
Durability-improving		
Frost action	Wood resins, proteinaceous materials, and synthetic detergents (ASTM C 260).	Strength loss.
Thermal cracking Alkali-aggregate expansion Acidic solutions Sulfate solutions	Fly ashes, and raw or calcined natural pozzolans (ASTM C 618); granulated and ground iron blast-furnace slag (ASTM C 989); condensed silica fume; rice husk ash produced by controlled combustion. (High-calcium and high-alumina fly ashes, and slag-portland cement mixtures containing less than 60% slag may not be sulfate resistant.)	Loss of strength at early ages, except when highly pozzolanic admixtures are used in conjunction with water-reducing agents.

should not be expected to compensate for the poor quality of concrete ingredients or poor mix proportioning.

TEST YOUR KNOWLEDGE

1. Why are plasticizing admixtures called *water reducing?* What is the distinction between normal and high-range water-reducing admixtures according to the ASTM Standard Specification?
2. Can you list and define the seven *types* of chemical admixtures, four *classes* of mineral admixtures, and three *grades* of iron blast-furnace slag which are used as admixtures for concrete?
3. After reviewing the ASTM C 618 and C 989 Standard Specifications and other published literature, write a critical note comparing the two standards.
4. What are the essential differences in composition and mode of action between the surfactants used for air entrainment and those used for water reduction?

5. Some manufacturers claim that application of water-reducing admixtures can lower the cement content and increase the consistency and strength of a reference concrete mixture. Explain why all three benefits may not be available at the same time.
6. Commercial lignin-based admixtures when used as water-reducing agents may exhibit certain side effects. Discuss the possible side effects and explain how they are corrected.
7. In their composition and mechanism of action, how do the superplasticizers differ from the normal water-reducing admixtures? Addition of 1 to 2 percent of a normal water-reducing agent to a concrete mixture may cause segregation and severe retardation. These effects do not take place in the superplasticized concrete. Explain why.
8. (a) When added to portland cement paste in very small amounts, calcium chloride acts as a retarder, but in large amounts it behaves as an accelerator. Can you explain the phenomenon?
   (b) Why doesn't calcium sulfate behave in the same manner as calcium chloride?
   (c) As an accelerator why isn't sodium chloride as effective as calcium chloride?
   (d) Mineral acids are accelerators for portland cement, but organic acids do not show a consistent behavior.
   (e) Formic acid is an accelerator, while gluconic acid is a retarder. Explain why.
9. (a) What type of admixtures would you recommend for concreting in: (i) hot weather, (ii) cold weather.
   (b) When used as an accelerator, what effect would calcium chloride have on the mechanical properties, dimensional stability, and durability of concrete?
10. State several important reasons why it is desirable to use pozzolanic admixtures in concrete.
11. (a) Why clays and shales have usually to be heat-treated to make them suitable for use as a pozzolan?
    (b) Name some of the commonly available industrial by-products which show pozzolanic or cementitious property when used in combination with portlant cement.
12. What do you know about the origin and characteristics of the following mineral admixtures: pumice, zeolitic tuff, rice husk ash, condensed silica fume?
13. Compare and contrast industrial fly ashes and ground iron blast-furnace slag with respect to mineralogical composition and particle characteristics.
14. Explain the mechanism by which mineral admixtures are able to improve the pumpability and finishability of concrete mixtures. In the amounts normally used, some mineral admixtures are water reducing, whereas others are not. Discuss the subject with the help of examples.
15. Discuss the mechanisms by which mineral admixtures improve the durability of concrete to acidic waters. Why all fly ash-portland cement or slag-portland cement combinations may not turn out to be sulfate-resisting?
16. What maximum strength levels have been attained in recently developed high-strength concrete mixtures? Explain the role played by admixtures in the development of these concretes.

SUGGESTIONS FOR FURTHER STUDY

ACI Committee 212 Report, "Chemical Admixtures for Concrete," *ACI Materials Jour.*, Vol. 86, No. 3, pp. 297–327, 1989.

Admixtures, Proc. Int. Congr. on Admixtures, The Construction Press, New York, 1980.

BERRY, E. E., and V. M. MALHOTRA, "Fly Ash in Concrete—A Critical Review," *J. ACI,* Proc., Vol. 2, No. 3, pp. 59–73, 1982; Canmet Publ. 85-3, 1986.

HELMUTH, R., *Fly Ash in Cement and Concrete,* Portland Cement Association, 1987.

LEA, F. M., *The Chemistry of Cement and Concrete,* Chemical Publishing Company, Inc., New York, 1971, pp. 302–10, 414–89.

MALHOTRA, V. M., ed., *Supplementary Cementing Materials for Concrete,* Canmet Publ. SP86–8E, 1987.

MALHOTRA, V. M., ed., *Use of Fly Ash, Silica Fume, Slag, and Other Mineral Byproducts in Concrete,* Proc. Symp., ACI, SP 79 (1983), SP 91 (1986), SP 114 (1989), and SP 132 (1992).

RAMACHANDRAN, V. S., ed., *Concrete Admixtures Handbook,* Noyes Publications, 1984.

RIXOM, M. R., ed., *Concrete Admixtures: Use and Applications,* The Construction Press, New York, 1978.

SWAMY, R. N., ed., *Cement Replacement Materials,* Surrey Univ. Press, 1986.

Superplasticizers in Concrete, Transportation Research Board, National Academy of Sciences, Washington, D.C., Transportation Research Record 720, 1979.

CHAPTER 9

Proportioning Concrete Mixtures

PREVIEW

In pursuit of the goal of obtaining concrete with certain desired performance characteristics, the selection of component materials is the first step; the next step is a process called ***mix proportioning*** by which one arrives at the right combination of the components. Although there are sound technical principles that govern mix proportioning, for several valid reasons the process is not entirely in the realm of science. Nevertheless, since mix proportions have a great influence on the cost and properties of concrete, it is important that engineers who are often called on to develop or approve mix proportions be familiar with the underlying principles and the commonly used procedures.

In this chapter the significance and objectives of concrete mix proportioning are given. General considerations governing cost, workability, strength, and durability are discussed, and the ACI 211.1 Standard *Practice for Selecting Proportions for Normal, Heavy Weight and Mass Concrete,* is described, with a sample computation to illustrate the procedures. Note that in this chapter only the U.S. customary units are used (see Chapter 1 for conversion table to S.I. units).

SIGNIFICANCE AND OBJECTIVES

The ***proportioning of concrete mixtures,*** also referred to as ***mix proportioning*** or ***mix designing,*** is a ***process*** by which one arrives at the right combination of cement, aggregates, water, and admixtures for making concrete according to given specifications. For reasons described below, ***this process is considered an art rather than a science.*** Although many engineers do not feel comfortable with matters that cannot be reduced to an exacting set of numbers, with an understanding of the underlying principles and with some practice, the art of proportioning concrete mixtures can be mastered. Given an opportunity, the exercise of this art is very rewarding because the effects of mix proportioning on the cost of concrete and several important properties of both fresh and hardened concrete can be clearly seen.

One purpose of mix proportioning is to obtain a product that will perform according to certain predetermined requirements, the most essential requirements being ***the workability of fresh concrete and the strength of hardened concrete at a specified age.*** Workability, which is discussed in more detail in Chapter 10, is the property that determines the ease with which a concrete mixture can be placed, compacted, and finished. Durability is another important property, but it is generally assumed that under normal exposure conditions durability will be satisfactory if the concrete mixture develops the necessary strength. Of course, under severe conditions, such as freeze-thaw cycles or exposure to sulfate water, the proportioning of concrete mixture will require special attention.

Another purpose of mix proportioning is to obtain a concrete mixture satisfying the performance requirements ***at the lowest possible cost;*** this involves decisions regarding the selection of ingredients that are not only suitable but are available at reasonable prices. The ***overall objective*** of proportioning concrete mixtures can therefore be summarized as selecting the suitable ingredients among the available materials and determining the most economical combination that will produce concrete with certain minimum performance characteristics.

The tools available to the engineer to achieve this objective are limited. An obvious constraint in concrete mix proportioning is that within a fixed volume you cannot alter one component independent of others; for example, in a cubic yard of concrete, if the aggregate component is increased, the cement paste component will decrease. With concrete-making materials of given characteristics, and given job conditions (i.e., structural design and equipment for handling concrete), the variables generally under the control of a mix designer are as follows: the cement paste/aggregate ratio in the mixture, the water/cement ratio in the cement paste, the sand/coarse aggregate ratio in the aggregates, and the use of admixtures.

The task of mix proportioning is complicated by the fact that certain desired properties of concrete may be oppositely affected by changing a specific variable. For example, the addition of water to a stiff concrete mixture with a given cement content will make the flowability of fresh concrete better but at the same time reduce the strength. In fact, workability itself is composed of two main components [i.e., consistency (ease of flow) and cohesiveness (resistance to segregation)], and both tend to be affected in an opposite manner when water is added to a given concrete mixture. The process of mix designing therefore boils down to the art of balancing the various conflicting effects, such as those described above.

GENERAL CONSIDERATIONS

Before discussing the specific principles underlying the procedures commonly used for mix proportioning, it is desirable to examine some of the general considerations governing the whole process. The considerations of cost, workability, strength, and durability of concrete are usually the most important, and will be discussed next.

Cost

A consideration that should be obvious is the choice of concrete-making materials that are technically acceptable, and at the same time economically attractive. In other words, when a material is available from two or more sources and a significant price differential exists, the least expensive source of supply is usually selected unless there are demonstrable technical reasons that the material will not be suitable for the job at hand.

In spite of the usually small differences in the price of aggregates from various local sources, the overall savings for a large project are worthy of examination. Assuming that a mix design calls for 3000 lb/yd^3 of total aggregate for a job requiring approximately 6 million cubic yards of concrete, and that the two sources capable of furnishing suitable aggregates have a 12-cent/ton price difference between them, a simple computation will show that a cost saving of over $1 million is possible if the less expensive aggregate is selected for use.

At times, for traditional or other reasons which may no longer be valid, some specifying agencies continue to require materials for concrete that are more expensive and perhaps unnecessary. For example, requiring the use of a low-alkali portland cement when the locally available cements are of the high-alkali type and the aggregates are essentially free from alkali-reactive minerals will increase the cost of concrete due to the extra haulage expense for the low-alkali cement. If the aggregate under consideration contains reactive particles, the use of mineral admixtures in combination with a high-alkali cement may turn out to be the more cost-effective alternative.

A *key consideration* that governs many of the principles behind the procedures for proportioning concrete mixtures is the recognition that *cement costs much more than aggregates* (e.g., 10 times or even more); therefore, all possible steps should be taken to reduce the cement content of a concrete mixture without sacrificing the desirable properties of concrete, such as strength and durability.

For the purposes of illustration, let us refer to the data in Fig. 3–7. Between concrete mixes 1 and 3, a reduction in the cement content from 530 to 460 lb per cubic yard of concrete at a given water/cement ratio (i.e., without compromising the strength of concrete) made it possible to lower the cost by $1.55 per cubic yard, provided that a 1-in. consistency (slump) instead of 6 in. was acceptable for the job. This may well be the case with lightly reinforced or unreinforced massive concrete structures. The economic implication of reduction in cement content can be enormous in projects requiring large amounts of concrete.

The scope for cost reduction can be enlarged further, without compromising the essential performance characteristics of a concrete mixture, *if cheaper materials are found to replace a part of the portland cement.* For instance, under most conditions, substitution of pozzolanic or cementitious by-products (such as fly ash or ground iron

General Considerations

blast-furnace slag) for portland cement is likely to produce direct savings in the cost of materials. Furthermore, at some point in the future every nation will have to consider the *indirect cost savings* resulting from resource preservation and reduced pollution when these industrial by-products are utilized properly, instead of being dumped into the environment.

Workability

Workability of fresh concrete determines the ease with which a concrete mixture can be handled without harmful segregation. In all likelihood, a concrete mixture that is difficult to place and consolidate will not only increase the cost of handling but will also have poor strength, durability, and appearance. Similarly, mixtures that are prone to segregate and bleed are more expensive to finish and will yield less durable concrete. Thus workability can affect both the cost and the quality of concrete mixtures.

As an important property of concrete, that is mostly dependent on mix design, there is a problem with workability. The term *workability represents many diverse characteristics of fresh concrete that are difficult to measure quantitatively*. Consequently, the proportioning of concrete mixtures for a desirable but not-fully-definable measure of workability remains an art as well as a science. This is another reason why a mere knowledge of mix design procedures is not sufficient without an understanding of the basic principles involved.

General considerations that guide the decisions affecting workability of concrete mixtures are as follows:

- The consistency of concrete should be no more than necessary for placing, compacting, and finishing.
- The water requirement for a given consistency depends mostly on the aggregate characteristics; therefore, wherever possible the cohesiveness and finishability of concrete should be improved by increasing the sand/coarse aggregate ratio rather than by increasing the proportion of fine particles in the sand.
- For concrete mixtures requiring high consistency at the time of placement, the use of water-reducing and set-retarding admixtures should be considered rather than the addition of more water at the job site; water that has not been accounted for in mix proportioning has frequently been responsible for the failure of concrete to perform according to the design specifications.

Strength and Durability

It was described in Chapter 2 that strength and permeability of hydrated cement paste are mutually related through the capillary porosity which is controlled by water/cement ratio and degree of hydration (Fig. 2–11). In general, with the exception of freeze-thaw resistance, since durability of concrete is controlled mainly by its permeability, it is not difficult to understand why there is a direct relationship between strength and durability. Consequently, *in routine mix design operations, only workability and strength are emphasized;* consideration of durability is ignored unless special environmental exposures require it.

With normally available cements and aggregates, structural concretes of consistency and strength that are adequate for most purposes—for instance, up to 6 in. of slump and 5000-psi 28-day compressive strength (which corresponds to an 0.5 water/cement ratio)—can be produced without any difficulty. When strength or durability considerations require a low water/cement ratio it is generally achieved not by increasing the cement content, but by lowering the water demand at a given cement content (through control of the aggregate grading and the use of water-reducing admixtures). This is not only more economical, but it also reduces the chances of cracking due to a high heat of hydration or drying shrinkage. To obtain high consistency, high strength, or control of the rate of strength development, the use of water-reducing and set-controlling admixtures is therefore often considered.

Ideal Aggregate Grading

From the foregoing considerations of cost, workability, strength, and durability, it may be concluded that the most dense aggregate-packing with a minimum content of voids will be the most economical because it will require the least amount of cement paste. This conclusion led to a large number of theoretical studies on the packing characteristics of granular materials. The objective of such studies was to obtain mathematical expressions or ideal grading curves which would help determine the ideal combinations of particles of different sizes that produce the minimum void space. Most of the theoretical expressions and curves thus developed are parabolic. It is not necessary to describe them here because, in practice, these ideal gradings simply do not produce workable concrete mixtures, besides being uneconomical.

In concrete technology, the idea of an ideal aggregate grading has now largely been abandoned. For practical purposes it is adequate to follow the grading limits specified by ASTM C 33 (Chapter 7), which are not only broad and therefore economically feasible, but are also based on practical experience rather than theoretical considerations. It is possible to make satisfactory concrete mixtures from almost any type of aggregate grading within the ASTM C 33 specification limits; aggregate gradings outside the limits may cause workability problems and may not be cost-effective (i.e., may produce large void space on compaction).

Specific Principles

When reviewing the following specific principles that are behind the procedures for selection of concrete mix proportions, it will be helpful to remember again that the underlying goal is to strike a reasonable balance between workability, strength, durability, and cost considerations.

Workability. As already stated, workability embodies certain characteristics of fresh concrete, such as consistency and cohesiveness. *Consistency,* broadly speaking, is a measure of the wetness of the concrete mixture, which is commonly evaluated in terms of slump (i.e., the wetter the mixture, the higher the slump). Since water content is a key factor affecting the cost economy, it should be noted that there is almost a direct proportionality between slump and water content for a given set of materials. For a given slump, the mixture water requirement generally decreases as: (1) the maximum size of a well-graded aggregate is increased; (2) the content of angular and rough-textured particles in the aggregate is reduced, and (3) the amount of entrained air in the concrete mixture is increased.

Cohesiveness is a measure of compactability and finishability, which is generally evaluated by trowelability and visual judgment of resistance to segregation. In trial mixtures when cohesiveness is judged as poor, it can usually be improved by taking one or more of the following steps: increase in sand/coarse aggregate proportion, partial replacement of coarse sand with a finer sand, and increase of cement paste/aggregates ratio (at the given water/cement ratio).

Since the slump affects the ease with which the concrete mixture will flow during placement, and the test for slump is simple and quantitative, most *mix design procedures rely on slump as a crude index of workability;* it is assumed that mixtures containing adequate cement content (with or without mineral admixtures) and well-graded aggregates will have a satisfactory degree of cohesiveness. It should be noted that in the laboratory several trial mixtures are usually necessary before arriving at a qualitative notion of workability that is considered satisfactory. Due to differences in equipment, further adjustment in mix proportions may be needed after experience with full-size field batches. This is why the importance of past experience is recognized by the mix-proportioning procedures.

It is worth mentioning here that there are no standard requirements for workability because the requirements needed for a particular placement vary, depending on the type of construction and the equipment used to transport and consolidate concrete. For example, the workability desired for a slip-formed unreinforced pavement will not be the same as for a congested reinforced column, and the workability desired for pumped concrete in a high-rise structure will not be the same as for mass concrete to be placed by a crane or a belt conveyor.

Strength. From the standpoint of structural safety, the strength of concrete specified by the designer is treated as the minimum required strength. Therefore, to account for variations in materials, methods of mixing, transporting, and placing of concrete, as well as the making, curing, and testing of the concrete specimens, ACI Building Code 318 requires a certain degree of strength overdesign, which is based on statistical considerations. In other words, depending on the variability of test results, the mix proportions selected must yield a mean or average strength which is higher than the minimum or the specified strength. The procedure for determining the average strength from a specified strength value is given in the Appendix at the end of this chapter. It should be noted that the average strength, not the specified strength, is used in mix design calculations.

Although other factors also influence strength, the tables and charts used for the purposes of mix proportioning assume that *strength is solely dependent on the water/cement ratio and the content of entrained air in concrete.* A more accurate relationship between the strength and water/cement ratio for a given set of materials and conditions may be available from past experience, or should be developed from trial mixtures. Depending on the moisture state of the aggregates, it is necessary to make corrections in the amounts of mixing water, sand, and coarse aggregate in order to assure that the water/cement ratio in the concrete mixture is correct.

Durability. As stated earlier, when concrete is subject to normal conditions of exposure, *the mix-proportioning procedures ignore durability,* because strength is considered to be an index of general durability. However, under conditions that may tend to shorten the service life of concrete, its durability may be enhanced by special considerations in mix proportioning. For example, entrained air is required in all exposed concrete in climates where freezing occurs. Concrete to be exposed to chemical attack by deicing salts or acidic or sulfate waters may require the use of water-reducing and mineral admixtures. In a given situation, although a higher water/cement ratio would satisfy the strength requirement, a lower water/cement ratio may have to be used when specified from the standpoint of exposure conditions.

PROCEDURES

Numerous procedures for computing concrete mix proportions are prevalent in the world. A comprehensive review of the British procedure is contained in *Properties of Concrete*.[1] To illustrate the principles already described, the two procedures used in the United States will be described here. The procedures are based on the report by ACI Committee 211.[2]

The *weight method* is considered less exact but does not require the information on the specific gravity of the concrete-making materials. The *absolute volume method* is considered more exact. Both procedures involve a sequence of nine steps which are given below, the first six steps being common. To the extent possible, the following background data should be gathered before starting the calculations:

- Sieve analysis of fine and coarse aggregate; fineness modulus
- Dry-rodded unit weight of coarse aggregate
- Bulk specific gravity of materials
- Absorption capacity, or free moisture in the aggregate
- Variations in the approximate mixing water requirement with slump, air content, and grading of the available aggregates
- Relationships between strength and water/cement ratio for available combinations of cement and aggregate

---

[1] A. M. Neville, *Properties of Concrete*, Pitman Publishing, Inc., Marshfield, Mass., 1981.

[2] Standard *Practice for Selecting Proportions for Normal, Heavy-Weight and Mass Concrete*, ACI 211.1 Report, ACI Manual of Concrete Practice, Part 1, 1991.

- Job specifications if any [e.g., maximum water/cement ratio, minimum air content, minimum slump, maximum size of aggregate, and strength at early ages (normally, 28-day strength is specified)]

Regardless of whether the concrete characteristics are prescribed by the specifications or left to the mix designer, the batch weights in pounds per cubic yard of concrete can be computed in the following sequence:

Step 1: Choice of slump. If the slump is not specified, a value appropriate for the work can be selected from Table 9–1. Mixes of the stiffest consistency that can be placed and compacted without segregation should be used.

Step 2: Choice of maximum size of aggregate. For the same volume of the coarse aggregate, using a large maximum size of a well-graded aggregate will produce less void space than using a smaller size, and this will have the effect of reducing the mortar requirement in a unit volume of concrete. Generally, the maximum size of coarse aggregate should be the largest that is economically available and consistent with the dimensions of the structure. In no event should the maximum size exceed one-fifth of the narrowest dimension between the sides of the forms, one-third the depth of slabs, or three-fourths of the minimum clear spacing between reinforcing bars.

Step 3: Estimation of mixing water and the air content. The quantity of water per unit volume of concrete required to produce a given slump is dependent on the maximum particle size, shape, and grading of the aggregates, as well as on the amount of entrained air; it is not greatly affected by the cement content of the concrete mixture. If data based on experience with the given aggregates are not available, assuming normally-shaped and well-graded particles, an estimate of the mixing water with or without air entrainment can be obtained from Table 9–2 for the purpose of computing the trial batches. The data in the table also show the approximate amount of entrapped air expected

TABLE 9–1 RECOMMENDED SLUMP FOR VARIOUS TYPES OF CONSTRUCTION

	Slump (in.)	
Types of construction	Maximum[a]	Minimum
Reinforced foundation walls and footings	3	1
Plain footings, caissons, and substructure walls	3	1
Beams and reinforced walls	4	1
Building columns	4	1
Pavements and slabs	3	1
Mass concrete	2	1

[a] May be increased 1 in. for methods of consolidation other than vibration.
Source: Reproduced with permission from the American Concrete Institute.

TABLE 9-2 APPROXIMATE MIXING WATER AND AIR CONTENT REQUIREMENTS FOR DIFFERENT SLUMPS AND NOMINAL MAXIMUM SIZES OF AGGREGATES

Slump, in.	Water, lb/yd³ of concrete for indicated nominal maximum sizes of aggregate							
	⅜ in.*	½ in.*	¾ in.*	1 in.*	1½ in.*	2 in.*†	3 in.†	6 in.†
	Non-air entrained concrete							
1 to 2	350	335	315	300	275	260	220	190
3 to 4	385	365	340	325	300	285	245	210
6 to 7	410	385	360	340	315	300	270	—
More than 7*	—	—	—	—	—	—	—	—
Approximate amount of entrapped air in non-air-entrained concrete, percent	3	2.5	2	1.5	1	0.5	0.3	0.2
	Air-entrained concrete							
1 to 2	305	295	280	270	250	240	205	180
3 to 4	340	325	305	295	275	265	225	200
6 to 7	365	345	325	310	290	280	260	—
More than 7*	—	—	—	—	—	—	—	—
Recommended averages total air content, percent for level of exposure:								
Mild exposure	4.5	4.0	3.5	3.0	2.5	2.0	1.5*††	1.0*††
Moderate exposure	6.0	5.5	5.0	4.5	4.5	4.0	3.5*††	3.0*††
Severe exposure††	7.5	7.0	6.0	6.0	5.5	5.0	4.5*††	4.0*††

*The quantities of mixing water given for air-entrained concrete are based on typical total air content requirements as shown for "moderate exposure" in the table above.

†The slump values for concrete containing aggregate larger than 1½ in. are based on slump tests made after removal of particles larger than 1½ in. by wet-screening.

**For concrete containing large aggregates that will be wet-screened over the 1½ in. sieve prior to testing for air content, the percentage of air expected in the 1½ in. minus material should be as tabulated in the 1½ in. column. However, initial proportioning calculations should include the air content as a percent of the whole.

††When using large aggregate in low cement factor concrete, air entrainment need not be detrimental to strength. In most cases mixing water requirement is reduced sufficiently to improve the water-cement ratio and to thus compensate for the strength-reducing effect of air-entrained concrete. Generally, therefore, for these large nominal maximum sizes of aggregate, air contents recommended for extreme exposure should be considered even though there may be little or no exposure to moisture and freezing.

in non-air-entrained concrete and recommend levels of air content for concrete in which air is to be purposely entrained for frost resistance.

Step 4: Selection of water/cement ratio. Since different aggregates and cements generally produce different strengths at the same water/cement ratio, it is highly desirable to develop the relationship between strength and water/cement ratio for the materials actually to be used. In the absence of such data, approximate and relatively conservative values for concretes made with Type I portland cement can be taken as shown in Table 9–3. Since the selected water/cement ratio must satisfy both the strength and the durability criteria, the value obtained from the table may have to be reduced depending on the special exposure requirements (Table 9–4).

Step 5: Calculation of cement content. The required cement content is equal to the mixing water content (Step 3) divided by the water/cement ratio (Step 4).

Step 6: Estimation of coarse-aggregate content. Economy can be gained by using the maximum possible volume of coarse aggregate on a dry-rodded basis per unit volume of concrete. Data from a large number of tests have shown that for properly graded materials, the finer the sand and the larger the size of the particles in coarse aggregate, the more is the volume of coarse aggregate that can be used to produce a concrete mixture of satisfactory workability. It can be seen from the data in Table 9–5 that, for a suitable degree of workability, the volume of coarse aggregate in a unit volume of concrete is dependent only on its maximum size and the fineness modulus of the fine aggregate. It is assumed that differences in the amount of mortar required for workability with

TABLE 9–3 RELATIONSHIPS BETWEEN WATER-CEMENT RATIO AND COMPRESSIVE STRENGTH OF CONCRETE

	Water-cement ratio, by weight	
Compressive strength at 28 days (psi)[a]	Non-air-entrained concrete	Air-entrained concrete
6000	0.41	—
5000	0.48	0.40
4000	0.57	0.48
3000	0.68	0.59
2000	0.82	0.74

[a] Values are estimated average strengths for concrete containing not more than the percentage of air shown in Table 9–2. For a constant water-cement ratio, the strength of concrete is reduced as the air content is increased. Strength is based on 6 by 12-in. cylinders moist-cured 28 days at 73.4 ± 3°F (23 ± 1.7°C) in accordance with Section 9(b) of ASTM C 31, for *Making and Curing Concrete Compression and Flexure Test Specimens in the Field.*

Source: Reproduced with permission from the American Concrete Institute.

TABLE 9-4 RECOMMENDATIONS FOR NORMAL WEIGHT CONCRETE SUBJECT TO SULFATE ATTACK

Exposure	Water soluble sulfate* (SO_4) in soil, percent	Sulfate* (SO_4) in water, ppm	Cement	Water-cement ratio, maximum†
Mild	0.00–0.10	0–150	—	—
Moderate†	0.10–0.20	150–1500	Type II, IP(MS), IS(MS)‡	0.50
Severe	0.20–2.00	1500–10,000	Type V§	0.45
Very severe	Over 2.00	Over 10,000	Type V + pozzolan or slag**	0.45

* Sulfate expressed as SO_4 is related to sulfate expressed as SO_3 as in reports of chemical analysis of cement as $SO_3 \times 1.2 = SO_4$.

† When chlorides or other depassivating agents are present in addition to sulfate, a lower water-cement ratio may be necessary to reduce corrosion potential of embedded items. Refer to Chapter 4.

‡ Or a blend of Type I cement and a ground granulated blast furnace slag or a pozzolan that has been determined by tests to give equivalent sulfate resistance.

§ Or a blend of Type II cement and a ground granulated blast furnace slag or a pozzolan that has been determined by tests to give equivalent sulfate resistance.

** Use a pozzolan or slag that has been determined by tests to improve sulfate resistance when used in concrete containing Type V cement.

Source: ACI Committee 201, Guide to Durable Concrete, *ACI Materials Jour.*, Vol. 88, No. 5, p. 553, 1991.

different aggregates, due to differences in particle shape and grading, are compensated for automatically by differences in dry-rodded void content.

The volume of aggregate, in cubic feet, on a dry-rodded basis, for 1 yd³ of concrete is equal to the volume fraction obtained from Table 9–5 multiplied by 27. This volume is converted to the dry weight of coarse aggregate by multiplying by its dry-rodded unit weight.

Step 7: Estimation of fine-aggregate content. At the completion of Step 6, all the ingredients of the concrete have been estimated except the fine aggregate; its quantity is determined by difference, and at this stage either the "weight" method or the "absolute volume" method can be followed.

According to the *weight method,* if the unit weight of fresh concrete is known from previous experience, then the required weight of fine aggregate is simply the difference between the unit weight of concrete and the total weights of water, cement, and coarse aggregate. In the absence of a reliable estimate of the unit weight of concrete, the first estimate for a concrete of medium richness (550 lb of cement per cubic yard, medium slump of 3 to 4 in.)

TABLE 9-5 VOLUME OF COARSE AGGREGATE PER UNIT OF VOLUME OF CONCRETE

Maximum size of aggregate (in.)	Volume of dry-rodded coarse aggregate[a] per unit volume of concrete for different fineness moduli of sand			
	2.40	2.60	2.80	3.00
3/8	0.50	0.48	0.46	0.44
1/2	0.59	0.57	0.55	0.53
3/4	0.66	0.64	0.62	0.60
1	0.71	0.69	0.67	0.65
1½	0.75	0.73	0.71	0.69
2	0.78	0.76	0.74	0.72
3	0.82	0.80	0.78	0.76
6	0.87	0.85	0.83	0.81

[a] Volumes are based on aggregates in dry-rodded condition as described in ASTM C 29, *Unit Weight of Aggregate*. These volumes are selected from empirical relationships to produce concrete with a degree of workability suitable for usual reinforced construction. For less workable concrete such as required for concrete pavement construction they may be increased about 10%. For more workable concrete, such as may sometimes be required when placement is to be by pumping, they may be reduced up to 10%.

Source: Reproduced with permission from the American Concrete Institute.

and approximately 2.7 aggregate specific gravity can be obtained from Table 9-6. Experience shows that even a rough estimate of the unit weight is adequate for the purpose of making trial batches.

In the case of the *absolute volume method* the total volume displaced by the known ingredients (i.e., water, air, cement, and coarse aggregate) is subtracted from the unit volume of concrete to obtain the required volume of fine aggregate. This in turn is converted to weight units by multiplying it by the density of the material.

Step 8: Adjustments for aggregate moisture. Generally, the stock aggregates are moist; without moisture correction the actual water/cement ratio of the trial mix will be higher than selected by Step 4, and the saturated-surface dry (SSD) weights of aggregates will be lower than estimated by Steps 6 and 7. The mix proportions determined by Steps 1 to 7 are therefore assumed to be on an SSD basis. For the trial batch, depending on the amount of free moisture in the aggregates, the mixing water is reduced and the amounts of aggregates correspondingly increased, as will be shown by the sample computations.

Step 9: Trial batch adjustments. Due to so many assumptions underlying the foregoing theoretical calculations, the mix proportions for the actual materials to be used must be checked and adjusted by means of laboratory trials consisting of small batches (e.g., 0.01 yd^3 of concrete). Fresh concrete should be tested for slump, workability (freedom from segregation), unit weight, and air content; specimens of hardened concrete cured under standard conditions should be tested for strength at the specified age. After several trials, when a mixture satisfying the desired criteria of workability and strength is obtained, the mix proportions of the laboratory-size trial batch are scaled up for producing fullsize field batches.

SAMPLE COMPUTATIONS

Job Specifications

Type of construction	Reinforced concrete footing
Exposure	Mild (below ground, not exposed to freezing or sulfate water)
Maximum size of aggregate	1½ in.
Slump	3 to 4 in.
Specified 28-day compressive strength	3500 psi

Characteristics of the Materials Selected

	Cement, Lone Star, Type 1	Fine aggregate, Felton, No. 2	Coarse aggregate, Fair Oaks, Gravel
Bulk specific gravity	3.15	2.60	2.70
Bulk density (lb/ft^3)	196	162	168
Dry-rodded unit weight (lb/ft^3)	–	–	100
Fineness modulus	–	2.8	–
Moisture deviation from SSD condition (%)	–	+2.5	+0.5

Steps 1 To 7: Computing Mix Proportions (SSD Basis, lb/yd^3)

Step 1. Slump = 3 to 4 in. (given).

Step 2. Maximum aggregate size = 1½ in. (given).

Step 3. Mixing water content (non-air-entrained concrete) = 300 lb. Approximate amount of entrapped air = 1 percent (Table 9–2).

Step 4. Average strength from equations in the Appendix (assuming 300 psi standard deviation from past experience) = 3500 + 1.34 × 300 = 3900 psi. Water/cement ratio (Table 9–3) = 0.58.

Step 5. Cement content = 300/0.58 = 517 lb.

Step 6. Volume fraction of gravel on dry-rodded basis (Table 9–5) = 0.71

Sample Computations

TABLE 9–6 FIRST ESTIMATE OF WEIGHT OF FRESH CONCRETE

Maximum size of aggregate (in.)	First estimate of concrete weight[a] (lb/yd^3)	
	Non-air-entrained concrete	Air-entrained concrete
3/8	3840	3690
1/2	3890	3760
3/4	3960	3840
1	4010	3900
1 1/2	4070	3960
2	4120	4000
3	4160	4040
6	4230	4120

[a] Values calculated for concrete of medium richness (550 lb of cement per cubic yard) and medium slump with aggregate specific gravity of 2.7. Water requirements based on values for 3 to 4 in. of slump in Table 9–2. If desired, the estimated weight may be refined as follows when necessary information is available: for each 10-lb difference in mixing water from the Table 9–2 values for 3 to 4 in. of slump, correct the weight per cubic yard 15 lb in the opposite direction; for each 100-lb difference in cement content from 550 lb, correct the weight per cubic yard 15 lb in the same direction; for each 0.1 by which aggregate specific gravity deviates from 2.7, correct the concrete weight 100 lb in the same direction.

Source: Reproduced with permission from the American Concrete Institute.

Dry-rodded volume of gravel = 0.71 × 27 = 19.17 ft^3
Weight of gravel = 19.17 × 100 = <u>1917 lb.</u>

Step 7. Using the *weight method:* unit weight of concrete (Table 9–6) = 4070 lb/yd^3

Weight of sand = 4070 − (300 + 517 + 1917) = <u>1336 lb.</u>

Using the *absolute volume method:*

Volume displaced by water   = 300/62.4   = 4.81 ft^3
Volume displaced by cement = 517/196   = 2.64 ft^3
Volume displaced by gravel = 1917/168   = 11.43 ft^3
Volume displaced by air   = 27 × 0.01   = <u>0.27</u> ft^3
                                total   19.15

Volume displaced by sand   = (27 − 19.15) = 7.85 ft^3
Weight of sand = 7.85 × 162 = <u>1272 lb.</u>

Since the absolute volume method is more exact, the proportions determined by this method will be used.

Step 8: Moisture Adjustment for the Laboratory Trial Batch

Material	SSD (lb/yd³)	SSD (lb/0.01 yd³)	Moisture correction (lb)	Mix proportions for the first trial batch (lb)
Cement	517	5.17		5.17
Sand	1272	12.72	12.72 × 0.025 = 0.3	13.02
Gravel	1917	19.17	19.17 × 0.005 = 0.1	19.27
Water	300	3.00	3 − (0.3 + 0.1)	2.60
Total	4006	40.06	←—— must be equal ——→	40.06

Step 9: Making The First Laboratory Trial and Adjusting The Proportions

Measured properties of fresh concrete from the first trial batch:

Slump = 4¾ in.
Workability = slight tendency to segregate and bleed
Unit weight = 148 lb/ft³ (3996 lb/yd³)
Air content = 1%

Action taken for the second trial batch: reduce the gravel by 1/4 lb and increase the sand by the same amount.

Batch weights for the second trial batch:

Cement = 5.17 lb
Sand   = 13.27 lb
Gravel = 19.02 lb
Water = 2.60 lb
         40.06 lb

Measured properties of fresh concrete from the second trial batch:

Slump = 4 in.
Workability = satisfactory
Unit weight = 148 lb/ft³
Air content = 1%

Three 3- by 6-in. cylinders were cast and moist cured at 73.4 ± 3°F.

Average 28-day compressive strength was 4250 psi, with less than 5 percent variation in strength between the individual cylinders.

Recalculated mix proportions for the full-size field batch are as follows.

	Present stock (lb/yd³)	Moisture correction (for conversion to SSD condition) (lb)	SSD basis (lb/yd³)
Cement	517		517
Sand	1327	1327 × 0.025 = 33	1294
Gravel	1902	1902 × 0.005 = 10	1892
Water	260	260 + (33 + 10)	303
Total	4006	←—— must be equal ——→	4006

APPENDIX: METHODS OF DETERMINING AVERAGE COMPRESSIVE STRENGTH FROM THE SPECIFIED STRENGTH[3]

ACI 322, *Building Code Requirements for Structural Plain Concrete*, and ACI 318, *Building Code Requirements for Reinforced Concrete*, specify that concrete shall be proportioned to provide an average compression strength (f'_{cr}) which is higher than the specified strength (f'_c) so as to minimize the probability of occurrence of strengths below f'_c.

When a concrete production facility has a suitable record of 30 consecutive tests of similar materials and conditions expected, the standard deviation can be calculated in accordance with the expression

$$S = \left[\frac{\Sigma(x_i - \bar{x})^2}{n-1} \right]^{1/2} \qquad (9\text{--}1)$$

where S is the standard deviation (psi), x_i the strength value from an individual test, $\bar{x}$ the average strength of n tests, and n the number of consecutive strength tests. When data for 15 to 25 tests are available, the calculated value of the standard deviation may be modified according to the following data:

Number of tests	Multiplication factor
15	1.16
20	1.08
25	1.03
30 or more	1.00

The required average compressive strength (f'_{cr}), which is to be used as the basis for calculating concrete mix proportions, shall be the larger of Eq. (9–2) or (9–3):

$$f'_{cr} = f'_c + 1.34S \qquad (9\text{--}2)$$

$$f'_{cr} = f'_c + 2.33S - 500 \qquad (9\text{--}3)$$

Equation (9–2) provides a probability of 1 in 100 that averages of three consecutive tests will be below the specified strength f'_c. Equation (9–3) provides a similar probability of individual tests being more than 500 psi below the specified strength.

When adequate data are not available to establish a standard deviation, the required average strength can be determined from the following:

Specified compressive strength, f'_c (psi)	Required average compressive strength, f'_{cr} (psi)
Less than 3000	$f'_c + 1000$
3000 to 5000	$f'_c + 1200$
Over 5000	$f'_c + 1400$

[3] Based on ACI Building Code 318.

Figure 9–1 shows a flowchart from the ACI Building Code Commentary (318R-89) outlining the mix selection and documentation procedure based either on field experience or trial mixtures.

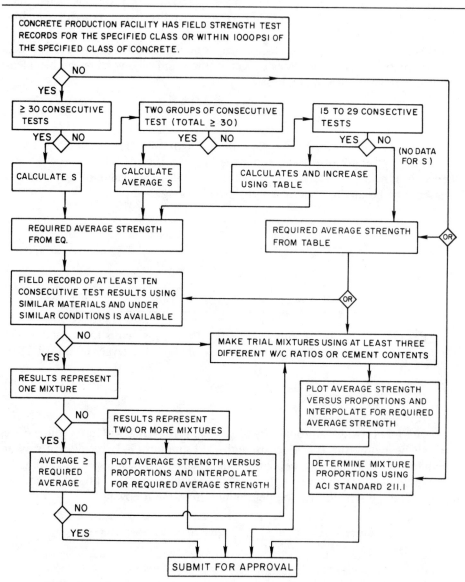

Figure 9-1 Flow chart for selection and documentation of concrete proportions. (Adapted from ACI 318R-89. Reproduced by permission.)

Suggestions for Further Study

TEST YOUR KNOWLEDGE

1. Explain why the process of proportioning concrete mixtures is still in the realm of art. Have you any ideas on how to make the currently used practice in the United States more scientific?
2. You find yourself the project manager for a concrete structure involving several million cubic yards of concrete. Briefly, what tips would you like to pass on to the engineer in charge of mix proportioning on the subject of materials cost reduction? In your answer emphasize the key ingredient in concrete from the standpoint of cost.
3. Why is it not necessary to take into account durability considerations in concrete mix proportioning when the concrete is subject to normal exposure conditions? Give examples of circumstances when durability must be considered in mix designing.
4. Theoretically derived ideal gradings of aggregates for maximum density should be the most economical, yet the practice is not followed. Can you explain why?
5. In mix designing, why is it desirable to use a minimum amount of water? For a given slump, how can you reduce the amount of water?
6. Describe the significance of workability of concrete and the factors affecting the property.
7. According to the ACI Building Code 318, selection of mix proportions should be based on the average strength, not the specified strength. Is this justified? Given a specified strength value, what procedures are used to determine the average strength?
8. With respect to the ACI 211.1, Standard Practice for Selecting Proportions for Normal Heavy-Weight and Mass Concrete, explain the principles underlying the following:
   (a) Estimation of water content.
   (b) Estimation of coarse aggregate content.
   (c) Estimation of fine aggregate content by the *weight method*.
   (d) Estimation of fine aggregate content by the *absolute volume method*.
9. (a) Briefly state the influence of maximum aggregate size (i.e., 19 mm vs. 38 mm) on the mixing water content and the cement content of a concrete mixture with a given water/cement ratio of 0.5.
   (b) Why is it important to control the aggregate gradation once the concrete mix design has been selected? How is this gradation control expressed in a specification?
10. Given the following SSD proportions (lb/yd^3) for a concrete mixture, compute the batch weights for the job when the sand contains 4% free moisture and the gravel has 1% effective absorption:
    cement = 500; water = 300; sand = 1350; gravel = 1900.
11. The proportions by mass for a concrete mixture are given as follows:
    cement = 1
    water = 0.53
    sand = 2.50
    gravel = 3.50
    If the unit weight is 149 lb/ft^3, compute the cement content.
12. Determine the SSD mix proportions of concrete required for an outdoor pavement subject to frequent freeze-thaw cycles. The following data are given:
    Specified 28-day compressive strength: 3000 psi
    Slump: 3 in.

Coarse aggregate: 1 in. max. size; 102 lb/ft^3 dry-rodded weight
Fine aggregate: 2.8 fineness modulus
Specified gravities of cement, coarse aggregate, and fine aggregate: 3.15, 2.72, and 2.70, respectively

SUGGESTIONS FOR FURTHER STUDY

ACI Committee 318, *Building Code Requirements for Reinforced Concrete, Building Code Commentary*, ACI 318R, 1983.

ACI Standard 211.1, *Standard Practice for Selecting Proportions for Normal, Heavyweight, and Mass Concrete*, ACI Manual of Concrete Practice, Part 1, 1991.

ACI Committee 211.2. "Standard Practice for Selecting Proportions for Structural Lightweight Concrete," *ACI Materials Jour.*, Vol. 87, No. 6, pp. 638–651, 1990.

Design and Control of Concrete Mixtures, Portland Cement Association, Skokie, Ill., 1979, pp. 55–65.

NEVILLE, A. M., *Properties of Concrete,* Pitman Publishing, Inc., Marshfield, Mass., 1981, pp. 648–712.

CHAPTER 10

Concrete at Early Ages

PREVIEW

The selection of proper materials and mix proportions no doubt are important steps in achieving the goal of producing a concrete that will meet given requirements of strength and durability in a structure; however, the goal may still remain elusive if adequate attention is not paid to the operations to which concrete is subjected at the early age. The term early age covers only an insignificant amount of time (e.g., first 2 days after production) in the total life expectancy of concrete, but during this period numerous operations are performed such as mixing, conveying to the job site, placement in forms, consolidation, finishing, curing, and removal of formwork. These operations are affected by the characteristics of fresh concrete: for instance, workability, setting time, and maturity or rate of strength gain. Obviously, the control of both the early-age operations and properties of fresh concrete is essential to assure that a finished concrete element will be structurally adequate for the purpose for which it was designed.

A detailed description of the operations and equipment used for batching, mixing, conveying, placing, consolidation, and finishing fresh concrete is beyond the scope of this book. Only the basic steps and their significance are described in this chapter. Also described are the significance and control of workability, slump loss, segregation and bleeding, plastic shrinkage, setting time, and temperature of concrete. As effective and economical tools of modern quality assurance programs, accelerated strength testing procedures, in situ and non-destructive test methods, and statistical quality control charts are briefly discussed.

DEFINITIONS AND SIGNIFICANCE

In the medical profession it is well recognized that in order to develop into a healthy person a newborn baby needs special attention during the early period of growth. Something similar applies to concrete, but in both cases there is no clear definition of *how early is the early age*. Concrete technologists agree that deficiencies acquired by freshly made concrete due to loss of workability at or before placement, segregation and bleeding during consolidation, and an unusually slow rate of maturity (strength gain) can impair a concrete permanently and reduce its service life.

Addressing the question of how early is the early age, S. G. Bergstrom of the Swedish Cement and Concrete Research Institute said:

> Time is not a very good measure of "early." The time when the concrete has reached a certain maturity, is dependent on so many factors: cement type, reactivity of the cement, temperature, admixtures, etc. The time factor is not significant in the general case if you are not specifying the case very carefully. Then of course the degree of hydration gives a much better indication, which however is not always available if we deal with the practical side. You can also use another more practical definition perhaps, giving the time the property you are interested in has reached the level you need. All times earlier than that level are evidently early ages; which means that the definition depends on the way you will use concrete. The form stripper would say that he needs about 15 MPa, whereas a slipformer does not need as much as that. These two have quite different concepts about early age. The answer is that there is no universal answer. When we try to define the area where we are going to work, we will, *as a rule of thumb and for the normal concrete, in normal situations, say about 24 hours, some say about 48 hours, but that is just to indicate the order of magnitude*....[1]

Since a normal concrete mixture (i.e., concrete made with a normal portland cement and subjected to normal handling operations and curing temperatures) generally takes 6 to 10 hr for setting and 1 to 2 days for achieving a desired strength level before formwork can be removed, the definition of early age includes on the one hand the freshly mixed concrete of plastic consistency and, on the other, 1- to 2-day-old hardened concrete that is strong enough to be left unattended (except continuation of moist curing, as will be discussed).

The early-age period in the life of concrete is insignificantly small compared to the total life expectancy, but during this period it is subjected to many operations which are not only affected by the properties of the material but also influence them. For instance, a mixture with poor workability will be hard to mix; on the other hand, too much mixing may reduce the workability. It is beyond the scope of this book to describe in detail the operations and the equipment used, but engineers should be familiar with the sequence of main operations, their interaction with characteristics of concrete at early ages, and the terminology used in the field practice.

---

[1] S. G. Bergstrom, Conclusions from the Symposium on Concrete at Early Ages, Paris, April 6–8, 1982, *RILEM Bulletin*.

Batching, Mixing, and Conveying

In general, the sequence of main operations is as follows: batching, mixing, and conveying the concrete mixture from the point where it is made to the job site; placing the plastic concrete at the point where it is needed; compacting and finishing while the mixture is still workable; finally, moist curing to achieve a desired degree of maturity before formwork removal. The operations described below are divided into separate categories only for the purpose of understanding their significance and the basic equipment involved; in practice, they may overlap. For example, in the truck-mixing method, the mixing and transporting operations are carried out simultaneously.

Finally, there are facets of concrete behavior at early ages that cannot be considered as intrinsic properties of the material but are important because of their effect on the long-term performance of a concrete structure. They include workability, rate of slump loss, segregation and bleeding, plastic shrinkage, setting time, and curing temperature. In practice, many of these characteristics and phenomena are interrelated; however, for the purposes of achieving a clear understanding of their significance and control, they will be discussed individually.

BATCHING, MIXING, AND CONVEYING

Most specifications require that *batching of concrete ingredients* be carried out by weight rather than by volume. This is because bulking of damp sand (p. 238) causes inaccuracies in measurement. Water and liquid admixtures can be batched accurately either by volume or weight. As discussed later, in many countries most concrete today is batched and mixed by ready-mixed concrete plants, where the batching is generally automatic or semiautomatic. For instance, in the United States during the period 1966–1980 the proportion of manually batched concrete materials diminished from 54 percent to less than 25 percent.

Abnormal handling and maturing characteristics of fresh concrete mixtures that are not uniform in appearance have often been attributed to inadequate *mixing*. Therefore, accurately proportioned concrete ingredients must be mixed thoroughly into a homogeneous mass. Depending on cost economy, type of construction, and amount of concrete required, the mixing operation can either be performed on-site or in a central off-site facility (ready-mixed concrete plant). On-site mixers can be either stationary or paving type; ready-mixed plants contain stationary mixers of sizes up to 12 yd^3 (9 m^3) which can be of the tilting or the nontilting type, or the open-top revolving blade or paddle type.

Ready-mixed concrete is defined as concrete that is manufactured for delivery to a purchaser in a plastic and unhardened state. During the last 50 years of its worldwide development, the ready-mixed concrete industry has experienced tremendous growth. For example, in the United States there are more than 5000 companies operating about 10,000 plants which furnish over two-thirds of the total concrete consumed in the country.[2] Most of the plants are equipped with automatic

---

[2] R. D. Gaynor, National Ready Mixed Concrete Association, Publication 157, March 1979.

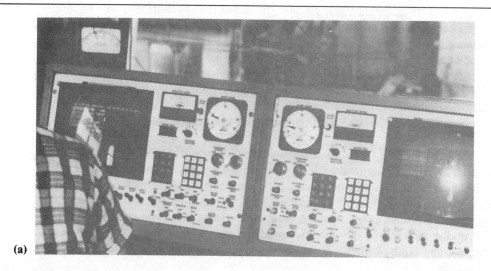

Figure 10–1 (a) Control panel for batching concrete in a modern ready-mixed concrete plant; (b) central mixing in a tilting-type stationary mixer; (c) in truck-mixing, concrete is mixed completely in the truck mixer. [(a), Photograph courtesy of Lone Star Industries, San Francisco, California; (b), (c), photographs courtesy of Portland Cement Association, Skokie, Ill.]

or semi-automatic batching systems and controls made possible by the use of microprocessors and computers (Fig. 10–1a). Truck mixing (Fig. 10–1c) rather than central mixing (Fig. 10–1b) is the commonly used method of mixing in the United States, although for the purpose of achieving better quality control the proportion of centrally mixed concrete increased from 15 percent to 27 percent, and truck-mixed concrete dropped from 83 percent to 71 percent during the period 1966–1977. Inclined-axis mixers of revolving drum type, either with rear or front discharge, are commonly used. In the past 10 to 15 years, for large or important work there has been a gradual change away from the prescriptive to performance or strength

Batching, Mixing, and Conveying

specifications. Also, increasingly, ready-mixed concrete producers are assuming technical responsibility for mix design and quality control.

Transportation of ready-mixed concrete to the job site should be done as quickly as possible to minimize stiffening to the extent that after placement, full consolidation and proper finishing become difficult. The causes and control of stiffening or loss of consistency, which is also referred to as slump loss, are discussed later. Under normal conditions there is usually a negligible loss of consistency during the first 30 min after the beginning of cement hydration. When concrete is kept in a slow state of agitation or is mixed periodically, it undergoes a small slump loss with time, but this usually does not present a serious problem for placing and consolidation of freshly made concrete within 1½ hr. However, as discussed next, attention must be paid to possible delays in transporting and placing concrete under hot and dry weather conditions.

A summary of the most common methods and equipment for transporting concrete is shown in Table 10–1. According to the Portland Cement Association:

> There have been few, if any, major changes in the principles of conveying concrete in the last 40 years. What has changed is the technology that led to development of better

TABLE 10–1 METHODS AND EQUIPMENT FOR HANDLING AND PLACING CONCRETE

Equipment	Type and range of work for which equipment is best suited	Advantages	Points to watch for:
Truck agitator	Used to transport concrete for all uses in pavements, structures, and buildings. Haul distances must allow discharge of concrete within 1½ hours, but limit may be waived.	Truck agitators usually operate from central mixing plants where quality concrete is produced under controlled conditions. Discharge from agitators is well controlled. There is uniformity and homogeneity of concrete on discharge.	Timing of deliveries to suit job organization. Concrete crew and equipment must be ready onsite to handle concrete in large batches.
Truck mixer	Used to mix and transport concrete to job site over short and long hauls. Hauls can be any distance.	No central mixing plant needed, only a batching plant since concrete is completely mixed in truck mixer. Discharge is same as for truck agitator.	Control of concrete quality is not as good as with central mixing. Slump tests of concrete consistency are needed on discharge. Careful preparations are needed for receiving the concrete.

TABLE 10–1 Continued

Equipment	Type and range of work for which equipment is best suited	Advantages	Points to watch for:
Nonagitating truck	Used to transport concrete on short hauls.	Capital cost of nonagitating equipment is lower than that of truck agitators or mixers.	Concrete slump should be limited. Possibility of segregation. Height is needed for high lift of truck body upon discharge.
Mobile continuous mixer	Used for continuous production of concrete at job site.	Combination materials transporter and mobile mixing system for quick, precise proportioning of specified concrete. One-man operation.	Trouble-free operation requires good preventive maintenance program on equipment. Materials must be identical to those in original mix-design proportioning.
Crane	The right tool for work above ground level.	Can handle concrete, reinforcing steel, formwork, and sundry items in high-rise, concrete-framed buildings.	Has only one hook. Careful scheduling between trades and operations are needed to keep it busy.
Buckets	Used on cranes and cableways for construction of buildings and dams. Convey concrete direct from central discharge point to formwork or to secondary discharge point.	Enable full versatility of cranes and cableways to be exploited. Clean discharge. Wide range of capacities.	Select bucket capacity to conform with size of the concrete batch and capacity of the placing equipment. Discharge should be controllable.
Barrows and buggies	For short flat hauls on all types of on-site concrete construction especially where accessibility to work area is restricted.	Very versatile and therefore ideal inside and on job sites where placing conditions are constantly changing.	Slow and labor intensive.
Chutes	For conveying concrete to lower level, usually below-ground level, on all types of concrete construction.	Low cost and easy to maneuver. No power required, gravity does most of the work.	Slopes range between 1 to 2 and 1 to 3 and chutes must be adequately supported in all positions. Arrange for discharge at end (downpipe) to prevent segregation.

Batching, Mixing, and Conveying

TABLE 10–1 Continued

Equipment	Type and range of work for which equipment is best suited	Advantages	Points to watch for:
Belt conveyors	For conveying concrete horizontally or to a higher level. Usually used between main discharge point and secondary discharge point. Not suitable for conveying concrete directly into formwork.	Belt conveyors have adjustable reach, traveling diverter, and variable speed both forward and reverse. Can place large volumes of concrete quickly when access is limited.	End-discharge arrangements needed to prevent segregation. Leave no mortar on return belt. In adverse weather (hot, windy) long reaches of belt need cover.
Pneumatic guns	Used where concrete is to be placed in difficult locations and where thin sections and large areas are needed.	Ideal for placing concrete in free-form shapes, for repairing and strengthening buildings, for protective coatings, and thin linings.	Quality of work depends on skill of those using equipment. Only experienced nozzlemen should be employed.
Concrete pumps	Used to convey concrete direct from central discharge point to formwork or to secondary discharge point.	Pipelines take up little space and can be readily extended. Deliver concrete in continuous stream. Mobile-boom pump can move concrete both vertically and horizontally.	Constant supply of fresh, plastic concrete is needed with average consistency and without any tendency to segregate. Care must be taken in operating pipeline to ensure an even flow and to clean out at conclusion of each operation. Pumping vertically, around bends, and through flexible hose will considerably reduce the maximum pumping distance.
Dropchutes	Used for placing concrete in vertical forms of all kinds. Some chutes are in one piece, while others are assembled from a number of loosely connected segments.	Dropchutes direct concrete into formwork and carry it down to bottom of forms without segregation. Their use avoids spillage of grout and concrete on the form sides, which is harmful when off-the-form	Dropchutes should have sufficiently large, splayed-top openings into which concrete can be discharged without spillage. The cross section of dropchute should be chosen to permit inserting into the formwork with-

TABLE 10-1 Continued

Equipment	Type and range of work for which equipment is best suited	Advantages	Points to watch for:
		surfaces are specified. They also will prevent segregation of coarse particles.	out interfering with steel reinforcing.
Tremies	For placing concrete under water.	Can be used to funnel concrete down through the water into the foundation or other part of the structure being cast.	Precautions are needed to ensure that the tremie discharge end is always buried in fresh concrete, so that a seal is preserved between water and concrete mass. Diameter should be 10 to 12 in. (200 to 300 mm) unless pressure is available. Concrete pumps can be used. Concrete mixture needs more cement, 6½ to 8 bags per cubic yard (363 to 446 kg/m^3), and greater slump, 6 to 9 in. (150 to 230 mm), because concrete must flow and consolidate without any vibration.
Screw spreaders	Used for spreading concrete over flat areas as in pavements.	With a screw spreader a batch of concrete discharged from bucket or truck can be quickly spread over a wide area to a uniform depth. The spread concrete has good uniformity before vibration for final compaction.	Screws are usually used as part of a paving train. They should be used for spreading before vibration is applied.

Source: Reproduced from *Design and Control of Concrete Mixtures, 12th Edition*, Portland Cement Association, Skokie, Ill., 1979, pp. 70–71.

machinery to do the work more efficiently. The wheelbarrow has become the power buggy; the bucket hauled over a pulley has become the hoist; and the horse drawn wagon is now the ready-mixed concrete truck. For some years, concrete was placed in reinforced concrete buildings by means of a tower and long chutes.... As concrete-framed buildings became taller, the need to bring reinforcement and formwork as well as concrete to higher levels led to the development of the tower crane—a familiar sight on the building skyline today.[3]

In choosing the method and equipment for transporting and placing concrete, a primary objective is to assure that concrete will not segregate. The causes, significance, and control of segregation (i.e., the tendency of the coarse aggregate to separate from the mortar) are discussed later.

PLACING, COMPACTING, AND FINISHING

After arrival at the job site the ready-mixed concrete should be *placed as near as possible to its final position* (Fig. 10–2). To minimize segregation, concrete should not be moved over too long a distance as it is being placed in forms or slabs. In general, concrete is deposited in horizontal layers of uniform thickness, and each layer is thoroughly compacted before the next is placed. The rate of placement is kept rapid enough so that the layer immediately below is still plastic when a new layer is deposited. This prevents cold joints, flow lines, and planes of weakness that result when fresh concrete is placed on hardened concrete.

Consolidation or *compaction* is the process of molding concrete within the forms and around embedded parts in order to eliminate pockets of entrapped air. This operation can be carried out by hand rodding and tamping, but almost universally is carried out now by mechanical methods, such as power tampers and vibrators that make it possible to place stiff mixtures with a low water/cement ratio or a high coarse-aggregate content; mixtures of high consistency should be consolidated with care because concrete is likely to segregate when intensely worked. Vibrators should only be used to compact concrete and not to move it horizontally, as this would cause segregation.

Vibration, either internal or external, is the most widely used method for compacting concrete. The internal friction between the coarse aggregate particles is greatly reduced on vibration; consequently, the mixture begins to flow and this facilitates consolidation. One purpose of using internal vibrators (described below) is to force entrapped air out of the concrete by plunging the vibrator rapidly into the mixture and removing it slowly with an up-and-down motion. The rapid penetration forces the concrete upward and outward, thereby helping the air to escape. As the vibrator is removed, bubbles of air rise to the surface.

---

[3] *Design and Control of Concrete Mixtures, 12th Edition,* Portland Cement Association, Skokie, Ill., 1979, p. 69.

Figure 10-2 Placement of concrete as near as possible to its final position prevents segregation. (From *Design and Control of Concrete, 12th Edition*, Portland Cement Association, Skokie, Ill., 1979.)

Internal or *immersion-type vibrators*, also called spud or poker vibrators, are commonly used for compacting concrete in beams, columns, walls, and slabs. Flexible-shaft vibrators usually consist of a cylindrical vibrating head, 19 to 175 mm in diameter, connected to a driving motor by a flexible shaft. Inside the head an unbalanced weight rotates at high speed, causing the head to revolve in a circular orbit. Small vibrators have frequencies ranging from 10,000 to 15,000 vibrations per minute and low amplitude, between 0.4 and 0.8 mm (deviation from the point of rest); as the diameter increases, the frequency decreases and the amplitude increases. An idealized representation of the sequence of actions during concrete consolidation by a high-frequency vibrator is shown in Fig. 10-3.

External or *form vibrators* can be securely clamped to the outside of the forms. They are commonly used for compacting thin or heavily reinforced concrete members. While the concrete mixture is still mobile, vibration of members congested with reinforcement helps to remove air and water that may be entrapped underneath the reinforcing bars, thus improving the rebar-concrete bond. Precasting plants generally use vibrating tables equipped with suitable controls so that the frequency and amplitude can be varied according to size of the members and consistency of the concrete. Surface vibrators such as vibrating screeds are used to consolidate concrete in floors and slabs up to 150 mm thick.

Revibration of concrete an hour or two after initial consolidation, but before setting, is sometimes needed in order to weld successive castings together. This helps to remove any cracks, voids, or weak areas created by settlement or bleeding, particularly around reinforcing steel or other embedded materials.

Flatwork such as slabs and pavements require proper *finishing* to produce dense surfaces that will remain maintenance-free. Depending on the intended use, some surfaces require only strike-off and screeding, whereas others may need finishing operations consisting of the *sequence of steps described below, which must be carefully coordinated with the setting and hardening of the concrete mixture*.

Screeding is the process of striking off excess concrete to bring the top surface

Placing, Compacting, and Finishing

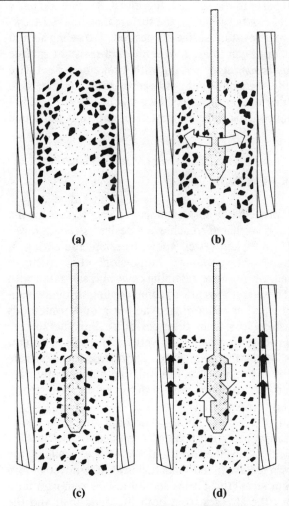

Figure 10-3 Idealized representation of the influence of a high-frequency vibrator on concrete consolidation. (a) The mix is introduced into the form. (b) The vibrator moves aggregate closer together at the form face and cement-sand mortar begins to move outward; air pockets collect on the faces of forms. (c) The mortar continues to move through the coarse aggregate toward the face of the form. (d) The movement of mortar toward the face is complete; as the operator moves the vibrator down and up, air bubbles move upward along the form face and out of the concrete. (Illustration courtesy *Concrete Construction*, Vol. 17, No. 11, 1972. By permission of Concrete Construction Publications, Inc., 426 South Westgate, Addison, Illinois 60101.)

to the desired grade. With a sawing motion a straight edge is moved across the surface with a surplus of concrete against the front face of the straight edge to fill in low areas. A *darby* or *bull-float* is used immediately after screeding to firmly embed large aggregate particles and to remove any remaining high and low spots. Bullfloating must be completed before any excess bleed water accumulates on the surface because this is one of the principal causes of surface defects, such as dusting or scaling, in concrete slabs. When the bleed-water sheen has evaporated and the concrete is able to sustain foot pressure with only slight indentation, the surface is ready for floating and final finishing operations. *Floating* is an operation carried out with flat wood or metal blades for the purposes of firmly embedding the aggregate, compacting the surface, and removing any remaining imperfections.

Floating tends to bring the cement paste to the surface; therefore, floating too early or for too long can weaken the surface. After floating, the surface may be steel-troweled if a very smooth and highly wear resistant surface is desired. Troweling should not be done on a surface that has not been floated. When a skid-resistant surface is desired, this can be produced by *brooming* or *scoring* with a rake or a steel-wire broom before the concrete has fully hardened (but has become sufficiently hard to retain the scoring). Various finishing operations are shown in Fig. 10–4. For additional durability and wear resistance, a surface treatment of the fully-hardened concrete may be considered.

CONCRETE CURING AND FORMWORK REMOVAL

The two objects of *curing* are to prevent the loss of moisture and to control the temperature of concrete for a period sufficient to achieve a desired strength level. When the ambient temperature is sufficiently well above freezing, the curing of pavements and slabs can be accomplished by ponding or immersion; other structures can be cured by spraying or fogging, or moisture-retaining coverings saturated with water, such as burlap or cotton. These methods afford some cooling through evaporation, which is beneficial in hot-weather concreting. Another group of methods are based on prevention of moisture loss from concrete by sealing the surface through the application of waterproof curing paper, polyethylene sheets, or membrane-forming curing compounds. When the ambient temperature is low, concrete must be protected from freezing by the application of insulating blankets; the rate of strength gain can be accelerated by curing concrete with the help of live steam, heating coils, or electrically heated forms or pads.

Formwork removal is generally the last operation carried out during the early-age period of concrete. The operation has great economic implication because on the one hand, removing forms quickly keeps the construction costs low, while on the other, concrete structures are known to have failed when forms were removed before the concrete had attained sufficient strength.[4] Forms should not be removed until concrete is strong enough to carry the stresses from both the dead load and the imposed construction loads. Also, concrete should be hard so that the surface is not injured in any way during form removal or other construction activities. Since the strength of a freshly hydrated cement paste depends on the ambient temperature and availability of moisture, it is better to rely on a direct measure of concrete strength rather than an arbitrarily selected time of form removal. Under standard temperature and moist-curing conditions, normal concretes made with Type I portland cement gain an adequate strength level (e.g., 500- to 1000-psi or 3.5- to 7-MPa compressive strength) in 24 hours, and those made with Type III cement in about 12 hours. For the safety of structures in cold weather, designers often specify a minimum compressive strength before concrete is exposed to freezing. In hot weather, moisture from fresh concrete may be lost by evaporation, causing interrup-

---

[4] See box on p. 341.

(a)

(b)

(c)

Figure 10–4 (a) Beam straight edges equipped with vibrators reduce the work of strike-off and consolidate the concrete; (b) bull-floating must be completed before any excess bleed water accumulates on the surface; (c) when the bleed-water sheen has evaporated and the concrete will sustain foot pressure with only slight indentation, the surface is ready for floating and final finishing operations. (From *Design and Control of Concrete, 12th Edition,* Portland Cement Association, Skokie, Ill., 1979.)

tion of the hydration and strength gain process; also surface cracking due to plastic shrinkage may occur, as described below.

WORKABILITY

Definition and Significance

Workability of concrete is defined in ASTM C 125 as the property determining the effort required to manipulate a freshly mixed quantity of concrete with minimum loss of homogeneity. The term *manipulate* includes the early-age operations of placing, compacting, and finishing. The effort required to place a concrete mixture is determined largely by the overall work needed to initiate and maintain flow, which depends on the rheological property of the lubricant (the cement paste) and the internal friction between the aggregate particles on the one hand, and the external friction between the concrete and the surface of the formwork on the other. *Consistency,* measured by slump-cone test or Vebe apparatus (described below), is used as a simple index for mobility or flowability of fresh concrete. The effort required to compact concrete is governed by the flow characteristics and the ease with which void reduction can be achieved without destroying the stability under pressure. Stability is an index of both the water-holding capacity (the opposite of bleeding) and the coarse-aggregate-holding capacity (the opposite of segregation) of a plastic concrete mixture. A qualitative measure of these characteristics is generally covered by the term *cohesiveness.*

It should be apparent by now that *workability is a composite property,* with at least two main components:

1. Consistency describes the ease of flow
2. Cohesiveness describes the tendency to bleed or segregate

Like durability, **workability is not a fundamental property of concrete;** to be meaningful it must be related to the type of construction and methods of placing, compacting, and finishing. Concrete that can readily be placed in a massive foundation without segregation would be entirely unworkable in a thin structural member. Concrete that is judged to be workable when high-frequency vibrators are available for consolidation, would be unworkable if hand tamping is used.

The *significance of workability* in concrete technology is obvious. It is one of the key properties that must be satisfied. Regardless of the sophistication of the mix design procedure used and other considerations, such as cost, a concrete mixture that cannot be placed easily or compacted fully is not likely to yield the expected strength and durability characteristics.

Measurement

The definition of workability, the composite nature of the property, and its dependence on the type of construction and methods of placing, compacting, and finishing are the reasons why no single test method can be designed to measure workability. The most universally used test, which measures only the consistency of concrete, is called the *slump test*. For the same purpose, the second test in order of importance is the *Vebe test*, which is more meaningful for mixtures with low consistency. The third test is the *compacting factor test*, which attempts to evaluate the compactability characteristic of a concrete mixture. The slump test is covered by ASTM C 143, and the other two by ACI Standard 211.3. Only brief descriptions of the equipment and procedures are given below.

Slump test. The equipment for the slump test is indeed very simple. It consists of a tamping rod and a truncated cone, 300 mm height and 100 mm diameter at the top, and 200 mm diameter at the bottom. The cone is filled with concrete, then slowly lifted. The unsupported concrete cone slumps down by its own weight; the decrease in the height of the slumped cone is called the *slump of concrete*. Details of the procedure based on ASTM C 143 are shown in Fig. 10–5.

The slump test is not suitable for measuring the consistency of very wet or very dry concrete. It is not a good measure for workability, although it is a fairly good measure of the consistency or flow characteristic of a concrete mixture. A main function of this test is to provide a simple and convenient method for controlling the batch-to-batch uniformity of ready-mixed concrete. For example, a more than normal variation in slump may mean an unexpected change in mix proportions or aggregate grading or moisture content. This enables the ready-mixed plant operator to check and remedy the situation.

Vebe test. The equipment for the test, which was developed by Swedish engineer V. Bährner, is shown in Fig. 10–6a. It consists of a vibrating table, a cylindrical pan, a slump cone, and a glass or plastic disk attached to a free-moving rod which serves as a reference end-point. The cone is placed in the pan, filled with concrete, and removed. The disk is brought into position on top of the concrete cone, and the vibrating table is set in motion. The time required to remold the concrete, from the conical to the cylindrical shape until the disk is completely covered with concrete, is the measure of consistency and is reported as the number of Vebe seconds.

Compacting factor test. This test, developed in Great Britain, measures the degree of compaction achieved when a concrete mixture is subjected to a standard amount of work. The degree of compaction, called the *compacting factor,* is measured by the density ratio (i.e., the ratio of the density actually achieved in the test to the density of the same concrete when in fully compacted condition). The apparatus consists essentially of two conical hoppers fitted with doors at the base and placed one above the other (Fig. 10–6b), and a 150- by 300-mm cylinder placed below the hoppers. The upper hopper, which is bigger than the lower, is filled with concrete and struck off without compacting. By opening the door at the bottom of the hopper, the concrete is allowed to fall by gravity into the lower hopper, which overflows. This assures that a given amount of concrete is obtained in a standard state of compaction

1. Stand on the two foot pieces of cone to hold it firmly in place during Steps 1 through 4. Fill cone mold 1/3 full by volume [2-5/8" (67 mm) high] with the concrete sample and rod it with 25 strokes using a round, straight steel rod of 5/8" (16 mm) diameter x 24" (600 mm) long with a hemispherical tip end. Uniformly distribute strokes over the cross section of each layer. For the bottom layer, this will necessitate inclining the rod slightly and making approximately half the strokes near the perimeter (outer edge), then progressing with vertical strokes spirally toward the center.

2. Fill cone 2/3 full by volume (half the height) and again rod 25 times with rod just penetrating into, but not through, the first layer. Distribute strokes evenly as described in Step 1.

3. Fill cone to overflowing and again rod 25 times with rod just penetrating into, but not through, the second layer. Again distribute strokes evenly.

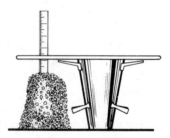

4. Strike off excess concrete from top of cone with the steel rod so the cone is exactly level full. Clean the overflow away from the base of the cone mold.

5. Immediately after completion of Step 4, the operation of raising the mold shall be performed in 5 ± 2 sec. by a steady upward lift with no lateral or torsional motion being imparted to the concrete. The entire operation from the start of the filling through removal of the mold shall be carried out without interruption and shall be completed within an elasped time of 2-1/2 minutes.

6. Place the steel rod horizontally across the inverted mold so the rod extends over the slumped concrete. Immediately measure the distance from bottom of the steel rod to the displaced original center of the top of the specimen. This distance, to the nearest 1/4 inch (6 mm), is the slump of the concrete. If a decided falling away or shearing off of concrete from one side or portion of the mass occurs, disregard the test and make a new test on another portion of the sample.

without the influence of human factor. The door of the lower hopper is released and the concrete falls into the cylinder. Excess material is struck off and the net weight of concrete in the known volume of the cylinder is determined, from which the density is easily calculated.

Factors Affecting Workability and Their Control

For obvious reasons, instead of workability it is more appropriate to consider how various factors affect consistency and cohesiveness, because these two components of workability may be oppositely influenced by changing a particular variable. In general, through their influence on consistency and/or cohesiveness, the workability

Workability

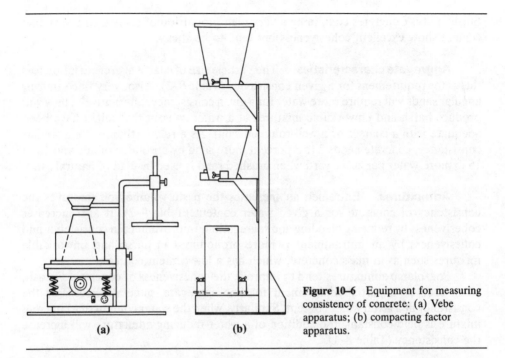

Figure 10–6 Equipment for measuring consistency of concrete: (a) Vebe apparatus; (b) compacting factor apparatus.

of concrete mixtures is controlled by water content, cement content, aggregate grading and other physical characteristics, admixtures, and factors affecting slump loss.

Water content. ACI 211.1, *Standard Practice for Proportioning Concrete Mixtures* (see Table 9–2), assumes that, for a given maximum size of coarse aggregate, the slump or consistency of concrete is a direct function of the water content; i.e., within limits it is independent of other factors such as aggregate grading and cement content. In predicting the influence of mix proportions on consistency it should be noted that of the three factors, that is, water/cement ratio, aggregate/cement ratio, and water content, only two are independent. For example, when the aggregate/cement ratio is reduced but the water/cement ratio is kept constant, the water content increases and consequently the consistency. On the other hand, when the water content is kept constant but the aggregate/cement ratio is reduced, the water/cement ratio decreases and the consistency would not be affected.

Concrete mixtures with very high consistency tend to segregate and bleed, therefore adversely affecting the finishability; mixtures with too low consistency may be difficult to place and compact, and the coarse aggregate may segregate on placement.

Cement content. In normal concrete, at a given water content, a considerable lowering of the cement content tends to produce harsh mixtures, with poor

326 Concrete at Early Ages   Chap. 10

finishability. Concretes containing a very high proportion of cement or a very fine cement show excellent cohesiveness but tend to be sticky.

Aggregate characteristics. The particle size of coarse aggregate influences the water requirement for a given consistency (Table 9–2). Also, very fine sands or angular sands will require more water for a given consistency; alternatively, they will produce harsh and unworkable mixtures at a water content that might have been adequate with a coarser or a well-rounded sand. As a rule of thumb, for a similar consistency, concrete needs 2 to 3 percent more sand by absolute volume and 10 to 15 lb more water per cubic yard when crushed sand is used instead of natural sand.

Admixtures. Entrained air increases the paste volume and improves the consistency of concrete for a given water content (Table 9–2). It also increases cohesiveness by reducing bleeding and segregation. Improvement in consistency and cohesiveness by air entrainment is more pronounced in harsh and unworkable mixtures such as in mass concrete, which has a low cement content.

Pozzolanic admixtures tend to improve the cohesiveness of concrete. Fly ash, when used as a partial replacement for fine aggregate, generally increases the consistency at a given water content. Similarly, when the water content of a concrete mixture is held constant, the addition of a water-reducing admixture will increase the consistency (Table 8–1).

SLUMP LOSS

Definitions

Slump loss can be defined as the loss of consistency in fresh concrete with elapsed time. Slump loss is a ***normal phenomenon*** in all concretes because it results from gradual stiffening and setting of hydrated portland cement paste, which is associated with the formation of hydration products such as ettringite and the calcium silicate hydrate (Chapter 6). Slump loss occurs when the free water from a concrete mixture is removed by hydration reactions, by adsorption on the surfaces of hydration products, and by evaporation.

Under normal conditions, the volume of hydration products during the first half-hour after the addition of water to cement is small and the slump loss is negligible. Thereafter, concrete starts losing slump at a rate determined mainly by elapsed time after hydration, temperature, cement composition, and admixtures present. Generally, the changes in the consistency of concrete up to the time of placement are closely monitored, and proper adjustments are made to assure sufficient consistency for placement and subsequent operations (e.g., compaction and finishing). To overcome the problems caused by slump loss, certain field practices have evolved such as starting with a higher initial slump of ready-mixed concrete than is needed at the job site in order to compensate for the expected slump loss, or adding extra water (within the permissible water/cement ratio) just before placement and remixing the concrete mixture thoroughly. The latter practice is known as *retempering*.

Under certain circumstances, concrete exhibiting an unusually large loss of slump during the first ½ or 1 hr may have the effect of making the mixing, conveying, placing, compacting, and finishing operations difficult or, at times, even impossible. In practice, a reference to a concrete showing the slump loss generally means a quick and unusually large loss of consistency that is beyond the expected or normal behavior. It should be noted that the slump measurements are taken up to the time before placement; however, operational problems can also arise when there is a severe loss of consistency during or immediately after placement. It has therefore been suggested[5] that the definition of the term slump loss be applied to an *unusual rate of stiffening* in fresh concrete (whether measured or not) that causes unwanted effects.

Significance

The premature stiffening of fresh concrete, depending on when the problem appears, may mean an increase in mixer drum torque, a requirement for extra water in the mixer or at the job site, hang-up of concrete within the drum of a truck mixer, difficulty in pumping and placing concrete, extra labor needed in placement and finishing operations, loss of production and quality of workmanship, and loss of strength, durability, and other properties when the retempering water is excessive or is not mixed thoroughly.

When job site inspection and quality control are lax, construction crews frequently adopt the bad practice of adding extra water to concrete, whether needed or not. Many failures of concrete to perform satisfactorily have been documented to the careless addition of the retempering water which was either poorly mixed or not accounted for in the mix design calculations. For example,[6] the removal of forms from an unusually large concrete placement revealed areas of severe honeycombing. Construction personnel indicated that quick setting had been experienced, primarily during periods of high ambient temperature. Petrographic analysis of cores revealed that areas of different water/cement ratio were present within a core, indicating that retempering water had been added owing to slump loss and that incomplete intermixing of the retempering water had occurred. The National Ready Mixed Concrete Association offers this advice: *A wasted load of questionable concrete may represent a tremendous bargain for the company, compared to its possible use and failure to perform.*

[5] B. Erlin and W. G. Hime, *Concr. Int.*, Vol. 1, No. 1, pp. 48–51, 1979.
[6] Ibid.

Causes and Control

The primary causes of slump-loss problems in concrete are: (1) the use of an abnormal-setting cement; (2) unusually long time for mixing, conveying, placement, compaction, or finishing operation; (3) high temperature of concrete due to excessive heat of hydration and/or the use of concrete-making materials that are stored at a high ambient temperature.

Typical data[7] on the influence of cement composition, elapsed time after hydration, and temperature, on the rate of slump loss in normal concrete mixtures are shown in Table 10–2. All concretes contained 517 lb/yd^3 Type I portland cement, 1752 lb/yd^3 coarse aggregate, and 824 lb/yd^3 fine aggregate. The water content was varied to obtain different initial slumps: approximately 7 in., 5 in., or 3 in. Cement A was a low-alkali cement (0.16 percent equivalent Na_2O) with 9 percent C_3A content, whereas cement B was high-alkali (0.62 percent equivalent Na_2O) with 10.6 percent C_3A content; both had similar SO_3 content and Blaine surface area. The following conclusions were drawn from the investigation:

1. In general, the amount of slump loss was proportional to the initial slump; the higher the initial slump, the higher the slump loss. For example, in the case of cement A, at the close of the 2-hr test at 70°F, concrete mix 1 (initial slump 7½ in.) lost 5¼ in., whereas concrete mix 3 (initial slump 5 in.) lost 3½ in., and

TABLE 10–2 EFFECT OF CEMENT COMPOSITION, ELAPSED TIME, AND TEMPERATURE ON SLUMP OF CONCRETE MIXTURES WITH DIFFERENT INITIAL SLUMPS

		Slump (in.)				
Concrete mix	Cement	Initial	30 min	60 min	90 min	120 min
		Concrete temperature 70°F				
1	A	7½	7	5½	3¾	2¼
2	B	7⅛	4¾	3¼	2½	1⅞
3	A	5	4⅜	3⅛	2¼	1½
4	B	5¼	3¼	2½	1¾	1¼
5	A	3⅝	3¼	2⅝	1⅞	1⅜
6	B	3½	2⅝	2	1½	⅞
		Concrete temperature 85°F				
7	A	7⅛	5⅜	4⅜	2⅝	1⅝
8	B	7½	5½	3½	2½	1⅜
9	A	5½	4½	3⅝	2⅝	1⅝
10	B	5½	4⅛	2¾	1⅞	1⅛
11	A	3½	3½	2½	1⅞	1⅛
12	B	3¾	2¼	1⅝	1⅜	¾

Source: Based on R. W. Previte, *J. ACI,* Proc., Vol. 74, No. 8, pp. 361–67, 1977.

[7] R. W. Previte, *J. ACI,* Proc., Vol. 74, No. 8, pp. 361–67, 1977.

concrete mix 5 (initial slump 3⅝ in.) lost 2¼ in. slump. Regardless of the initial slump, the final slump values after 2 hours of hydration were of the order of 1½ to 2 in. In such a case the method of compensating for the expected slump loss by designing for a higher initial slump is not recommended because the retempering water required at the job site may have the effect of pushing up the water/cement ratio of the concrete mixture with a high initial slump to an undesirable level.

2. In general, early slump loss tends to be directly proportional to the temperature of concrete. For example, a comparison of the 7-in. slump concretes made with cement A at two different temperatures [i.e., 70°F (concrete mix 1) and 85°F (concrete mix 7)] showed that at 30, 60, and 90 min elapsed times, the former lost ½ in., 1⅞ in., and 3¾ in. slump, while the latter lost 1¾ in., 2¾ in., and 4½ in., respectively.

3. In regard to the effect of cement composition, greater slump loss rates were observed for all test conditions in the case of concretes made with the cement containing higher C_3A and high-alkali content (cement B). For instance, at 70°F and 30, 60, and 90 min elapsed times, concrete mix 1 lost ½ in., 1⅞ in., and 3¾ in., compared to 2⅜ in., 3⅞ in., and 4⅝ in., respectively, for concrete mix 2.

Erlin and Hime[8] reported case histories of unusual slump loss attributed to cement composition or cement-admixture interaction. During slip-form construction of a concrete silo, surface irregularities were observed when a light-colored portland cement was used; such irregularities did not occur when a darker cement was used in the initial stages of construction. Workers had noticed higher pumping pressures at the time of placing the concrete made with the light-colored cement. Laboratory analysis revealed that this cement contained calcium sulfate largely in the form of dehydrated gypsum, and was severely false-setting (see Fig. 6-8). This created a condition that caused the concrete surface to *tear* when the forms were slipped.

In a second case,[9] during transit in a ready-mix truck the concrete set so severely that it had to be blasted loose. Laboratory tests showed that the concrete contained two or three times the normal dose of an admixture containing triethanolamine, which is an accelerator. With additions of the admixture, the cement stiffened rapidly and produced considerable heat (i.e., a flash set). From the analysis of the cement it was found that calcium sulfate was present mostly in the form of insoluble natural anhydrite. Thus imbalance in the reactions involving sulfate and aluminate phases, as discussed earlier (Fig. 6-8), led to rapid setting. In another incident, due to the presence of a glucoheptanate-type coloring agent in an admixture, retardation of the cement was so severe that no stiffening and setting occurred

---

[8] B. Erlin and W. G. Hime, *Concr. Int.*, Vol. 1, No. 1, pp. 48–51, 1979.
[9] Ibid.

even at 24 hr; therefore, the concrete had to be removed the next day. Some water-reducing agents, especially the high-range type or superplasticizers, tend to accelerate the slump loss. This is because an efficient dispersion of the cement-water system enhances the rate of formation of the hydration products.

According to Tuthill,[10] problems attributed to slump loss often arise at the very start of a placing operation if mixing is permitted before the formwork is positively ready to receive the concrete, or if the first batches are on the low side of the slump range and are judged too dry to make a safe start without delay, where there is no newly placed concrete into which to work them. Either of these two common problems cause concrete to stay in trucks or buckets, losing slump with time. Delays from mixing to placement of concrete can have a serious effect on production rates, aside from the direct time loss, especially in operations such as pumping, tunnel lining, slip-formed paving, and tremie concreting, which depend heavily on a reasonably constant degree of concrete consistency.

Slump-loss problems occur most often in hot weather; the higher the temperature at which concrete is mixed, handled, and placed, the more likely it is that slump loss is the cause of operating problems. ACI Committee 305 cautions that difficulties may be encountered with concrete at a placing temperature approaching 32°C, and every effort should be made to place it at a lower temperature. In hot and dry weather it is recommended that aggregate be stored in shaded areas and be cooled by sprinkling water. According to Tuthill,[11] the use of chipped ice as a partial or complete replacement for mixing water is the best way to bring down the concrete temperature; each 3 kg of ice will reduce the temperature of 1 m^3 of concrete about 0.7°C.

In conclusion, elimination of every possible delay in concrete handling operations, keeping the temperature of concrete as close to the range 10 to 21°C as possible, and a laboratory check on the stiffening and setting characteristics of the cement (with or without the admixtures selected for use) are the necessary preventive measures to control slump loss problems.

SEGREGATION AND BLEEDING

Definitions and Significance

Segregation is defined as separation of the components of fresh concrete so that they are no longer uniformly distributed. There are *two kinds of segregation.* The first, which is characteristic of dry concrete mixtures, consists of separation of the mortar from the body of concrete (e.g., due to overvibration). Bleeding, as explained next, is the second form of segregation and is characteristic of wet concrete mixtures.

Bleeding is defined as a phenomenon whose external manifestation is the

---

[10] L. H. Tuthill, *Concr. Int.,* Vol. 1, No. 1, pp. 30–35, 1979.

[11] Ibid.

Segregation and Bleeding

appearance of water on the surface after a concrete has been placed and compacted but before it has set (i.e., when sedimentation can no longer take place). Water is the lightest component in a concrete mixture; thus bleeding is a form of segregation because solids in suspension tend to move downward under the force of gravity. Bleeding results from the inability of the constituent materials to hold all the mixing water in a dispersed state as the relatively heavy solids settle.

It is important to reduce the tendency for segregation in a concrete mixture because full compaction, which is essential to achieve the maximum strength potential, will not be possible after a concrete has segregated. The bleeding phenomenon has several manifestations. First, only some of the bleeding water reaches the surface; a large amount of it gets trapped under larger pieces of aggregate[12] and horizontal reinforcing bars, when present. If the loss of bleed water were uniform throughout a concrete mixture, and if before revibrating concrete the bleed water appearing on the surface was removed by processes such as vacuum extraction, the quality of concrete would be improved as the result of a reduction in the water/cement ratio. In practice, however, this does not happen. Usually, bleed water pockets under coarse aggregate and reinforcing steel are numerous and larger in the upper portion of the concrete, which is therefore weaker than the concrete in the lower portion of the structure.

Laitance, associated with the external manifestation of bleeding, is caused by the tendency of water rising in the internal channels within concrete, carrying with it very fine particles of cement, sand, and clay (present as a contaminant in aggregate) and depositing them in the form of a scum at the concrete surface. The laitance layer contains a very high water/cement ratio and is therefore porous, soft, and weak. When a floor slab or a pavement suffers from laitance, instead of having a hard durable surface the concrete will present a soft surface that is prone to ***dusting***. The hydration products in the porous cement paste of the laitance layer are easily carbonated in air. Therefore, if laitance occurs at the top of a casting or lift, poor bond to the next lift will result. Thus laitance on old concrete should always be removed by brushing and washing or by sand blasting before new concrete is placed. Bleeding also plays a significant role in plastic shrinkage cracking, which is discussed later.

Measurement

There are no tests for the measurement of segregation; visual observation and inspection of cores of hardened concrete are generally adequate to determine whether segregation is a problem in a given situation. There is, however, an ASTM Standard test for the measurement of rate of bleeding and the total bleeding capacity of a concrete mixture. According to ASTM C 232, a sample of concrete is placed and consolidated in a container 250 mm diameter and 280 mm height. The bleed water accumulated on the surface is withdrawn at 10-min intervals during the first

---

[12] For internal manifestations of bleeding, also see Fig. 2–4.

40 min, and thereafter at 30-min intervals. Bleeding is expressed in terms of the amount of accumulated water as the percentage of net mixing water in the sample.

Causes and Control

A combination of improper consistency, excessive amount of large particles of coarse aggregate with either too high or too low density, presence of less fines (due to low cement and sand contents or the use of a poorly graded sand), and inappropriate placing and compacting methods are generally the causes for segregation and bleeding problems in concrete. Obviously, the problems can be reduced or eliminated by paying attention to mix proportioning and to handling and placement methods.

Segregation in very dry mixtures can sometimes be reduced by increasing the water content slightly. In most cases, however, proper attention to aggregate grading is called for. This may involve a lowering of the maximum size of coarse aggregate and the use of more sand or a finer sand. Increase in the cement content and the use of mineral admixtures and air entrainment are the commonly employed measures in combating the tendency of concrete mixtures to bleed. It is interesting to point out that high C_3A-high alkali cements, which show greater slump loss, tend to reduce bleeding. When a concrete mixture has to be dropped from a considerable height (such as in tremie concreting) or discharged against an obstacle, the material should be highly cohesive and extra care is necessary during placement.

EARLY VOLUME CHANGES

Definitions and Significance

After fresh concrete has been in deep forms, such as the forms for a column or a wall, after a few hours it will be observed that the top surface has subsided. The tendency toward subsidence is also confirmed by the presence of short horizontal cracks. This reduction in volume of fresh concrete is known as **prehardening** or **presetting shrinkage,** or **plastic shrinkage,** since the shrinkage occurs while the concrete is still in the plastic state. In the United States, the term *plastic shrinkage* is usually used in the case of concrete slabs only, as discussed below.

As a result of prehardening shrinkage, cracks develop over obstructions to uniform settlement: for instance, reinforcing bars and large aggregate particles. In slabs, rapid drying of fresh concrete causes plastic shrinkage when the rate of loss of water from the surface exceeds the rate at which the bleed water is available. If at the same time the concrete near the surface has become too stiff to flow but is not strong enough to withstand the tensile stress caused by restrained shrinkage, cracks will develop. Typical plastic shrinkage cracks (Fig. 10–7) are parallel to one another and are 0.3 to 1 m apart and 25 to 50 mm deep.

Early Volume Changes

Figure 10–7 Plastic shrinkage cracking in freshly placed concrete. (Photograph from *Concrete in Practice,* Brochure 5, Courtesy National Ready Mixed Concrete Association, 1979.)

Causes and Control

A variety of causes contribute to plastic shrinkage in concrete: for example, bleeding or sedimentation, absorption of water by subgrade or forms or aggregate, rapid water loss by evaporation, reduction in the volume of the cement-water system, and bulging or settlement of the formwork. The following conditions, singly or collectively, increase the rate of evaporation of surface moisture and enhance the possibility of plastic-shrinkage cracking: high concrete temperature, low humidity, and high wind velocity. When the rate of evaporation exceeds 0.2 lb/ft^2 per hour (1 kg/m^2 per hour), precautionary measures are necessary to prevent the plastic-shrinkage cracking. The Portland Cement Association has developed charts (Fig. 10–8) for determining when the precautionary measures, listed as follows, should be taken:

- Moisten the subgrade and forms.
- Moisten aggregates that are dry and absorptive.
- Erect temporary windbreaks to reduce wind velocity over the concrete surface.
- Erect temporary sunshades to reduce concrete surface temperature.
- Keep the fresh concrete temperature low by cooling the aggregate and mixing water.
- Protect concrete with temporary coverings such as polyethylene sheeting during any appreciable delay between placing and finishing.
- Reduce the time between placing and start of curing by eliminating delays during construction.
- To minimize evaporation, protect the concrete immediately after finishing by wet burlap, fog spray, or a curing compound.

Settlement cracks and plastic shrinkage cracks in flatwork can be eliminated by revibration of concrete when it is still in the plastic state. Revibration is also known

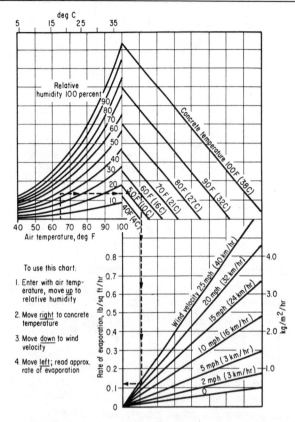

Figure 10-8 Estimating the rate of moisture evaporation from a concrete surface. (From *J. ACI*, Proc., Vol. 74, No. 8, p. 321, 1977.)

This chart, developed by Portland Cement Association, provides a graphic method of estimating the loss of surface moisture for various weather conditions. If the rate of evaporation approaches 0.2 lb/ft² (1 kg/m²) per hour, it is recommended that precautions be taken against plastic shrinkage cracking.

to improve the bond between concrete and reinforcing steel and to enhance the strength of concrete by relieving the plastic shrinkage stresses around the coarse aggregate particles.

SETTING TIME

Definitions and Significance

The reactions between cement and water are the primary cause of the setting of concrete although for various reasons, as will be discussed later, the setting time of concrete does not coincide with the setting time of the cement with which the concrete is made. As described in Chapter 6, the phenomena of stiffening, setting,

Setting Time

and hardening are physical manifestations of the progressive hydration reactions of cement. Also, the initial and the final setting times of cement are the points, defined arbitrarily by the method of test, which determine the onset of solidification in a fresh cement paste. Similarly, **setting of concrete** is defined as the onset of solidification in a fresh concrete mixture. Both the initial and the final setting times of concrete are arbitrarily defined by a test method such as the penetration resistance method (ASTM C 403), which is described below.

The **initial setting time** and the **final setting time,** as measured by penetration resistance methods, do not mark a specific change in the physical-chemical characteristics of the cement paste; they are purely functional points in the sense that the former defines the limit of handling and the latter defines the beginning of development of mechanical strength. Fig. 10–9 illustrates that initial set and final set of concrete measured by ASTM C 403 do not have to coincide exactly with the periods marking the end or the complete loss of workability and the beginning of mechanical strength. Instead, the initial set represents approximately the time at which fresh concrete can no longer be properly mixed, placed, and compacted; the final set represents approximately the time after which strength begins to develop at a significant rate. Obviously, a knowledge of the changes in concrete characteristics, as defined by the initial and final setting times, can be of considerable value in scheduling concrete construction operations. The test data can also be useful in comparing the relative effectiveness of various set-controlling admixtures.

Measurement and Control

The most commonly used method is the ASTM C 403, *Test for Time of Setting of Concrete Mixtures by Penetration Resistance,* which provides a standard procedure for measurement of setting time of concrete with a slump greater than zero by testing mortar sieved from the concrete mixture. Briefly, the test consists of removing the

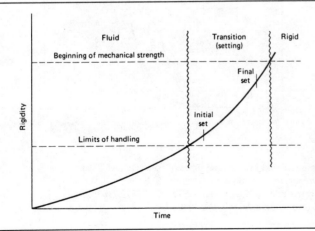

Figure 10–9 The progress of setting and hardening in concrete. (From S. Mindess and J. F. Young, *Concrete,* © 1981. Reprinted by permission of Prentice-Hall, Inc., Englewood Cliffs, N.J.)

mortar fraction from the concrete, compacting it in a standard container, and measuring the force required to cause a needle to penetrate 25 mm into the mortar. The times of set are determined from the rate of solidification curve, obtained from a linear plot of data with elapsed time as the abscissa and penetration resistance as the ordinate. Initial and final set are defined as times at which the penetration resistances are 500 psi (3.5 MPa) and 4000 psi (27.6 MPa), respectively. It should be noted that these arbitrarily chosen points do not indicate the strength of concrete; in fact, at 500 psi (3.5 MPa) penetration resistance the concrete has no compressive strength, while at 4000 psi (27.6 MPa) penetration resistance the compressive strength may be only about 100 psi (0.7 MPa).

The principal factors controlling the setting times of concrete are cement composition, water/cement ratio, temperature, and admixtures. Cements that are quick setting, false setting, or flash setting will tend to produce concretes with corresponding characteristics. Since the setting and hardening phenomena in a hydrating cement paste are influenced by the filling of void space with the products of hydration, the water/cement ratio will obviously affect setting times. Thus the setting-time data on a cement paste do not coincide with the setting times of concrete containing the same cement because the water/cement ratios in the two cases are usually different. In general, the higher the water/cement ratio, the longer the time of set.

The effects of cement composition, temperature, and retarding admixtures on typical rates of setting obtained by ASTM C 403 test are shown in Fig. 10–10. When a concrete mixture was made and stored at 10°C instead of 23°C, the initial and the final setting times were retarded approximately by 4 and 7 hr, respectively. With cement B and a set-retarding admixture, the retarding effect of the admixture was found to be greater at the higher temperature.

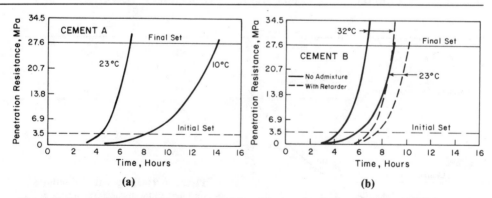

Figure 10–10 (a) Effect of temperature on initial and final setting times of concrete (ASTM C 403); (b) effect of a retarding admixture on setting times of concrete (ASTM C 403). (Reprinted with permission, from J. H. Sprouse and R. B. Peppler, ASTM STP 169B, 1978, pp. 105–21. Copyright, ASTM, 1916 Race Street, Philadelphia PA 19103.)

TEMPERATURE OF CONCRETE

Significance

Among other problems, as will be described below, in hot weather unprotected concrete is subject to plastic shrinkage cracking during early ages. On the other hand, in cold weather insufficient curing of concrete may seriously impede the rate of strength development. Premature removal of formwork (i.e., before concrete acquires sufficient *maturity* or strength) has in the past led to disastrous consequences in terms of both economic and human costs (see box, p. 341). The problem usually arises from construction decisions based on laboratory-cured cylinders, when the actual curing history of the in-place field concrete has been far different. It seems, therefore, that engineers should have a general understanding of possible effects of both lower- and higher-than-normal curing temperatures on the properties of concrete at early ages, and methods of evaluating and controlling these properties.

Cold-Weather Concreting

It is generally accepted that there is little cement hydration and strength gain if concrete is frozen and kept frozen below $-10°C$. Therefore, fresh concrete must be protected against disruptive expansion by freezing until adequate strength has been gained,[13] and the degree of saturation of the concrete has been sufficiently reduced by some progress in the hydration process. Without external heat sources, heat of cement hydration in large and well-insulated concrete members may be adequate to maintain satisfactory curing temperatures, provided that concrete has been delivered at a proper temperature, and coldness of frozen ground, formwork, and reinforcing bars have been taken into consideration.

In recommended practice for cold-weather concreting, ACI Committee 306R gives placement temperatures for normal-weight concrete (Table 10–3). It may be noted that lower concrete temperatures are permitted for massive sections because with these the heat generated during hydration is dissipated less rapidly than from flatwork. Also, since more heat is lost from concrete during conveying and placing at lower air temperatures, the recommended concrete temperatures are higher for colder weather (see lines 1, 2, and 3 in Table 10–3).

Insufficient curing of concrete can be detrimental to properties other than strength. But strength is at the center of most of the decision making because form stripping, prestressing, and other such operations in concrete construction are guided by the strength of concrete on hand. Strength is also the criterion when durability of concrete in early exposure to aggressive waters is of concern. The traditional method for determining safe stripping times is to test laboratory-cured concrete cylinders and strip the forms when the cylinders reach the specified

---

[13] A minimum compressive strength of 500 psi (3.5 MPa) prior to freezing is stated in ACI 306R as a criterion for preventing frost damage.

TABLE 10-3 RECOMMENDED CONCRETE TEMPERATURE FOR COLD-WEATHER CONSTRUCTION: AIR-ENTRAINED CONCRETE[a]

Line	Condition		Sections less than 12 in. (300 mm) thick		Sections 12–36 in. (300 mm–0.9 m) thick		Sections 36–72 in. (0.9–1.8 m) thick		Sections over 72 in. (1.8 m) thick	
			°F	°C	°F	°C	°F	°C	°F	°C
1	Minimum temperature fresh concrete *as mixed* in weather indicated, °F (°C)	Above 30°F (−1°C)	60	16	55	13	50	10	45	7
2		0°F to 30°F	65	18	60	16	55	13	50	10
3		Below 0°F (−18°C)	70	21	65	18	60	16	55	13
4	Minimum temperature fresh concrete *as placed* and *maintained*		55	13	50	10	45	7	40	5
5	Maximum allowable *gradual* drop in temperature in first 24 hr after end of protection		50	28	40	22	30	17	20	11

[a] For durability and safe stripping strength of *lightly stressed* members. ACI 306 recommends 1 to 3 day's duration of the temperatures shown in the table, depending on whether the concrete is conventional or the high-early-strength type. For *moderately and fully stressed* members, longer durations are recommended. Also, for concrete that is *not air entrained* it is recommended that protection for durability should be at least twice the number of days required for air-entrained concrete.

Source: Adapted from ACI 306-78.

Temperature of Concrete

strength. This procedure has led to problems when the curing history of the cylinder in the laboratory is considerably different from the curing history of the in-place concrete. In the report of ACI Committee 306, the *maturity method* is recommended as an alternative to using laboratory or field-cured cylinders.

The maturity method. Since degree of cement hydration depends on both time and temperature, the strength of concrete may be evaluated from a concept of maturity which is expressed as a function of the time and the temperature of curing. To use the concept for estimating concrete strength, it is assumed that for a particular concrete mixture, concretes of the same maturity will attain the same strength, regardless of the time-temperature combinations leading to maturity. Also, $-10°C$ or $14°F$ is generally assumed as the datum temperature below which there is no strength gain: thus the expression

$$\text{Maturity Function} = \int_o^t (T - T_o) \, dt$$
$$M(t) = \Sigma \, (T_a - T_o)\Delta t$$

where $M(t)$ is the temperature-time factor at age t (degree-days or degree-hours), Δt, T_a and T_o are time interval, average concrete temperature during the time interval t, and the datum temperature respectively. Before construction begins, a calibration curve is drawn plotting the relationship between concrete compressive strength and the value M for a series of test cylinders made from the particular concrete mixtures but with different combinations of time and curing temperature (Fig. 10–11).

Some researchers have reported good correlation between maturity and compressive strength of concrete, whereas others have questioned the validity of the maturity concept. For instance, it is pointed out that the maturity concept does not take into consideration the influences of humidity of curing, heat of hydration, variable cement composition, and curing temperature during the early age, which contrary to the assumption under the maturity concept, seem to exercise a disproportional effect on strength with time. The data in Fig. 10–11 show that the plots showing the strength-maturity relation were significantly different from each other when a concrete mixture was cured at 30°F instead of at 70 or 100°F during the early age.

Control of concrete temperature. From Table 10–3 it should be noted that for cold-weather concreting, making fresh concrete mixtures at a temperature above 70°F (21°C) is not recommended. The higher temperatures do not necessarily offer better protection: first, because at higher temperatures the rate of heat loss is greater, and second, the water requirement for the same consistency is more. Depending on the ambient temperature and transport time from the production to the job site, the temperature of concrete *as mixed* is maintained at not more than 10°F (5.6 °C) above the minimum recommended in Table 10–3. As discussed below, the temperature of fresh concrete is usually controlled by adjusting the temperatures of mixing water and aggregates.

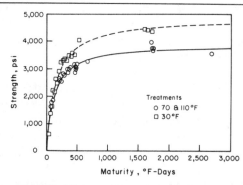

Figure 10–11 Strength-maturity relation in concrete. (From N. J. Carino, H. S. Lew, and C. K. Volz, *J. ACI, Proc.*, Vol. 80, No. 2, 1983.)

The strength of hardening concrete can be predicted at any age by computing the maturity, based on the temperature-time history of the concrete. Thus the maturity concept is recommended as an alternative to using direct strength testing of field-cured cylinders. The limitations of the concept should be noted when it is used for the removal of formwork after the concrete has gained sufficient strength to support self-weight and other construction loads. The data in the figure show that the strength-maturity relation of a given concrete mixture is affected by the temperature at early ages.

Of all the concrete-making components, mixing water is the easiest to heat. Also, it makes more practical sense to do so because water can store five times as much heat as can the same mass of cement or aggregate. Compared to a specific heat of 1.0 for water, the average specific heat for cement and aggregates is 0.22. At temperatures above freezing, it is rarely necessary to heat aggregates; at temperatures below freezing, often only the fine aggregates need to be heated to produce concrete at the required temperature, which is generally accomplished by circulating hot air or steam through pipes embedded in the aggregate stockpile.

Concrete temperature can be measured directly by a mercury thermometer or a bimetallic thermometer. It can also be estimated using the expression

$$T = \frac{0.22(T_a W_a + T_c W_c) + T_w W_w + T_{wa} W_a}{0.22(W_a + W_c) + W_w + W_{wa}} \tag{10-1}$$

where T is temperature of the fresh concrete in °F; T_a, T_c, T_w, and T_{wa} are temperatures of aggregates, cement, mixing water, and free moisture in aggregates, respectively; and W_a, W_c, W_w, and W_{wa} are weights (in pounds) of aggregates, cement, mixing water, and free moisture in aggregates, respectively. The formula remains the same in SI units except that °F is changed to °C and pounds to kilograms.

Hot-Weather Concreting

For the purposes of construction problems in normal structural concrete, ACI Committee 305 defines *hot weather* as any combination of high air temperature, low relative humidity, and wind velocity tending to impair the quality of fresh or hardened concrete or otherwise resulting in abnormal properties. In addition to the increase in slump loss and plastic-shrinkage cracking, and the decrease of setting time in fresh concrete (already described) hot weather increases the mixing water

> In Kiev, capital of the industrial Ukraine, workers were in a bind to get a building up in the allotted time. The newspaper *Rabochaya Gazeta* said the construction crews fiddled with the architect's plans to cut down the work and then produced a building in record time. When the workers eagerly swung the roof into place, the structure neatly collapsed in a heap. They had left out that part that says "allow the concrete to dry [*cure.*"]
>
> *Source:* UPI report published in the San Francisco *Sunday Examiner and Chronicle,* January 4, 1976
>
> On 27 April 1978 a cooling tower under construction at Willow Island in West Virginia, collapsed—killing 51 workers. The contractor was using a slip-formed construction process involving a multilayer scaffold that raises itself up the wall by its own power after anchoring into the hardened concrete of the previous day's work. According to an investigation by the Office of Safety and Health Administration, the accident "could have been prevented if proper engineering practices had been followed." Investigation findings cited that one of the key factors contributing to the collapse was "a failure to make field tests to be sure that the concrete had *cured sufficiently* before the support forms were removed."
>
> *Source:* Based on a report by Eugene Kennedy published in the *San Francisco Examiner and Chronicle,* December 3, 1978

requirement for a given consistency (Fig. 10–12) and creates difficulty in holding air in an air-entrained concrete. Retempering of fresh concrete is frequently necessary in hot weather; at times this causes adverse effects on strength, durability, dimensional stability, and appearance of the hardened concrete. Also, in the range 40 to 115°F (4 to 46 °C), concretes placed and cured at higher temperatures normally develop greater early strengths, but at 28 days and later ages the strengths are lower (Fig. 3–10).

Control of concrete temperature. As explained earlier, since the mixing water has the greatest effect per unit weight of any of the ingredients on the temperature of concrete, the use of cooled mixing water and/or ice offers the best way of lowering the temperature of concrete. The expression for determining the temperature of concrete in cold weather by using hot water (p. 340) can be employed for calculating how much cold water will be needed to lower the temperature of a concrete by a given amount. Alternatively, charts such as that shown in Fig. 10–13a can be used. The data in Fig. 10–13a pertain to a nominal concrete mixture containing 564 lb/yd^3 (335 kg/m^3) cement, 284 lb/yd^3 (170 kg/m^3) water, and 3100 lb/yd^3 (1830 kg/m^3) aggregate.

The use of shaved or chipped ice as all or part of the required mixing water is the most effective way of reducing the concrete temperature, because on melting ice absorbs 144 Btu/lb (80 cal/g) heat. Figure 10–13b illustrates possible reductions in concrete temperature by substitution of varying amounts of ice at 32°F for mixing water at the temperatures shown. It can be seen from Fig. 10–13 that with normal

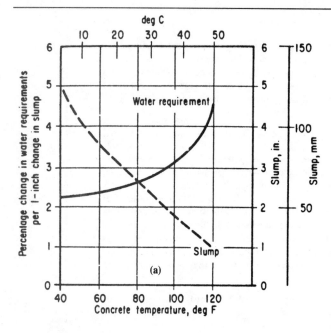

Figure 10–12 (a) Effect of concrete temperature on slump and on water required to change slump (average data for Type I and II cements). (Report of ACI Committee 305 on Hot Weather Concreting, *ACI Materials Jour.*, Vol. 88, No. 4, p. 422, 1991.)

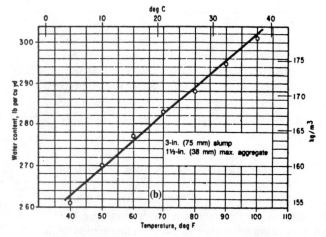

Figure 10–12 (b) Effect of ambient temperature increase on the water requirement of concrete. (Report of ACI Committee 305 on Hot Weather Concreting, *ACI Materials Jour.*, Vol. 88, No. 4, p. 422, 1991.)

The water requirement of a concrete mixture increases with an increase in the temperature of concrete. As shown in the figure, if the temperature of fresh concrete is increased from 50°F to 100°F, the water requirement increases by about 33 lb/yd³ for maintaining the same 3-in. slump. This increase in water content can reduce the 28-day compressive strength of concrete by 12 to 15 percent.

Temperature of Concrete

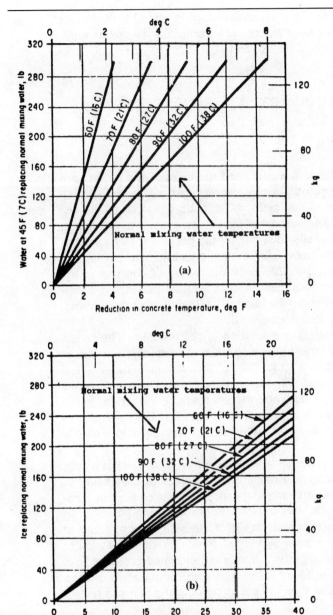

Figure 10–13 Determination of reduction in concrete temperature: (a) by adding cooled water; (b) by adding ice. (From ACI Committee 305, "Hot Weather Concreting," *ACI Materials Jour.*, Vol. 88, No. 4, p. 423, 1991.

Part (a) shows the effect of cooled (45°F) mixing water and part (b) shows effect of ice in mixing water on concrete temperature. Normal mixing water temperatures are shown on the curves. The data are applicable to average mixes made with typical natural aggregates. A comparison of the two figures shows that the use of ice as part of the mixing water is highly effective in reducing concrete temperature because on melting alone, ice absorbs heat at the rate of 144 Btu/lb (80 cal/g).

mixing water at 100°F there will be a 6°F temperature reduction when 120 lb of water at 45°F replaces the mixing water; the same amount of ice replacing the mixing water would have reduced the temperature of the concrete by 23°F.

TESTING AND CONTROL OF CONCRETE QUALITY

Methods and Their Significance

Engineers representing the owner, the designer, and the builder of a project are frequently required to develop or approve a *quality assurance* program which, among other things, involves the selection of test methods, statistical analysis of the test results, and follow-up procedures. The objective of such a program is to assure that a finished concrete element is structurally adequate for the purpose for which it was designed. The size of concrete structures being designed and built today and the speed of modern construction (e.g., over 200 m^3/hr placement of concrete in hydroelectric projects) require, for obvious reasons, that decision making on acceptance or rejection of concrete quality not be left to the 28-day compression test, which continues to be the basis of design specifications.

Accelerated strength testing offers one solution to the problem. The procedures are being used increasingly on large projects to make a preliminary assessment 1 or 2 days after placing concrete as to whether the product will reach the required strength level. A low value from an accelerated strength test can warn the contractor of a potential problem and provide an early opportunity for remedial action. In the case where substandard concrete has been placed, it is easier and less expensive to remove it when the concrete is a few days old rather than 28 days old and probably covered with a superstructure.

A criticism against the testing of concrete samples drawn from batches before their placement is that the test specimens may not truly represent the quality of concrete in a structure, due possibly to sampling errors and differences in compaction and curing conditions. Also, on large projects the cost of strength testing can be considerable. As an alternative approach to direct strength testing, many in situ/nondestructive test methods have been developed, which provide an excellent means of control of in-place concrete quality. Although in situ/nondestructive tests are not accepted as a complete substitute for direct strength tests, when used in conjunction with core strength tests or standard compression tests, they can reduce the cost of testing for quality control.

In large-scale industrial productions an effective and economical system of quality control has to rely on statistical methods of data processing and decision making. A primary statistical tool in concrete quality control programs is the use of control charts which graphically show the results of tests and also contain limit lines indicating the need for action when the plotted data approach these limit lines.

Accelerated Strength Testing

Based on reports by Philleo,[14] and Malhotra,[15] a brief review of the three test procedures covered by ASTM C 684 follows.

Procedure A (warm-water method). This is the simplest of the three methods and consists of curing standard cylinders (in their molds) in a water bath maintained at 35°C for 24 hr. A limitation of the method is that strength gain, compared to the 28-d moist-cured concrete at normal temperature, is not high. In the mid-1970s, the U.S. Corps of Engineers[16] conducted an extensive study on the evaluation of the warm-water method, from which it was concluded that accelerated strength testing with this method is a reliable means of routine quality control for concrete.

Procedure B (boiling-water method). This method consists of normal curing the cylinders for 24 hr, then curing in a boiling-water bath at 100°C for 3½ hr, and testing 1 hr later. The method is the most commonly used of the three procedures because compared with the 24-hr warm-water method, the strength gain at 28½ hr is much higher and cylinders can be transported to a central laboratory for strength testing, thus eliminating the need for an on-site laboratory. In the early 1970s the method was used successfully to develop concrete mix proportions in preliminary laboratory studies and to check field concrete in the construction of a large number of dikes, spillways, and a huge underground power station for the Churchill Falls Project in Labrador, Canada.

Procedure C (autogenous method). In this method, test cylinders immediately after casting are placed in insulated containers and are tested 48 hr later. No external heat source is provided, acceleration of strength gain being achieved by heat of hydration of the cement alone. Again, the strength gain at the end of the curing period is not high; also, the method is judged to be the least accurate of the three. It was used as an integral part of the quality control program in the construction of the CN Communication Tower located in Toronto, Canada. The project, completed in 1974, involved placing 30580 m^3 of slip-formed concrete to a height of 475 m. It is believed that the accelerated strength testing played an important role in the quality control of concrete and in the overall structural safety of the world's tallest free-standing structure (Fig. 10-14).

---

[14] R. E. Philleo, in *Progress in Concrete Technology,* ed. V. M. Malhotra, CANMET, Ottawa, 1980, pp. 729-48.

[15] V. M. Malhotra, *Concr. Int.,* Vol. 3, No. 11, pp. 17-21, 1981.

[16] J. F. Lamond, *J ACI,* Proc., Vol. 76, No. 4, pp. 499-512, 1979.

Figure 10-14 The CN communication tower, Toronto, Canada, 1974. [Photograph courtesy of CANMET, Ottawa.]

This slender, tapering tower is a beautiful landmark in concrete. The 1815-ft tower is the world's tallest free-standing structure, containing 39,800 yd³ of slip-formed concrete to a height of 1590 ft. Post-tensioning of concrete not only permitted a substantial reduction in the foundation requirement but also ensured that concrete will remain free from cracks, which is important for a structure exposed to considerable variations in ambient temperatures and humidity. With the slip-formed concrete rising almost at 20 ft/day, accelerated strength testing of concrete, based on the autogenous curing method, was a bold and necessary step for maintaining the construction schedule.

In Situ and Nondestructive Testing

The in situ tests can be classified into two categories: first, those that attempt to measure some property of concrete from which an estimate of strength, durability, and elastic behavior of the material may be obtained; and second, those that attempt to determine position, size, and condition of reinforcement; areas of poor consolidation, voids, and cracks; and the moisture content of in-place concrete. In regard to the tests providing an estimate of strength, Malhotra[17] suggests that unless comprehensive laboratory correlations on the field materials and mix proportions have been established between the strength parameters to be predicted and the results of in situ/nondestructive tests (NDT), the latter should not be used to predict strength. As a part of an overall quality assurance program for large projects, these methods have, however, proven to be unquestionably valuable.

[17] V. M. Malhotra, *Proc. CANMET/ACI Conf. on In Situ/Nondestructive Testing of Concrete*, ACI SP-82, pp. 1–16, 1984.

Testing and Control of Concrete Quality

A brief description and the significance of the commonly known in situ and nondestructive tests for assessment of concrete quality, which is based on a review by Malhotra,[18] is as follows.

Surface hardness methods. The surface hardness method consists essentially of impacting the concrete surface in a standard manner, using a given energy of impact, and measuring the size of indentation or rebound. The most commonly used method employs the *Schmidt rebound hammer,* which consists of a spring-controlled hammer that slides on a plunger. When the plunger is pressed against the surface of the concrete, it retracts against the force of the spring; when completely retracted, the spring is automatically released. The hammer impacts against the concrete surface and the spring-controlled mass rebounds, taking a rider with it along a guide scale which is used for the hammer rebound number. A standard procedure is described in detail in ASTM C 805.

The Schmidt hammer is simple and the method provides a quick and inexpensive means of checking uniformity of in-place hardened concrete, but the results of the test are affected by smoothness, degree of carbonation and moisture condition of the surface, size and age of specimen, and type of coarse aggregate in concrete. According to Malhotra, using a properly calibrated hammer, the accuracy of predicting concrete strength in laboratory specimens is ± 15 to 20 percent, and in a concrete structure it is ± 25 percent.

Penetration resistance techniques. The equipment to determine the penetration resistance of concrete consists essentially of powder-activated devices; a currently used system is known as the *Windsor probe.* In this system a powder-activated driver is used to fire a hardened-alloy probe into the concrete. The exposed length of the probe is a measure of the penetration resistance of concrete. Again, due to the small area under test the variation in the probe-test results is higher when compared with the variation in standard compressive strength tests on companion specimens. But this method is excellent for measuring the relative rate of strength development of concrete at early ages, especially for the purpose of determining stripping time for formwork. A standard test procedure is described in ASTM C 803.

Pullout tests. A pullout test consists of pulling out from concrete a specially shaped steel insert whose enlarged end has been cast into the fresh concrete. The force required for pullout is measured, using a dynamometer. Because of its shape, the steel insert is pulled out with a cone of concrete; therefore, the damage to the concrete surface has to be repaired after the test. If the test is being used for determining the safe form-stripping time, the pullout assembly need not be torn out of concrete; instead, when a pre-determined pullout force has been reached on the gage, the test may be terminated and the forms be removed safely.

During the test the concrete is in shear/tension, with generating lines of the

---

[18] Ibid.

cone running at approximately 45° to the direction of the pull. The pullout strength is of the order of 20 percent of the compressive strength, and is probably a measure of the direct shear strength. Like the penetration resistance test, the pullout test is an excellent means of determining the strength development of concrete at early ages and safe form-stripping times. Also, the technique is simple and the procedure is quick. The main advantage of pullout tests is that they attempt to measure directly the in situ strength of concrete. The major drawback is that unlike most other in situ tests, the pullout test has to be planned in advance, although new techniques are being developed to overcome this difficulty. A suitable test procedure is described in ASTM C 900.

Ultrasonic pulse velocity method. The ultrasonic pulse velocity method consists of measuring the time of travel of an ultrasonic wave passing through the concrete. The times of travel between the initial onset and reception of the pulse are measured electronically. The path length between transducers, divided by the time of travel, gives the average velocity of wave propagation. The relationships between pulse velocity and strength are affected by a number of variables, such as the age of concrete, moisture condition, aggregate/cement ratio, type of aggregate, and location of reinforcement. The method is therefore recommended for the purpose of quality control only; generally, attempts to correlate pulse velocity data with concrete strength parameters have not been successful. A suitable apparatus and a standard procedure are described in ASTM C 597.

Maturity meters. As described earlier, the basic principle behind the maturity concept is that the strength of concrete varies as a function of both time and temperature. Consequently, maturity meters have been developed to provide an estimate of concrete strength by monitoring the temperature of concrete with time.

Methods for assessing properties other than strength. Cover meters and pachometers are magnetic devices based on the principle that the presence of steel affects the field of an electromagnet. They are useful for determining the depth of concrete cover and the position of reinforcing bars. Electrical methods are gaining increasing acceptance as a tool for evaluation of in-place concrete (e.g., to determine reinforcement corrosion or the moisture content of concrete). Radiographic methods are available to reveal the position and condition of reinforcement, air voids, segregation, and cracking. Pulse-echo techniques are being developed to delineate voids and internal discontinuities in concrete.

Core Tests

In situ/NDT methods provide an effective way of obtaining a considerable number of preliminary test data at relatively little cost. When these tests indicate internal cracking or zones of weaker concrete, it is necessary to perform direct strength testing on cores obtained by a rotary diamond drill (ASTM C 42). The core strengths

Testing and Control of Concrete Quality

are generally lower than those of standard-cured concrete cylinder, especially in high-strength concretes. In the case of a high cement content and a correspondingly high heat of hydration, large sections of in-place concrete are vulnerable to microcracking in the transition zone between the coarse aggregate and the hydrated cement paste. Consequently, the ratio of core strength to cylinder strength decreases as the strength of the concrete increases. The strength of the core will also depend on its position in the structure. Generally, due to the differential bleeding effect, the cores taken near the top of a structural element are weaker than those from the bottom.

Quality Control Charts

As stated earlier, with high production rates of centralized ready-mixed plants or on-site concrete plants in large projects, the effective and economical system of quality control must be based on statistical methods. Statistical procedures are based on the laws of probability, and for these laws to properly function, the first requirement is that the data be gathered by random sampling. The second important statistical concept is that of the frequency distribution according to the bell-shaped normal-distribution Gaussian curve (Fig. 10–15a). A detailed discussion of statistical symbols and their definitions is outside the scope of this book; those interested should refer to any standard textbook on statistics, or ASTM Special Technical Publication 15D (1976).

Statistical quality control utilizes control charts that show graphically the results of a continuous testing program. The charts contain upper- and lower-limit lines that indicate the need for action when the plotted curve approaches or crosses them. The limit lines relate to the normal-distribution curve; in fact, a control chart may be considered as a normal-distribution curve laid on its side (Fig. 10–15b). Figure 10–15c illustrates the use of control charts in concrete quality control operations.

Based on the report of ACI Committee 214, typical quality control charts for continuous evaluation of strength test data of concrete are shown in Fig. 10–16. Figure 10–16a is a plot for *individual strength values*; the line for required average strength, σ_{cr}, is obtained from the expression $\sigma_{cr} = \sigma_c' + ts$, where σ_c' is the specified design strength, t a constant, and s the standard deviation. The chart gives an indication of the range or scatter between individual test values, and the number of low values. Unless the trend of individual low values persists, occasional low values may not be significant because they may represent chance variations rather than any problems with materials or testing method. Figure 10–16b is a plot of the *moving average for strength*; each point represents the average of the previous five sets of strength tests (each set of strength tests normally represents data from 3 test cylinders). This chart tends to smooth out chance variations and can be used to indicate significant trends due to variations in materials and processes that affect strength. Figure 10–16c is a plot of the *moving average for range,* where each point represents the average of the ranges of the 10 previous sets of strength tests. The chart provides

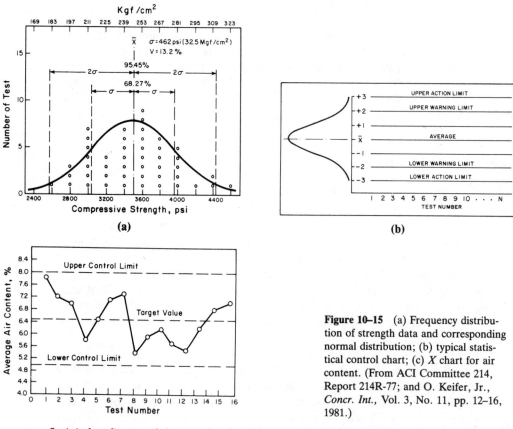

Figure 10–15 (a) Frequency distribution of strength data and corresponding normal distribution; (b) typical statistical control chart; (c) X chart for air content. (From ACI Committee 214, Report 214R-77; and O. Keifer, Jr., *Concr. Int.*, Vol. 3, No. 11, pp. 12–16, 1981.)

Statistical quality control charts are based on frequency distribution predicted by the normal-distribution curve. On a typical control chart, the upper and lower control limits may be derived from the normal-distribution curve laid on its side.

a control on the reproducibility of the test procedures; when the range chart indicates poor reproducibility between different sets of data, it is time to check the testing procedures.

EARLY AGE CRACKING IN CONCRETE

In designing reinforced concrete elements it is assumed that concrete would crack due to thermal and humidity cycles, however, by careful design and detailing, the cracks can be controlled and crack-widths can be limited. While, in principle at least, the thermal and drying shrinkage cracks can be predicted and controlled, extensive cracking can develop due to other causes. It is not easy to distinguish the different crack configurations and this often requires a number of laboratory tests and com-

Early Age Cracking in Concrete

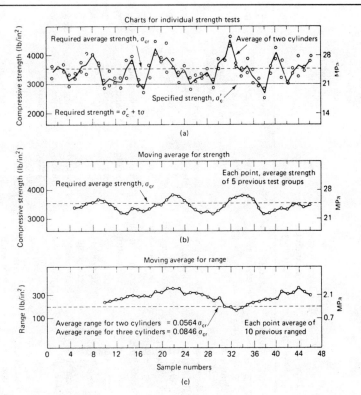

Figure 10–16 Typical quality control charts for concrete strength. (From ACI Committee 214, Report 214R-77; and O. Keifer, Jr., *Concr. Int.*, Vol. 3, No. 11, pp. 12–16, 1981.)

pilation of complete history of the project including concrete mix design, placement conditions, curing methods, formwork removal, and loading history. Based on a report by the Concrete Society of U.K., crack types are illustrated in Fig. 10–17 and their classification with possible causes and prevention methods are listed in Table 10–4. Most of the causes responsible for non-structural cracking have been described earlier in this Chapter and in Chapter 5. Two other types of non-structural cracks, which have not been discussed, will be briefly reviewed next.

Plastic settlement cracks occur when the bleeding and settlement are high and there is some restraint to the settlement. It should be noted that these cracks are independent of evaporation and superficial drying. Methods used to prevent the settlement cracks include the following: reduction of bleeding, reduction of tendency for settlement by providing adequate restraint, revibration of concrete.

Hair-line, discontinuous surface cracking, also called **crazing**, can appear in hardened concrete after several weeks. These cracks are particularly observed during rainy periods when they absorb moisture and polluents from the atmosphere,

TABLE 10-4 CLASSIFICATION OF CRACK TYPES*

Type of cracking	Letter (see Figure 10-17)	Subdivision	Most common location	Primary cause (excluding restraint)	Secondary causes/ factors	Remedy (assuming basic redesign is impossible) in all cases reduce restraint	Time of appearance
Plastic settlement	A	Over reinforcement	Deep sections	Excess bleeding	Rapid early drying conditions	Reduce bleeding (air entrainment) or revibrate	Ten minutes to three hours
	B	Arching	Top of columns				
	C	Change of depth	Trough and waffle slabs				
Plastic shrinkage	D	Diagonal	Roads and slabs	Rapid early drying	Low rate of bleeding	Improve early curing	Thirty minutes to six hours
	E	Random	Reinforced concrete slabs				
	F	Over reinforcement	Reinforced concrete slabs	Ditto plus steel near surface			
Early thermal contraction	G	External restraint	Thick walls	Excess heat generation	Rapid cooling	Reduce heat and/or insulate	One day to two or three weeks
	H	Internal restraint	Thick slabs	Excess temperature gradients			
Long-term drying shrinkage	I		Thin slabs (and walls)	Inefficient joints	Excess shrinkage Inefficient curing	Reduce water content Improve curing	Several weeks or months
Crazing	J	Against formwork	'Fair faced' concrete	Impermeable formwork	Rich mixes	Improve curing and finishing	One to seven days sometimes much later
	K	Floated concrete	Slabs	Over-trowelling	Poor curing		
Corrosion of reinforcement	L	Natural	Columns and beams	Lack of cover	Poor quality concrete	Eliminate causes listed	More than two years
	M	Calcium chloride	Precast concrete	Excess calcium chloride			

*Adapted from Concrete Society of U.K., Tech. Report No. 22, 1985.

TABLE 10-5 RELATIVE EFFECTS OF MATERIAL CHARACTERISTICS, MIX PROPORTIONS, AND EARLY-AGE OPERATIONS ON THE PROPERTIES OF CONCRETE

Properties	Type of portland cement	Aggregate characteristics	Type of admixture	Mix proportions	Placing and compaction	Surface treatment	Curing conditions (temperature and humidity)
Workability							
Consistency	M	L	L	L	n	n	c
Cohesiveness	M	L	L	L	M	n	c
Setting time	L	n	L	M	n	-	c
Strength							
Early	L	n	L	L	L	n	L
Ultimate	n	n	M	L	L	n	L
Permeability	n	L	L	L	L	L	L
Shrinkage							
Plastic	n	n	n	M	M	n	L
Drying	n	L	M	L	n	n	L
Thermal	L	L	L	M	L	M	L
Surface appearance	n	n	n	L	M	L	M
Frost resistance	n	M	L	L	M	M	L
Abrasion resistance	n	L	n	L	L	M	L
Coefficient of thermal expansion	n	L	n	L	n	n	n

L, large effect; M, moderate effect; n, no or negligible effect; c, not applicable since curing starts after the removal of formwork.

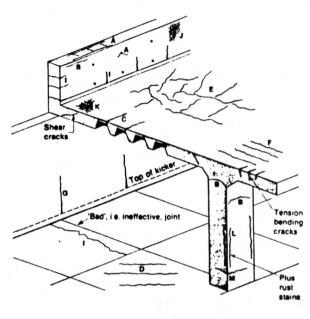

Figure 10–17 Cracks in a hypothetical concrete structure. (From Concrete Society, *Constructural Cracks in Concrete,* Report of a Society Working Party, The Concrete Society, Technical Report, No. 22, 1985.)

giving the disagreeable impression of damage to concrete. In reality, the cracks are quite superficial, perhaps not more than a fraction of a millimeter deep and do not cause structural problems with the exception of opening up later and providing a passage for aggressive agents. Crazing usually occurs as a result of inadequate finishing and curing particularly in the presence of high humidity gradients between the surface and the bulk of concrete. The use of smooth and impermeable formwork surfaces (steel, plastic), or over-trowelling of rich concrete mixtures tends to concentrate the cement paste at the concrete surface which cracks easily due to drying shrinkage, producing crazing.

A review of **structural cracks** due to insufficient reinforcement or higher loads than designed, is beyond the scope of this book.

CONCLUDING REMARKS

In this chapter it is shown that various early-age operations, such as placing and compaction, surface treatment, and curing, have an important effect on the properties of concrete. In Chapters 6 to 9 a similar conclusion was reached; that is, the characteristics of cement, aggregate, and admixtures also have an important effect on the properties of concrete. To keep the various factors influencing the properties of concrete in proper perspective, it should be interesting to see at one glance their relative significance with respect to some of the major properties of concrete. Such an attempt is presented in Table 10–5.

Suggestions for Further Study

The information in Table 10–5 is qualitative only, nevertheless it is useful for educational purposes. For instance, it may surprise some engineers to discover that the type of cement influences mainly the setting time, the early strength, and the heat of hydration (thermal shrinkage of concrete). On the other hand, mix proportions, placing and compaction, and curing conditions have a far-reaching effect on several important properties of concrete, such as ultimate strength, permeability, plastic shrinkage, and drying shrinkage.

TEST YOUR KNOWLEDGE

1. Explain the operations covered by the following terms, and discuss the significance of these operations: retempering, revibration, screeding, bullfloating, scoring.
2. What is the principle behind consolidation of concrete mixtures by vibration? Describe the sequence of actions that take place in a fresh concrete mixture when it is exposed to a high-frequency vibrator.
3. Explain the two important objects of curing and how they are achieved in: (a) cold-weather concreting and (b) hot-weather concreting.
4. How would you define workability? Is workability a fundamental property of fresh concrete? If not, why? What are the principal components of workability and their significance in the concrete construction practice?
5. Define the following phenomena, and give their significance and the factors affecting them: slump loss, segregation, bleeding.
6. Suggest at least two methods by which you can reduce "bleeding" of a concrete mixture.
7. With the help of a sketch briefly describe the "Vebe Test." What is the objective of this test, and for which concrete types it is more suitable when compared to a commonly used test method?
8. What are harmful manifestations of plastic shrinkage of concrete in: (a) reinforced columns and (b) slabs? Assuming that the air temperature is 70°F, the concrete temperature is 75°F, and the wind velocity is 20 miles/hr, determine the rate of evaporation. If this rate is too high from the standpoint of risk of plastic-shrinkage cracking, what precautionary measures would you take? Alternatively, determine the temperature to which concrete must be cooled to reduce the rate of evaporation to a safe limit.
9. Why may the setting time of concrete be substantially different from the setting time of the cement from which the concrete is made? Define the initial and the final setting times as measured by the penetration resistance method (ASTM C 403). What is their significance in the concrete construction practice?
10. With the help of suitable curves show the effect of both accelerating and retarding admixtures on the setting time of a concrete mixture.
11. Briefly discuss the effect of temperature on the setting time of concrete. What is the most efficient way of reducing the temperature of a fresh concrete mixture? Explain why.
12. In the ACI 306R (*Recommended Practice for Cold-Weather Concreting*), explain why: (a) higher concrete temperatures are required for colder weather and (b) lower concrete temperatures are permitted for massive concrete members.
13. Explain the maturity concept, its application, and its limitations.

14. (a) For a concrete mixture containing 600 lb of cement, 3050 lb of aggregate (SSD condition), and 310 lb of mixing water, calculate the temperture of concrete in summer, assuming that the cement and the aggregate are at 85°F and the water has been cooled to 40°F. (b) For the concrete mixture in part (a) calculate the temperature of concrete in winter, assuming that the cement and the aggregate are at 40°F and the water has been heated to 150°F.
15. You have recently taken charge of a large project. Write a short note for the attention of the owner on the subject of a concrete quality assurance program, explaining briefly the advantages, disadvantages, and testing costs of the three accelerated testing procedures and the various nondestructive test methods.
16. What are the principles behind the following test procedures: Schmidt hammer test, Windsor probe test, pullout test, pulse-velocity test? Explain which you would recommend for deciding the formwork removal time.
17. Describe the essential elements of statistical quality control charts. In the case of concrete strength data, explain why moving-average and moving-range charts are more useful than those containing a plot of individual strength values.

SUGGESTIONS FOR FURTHER STUDY

Report of ACI Committee 228. "In-Place Methods for Determination of Strength of Concrete," *ACI Materials Jour.*, Vol. 85, No. 5, pp. 446–71, 1988.

Report of ACI Committee 214, "Recommended Practice for Evaluation of Strength Test Results of Concrete," *ACI Manual of Concrete Practice*, Part 2, 1991.

Report of ACI Committee 305, "Hot Weather Concreting," *ACI Materials J.*, Vol. 88, No. 4, pp. 417–36, 1991.

Report of ACI Committee 306, "Cold Weather Concreting," *ACI Manual of Concrete Practice*, Part 2, 1991.

ASTM, *Significance of Tests and Properties of Concrete and Concrete-Making Materials*, STP 169B, 1978, Chaps. 7, 9, 13, and 15.

Design and Control of Concrete Mixtures, 13th Edition, Portland Cement Association, Skokie, Ill., 1988.

In Situ/Non-destructive Testing of Concrete, Proc. CANMET/ACI Conf., ACI SP-82, 1984.

MINDNESS, S., and J. F. YOUNG, *Concrete*, Prentice Hall, Inc., Englewood Cliffs, N.J., 1981, Chaps. 8, 11, and 17.

NEVILLE, A. M., *Properties of Concrete*, Pitman Publishing, Inc., Marshfield, Mass., 1981, Chaps. 4, 8, and 10.

POWERS, T. C., *The Properties of Fresh Concrete*, John Wiley & Sons, Inc., New York, 1968.

TATTERSALL, G. H., *The Workability of Concrete*, Cement and Concrete Association, Wexham Springs, Slough, U.K., 1976.

CHAPTER 11

Progress in Concrete Technology

PREVIEW

Normal concrete made with portland cement and conventional natural aggregate suffers from several deficiencies. Attempts to overcome these deficiencies have resulted in the development of special concretes, which represent advances in the concrete technology and are the subject of this chapter. Theoretical considerations underlying special concretes, their mix proportions, properties, and applications will be described. The major problems in the use of concrete as a material of construction and the ways found to tackle them successfully are summarized in the following paragraphs.

Compared to steel, the **low strength/weight ratio** for concrete presents an economic problem in the construction of tall buildings, long-span bridges, and floating structures. To improve the strength/weight ratio, two approaches suggest themselves: either the density of the material should be lowered, or the strength should be increased. The first approach has been practiced successfully for the last 70 years. Structural lightweight aggregates are commonly used throughout the world to produce **lightweight concretes** with about 100 lb/ft^3 (1600 kg/m^3) unit weight and 4000 to 6000 psi (25 to 40 MPa) compressive strength.

In accordance with the second approach, during the 1970s, normal-weight 150 lb/ft^3 (2400 kg/m^3) **high-strength concretes** with 9000 to 12,000 psi (60 to 80 MPa) compressive strengths have been produced industrially by using normal or superplasticizing water-reducing admixtures and pozzolans. However, it appears that the limiting strength/weight ratio for concrete has not yet been reached. Developments are under way to produce stronger lightweight aggregates so that light weight and high strength may be achieved in concrete simultaneously. Application of superplasticizing admixtures is not limited to the production of high-strength concretes. As structures become larger and designs become more complex, massive placements of concrete mixtures are needed for heavily reinforced structural members. Superplasticized concrete now fulfills this need for **high-workability or flowing-concrete** mixtures, without a high water/cement ratio and segregation. These concretes with high workability, even under hot-weather conditions, do not require mechanical consolidation. They have extended the use of concrete to new frontiers.

Shrinkage of concrete on drying frequently leads to cracking; this is recognized in concrete design and construction practice, especially in regard to pavements, floors, and relatively thin structural members. To counteract this problem **shrinkage-compensating concrete** containing expansive cements or expansive admixtures has been employed successfully for the last 20 years.

Again, compared to other building materials the toughness of concrete is very low and therefore the impact resistance is poor. This characteristic has been substantially improved by using the concept of microlevel reinforcement. **Fiber-reinforced concrete** containing steel, glass, or polypropylene fibers has been employed successfully in situations where resistance to impact is important.

Imperviousness in materials is important for durability to moisture and to strong chemical solutions. Three types of **concretes containing polymers** have been developed which show very low permeability and excellent chemical resistance. Overlays composed of these concretes are suitable for the protection of reinforcing steel from corrosion in industrial floors and bridge decks; the concretes are also useful for rehabilitation of deteriorated pavements.

Heavyweight concrete made with high-density minerals is about 50 percent heavier than normal concrete containing conventional aggregate; this type of concrete is used for **radiation shielding** in nuclear power plants when limitations of usable space require a reduction in the thickness of the shield.

Mass concrete for dams and other large structures has been around for some time, but methods selected to control the temperature rise have had a considerable influence on the construction technology during the last 30 years. Precooling of concrete materials has virtually eliminated the need for expensive post-cooling operations and has made faster construction schedules possible. Dams that are less than 100 m high can be built with **roller-compacted concrete,** using ordinary earth-moving equipment, at speeds and costs that were unimaginable only 10 years ago.

STRUCTURAL LIGHTWEIGHT CONCRETE

Definitions and Specifications

Structural lightweight concrete is structural concrete in every respect except that for reasons of overall cost economy, the concrete is made with lightweight cellular aggregates so that its unit weight is approximately two-thirds of the unit weight of concrete made with typical natural aggregates. Since light weight, and not strength, is the primary objective, the specifications limit the maximum permissible unit weight of concrete. Also, since highly porous aggregates tend to reduce concrete strength greatly, the specifications require a minimum 28-day compressive strength to ensure that the concrete is of structural quality.

ACI 213R-87, *Guide for Structural Lightweight Aggregate Concrete,*[1] defines

---

[1] Report of ACI Committee 213, *ACI Materials J.,* Vol. 87, No. 3, pp. 638–51, 1987.

structural lightweight aggregate concretes as concretes having a 28-day compressive strength in excess of 2500 psi (17 MPa) and a 28-day, air-dried unit weight not exceeding 115 lb/ft^3 (1850 kg/m^3). The concrete may consist entirely of lightweight aggregates or, for various reasons, a combination of lightweight and normal-weight aggregates. From the standpoint of workability and other properties, it is a common practice to use normal sand as fine aggregate, and to limit the nominal size of the lightweight coarse aggregate to a maximum of 19 mm. According to ASTM C 330, fine lightweight and coarse lightweight aggregates are required to have dry-loose weights not exceeding 70 lb/ft^3 (1120 kg/m^3) and 55 lb/ft^3 (880 kg/m^3), respectively. The specification also contains requirements with respect to grading, deleterious substances, and concrete-making properties of aggregate, such as strength, unit weight, drying shrinkage, and durability of concrete containing the aggregate.

ASTM C 330 Standard Specifications requirements for compressive and tensile strengths and unit weight of structural lightweight concrete are shown in Table 11-1.

Mix-Proportioning Criteria

For various reasons the absolute volume method, which is the basis of the ACI method of proportioning normal-weight concrete mixtures (Chapter 9), is not useful for designing lightweight concrete mixtures. First, the relation between strength and water/cement ratio cannot be effectively employed because it is difficult to determine how much of the mixing water in concrete will be absorbed by the aggregate. The difficulty is caused not only by the large amounts (10 to 20 percent) of water absorption by porous aggregate, but also by the fact that some aggregates continue to absorb water for several weeks. Therefore, reliable estimates of moisture deviation from the saturated-surface dry (SSD) condition and of the SSD bulk specific

TABLE 11-1 REQUIREMENTS FOR STRUCTURAL LIGHTWEIGHT CONCRETE[a]

Air-dried, 28-day unit weight, max. [lb/ft^3 (kg/m^3)]	28-day splitting tensile strength, min. [psi (MPa)]	28-day compressive strength, min. [psi (MPa)]
All lightweight aggregates		
110 (1760)	320 (2.2)	4000 (28)
105 (1680)	300 (2.1)	3000 (21)
100 (1600)	290 (2.0)	2500 (17)
Combination of normal sand and lightweight aggregate		
115 (1840)	330 (2.3)	4000 (28)
110 (1760)	310 (2.1)	3000 (21)
105 (1680)	300 (2.1)	2500 (17)

[a] The compressive strength and unit weight shall be the average of three specimens, and the splitting tensile strength shall be the average of eight specimens.

Source: Reprinted, with permission, from the Annual Book of ASTM Standards, Section 4, Vol. 04.02. Copyright, ASTM, 1916 Race Street, Philadelphia, PA 19103.

gravity are very difficult. Also, unlike normal-weight aggregates, the bulk specific gravity of lightweight aggregates can vary widely with grading.

Workability considerations in freshly made lightweight aggregate concretes require special attention because with high-consistency mixtures the aggregate tends to segregate and float on the surface. To combat this tendency, it is often necessary to limit the maximum slump and to entrain air (irrespective of whether durability of concrete to frost action is a consideration). Approximately, 5 to 7 percent air entrainment is generally required to lower the mixing water requirement while maintaining the desired slump and reducing the tendency for bleeding and segregation. Consequently, structural engineers' specifications for lightweight concrete usually include minimum permissible values for compressive strength, maximum values for unit weight and slump, and both minimum and maximum values for air content.

For the purpose of mix design, the compressive strength of lightweight aggregate concrete is usually related to cement content at a given slump rather than to the water/cement ratio. In most cases, the compressive strength at a given cement and water content can be increased by reducing the maximum size of coarse aggregate and/or partial replacement of lightweight fine aggregate with a good-quality natural sand. According to ACI 213R-79, the approximate relationship between average compressive strength and cement contents for both all-lightweight and sanded-lightweight concretes is shown in Table 11-2. It should be noted that complete replacement of the lightweight fines will increase the unit weight by about 10 lb/ft^3 (160 kg/m^3) at the same strength level.

With some lightweight aggregates, it may be possible to use the ACI 211.1 volumetric method for proportioning normal-weight concrete mixtures and adjust the proportions by trial and error until the requirements of workability of fresh concrete and the physical properties of hardened concrete are satisfactorily met. In this case it is often convenient to start with equal volumes of fine and coarse aggregate and to make adjustments as necessary for achieving the desired slump with minimum segregation. The problems of mix proportioning with the lightweight aggregates and the methods to get around them are described in ACI Standard 211.2 (*Standard Practice for Selecting Proportions for Structural Lightweight Concrete,*

TABLE 11-2 APPROXIMATE RELATIONSHIP BETWEEN AVERAGE COMPRESSIVE STRENGTH AND CEMENT CONTENT

Compressive strength	Cement [lb/yd^3 (kg/m^3)]	
psi (MPa)	All-lightweight	Sanded-lightweight
2500 (17.24)	400–510 (240–305)	400–510 (240–305)
3000 (20.68)	440–560 (260–335)	420–560 (250–335)
4000 (27.58)	530–660 (320–395)	490–660 (290–395)
5000 (34.47)	630–750 (375–450)	600–750 (360–450)
6000 (41.37)	740–840 (440–500)	700–840 (420–500)

Source: Report of ACI Committee 213R-87.

p. 307), which should be consulted when a detailed and more accurate procedure is needed.

Properties

Workability. The properties of fresh concrete made with lightweight aggregate and the factors affecting them are essentially the same as in normal-weight concrete. Due to low density and the characteristic rough texture of porous aggregates especially in the crushed state, the workability of concrete needs special attention. In general, placing, compacting, and finishing lightweight aggregate concretes requires relatively less effort; therefore, even 50 to 75 mm slump may be sufficient to obtain workability of the type that is usually shown by 100 to 125 mm-slump normal-weight concrete.

With lightweight aggregate concrete mixtures, high slump and overvibration are the two causes generally responsible for drawing the heavier mortar away from the surface, where it is needed for finishing. This phenomenon, called **floating of the coarse aggregate,** is the reverse of that for normal-weight concrete, where segregation results in an excess of mortar at the surface. ACI 213R-87 recommends a maximum of 100 mm of slump for achieving a good surface in floors made with lightweight aggregate concrete. Slump loss can be a serious problem when aggregate continues to absorb a considerable amount of water after mixing. This problem can be controlled by batching the aggregate in a damp condition.

Unit weight. Next to workability, unit weight and strength are the two properties generally sought from structural lightweight concrete. With given materials, it is generally desired to have the highest possible strength/unit weight ratio with the lowest cost of concrete. Specifications limit the air-dried unit weight of concrete to a maximum of 115 lb/ft^3 (1840 kg/m^3), but there is no minimum limit. However, it is a common experience that when a highly porous aggregate larger than 19 mm maximum size is used, the unit weight of concrete can be reduced to less than 90 lb/ft^3 (1440 kg/m^3), but the product may not be able to meet the minimum 2500-psi (17 MPa) 28-day compressive strength requirement for structural lightweight concrete.

The use of normal sand to control the properties of hardened concretes tends to increase the unit weight, although this tendency is partially offset from the opposite effect of entrained air, which is invariably prescribed for improving the workability. Most structural lightweight concretes weigh between 100 to 110 lb/ft^3 (1600 to 1760 kg/m^3); however, job specifications in special cases may allow higher than 115 lb/ft^3 (1840 kg/m^3).

Strength. Design strengths of 3000- to 5000-psi 28-day compressive strengths are common, although by using a high cement content and good-quality lightweight aggregate of small size (i.e., 9 or 13 mm maximum) it has been possible, in some precast and prestressing plants, to produce 6000- to 7000-psi (40 to 48 MPa)

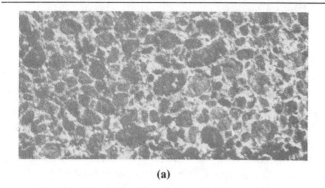

(a)

(b)

Figure 11-1 Fracture surface from concrete cylinders after splitting tension test: (a) concrete made with lightweight aggregate; (b) concrete made with rounded flint aggregate. (Photographs courtesy P. Nepper-Christensen, Aalborg Portland Co., Aalborg, Denmark.)

In lightweight aggregate concrete, the fracture passes through the cellular aggregate particles because both the transition zone and the cement paste are generally stronger. On the other hand, in normal-weight concrete the aggregate particles are dense and strong, and fracture is usually in the transition zone or the bulk cement paste, not through the aggregate.

concrete. Lightweight aggregates with controlled microporosity have been developed to produce 10,000- to 11,000-psi (70 to 75 MPa) lightweight concretes which generally weigh 115 to 125 lb/ft^3 (1840 to 2000 kg/m^3).

The splitting tensile strength of concrete cylinders (ASTM C 496) is a convenient relative measure for tensile strength. The data in Table 11-1 show that, like normal-weight concrete, this ratio between the splitting tensile strength and compressive strength decreases significantly with the increasing strength of lightweight concrete. The modulus of rupture of continuously moist-cured lightweight concrete also behaves in the same manner; tests on dried specimens show that the data are extremely sensitive to the moisture state. Examination of fractured specimens of lightweight aggregate concrete after the splitting tension test clearly reveals that unlike normal-weight concrete, the aggregate, and not the transition zone, is generally the weakest component in the system (Fig. 11-1). Holm et al.[2] have presented scanning electron micrographic evidence to show that due to the pozzolanic reaction,

[2] T. A. Holm, T. W. Bremner, and J. B. Newman, *Concr. Int.,* Vol. 6, No. 6, pp. 49–54, 1984.

the aggregate-cement paste bond at the surface of lightweight aggregates was stronger than the aggregate particles.

Dimensional stability. In the ACI Building Code 318, the elastic modulus of normal-weight or structural lightweight concrete is computed using the equation $E_c = 33W^{1.5}\sqrt{f_c}$ psi (Chapter 4). The values of the elastic modulus thus obtained may deviate from those obtained experimentally (ASTM C 469) by ±15 to 20 percent. From a large number of test specimens Schideler[3] found that the static elastic moduli for 3000- and 6000-psi concretes containing expanded-clay aggregate were 1.5×10^6 and 2.0×10^6 psi, respectively. A concrete with total replacement of lightweight sand with a normal sand generally gives 15 to 30 percent higher elastic modulus. Experimental studies indicate that the ultimate compressive strain of most lightweight concretes may be somewhat greater than the value 0.003 assumed for design purposes.

Compared to normal-weight concrete, concretes made with lightweight aggregate exhibit a higher moisture movement (i.e., a higher rate of drying shrinkage and creep), somewhat higher ultimate drying shrinkage (typically 800×10^{-6} in./in.), and considerably higher creep (typically 1600×10^{-6} in./in.). It seems that low strength and low elastic modulus have more pronounced effects on creep than on drying shrinkage. The effects of replacement of fine lightweight aggregate with normal sand on creep and drying shrinkage are shown in Fig. 11–2. In view of the comparatively low tensile strength of lightweight concrete (300 to 400 psi), a considerable tendency for the drying-shrinkage cracking is expected; this tendency is offset to some degree by a lower modulus of elasticity and a greater extensibility of lightweight concrete. Furthermore, Kulka and Polivka[4] point out that in a structure the creep and the shrinkage of lightweight concrete (this is also true of normal-weight concrete) would be much smaller than the values obtained on specimens in the laboratory. For example, in the design of a lightweight-concrete bridge structure, the shrinkage and creep values were reduced from the laboratory test data by about 15 to 20 percent due to the size effect of the member, 10 to 20 percent due to humidity of the environment, and about 10 to 15 percent due to reinforcement (overall about 50 percent reduction).

Durability. The freeze-thaw resistance of air-entrained lightweight concrete is similar to that of normal-weight concrete. Air entrainment is especially helpful when aggregates are close to saturation and concrete will be subject to frost action. The hydraulic pressure caused by the expulsion of water from the aggregate can be accommodated by the entrained air present in the cement mortar, thus preventing damage to the concrete.

Although air-dried lightweight aggregate concrete tends to show a greater degree of moisture absorption, this does not mean high permeability. In fact, the permeability of lightweight concrete is low, and therefore its durability to aggressive

---

[3] J. J. Schideler, *J. ACI*, Proc., Vol. 54, pp. 299–329, 1957.
[4] F. Kulka and M. Polivka, *Consult. Eng.*, Vol. 42, No. 12, 1978.

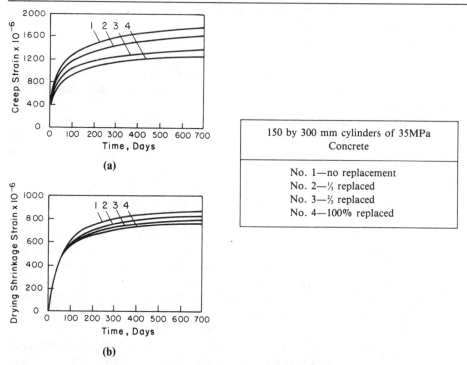

Figure 11–2 Effect of replacing lightweight fines by natural sand on (a) creep strain; (b) drying shrinkage strain. (From D. F. Orchard, *Concrete Technology*, Vol. 1, Elsevier Applied Science Publishers Ltd., Barking, Essex, U.K., 1979.)

chemical solutions is generally quite satisfactory. Holm et al.[5] reported that in 1973, concrete cores taken from a 3.25-mile, 5.2 km) long lightweight concrete deck (which was then 20 years old) revealed little evidence of microcracking, whereas from a parallel location, cores of concrete of the same age but containing a quartz gravel as aggregate were found to contain major cracks. Similarly, in 1980, cores of lightweight concrete from a World War I freighter, the USS *Selma,* which had been exposed to a marine environment for 60 years, also revealed little microcracking. The principal reason for low permeability and excellent durability of lightweight concrete is the general absence of microcracks in the aggregate-cement paste transition zone, which, according to Holm et al., is due to similarities of elastic moduli between the lightweight aggregate and the mortar fraction. Also, due to the pozzolanic reaction between the reactive aluminum silicates of the thermally activated lightweight aggregate particle and calcium hydroxide of the cement paste, the transition zone is normally dense and strong.

Lightweight aggregates are porous and therefore more easily friable than are normal-weight rocks and minerals. Consequently, concrete containing lightweight

[5]T. A. Holm, T. W. Bremner, and J. B. Newman, *Concr. Int.,* Vol. 6, No. 6, pp. 49–54, 1984.

aggregates generally shows *poor resistance to heavy abrasion*. Replacement of lightweight fines with natural sand improves the abrasion resistance.

The *coefficient of thermal expansion* of expanded-shale lightweight aggregate concretes is of the order of 6 to 8×10^{-6} per °C. In regard to resistance to heat flow, although insulating lightweight concretes (not described here) are better than structural lightweight concrete because of lower aggregate density, the *thermal conductivity* of the latter is still about half as much as that of normal-weight concrete, and therefore the fire endurance is considerably better. Typical values of thermal conductivity for concretes containing all expanded-clay aggregate and those containing expanded-clay coarse aggregate with natural sand, are 0.22 and 0.33 Btu/ft^2-hr-°F/ft., respectively.

Applications

According to ACI 213R-87, the use of lightweight aggregate concrete in a structure is usually predicated on a lower overall cost of the structure. While lightweight concrete will cost more than normal-weight concrete per cubic yard, the structure may cost less as a result of reduced dead weight and lower foundation costs. Wilson[6] cites several examples, including the following, to demonstrate that application of lightweight concrete can result in lower costs for foundations, reinforcing steel, and construction.

The construction of the lightweight concrete bridge deck for the San Francisco-Oakland Bay Bridge in 1936 resulted in a $3 million saving in steel. Since then, numerous lightweight concrete bridge decks have been built throughout the world. Strength is not a major consideration in floor slabs; therefore, a large amount of lightweight aggregate is used to reduce concrete dead weight in floors of high-rise buildings. An example of this application is the Lake Point Tower in Chicago, Illinois, which was built in 1968 and has 71 stories. The floor slabs from the second to the seventieth level and in the garage area were made of cast-in-place concrete with a density of 1730 kg/m^3 and a 7-day compressive strength (150- by 300-mm cylinders) of 20 to 22 MPa. The Australian Square, Sydney, Australia, completed in 1967, is a circular tower (50 stories) 184 m high by 42.5 m in diameter. A 13 percent saving in construction costs was achieved through the use of 31,000-m^3 lightweight concrete in the beams, columns, and floors above the seventh-floor level. The concrete had an average compressive strength of 34.3 MPa and an average density of 1792 kg/m^3 at 28 days. One Shell Plaza, Houston, Texas, is a 1969 all-lightweight concrete structure of 52 stories, containing a 70- by 52- by 2.5-m lightweight concrete pad 18 m below grade. The concrete, with a density of 1840 kg/m^3, had a 41.2-MPa 28-day compressive strength for shear walls, columns, and mat foundation, and 31.3 MPa for the floor structures. If normal concrete had been used, only a 35-story structure could have been safely designed due to the limited bearing capacity of the soil.

[6] H. S. Wilson, in *Progress in Concrete Technology,* ed. V. M. Malhotra, CANMET, Ottawa, 1981, pp. 141–87.

Kulka and Polivka[7] state that the basic economy of lightweight concrete can be demonstrated by the savings in reinforcing. In ordinary reinforced concrete the economic advantage is not as pronounced as in prestressed concrete. The prestressing force in most cases is computed strictly from the dead load of the structure; consequently, a weight reduction of 25 percent results in a substantial reduction in the weight of prestressing tendons. Among other advantages of reduction in weight of concrete is the superior resistance of shear elements to earthquake loading since seismic forces are largely a direct function of the dead weight of a structure.

It should be noted that major applications of lightweight concrete throughout the world continue to remain in the production of precast concrete elements and prefabricated panels. Due to lower handling, transportation, and construction costs, the lightweight concrete products are ideally suited for this type of construction (Fig. 11-3).

Although expanded clay and shale aggregates are most suitable for the production of structural-quality lightweight concrete, the escalation of fuel costs in the 1970s has priced these aggregates out of many markets. Consequently, there is renewed interest in finding natural lightweight aggregate of good quality. According to Bryan,[8] there are ample reserves of a suitable volcanic rock called rhyolite in Arizona, Colorado, California, Nevada, New Mexico, Oregon, and Utah. The

Figure 11-3 Structural lightweight concrete wall panels, 16 ft wide and 27 ft high, are light enough to be handled by a small crane. (Photographs courtesy Expanded Shale, Clay, and Slate Institute, Bethesda, Md.)

---

[7] F. Kulka and M. Polivka, *Consult. Eng.*, Vol. 42, No. 12, 1978.
[8] D. L. Bryan, *J. ACI*, Proc., Vol. 74, No. 11, p. N23, 1977.

weight of air-dried rhyolite concretes at 28 days was between 110 lb/ft^3 (1760 kg/m^3) and 125 lb/ft^3 (2000 kg/m^3); compressive strengths with a 611-lb (280 kg) cement content were well over 4000 psi (28 MPa). Concrete up to 6000 psi (40 MPa) has been designed and used in construction in the Reno, Nevada, area. Reported to be the world's largest casino, the MGM Grand Hotel in Reno built in 1978 contains over 65,000 yd^3 (50,000 m^3) of lightweight concrete with rhyolite aggregate.

HIGH-STRENGTH CONCRETE

Definition

For mixtures made with normal-weight aggregates, high-strength concretes are considered to be those which have compressive strengths in excess of 6000 psi (40 MPa). Two arguments are advanced[9] to justify this definition of high-strength concrete:

1. The bulk of the conventional concrete is in the range 3000 to 6000 psi. To produce concrete above 6000 psi, more stringent quality control and more care in the selection and proportioning of materials (plasticizers, mineral admixtures, type and size of aggregates, etc.), are needed. Thus, to distinguish this specially formulated concrete, which has a compressive strength above 6000 psi, it should be called *high strength*.
2. Experimental studies show that in many respects the microstructure and properties of concrete with a compressive strength above 6000 psi are considerably different from those of conventional concrete. Since the latter is the basis of current concrete design practice (e.g., the empirical equation for estimating the elastic modulus from compressive strength), the designer will be alerted if concrete of higher than 6000 psi is treated as a separate class.

Significance

Based on 1982 prices for concrete and steel in the Chicago area, the data in Fig. 11–4 show that the use of the highest-possible strength concrete and minimum steel offers the most economical solution for columns of high-rise buildings. According to a recent report, the increase of only 3.1 times in price for an increase of 4.7 times in load-carrying capacity clearly demonstrates the economy of using high-strength concrete in multistory building columns.[10] Consequently, in the United States during the past 20 years, high-strength concrete has been used mainly for constructing the reinforced concrete frames of buildings 30 stories and higher. The columns in the upper third of the building are made with conventional concrete (4000 to 5000 psi),

---

[9] S. P. Shah, *Concr. Int.*, Vol. 3, No. 5, pp. 94–98, 1981.
[10] J. Moreno, *Concr. Int.*, Vol. 12, No. 1, pp. 35–39, 1990.

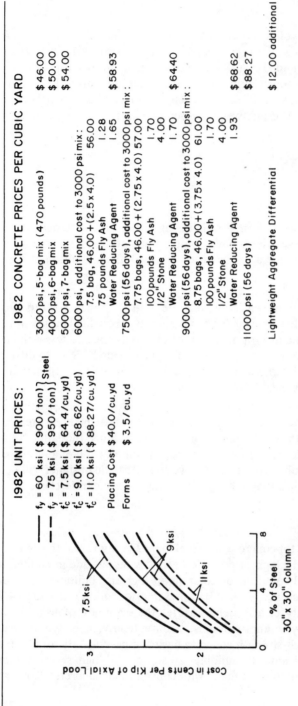

Figure 11-4 Relation between concrete strengths and column costs. (From J. Moreno, *Proc. ACI Seminar on High-Strength Concrete*, ACI Annual Convention, Los Angeles, 1983, pp. 52–68.)

but the size of the column can become cumbersome when conventional concrete is used in the lower two-thirds of the building.

Several prominent architects and engineers in the New York area who have been involved in the design of high-rise buildings seem to believe that the choice of steel frame versus reinforced concrete frame for a high-rise building does not depend on the cost of the frame alone.[11] With concrete, one can "fast track," that is, start construction even when the design of the superstructure is only partially done and get the building completed sooner (thus occupy it sooner). With a steel frame, on the other hand, the entire design must be completed before, and one may have to wait for several months before the steel structure is fabricated and shipped to the job site. Construction speed is a major consideration in the era of high interest rates; therefore, getting a building completed and occupied as quickly as possible by rent-paying tenants is important to a developer. Ten years ago in Manhattan, tall office buildings were almost all steel-frame; today perhaps as many as 25 percent of all new office buildings have concrete frames.

In the precast and prestressed concrete industries, the use of high-strength concrete has resulted in a rapid turnover of molds, higher productivity, and less loss of products during handling and transportation. Since their permeability is very low, high-strength concretes also find application where durability of concrete is adversely affected due to abrasion, erosion, or various chemical attacks listed in Fig. 5–10.

Materials and Mix Proportions

General considerations. The porosity of the three phases of concrete (i.e., aggregate, the cement paste, and the transition zone) is the most important strength-determining factor in high-strength concrete. In Chapter 3, it was described how the properties of concrete-making materials, mix proportions, and consolidation of a fresh concrete mixture influenced the porosities of cement paste and transition zone. When workability is adequate, it appears that the water/cement ratio holds the key to the porosities of both the hydrated cement paste and transition zone. Furthermore, with a low water/cement ratio it is generally observed that considerably high strength gains are achieved for very small decreases in the water/cement ratio. For instance, a study showed that for water/cement ratios of 0.38, 0.36, and 0.34, compressive strengths of concrete were 6000, 7500, and 9000 psi, respectively.[12] There is, however, one problem. With the decreasing water content, fresh concrete becomes more and more difficult to mix, place, and consolidate.

For the production of high-strength concrete, the opposing effects of water/cement ratio on consistency and strength of concrete cannot be harmonized without the use of water-reducing admixtures. This explains why the development of superplasticizing admixtures during the last 10 years has helped to increase the production

[11] "Structural Trends in New York City Buildings," *Civil Eng.-ASCE Mag.*, p. 30, January 1983.

[12] *High-Strength Concrete in Chicago High-Rise Buildings*, Chicago Committee on High-Rise Buildings, Report 5, February 1977, p. 29.

and use of high-strength concrete. In fact, without the excellent workability made possible by superplasticizers, it would not have been practical to produce heavily reinforced members of very-high-strength concrete (above 10,000 psi or 70 MPa compressive strength), which have water/cement ratios on the order of 0.3 or less.

As discussed in Chapter 3, with normal-weight conventional concrete containing strong aggregates of 25 or 38 mm maximum size and water/cement ratios in the range 0.4 to 0.7, it is generally the transition zone that is the weakest component of the system. At a given water/cement ratio, the strength of a concrete mixture can be increased significantly by simply reducing the maximum size of the coarse aggregate particles because this has a beneficial effect on the strength of the transition zone. Therefore, in proportioning high-strength concrete mixtures it is customary to limit the maximum size of the aggregate to 19 mm or lower.

The requirements of low water/cement ratio and small aggregate size mean that the cement content of the concrete mixtures will be high, generally above 650 lb/yd^3 (385 kg/m^3). Cement contents of approximately 1000 lb/yd^3 (600 kg/m^3) and even higher have been investigated, but found undesirable. With the increasing proportion of cement in concrete a strength plateau is reached, that is, there will be no more increase in strength with further increases in the cement content. This is probably due to the inherent inhomogeneity of the hydrated portland cement paste, in which the presence of large crystals of calcium hydroxide represents weak areas of cleavage under stress. Such inhomogeneous and weak areas in the transition zone are vulnerable to microcracking even before the application of an external load. This can happen as a result of thermal-shrinkage or drying-shrinkage stresses when considerable differences may arise between the elastic response of the cement paste and that of the aggregate. It should be noted that increases in cement content also means increases in cost, heat of hydration, and drying shrinkage of concrete.

When the inhomogeneity of the hydrated portland cement paste becomes strength limiting in concrete, the obvious solution is to modify the microstructure so that the components causing the inhomogeneity are eliminated or reduced. In the case of portland cement products, an inexpensive and effective way to achieve this is by incorporation of pozzolanic materials into the concrete mixture. As described earlier (page 209), pozzolanic admixtures, such as fly ash, react with calcium hydroxide to form a reaction product that is similar in composition and properties to the principal hydration product of portland cement; also, the pozzolanic reaction is accompanied by a reduction in large pores—an effect that is equally important for the enhancement of strength of the system.

In addition to a more homogeneous end product and a possible cost reduction when a part of the cement in a concrete mixture is replaced by a pozzolan, a major benefit can accrue in the form of a lower temperature rise due to less heat of hydration. Because of high cement contents, the users of high-strength concrete frequently experience thermal cracking in large structural elements. Thus, in certain cases, minimizing the risk of thermal cracking is itself enough of a justification for the partial substitution of cement by a pozzolan.

At a given water/cement ratio, when Class F fly ash is used as a partial replacement for portland cement, the early strength (3- and 7-day) of concrete cured at normal temperatures may be reduced almost in direct proportion to the amount of fly ash present in the total cementing material (cement + fly ash). However, concretes containing Class C fly ash or ground blast-furnace slag show a significant strength gain within the first 7 days of hydration; highly reactive pozzolans such as condensed silica fume and rice husk ash can make a strength contribution even at 3 days. Of course, under conditions of acceleration of cement hydration reactions, such as those present in the production of steam-cured precast concrete members, the differences in pozzolanic activity should not have any significant effect on the strength of the product.

It should be obvious that low strength at early age need not occur if, instead of cement, a part of the fine aggregate in concrete were replaced with a pozzolan. In fact, substantial strength gains, even at early ages, are observed by partial substitution of fly ash or ground blast-furnace slag for fine aggregate, provided it is not accompanied by an increase in water requirement of the concrete mixture. It should be noted that this approach will neither offer an economic advantage nor reduce the risk of thermal cracking when large elements are involved (since no reduction in the cement content is made); however, it is worthy of consideration if early strength is most important. The published literature contains numerous reports showing that for best overall results it may be advantageous in some cases to do both, that is, replace a part of the portland cement and a part of the fine aggregate with pozzolan. A method of proportioning high-strength concrete mixtures is suggested by Mehta and Aitcin.[13]

Typical concrete mixtures. The principles underlying the production of high-strength concrete are illustrated by the composition of the three concrete mixtures shown in Table 11-3. A major difference between the concrete for the Water Tower Place and the Texas Commerce Tower is that the former contains more cement and about 10 percent Class F fly ash by weight of the total cementing material, whereas the latter contains considerably less cement and about 20 percent Class C fly ash. Both concretes produced 113 mm slump, and 9400 to 9770-psi (65 to 67 MPa) 28-day compressive strength in the laboratory-cured specimens. The 139 tests on field concrete for the Texas Commerce Tower averaged 8146 psi (56 MPa), which is close to the average strength recommended by ACI 318 for a 7500-psi (52 MPa) specified strength. Under ideal field conditions, the strength of high-strength job concrete may be approximately 90 percent of the strength of the laboratory-cured specimens.

Very-high-strength concrete mixtures are being field tested at several places in the United States and Canada. Due to the presence of a highly reactive pozzolan such as silica fume, a high cement content, and a very low water/cement ratio that is made possible by the use of a large dosage of superplasticizer, the compressive strengths of the ultra-high-strength concrete mixture (Table 11-3) at 28 and 120 days were 16,170 psi (110 MPa) and 18,020 psi (125 MPa), respectively.

---

[13] P. K. Mehta and P. C. Aitcin, *Cement, Concrete, and Aggregates Journ.* ASTM, Vol. 12, No. 2, pp. 70–78, 1990.

TABLE 11-3 TYPICAL MIX PROPORTIONS FOR HIGH-STRENGTH CONCRETE (LB/YD^3)

Material	Water Tower Place, Chicago (1975)[a]	Texas Commerce Tower, Houston (1980)[b]	Test mixture for ultra-high-strength concrete (1984)[c]
Cement, Type I	846	658	1000
Pozzolan			
Class F fly ash	100	—	—
Class C fly ash	—	167	—
Silica fume	—	—	200
Crushed stone			
$\frac{3}{4}$ in., max.	—	1923	—
$\frac{5}{8}$ in., max.	1800	—	—
$\frac{1}{2}$ in., max.	—	—	1682
Sand	1025	974	905
Water	300	272	266
Water-reducing admixture ASTM C 494			
Type A (oz)	25	25	—
superplasticizer	—	—	High dosage[d]
Water/(cement + pozzolan)	0.32	0.33	0.22

[a] *High-Strength Concrete in Chicago High-Rise Buildings,* Chicago Committee on High-Rise Buildings, Report 5, February 1977, p. 45.

[b] J. E. Cook, "Research and Application of High-Strength Concrete Using Class C Fly Ash," *Concr. Int.,* Vol. 4, No. 7, pp. 72–80, 1982.

[c] J. Wolsiefer, "Ultra High-Strength Field Placeable Concrete with Silica Fume Admixture," *Concr. Int.,* Vol. 6, No. 4, pp. 25–31, 1984.

[d] The exact dosage is unspecified. However, the author states that the use of a high cement factor and fine silica fume pozzolan makes it mandatory to use high-range water reducers; the admixture is used at a very high dosage to compensate for the increased water demand of the fine materials and to allow placement of the concrete products at reasonable field slumps.

Superplasticized concrete. Although normal water-reducing admixtures can be used for making high-strength concrete mixtures, as shown in Table 11-3, concretes with a very high consistency (200 to 250 mm slump) and more than 10,000-psi (70 MPa) strength (less than an 0.3 water/cement ratio) are more easily produced by the application of high-range or superplasticizing admixtures. It is generally known that the addition of 0.5 to 1.5 percent of a conventional superplasticizer by weight of cement to a concrete mixture with 50- to 75-mm slump will cause a dramatic increase in the consistency (200 to 250 mm slump); however, this high-consistency concrete tends to revert back to the original consistency within 30 to 60 min (Fig. 11-5a).

When the time duration between mixing and placing of concrete is relatively short, the rapid slump loss may not be a problem. On the other hand, it is even beneficial in the precast concrete industry, where a stiff consistency is desired soon

High-Strength Concrete

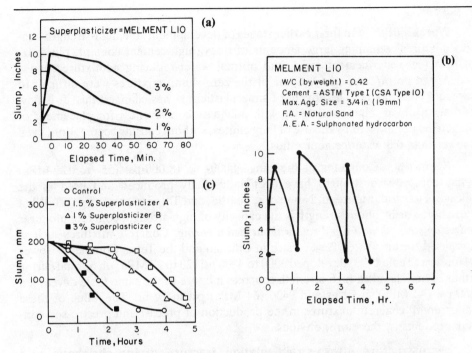

Figure 11–5 Effect of a superplasticizing admixture on concrete consistency: (a) slump loss with elapsed time; (b) slump recovery by repeated dosage; (c) effect of the superplasticizer composition on slump loss. [(a), (b), From V. M. Malhotra, ed., *Progress in Concrete Technology,* CANMET, Ottawa, 1980; (c), from M. Collepardi, M. Coradi, and M. Valente, *Transportation Research Board,* National Research Council, Washington, D.C., Transportation Research Record 720, p. 8, 1979.]

after placement so that steam can be injected. However, in the ready-mixed concrete industry, an unforeseen mishap would lead to serious problems in handling concrete with a high-slump-loss tendency. There are two ways to handle this problem. Researchers have shown that large increases in the slump of superplasticized concrete can be maintained for several hours by repeated dosages of the superplasticizing admixture (Fig. 11–5b).[14] Caution against segregation is recommended, however, when the second or third dosages of the admixture are added after the slump loss. The second approach involves modification of the admixture composition with a retarding agent so that the original high consistency may be maintained for 2 to 3 hr (Fig. 11–5c).[15] Ready-mixed concretes containing low-slump-loss superplasticizers have found industrial applications in hot-weather concreting.

[14] V. M. Malhotra, ed., *Progress in Concrete Technology,* CANMET, Ottawa, 1980, pp. 367–419.

[15] M. Collepardi, M. Corradi, and M. Valente, *Transportation Research Record,* No. 720, Transportation Research Board National Academy of Sciences, Washington, D.C., 1979, pp. 7–12.

Properties

Workability. In their earlier stages of development, the high-strength concrete mixtures containing large amounts of fines (high cement content, plus pozzolan) a low water/cement ratio, and normal water-reducing admixtures had a tendency to be *stiff and sticky*. Some of the zero-slump mixtures were difficult to place and consolidate. The advent of superplasticizers has changed this. It is now possible to obtain high consistency with adequate cohesiveness for placement of concrete by pumping or by the use of long chutes, without causing segregation, even at lower than 0.3 water/cement ratios.

Strength. Concretes in the range 8000- to 18,000-psi (55- to 120-MPa) 28-day compressive strength have been industrially produced and used in the Chicago, Houston, Montreal, New York, Seattle, and Toronto areas (p. 377). What is also noteworthy about strength is the capacity of high-strength concrete mixtures to *develop strength at a rapid rate without steam curing*. Concrete mixtures such as those used for the Water Tower Place in Chicago and the Texas Commerce Tower in Houston (Table 11–3) developed 3000 to 4000 psi (20 to 27 MPa) on normal curing within 24 hr. The ultra-high-strength concrete mixture of the same table developed 6260 psi (42 MPa) in 12 hr and 9430 (64 MPa) psi in 24 hr. The value of these high-strength concrete mixtures in the production of precast and prestressed concrete products is, therefore, obvious.

Microstructure, stress-strain relation, fracture, drying shrinkage, and creep. From the general principles behind the design of high-strength concrete mixtures, it is apparent that high strengths are made possible by reducing porosity, inhomogeneity, and microcracks in the hydrated cement paste and the transition zone. Since the existence of numerous microcracks in normal concrete is fundamental to stress-strain relations, creep, and fracture behavior (Chapters 3 and 4), it should not be difficult to understand why, as a result of a reduction in the size and number of microcracks, a high-strength concrete would behave in fundamentally different ways from normal concrete.

From an experimental study of progressive microcracking in concretes with uniaxial compressive strengths in the range 4500 to 11,000 psi (30 to 75 MPa), researchers at Cornell University[16] have drawn the following conclusions:

1. Compared to normal-strength concrete, high-strength concrete behaves more like a homogeneous material. For high-strength concretes, the stress-strain curves are steeper and more linear to a higher stress-strength ratio than in normal-strength concretes, because of a decrease in the amount and extent of microcracking in the transition zone.* Thus the high-strength concrete shows

---

[16] R. L. Carrasquillo, F. O. Slate, and A. H. Nilson, *J. ACI*, Proc., Vol. 78, No. 3, pp. 179–86, 1981; A. S. Ngab, F. O. Slate, and A. H. Nilson, *J. ACI*, Proc., Vol. 78, No. 4, pp. 262–68, 1981.

*The authors used the term "bond" instead of "transition zone." The latter term is used throughout this text to describe the contact area between the cement paste and coarse aggregate, since it is not a bond in the true physical-chemical sense.

a more brittle mode of fracture and less volumetric dilation. The study indicated that these concretes can be loaded to a higher stress-strength ratio without initiating a self-propagating mechanism leading to disruptive failure; that is, the sustained-load strength is a higher percentage of the short-term strength.

2. The amount of microcracking in high-strength concrete associated with shrinkage, short-term loading, and sustained loading is significantly less than in normal-strength concrete. The substantial time-dependent increase found in the latter is far less than in the former, explaining in part the much reduced creep in high-strength concretes.

3. From the above it is obvious that high-strength concretes (with f_c from 6000 to about 9000 psi) behave in fundamentally different ways from normal-strength concrete.

Based on the foregoing conclusions, it was suggested that many of the current code provisions which are applicable to normal-strength concrete should be modified when applied to high-strength concrete. For instance, instead of the ACI formula for computing the static modulus of elasticity for normal concrete from the 28-day compressive strength (Chapter 4), the following formula (psi units) was found to be more applicable to high-strength concrete: $E_c = 40,000\sqrt{f_c'} + (1 \times 10^6)$. It was recommended that the flexural strength values should be calculated by the formula $f_r = 11.7 \sqrt{f_c'}$. In the range 6000 to 15,000 psi (40 to 100 MPa), Yamamoto and Kobayashi[17] (Fig. 11–6) proposed the following relationship for predicting the splitting tensile strength (in MPa units): $f_t = 0.06f_c' + 0.8$.

It may be noted that the experimentally determined values for the static elastic modulus in compression for the three concrete mixtures (from left to right) shown in Table 11–3 were 5.4, 5.7, and 6.25×10^6 psi, respectively. When tested according to ASTM C 512, the 1-year shrinkage and creep strains for ultra-high-strength concrete containing the silica fume and superplasticizer were 315 and 480×10^{-6}, respectively, for a sustained load of 2500 psi applied at 28 days after curing. The data appear logical in view of the high elastic modulus, low porosity, and fewer microcracks with this type of concrete.

Durability. Many researchers have found that primarily due to low permeability, high-strength concretes exhibit excellent durability to various physical and chemical agents that are normally responsible for concrete deterioration (Chapter 5). So far, industrial applications of high-strength concrete are limited to structural members that are not exposed to freeze-thaw cycles. Superplasticized concretes generally suffer more than normal concrete from loss of entrained air and increased void spacing. However, in spite of the void spacings exceeding 200 μm, the freeze-thaw durability of superplasticized concretes, when tested in accordance with ASTM Standard C 666, is usually not impaired. The results of various durability

---

[17] Y. Yamamoto and M. Kobayashi, *Concr. Int.*, Vol. 4, No. 7, pp. 33–40, 1982.

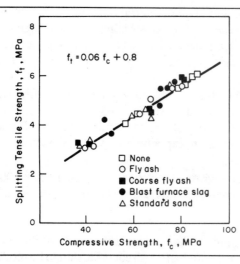

Figure 11-6 Relation between compressive and splitting tensile strengths of high-strength concrete containing mineral admixtures. (From Y. Yamamoto and M. Kobayashi, *Concr. Int.*, Vol. 4, No. 7, pp. 33–40, 1982.)

tests on the ultra-high-strength concrete mixture shown in Table 11-3 are reported by Wolsiefer[18] and summarized as follows.

With entrainment of 6.7 percent air, the compressive strength of concrete dropped to 11,000 psi (75 MPa); after 300 cycles of freeze-thaw (ASTM C 666), the expansion, weight loss, and durability factor were 0.008, 0.13, and 97 percent, respectively. The abrasion resistance test with the revolving disk machine (ASTM C 779) showed only an average of 1.5 mm depth of wear after 60 min of abrading. The deicer scaling test (ASTM C 672) showed no scaling after 50 cycles and slight scaling after 500 cycles. The chloride permeability test by ponding 3 percent NaCl solution on a 3-month-old test specimen (360 kg/m^3 cement, 76 kg/m^3 silica fume, and an 0.27 water/cementitious ratio) showed an average absorbed chloride ion content of 0.003 compared to 0.028 for the control concrete specimen (385 kg/m^3 cement, an 0.38 water/cement ratio, and no pozzolan). Specimens exposed to 4 MPa water pressure were virtually impermeable. Very-high-strength concrete (70 MPa or more) typically shows 100–1,000 coulombs of current passing through the specimen in a 6 h standard test (AASHTO T-277-83), which is rated as "very low."

Due to the high cement content thermal cracking can be a durability problem in high-strength concrete structures. ACI Committee 363 indicates the expected temperature rise to be 10–14°C for every 100 kg/m^3 cement content. A recent study showed 54°C temperature rise at the center of a 1220 mm cube of concrete made with 560 kg/m^3 cement.[19]

It seems that superplasticized, low water/cement ratio, high-strength concretes containing a high cement content and a good-quality pozzolan should have a great

---

[18] J. Wolsiefer, *Concr. Int.*, Vol. 6, No. 4, pp. 25–31, 1984.

[19] R. G. Burg and B. W. Ost, "Engineering Properties of Commercially Available High-Strength Concretes," *PCA Res. & Der. Bull. RD104T*, 1992.

potential where impermeability or durability, not strength, is the main consideration. Such applications include floors in the chemical and food industry, and bridge-deck overlays that are subject to severe chemical and physical processes of degradation.

Applications

So far, most of the structural applications of high-strength concrete in the United States have been limited to the metropolitan areas of Chicago, New York, and Houston. Beginning in 1965, with 7500-psi concrete for the Columns of Lake Point Tower, the Chicago area has high-strength concrete in more than 40 high-rise projects.[20] For commercial buildings up to 50 stories high, 9000-psi concrete has become more popular since 1972 because from an architectural standpoint it is claimed that the minimum column dimensions obtained with this strength provide the necessary aesthetic proportions. Twenty-eight lower stories of the 79-story Water Tower building (Fig. 11–7), contain columns of 9000-psi concrete. In 1982 a 100 MPa concrete was used for the Mercantile Building in Chicago. Again in Chicago, in 1989 six stories of columns for a high-rise building at 225 W. Wacker Dr. were made with a 96 MPa concrete containing silica fume. This project also has an experimental column made with a 117 MPa concrete. The Nova Scotia Plaza Building in Toronto and the Two Union Square Building in Seattle reportedly contain 95 MPa and 120 MPa concrete elements, respectively.

Constructed in 1979, the 53-story Helmsley Palace Hotel was the first structure in New York using high-strength concrete[21] (8000 psi). The concrete was used to reduce the size of columns. Prior to this building, in which a flat-plate concrete frame is used, New York buildings higher than 35 stories almost always had structural steel frames. Superplasticized, 8000-psi concrete was used for the 46-story, 101 Park Avenue building, which has a rigid concrete core and reinforced concrete frame of the beam-and-girder type. The high-strength concrete was used to form the core and the 46-in.-diameter columns. A 38-story concrete frame building at 535 Madison Avenue used superplasticized 8500-psi concrete for the purpose of reducing the column sizes.

The 75-story Texas Commerce Tower in Houston (Fig. 11–8), constructed in 1980, contains 94,000 yd^3 of concrete, of which 35 percent was high-strength. The following description of the structure by Cook illustrates the economic options and the type of flexibility enjoyed by the structural designer when concrete mixtures with a wide range of strengths are available for use:

> The foundation mat required 15,200 cu yd of 6000 psi concrete. The contractor elected to place the mat in two pours; the first placement consisted of 8300 cu yd over a 15 hr period, two weeks later the second placement of 6900 cu yd was made in 11 hr. Construction of the composite frame utilized 7500 psi concrete for spandrells, columns,

---

[20] Moreno, *Proc. ACI Seminar on High-Strength Concrete.*
[21] "Structural Trends in New York City Buildings."

Figure 11-7 Water Tower Place, Chicago, Illinois, 1976. (Photograph by permission of Concrete Construction Publications, Inc., 426 South Westgate, Addison, Illinois 60101.)

Chicago's Water Tower Place, rising 859 ft 2 in. above street level—literally a city within the city—contains residential condominiums, offices, a shopping center, and a luxury hotel. Concrete of seven different strengths was used. From the basement stories to the twenty-fifth floor, columns are made of 9000-psi concrete; above the twenty-fifth floor, the concrete strength in the columns was progressively reduced from 7500 psi to 4000 psi. Water-reducing admixtures and fly ash were used in the production of high-strength concrete mixtures. Floor slabs were placed with 4000-psi structural lightweight concrete, which reduced the slab dead load by more than a third. Built at a cost of $195 million, the building contains about 160,000 yd^3 of concrete and 12,000 tons of reinforcing steel.

and shear walls up to the 8th floor level; 6000 psi concrete from the 8th to the 30th floor; 5000 psi from the 30th to the 60th; and 4000 psi from the 60th floor to the roof.[22]

As stated earlier, an important advantage of the use of pozzolans in high-strength concrete is that relatively less heat of hydration is evolved per unit strength; therefore, the risk of thermal cracking is reduced. The construction of the New Tjörn

---

[22] J. E. Cook, *Concr. Int.*, Vol. 4, No. 7, p. 72, 1982.

Figure 11-8 Texas Commerce Tower, Houston, Texas, 1980. (Photograph courtesy of J. E. Cook, Gifford-Hill and Company, Inc., Dallas, Texas.)

Approximately 30,000 yd³ of high-strength concrete was used to make spandrell beams, columns, and shear walls up to the thirtieth floor of this 75-story building. A unique feature of the concrete mixture was the use of a high-calcium fly ash, which permitted considerable reduction in the cement content.

Bridge (Fig. 11-9), completed in November 1981, involved the first known use of condensed silica fume in high-strength concrete when heat of hydration was the main consideration. According to Rickne and Svensson:

> The principal reason for the use of silica fume was that this permitted a reduction of the cement content, without loss of strength. It was important to limit the cement content, since the bridge contains some very large concrete sections (principally the pylons) with concrete of very high quality, K50 with cube strength 50 MPa. A high cement content could give rise to excessive heat generation, with consequent risk of cracking.[23]

[23] S. Rickne and C. Svensson, *Nordisk Betong* (Stockholm), No. 2-4, pp. 213-16, 1982.

Figure 11-9 From disaster to inauguration: The new Tjörn bridge in Sweden was designed and built in less than 2 years. (Photograph from S. Rickne and C. Svensson, *Nordisk Betong* (Stockholm), No. 2–4, pp. 213–16, 1982.)

Early on the morning of Friday, January 18, 1980, an off-course bulk carrier collided with the 20-year-old steel arch bridge, bringing the arch span and the deck slab down into the water. Since the broken bridge served as a major communication link between two islands, a new bridge was urgently needed. In June the contract was awarded for the construction of a 654-m-long cable bridge with concrete pylons 100 m high and 4 m by 4.5 m in cross section. To reduce the risk of thermal cracking in the massive structure of 50-MPa concrete (high cement content), a part of the cement was replaced by silica fume and a further reduction in the temperature rise was achieved by the circulation of cold air through pipes embedded in concrete. The extremely short construction time of 15 months has aroused international attention because the time normally required for the construction of a structure of this size is twice as long.

The composition of K50 concrete used in the slip-formed pylons of the New Tjörn Bridge was: cement, 370 kg/m^3; silica fume, 37 kg/m^3; fine aggregate (0 to 8 mm), 785 kg/m^3; coarse aggregate (8 to 32 mm), 970 kg/m^3; and water, 205 kg/m^3. In difficult sections, retarding admixtures were used. The average value of actual concrete strengths obtained was 62 MPa. Reduction in cement content due to the use of silica fume, and artificial cooling of concrete by circulation of cold air through pipes embedded in concrete, helped to lower the peak concrete temperature by 10 to 12°C.

Holland[24] reported an interesting case in which the use of high-strength concrete mixture containing silica fume was considered for the purpose of long-time durability to abrasion-erosion. Concrete in a stilling basin at the Kinzua Dam, Pennsylvania, including a fiber-reinforced concrete (FRC) section, was severely eroded and was in need of rehabilitation (Fig. 11-10a). In 1983, the U.S. Army Corps of Engineers investigated several concrete mixtures with various aggregate types and admixtures. Typical abrasion-erosion data (Fig. 11-10b) from a special test developed by the Corps showed that a superplasticized, low water/cement ratio concrete containing silica fume showed considerably less erosion loss. Consequently, about 2500 yd^3 of concrete containing 657 lb/yd^3 of Type I portland cement, 118 lb/yd^3 of silica fume (18 percent by weight of cement), 1383 lb/yd^3 of sand, 1638 lb/yd^3 of crushed limestone, and 219 lb/yd^3 of water (0.283 ratio between water and cement plus silica fume) were used for the rehabilitation work. Fresh concrete had about a 9-in. slump, and the hardened concrete cylinders showed over 13,000- and 15,000-psi compressive strengths at 28 and 90 days, respectively. It may be noted that high durability of high-strength concrete to seawater is one of the reasons the material is increasingly being selected for the construction of islands in the oceans (Fig. 11-11).

HIGH-WORKABILITY CONCRETE

Definition and Significance

For want of a standard definition, high-workability concrete may be considered as the concrete of a flowing consistency (180 to 230 mm slump), which can be placed and compacted with little or no effort and which is, at the same time, cohesive enough to be handled without segregation and bleeding. Architects and engineers who have worked with superplasticized concretes to achieve high strength or high workability believe that the advent of superplasticizers has started a new era in the construction practice by extending the use of concrete to structures with complex designs and highly demanding applications. Superplasticizing chemicals (Chapter 8) are expensive; their use may increase the materials cost of a concrete mixture by about $5 per cubic yard, but the increased cost is easily offset by savings in labor cost

---

[24] T. C. Holland, Misc. Paper SL-83-16, U.S. Army Waterways Exp. Station, Vicksburg, Miss., 1983.

(a)

No, these are not craters on the moon. This is the result of abrasion-erosion damage to the concrete stilling basin at the Kinzua Dam in Pennsylvania.

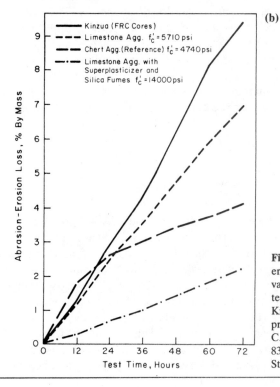

Figure 11–10 (a) Concrete damaged by erosion at the Kinzua Dam, Pennsylvania; (b) results of abrasion-erosion tests for concretes investigated for the Kinzua Dam stilling basin rehabilitation project. [(a), photograph courtesy of T. C. Holland; (b), from Misc. Paper SL 83–16, U.S. Army Waterways Exp. Station, 1983.]

Figure 11–11 (a) Floating concrete container terminal, Valdez, Alaska; (b) one-half section of the terminal under tow from Tacoma, Washington, to Valdez. (From *Concr. Construct.*, Vol. 28, No. 2, 1983; photograph by permission of Concrete Construction Publications, Inc., 426 South Westgate, Addison, Illinois 60101 and Concrete Technology Corporation, Tacoma, Wash.)

Problem: How will you construct in the middle of nowhere, at a reasonable cost, a high-capacity marine terminal which requires low maintenance, can be rapidly deployed, is suitable for deep water and poor soil conditions, is self-adjusting to even the tidal variations, and may be moved easily to another site should market conditions warrant it?

Solution: Build prestressed concrete floating dock sections at a dry-dock site where construction materials are readily available, then tow the sections to the deployment site and install the floating terminal.

Case History: A 100- by 700-ft floating container-terminal was needed off the coast of Alaska at Valdez. About 1600 miles away at Tacoma, Washington, two 100- by 350-ft concrete sections were fabricated from 468 prestressed concrete panels which served as form-work for the final 12-in.-thick cast-in-place concrete deck. Additionally, 266 concrete wall panels were made. Thus over 700 precast units were made of high-strength (8000-psi) concrete by accelerated curing, then trucked to the graving dock for assembly before casting of the concrete bottom. The hull bottom and bulkheads were post-tensioned together longitudinally and transversely. The world's first floating prestressed concrete container dock, the largest of its kind made of any material, went into service late in 1982; it will provide year-round service to ships with up to 50,000-ton capacity and barges up to 1500 tons.

TABLE 11-4 TYPICAL MIX PROPORTIONS OF HIGH-WORKABILITY CONCRETE MIXTURES[a]

		Mix proportions (lb/yd^3)					
No.	Type	Type I cement	Fine aggregate	Coarse aggregate	Water	Water-reducing admixture	Water/cement ratio
1	Control mix 75 mm slump	524	1428	1738	241	None	0.47
2	Control mix 230 mm slump	505	1378	1676	288	None	0.58
3	High-workability 230 mm slump	532	1423	1731	241	0.3%[b] melanine sulfonate	0.47
4	High-workability 230 mm slump	530	1415	1722	243	0.25%[b] naphthalene sulfonate	0.47

[a] Air contents of freshly mixed concretes were in the range of 5.8 to 6.7%; hardened concretes showed about 4% air.

[b] Admixtures are aqueous solutions, but their proportion is expressed as percent solids by weight of concrete.

Source: Adapted from S. H. Gebler, *Res. Dev. Bull. RD081.01T*, Portland Cement Association, Skokie, Ill., 1982.

that are made possible by maximum utilization of on-site time. Hewlett and Rixom[25] summarize the possible benefits from the use of flowing concrete as follows:

- Placing of concrete with reduced vibration effort in areas of closely bunched reinforcement and in areas of poor access (by using flowing concrete, the need to cut or adopt formwork to obtain vibrator access may be obviated)
- The capability of placing—very rapidly, easily, and without vibration—concrete for bay areas, floor slabs, roof decks, and similar structures
- The very rapid pumping of concrete
- Placing concrete by means of tremie pipe
- The production of uniform and compact concrete surfaces

Mix Proportions

As stated earlier, consistencies on the order of 175 to 230 mm slump can easily be achieved by superplasticizer addition to a concrete mixture with 50 to 75 mm slump. However, for such concrete to remain cohesive during handling operations, special attention has to be paid to mix proportioning. Hewlett and Rixom[26] suggest that a good starting point is to design mixtures that are suitable for pumping. Conventional mix-proportioning procedures described in Chapter 9 may be used to determine the

---

[25] P. Hewlett and R. Rixom, *J. ACI*, Proc., Vol. 74, No. 5, pp. N6–11, 1977.
[26] Ibid.

quantities of given materials, needed to yield the specified strength with a 75 mm slump. To avoid segregation on superplasticizer addition, a simple approach consists of increasing the sand content at the cost of the coarse aggregate content by 4 to 5 percent. For coarse sands, the adjustment in the sand/coarse aggregate ratio may involve increases of up to 10 percent in the sand content. Typical mix proportions of flowing concretes, investigated by Gebler[27] at the Construction Technology Laboratories of the Portland Cement Association, are shown in Table 11–4.

Important Properties

Compared to control concrete mixtures, the properties of freshly mixed as well as hardened concretes of the flowing type containing superplasticizers (Table 11–4) can be summarized as follows:

1. When added to a control concrete mixture with a 75 mm slump, both melamine sulfonate and naphthalene sulfonate superplasticizers were able to produce concretes with 215 to 230 mm slump. Mixtures containing the superplasticizers, however, reverted back to the original 75 mm slump within 30 to 60 min. Under the same conditions the high-slump control concrete, without any water-reducing admixture, lost slump at a much slower rate (Fig. 11–12).
2. The data in Table 11–5 show that the superplasticized concretes (mixes 3 and 4) exhibited only slight segregation and bleeding, compared to the high-slump control concrete (mix 2). Compared to the low-slump concrete with an equivalent water/cement ratio (mix 1), the superplasticized concretes showed a slight retardation of setting times.
3. Compressive and flexural strengths of superplasticized concrete mixtures (3 and 4) were not significantly different, compared to the 75 mm-slump control concrete prepared at an equivalent water/cement ratio (mix 1).
4. Although immediately after mixing the air content of the superplasticized mixtures was about 6 percent, it dropped to about 4 percent after the slump loss, and the spacing factor increased to 0.2 mm compared to 0.15 mm for the control mix (mix 1). Superplasticized concrete specimens compacted by standard rodding showed satisfactory durability to 300 cycles of freezing and thawing in water, but the specimens compacted by external vibration showed some reduction in freeze-thaw durability. Also, the flowing concretes containing superplasticizers performed satisfactorily in deicer scaling and chloride ion penetration tests.
5. Flowing concretes made with superplasticizing admixtures showed essentially no difference in drying shrinkage compared to that of the 75-mm-slump control concrete (with an equivalent water/cement ratio). However, compared to the 230 mm-slump control concrete (with an equivalent consistency), the drying shrinkage of the superplasticized concretes was less.

[27] S. H. Gebler, *Res. Dev. Bull. RD081.0IT*, Portland Cement Association, Skokie, Ill., 1982.

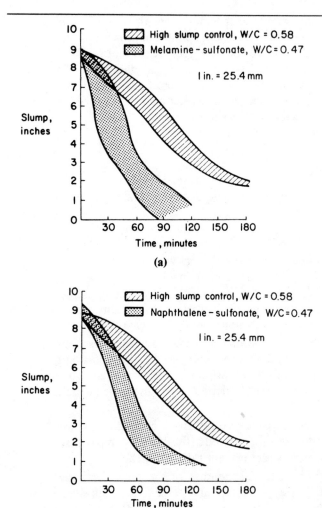

Figure 11-12 Range of slump loss in high-slump concretes: (a) naphthalene-sulfonate used as superplasticizer; (b) melamine-sulfonate used as superplasticizer. (From S. H. Gebler, *Bulletin RD081-01T,* Portland Cement Association, Skokie, Ill., 1982.)

TABLE 11-5 BLEEDING CHARACTERISTICS AND SETTING TIMES OF THE MIXTURES SHOWN IN TABLE 11-4

No.	Bleeding (ASTM C232)		Setting time (ASTM C403)	
	Percent	ml/cm² of surface	Initial (hr: min)	Final (hr: min)
1	1.09	0.031	4:14	5:46
2	3.27	0.143	5:06	7:10
3	1.59	0.060	4:52	6:28
4	1.50	0.059	4:55	6:20

Source: Adapted from S. H. Gebler, *Res. Dev. Bull. RD081.01T,* Portland Cement Association, Skokie, Ill., 1982.

Applications

To support the 73-story Raffles City building in Singapore, a monolithic foundation mat containing 14,300 yd^3 (11,000 m^3) of 35 MPa, 250-mm-slump concrete was recently placed in 43 hr, despite ambient temperatures in the range 29 to 32°C (Fig. 11–13). To obtain a sufficient rate of delivery, concrete had to be transported from three different batch plants; two were 25 min away, and one 45 min away. The concrete containing a low-slump-loss superplasticizer was placed with 0 to 250 mm slump despite the hot weather. The impact on the construction practice of the use of superplasticizers is evident from the following description of the project.

> Distribution of the concrete to all parts of the mat without pumps was accomplished by a unique method. In preparation for the pour, midway between street level and the top of the mat, three boxes with hand-operated gates were built. Two 12-in. diameter PVC pipes led down to each box from street level, and from each gate a chute led down to the mat. . . . as trucks arrived from the batch plants, they could empty at full speed into the pipes. The concrete flowed to the distribution boxes where it was directed wherever needed by workers who simply opened and closed the gates. Even though the

Figure 11–13 Raffles City building, Singapore, 1983. (Photograph courtesy of Modern Advanced Concrete, Inc., Treviso, Italy.)

The tallest concrete building in Asia, containing highly reinforced elements, utilized flowing-concrete mixtures with a superplasticizing retarder to counteract the high ambient temperatures, up to 32°C.

mat had an exceptionally high density of rebar, the superplasticized concrete flowed freely to all areas of the mat, with no pumping and almost no vibration. Despite ambient temperatures of 29 to 32°C, the concrete was placed without using ice.[28]

A high-strength, high-workability, superplasticized concrete containing a non-chloride accelerator was used successfully for a cold-weather concreting application in New York City. This application in the construction of Trump Tower (Fig. 11–14), a 68-story building, has some unusual design features:

> The lower 17 floors are designed for office and retail use, each providing space unbroken by columns. The 18th and 19th floors contain enough concrete (3000 cu yd) to make a moderate-size foundation mat. It is here, substantially set back from the perimeter of the office floors, that the structural columns of the residential space begin, rising to the top of the building at the 68th floor. Directly beneath them is office space. The entire weight of the residential floors is transferred to the outside columns of the office floors by a "sky foundation."
>
> Without the 230 mm-slump superplasticized concrete, construction of densely reinforced concrete elements might not have been economically feasible. Concrete haulage time from the control batch plant to the job site was about one hour; discharge and placement required about 1/2 hr with another hour for leveling and finishing the slabs. Concrete for some of the slabs, columns, and shear walls was placed when ambient temperature was $-8°C$, with a wind chill factor to $-18°C$. Concrete temperature had been raised to about 16°C by using steam and boiling water. The compressive strength of the concrete used for columns and shear walls was 8000 psi; 6000 psi concrete was used in other walls.[29]

Whereas *high-performance concrete* which combines high-workability, high-strength and high-durability characteristics are discussed in Chapter 13, an impressive demonstration of the use of **high-strength and high-workability concrete** mixtures is given by the massive oil storage tanks and drilling platforms which have been installed since 1973 in the British and the Norwegian sectors of the North Sea. The structures composed of prestressed and heavily reinforced concrete elements are exposed to unusual conditions; they are required to withstand not only the risk of corrosion by seawater but also extraordinary loads due to wave action.

Gerwick and Hognestad[30] have provided the design and construction details of the first structure, Ekofisk I, which contains a central oil reservoir about 45 m square and 70 m high. In plan, the tank is nine-lobed—a shape selected to provide maximum strength in the 0.5-m-thick reservoir walls. A unique feature is a concrete breakwater (Fig. 11–15) designed to damp wave pressures due to constant water movement. A concrete mixture with a low water/cement ratio and a high cement content (9000 to 10,000 psi compressive strength) was employed for making the precast, unreinforced, concrete spools of the breakwater.

A different design was adopted by the Norwegian Contractors Company,

---

[28] *Il Calcestruzzo Oggi*, (*Modern Advanced Concrete*, Treviso, Italy), Vol. 14, No. 1, November 1982–May 1983.

[29] *Il Calcestruzzo Oggi*, Vol. 13, No. 2, November 1981–May 1982.

[30] B. C. Gerwick, Jr., and E. Hognestad, *Civ. Eng.-ASCE Mag.*, August 1973.

Figure 11-14 Trump Tower building, New York City, 1981. (Photograph courtesy of Modern Advanced Concrete, Inc., Boca Raton, Florida.)

Trump Tower is the tallest reinforced concrete frame building in New York City. The heavily reinforced concrete elements of this 68-story building were made with a high-strength, high-workability concrete mixture containing a superplasticizing agent and a nonchloride accelerator. The accelerator was used due to the low ambient temperatures at the time of construction.

Figure 11–15 Ekofisk oil storage tanks and breakwater. (From B. C. Gerwick, Jr., and E. Hognestad, *Civil Eng.*, ASCE Environmental Design/Eng. Const., August 1973; photograph courtesy of B. C. Gerwick.)

Engineers were confronted with the following question in 1971: "In 70-m water in the middle of the North Sea (170 km from the coast of Norway), how would you go about building a 2-acre island, and beneath it a tank for the storage of 1 million barrels (159,000 m^3) of crude oil?" The structure was required to withstand extraordinary loads, including the 60-ton/m^2 pressure of the 100-year high (24 m) design wave. Ekofisk I (shown in the photograph) provided the answer. Completed by June 1973, this prestressed concrete structure is about 90 m high and rests on a 99-diameter roughly circular base.

An interesting element of the structure is a perforated breakwater designed to resist and dampen the wave energy. The 8634 breakwater holes are precast, unreinforced, concrete spools, 0.3 to 1.3 m in diameter. While the concrete mixture used for the slip-formed walls of the tank had an average 28-day compressive strength of 43 MPa, the breakwater spools were made from a stronger concrete mixture (62 to 69 MPa) which is expected to be highly erosion resistant.

The structure was prefabricated onshore, assembled at a dry dock, and then towed some 480 km by four of the world's most powerful tugs, totaling more than 40,000 hp.

High-Workability Concrete

TABLE 11-6 MIX PROPORTIONS AND PROPERTIES OF CONCRETE USED FOR THE CONSTRUCTION OF CONDEEP PLATFORMS IN NORWAY

Platform	Beryl A (1975)	Statfjord C (1984)
Water depth (m)	118	145
Number of cells/number of towers	19/3	24/4
Storage capacity (million barrels)	0.9	2
Maximum deck weight (tons)	28,000	50,000
Volume of concrete (m^3)	62,000	135,000
Mix proportions (kg/m^3)		
Cement	430	380
Sand	900	1030
Coarse aggregate		
32 mm, max.	900	
20 mm, max.		845
Water	175	160
Admixture type	Normal water-reducing lignosulfonate type	Superplasticizer, naphthalene sulfonate
Slump (mm)	120	220
Mean 28-day compressive strength (MPa)	55	62

which has built seven of the 15 offshore concrete platforms in the North Sea. Known as Condeep platforms, typically there is a caisson of 19 or 24 oil-storage cells (which rest on the sea floor) with three or four shafts supporting the deck. A photograph of the largest Condeep platform, Statfjord B (18,000 m^2 base area, 140,000 m^3 total concrete used), which was completed in 1981, is shown in Fig. 1-5. Some of the design and concrete quality data for the earliest and the newest installed Condeep platforms, Beryl A and Statfjord C, respectively, are given in Table 11-6, which is based on a report by Moksnes.[31] This author has documented an interesting case study which shows how construction costs can be reduced when close communication exists between the project engineer and the materials suppliers. In July 1979, observations from the caisson slip-forming operation on Statfjord B showed that the type of cement (ASTM Type II portland cement with 5.5 percent C_3A and 330 m^2/kg Blaine surface area) and mix proportions used for making concrete gave a maximum slip rate of just under 2 m/day. For slip-forming of the shafts it was considered desirable to increase the slip rate to more than 3 m/day. This was achieved mainly by finer grinding of the cement to 400 m^2/kg Blaine, which reduced the initial setting time from 9-1/2 hr to 7-1/2 hr (*Penetration Resistance Test*, ASTM C 403). The finer cement was used for slip-forming the shafts in Statfjord B and C.

A comparison of the mix proportions and properties between the Beryl A and Statfjord C concretes (Table 11-6) shows other interesting features. First, the application of the superplasticizing admixture in the latter produced a great improvement

[31] J. Moksnes, *Nordisk Betong (Stockholm)*, No. 2-4, pp. 102-5, 1982.

in the consistency (220 mm of slump instead of 120 mm for the Beryl A concrete, which contained the normal water-reducing admixture). Second, a combination of the smaller coarse aggregate size and better workability (compactability) gave 13 percent higher compressive strength for the Statfjord C concrete, in spite of a significant reduction in the cement content that was made possible by the application of the superplasticizer. Obviously, in a massive structure the reduction of cement content by 50 kg/m^3 would have beneficial effects in regard to both cost economy and risk of thermal cracking.

SHRINKAGE-COMPENSATING CONCRETE

Definition and the Concept

According to ACI Committee 223,[32] **shrinkage-compensating concrete** is an expansive cement concrete which, when properly restrained by reinforcement or other means, will expand an amount equal to or slightly greater than the anticipated drying shrinkage. Because of the restraint, compressive stresses will be induced in the concrete during expansion. Subsequent drying shrinkage will reduce these stresses. Ideally, a residual compression will remain in the concrete, eliminating the risk of shrinkage cracking.

A graphical representation of the concept in Fig. 11–16 compares the behavior of portland cement concrete with a Type K expansive cement concrete[33] during the early moist-curing and subsequent air-drying periods. Briefly speaking, this is how the concept underlying the use of shrinkage-compensating concrete works to reduce or eliminate the drying-shrinkage cracking in a reinforced concrete element. As the Type K cement hydrates, large amount of ettringite are formed. When the concrete sets and develops strength, it will bond to the reinforcement and at the same time start expanding if sufficient quantities of curing water are present. (See p. 147 for mechanisms of expansion by ettringite.) Since the concrete is bonded to steel, its expansion under the restraining influence of the steel will induce tension in the latter while the *concrete itself goes into compression*. At the end of moist curing, when the element is exposed to drying conditions, it will shrink like a normal portland cement concrete. However, the shrinkage will first relieve the *precompression* before inducing tensile stress in the concrete. By preventing the buildup of high tensile stress, the risk of cracking of concrete due to drying shrinkage is thus reduced.

Significance

Shrinkage on drying is a fundamental property of the calcium silicate hydrate (see Chapter 2) which is the principal hydration product of portland cements. Therefore,

[32] Report of ACI Committee 223, Manual of Concrete Practice, Part 1, 1991.

[33] The properties and applications of shrinkage-compensating concretes described in this book are based on Type K expansive cements (Chapter 6), which is the only type of expansive cement being produced in the United States.

Shrinkage-Compensating Concrete

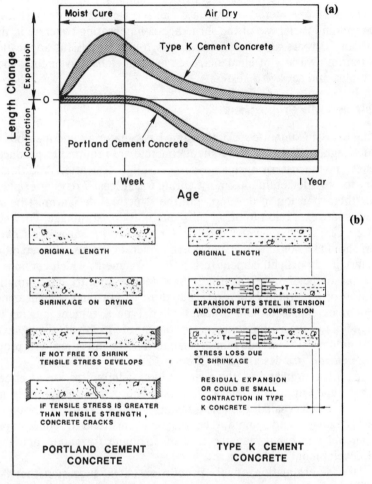

Figure 11–16 (a) Comparison of length change characteristics between portland cement and Type K cement concrete; (b) illustration showing why Type K cement concrete is resistant to cracking from drying shrinkage. (From J. V. Williams, Jr., *Concr. Int.*, Vol. 3, No. 4, p. 58, 1981.)

as described in Chapter 4, small and medium-size concrete elements are prone to undergo drying shrinkage, and they usually crack to relieve the induced stress when it exceeds the strength of the material. Cracks in concrete are unsightly; by increasing the permeability of concrete the cracks also become instrumental in reducing durability of structures exposed to aggressive waters. In some cases, they may even threaten the structural safety of a concrete member.

Current design and construction practices assume that concrete will crack, and try to get around the problem in many ways, such as selection and proportioning of concrete mixtures that will shrink less, provision of adequate joints in floor slabs or pavements, and reinforcement of concrete elements with steel. The advent of expansive cement has offered an alternative and cost-effective approach. As a result, since the first industrial use of the shrinkage-compensating concrete in the United States about 20 years ago, the material has found successful application in the construction of a variety of elements, such as floor slabs, pavements, roofs, water storage tanks, and sewage digesters.

Materials and Mix Proportions

According to ACI Committee 223, the same basic materials and methods necessary to produce high-quality portland cement concrete are required to produce satisfactory results in the use of shrinkage-compensating concrete. Additional care is necessary to provide continuous moist curing for at least 7 days after placement in order for the expansion to develop, and the structural design must be such as to ensure adequate expansion to offset subsequent drying shrinkage. For details, ACI Committee 223, *Recommended Practice for the Use of Shrinkage-Compensating Cements*, should be consulted. Expansion in concrete can be achieved either by the use of a modified portland cement (e.g., Type K cement, which contains $C_4A_3\bar{S}$ as the principal source of the reactive aluminate needed for ettringite formation) or by the addition of a suitable expansive admixture to portland cement concrete. To assure adequate expansion and restraint when Type K cement is being used, it is recommended to have a minimum cement content of 515 lb/yd^3 (305 kg/m^3) concrete with a minimum 0.15 percent reinforcement. The composition and properties of expansive cements are described in Chapter 6.

In regard to admixtures, the air-entraining admixtures are as effective with shrinkage-compensating concrete as with portland cement concrete in improving freeze-thaw and deicer salt durability. Calcium chloride, excessive amounts of fly ash and other pozzolans, and some water-reducing agents tend to reduce expansion by causing an imbalance between the rate of ettringite formation and the rate of strength development in the concrete.

For the determination of mix proportions, it is suggested that ACI 211.1 (Chapter 9) procedures be used except that compared to portland cement concrete, a slightly higher water/cement ratio may be employed in shrinkage-compensation concrete for achieving the same strength level. Owing to the relatively large amount of water needed for ettringite formation, and the water-imbibing property of ettringite, approximately 10 percent additional water may be used with a Type K expansive cement concrete, without strength impairment, in order to produce a consistency similar to that of a Type I portland cement concrete having the same cement content. In addition to the report of ACI 223, several other reports, such as the one by Williams,[34] contain useful information for mix proportioning and design and construction procedures for shrinkage-compensating concrete structures. Typical mix proportions and properties of shrinkage-compensating concrete used in a floor slab described by Hoffman and Opbroek[35] are given in Table 11–7.

[34] J. V. Williams, Jr., *Concr. Int.*, Vol. 3, No. 4, pp. 57–61, 1981.

[35] M. W. Hoffman and E. G. Opbroek, *Concr. Int.*, Vol. 1, No. 3, pp. 19–25, 1979.

TABLE 11-7 TYPICAL MIX PROPORTIONS AND PROPERTIES OF SHRINKAGE-COMPENSATING CONCRETE USED IN THE CONSTRUCTION OF A FLOOR SLAB

Mix proportions (no admixtures were used)	
Type K cement	588 lb/yd^3
Coarse aggregate (1 in., max.)	1790
Fine aggregate	1287
Water	312
Water/cement ratio	0.53
Properties	
Slump	4¾ in.
Specified compressive strength	4000 psi
Average 28-day strength	6034 psi

Source: Based on M. W. Hoffman and E. G. Opbroek, ACI, *Concr. Int.,* Vol. 1, No. 3, pp. 19–25, 1979.

Properties

Workability. Because of the water-imbibing characteristic of ettringite, which forms in relatively large quantities during very early stages of hydration, the concrete mixtures tend to be stiff but highly cohesive. Compared to normal portland cement, the use of a somewhat higher water/cement ratio (without the possibility of strength impairment) than recommended by the standard w/c-strength relationships for normal portland cement concrete is therefore permitted with expansive cements for achieving a reasonable consistency. Slumps in the range of 100 to 150 mm are recommended for most structural members, such as slabs, beams, reinforced walls, and columns. Because it is more cohesive or "fat" than portland cement concrete and has less tendency to segregate, the Type K shrinkage-compensating concrete is reported to be especially suitable for placement by pumping.

Slump loss. Slump loss under hot (concrete temperatures 32°C or higher) and dry conditions is more serious a problem in shrinkage-compensating concrete than in normal portland cement concrete. As a result of slump loss, excessive retempering of concrete on the job site will not only reduce the strength but also the expansion, which defeats the purpose for which the concrete is used. At higher than 27 to 29°C ambient temperatures, unless the concrete is cooled, both the amount of ettringite formed and the rate of its formation may be large enough to cause severe slump loss and quick setting. During construction of Houston's wastewater and sludge treatment plant,[36] completed in 1984, the contractors experienced extended periods of 38°C ambient temperatures, which necessitated the use of 50 to 125 lb (22 to 27 kg) of ice per cubic yard of Type K concrete in order to lower the concrete temperature to between 24 and 29°C at the job site.

---

[36] "Houston Plant Uses Type K Concrete," *Concr. Int.,* Vol. 3, No. 4, pp. 42–47, 1981.

Plastic shrinkage. Because of lack of bleeding and quicker stiffening and setting of concrete under hot, dry, and windy conditions, plastic shrinkage cracking is another problem for which extra precautions must be taken when using the shrinkage-compensating concrete. When fresh concrete is likely to be in contact with an absorptive surface such as a dry soil or an old concrete, the base should be thoroughly saturated by soaking it the evening before placement. Special precautions should be taken to avoid placement delays at the job site when using ready-mixed concrete. For slabs, fog spraying or covering the surface with wet blankets soon after placement is desirable in order to prevent rapid moisture loss.

Strength. The development of compressive, tensile, and flexural strength in shrinkage-compensating concrete is generally influenced by the same factors as portland cement concrete. Polivka and Willson[37] found that for a given water/cement ratio (in the range 0.4 to 0.65), the compressive strengths of Type K cement concrete were significantly higher than that of Type I portland cement concrete (Fig. 11–17a). In the case of shrinkage-compensating concrete, a denser cement paste matrix and a stronger transition zone between the cement paste and the coarse aggregate are the factors responsible for strengths higher than those of a portland cement concrete made with an equivalent water/cement ratio.

Volume changes. The drying-shrinkage characteristics of a shrinkage-compensating concrete are comparable to those of a corresponding portland cement concrete; the rate and the magnitude of shrinkage in both the cases are affected by the same factors, such as aggregate content and type, and water content. However, in the case of shrinkage-compensating concrete, the influence of the water/cement ratio on expansion during the early moist-curing period is quite important. Polivka and Willson's data (Fig. 11–17b) showed that with a water/cement ratio of 0.53 or less, the magnitude of the initial expansions was such that there was always some residual expansion even after 2 months of drying shrinkage. The magnitude of expansion decreased substantially with the increase in water/cement ratio (e.g., 0.76). Since the degree of needed precompression in shrinkage-compensating concretes may reduce considerably with water/cement ratios above 0.6, it is recommended that low water/cement ratios be used even when this is not needed from the standpoint of strength.

Everything else remaining the same, curing conditions are equally important for determining the overall volume change in shrinkage-compensating concrete. Probably for the reason cited in Chapters 5 and 6 (i.e., the role of water in ettringite expansion), Kesler[38] found (Fig. 11–17c) that the expansion of shrinkage-compensating concrete made with a Type M cement was significantly reduced when wet burlap or water curing was replaced by a curing procedure based on moisture retention (polyethylene sheet covering); only a little expansion occurred when the sealed-in

---

[37] M. Polivka and C. Willson, ACI SP-38, 1973, pp. 227–37.

[38] C. E. Kesler, *ASCE, Proc. J. Constr. Div.*, Vol. 102, No. C01, pp. 41–49, 1976.

Shrinkage-Compensating Concrete

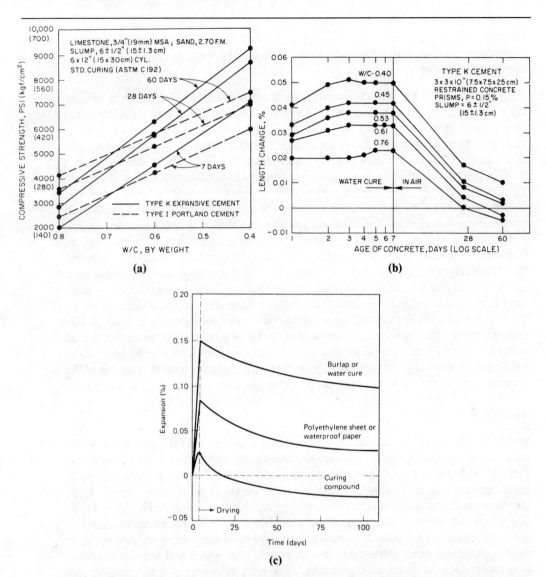

Figure 11-17 Factors affecting the properties of expansive cement concrete: (a) effect of water/cement ratio on strength; (b) effect of water/cement ratio on expansion; (c) effect of curing conditions on expansion. [(a), (b), From M. Polivka and C. W. Willson, *Expansive Cement*, ACI SP-38, 1973; (c), from C. E. Kesler, *Proc. ASCE J. Const. Div.*, Vol. 102, No. CO1, 1976.]

mixing water was the sole source of curing water. Available data on modulus of elasticity, creep, and Poisson's ratio show that for shrinkage-compensating concretes these properties are within the same range as those of portland cement concretes of comparable quality.

Durability. For several reasons, such as the restrained expansion of concrete, lack of bleeding, and little or no microcracking by drying shrinkage, the shrinkage-compensating concrete provides a more dense and essentially impermeable mass than does portland cement concrete of an equivalent water/cement ratio in the range 0.4 to 0.6. Consequently, laboratory and field experience has shown that, in general, Type K cement concretes possess a significantly higher resistance to abrasion, erosion, and chemical attack by aggressive solutions. Tests have also shown that when Type K expansive cements are made with an ASTM Type II or V portland cement clinker, the concrete shows a durability to sulfate attack that is similar to concrete made with a sulfate-resisting portland cement. This has been an important consideration in the use of Type K cement concrete for sanitary structures.

Properly designed, placed, and cured shrinkage-compensating concrete mixtures offer as much resistance to freezing and thawing, and deicer scaling, as portland cement concretes of comparable quality. Generally, a given amount of an air-training admixture will produce a volume of air and void spacing in a shrinkage-compensating concrete which is comparable to that of portland cement concrete, all other conditions being the same. For quick reference, a general comparison of the properties of shrinkage-compensating concrete with portland cement concrete of the same water/cement ratio is presented in Table 11-8.

Applications

Since the 1960s, expansive cements have been used in several countries for the purpose of producing both shrinkage-compensating and self-stressing concretes. Reviews of industrial applications in Japan,[39] the USSR,[40] and the United States[41] have appeared in the published literature. A description of some recent U.S. structures containing shrinkage-compensating concretes is presented in Table 11-9. It appears that most of the applications have been in structural elements, such as slabs, pavements, prestressed beams, and roofs. For obvious reasons, several hundred applications are reported for the construction of water and sewage-handling structures, such as water storage tanks, digesters, filtration plants, equalization basins, sludge removal facilities, spillways, cooling tower basins, and swimming pools.[42] Two of the recent applications are described below.

[39] S. Nagataki, *Proc. Cedric Willson Symp. on Expansive Cement Concrete*, ACI SP-64, 1977, pp. 43–79.

[40] V. V. Mikhailov and S. L. Litver, *Expansive and Stressing Cement* (translation from Russian), Sandia Laboratories, SAND 76-6010, 1975.

[41] P. K. Mehta and M. Polivka, *Proc. Sixth Int. Congr. on Chemistry of Cements*, Moscow, 1974; and V. M. Malhotra, ed., *Progress in Concrete Technology*, CANMET, Ottawa, 1980, pp. 229–92.

[42] J. V. Williams, Jr., *Concr. Int.*, Vol. 3, No. 4, pp. 57–61, 1981.

Shrinkage-Compensating Concrete

TABLE 11–8 COMPARISON OF PROPERTIES OF SHRINKAGE-COMPENSATING CONCRETE WITH PORTLAND CEMENT CONCRETE

Type of property	Characteristics of shrinkage-compensating concrete relative to portland cement concrete of similar water/cement ratio
Workability	
Consistency	Stiffer
Cohesiveness	Better
Time of set	Quicker
Strength	Better
Impermeability	Better
Drying shrinkage	Similar
Creep	Similar
Elastic modulus	Similar
Overall dimensional stability	Better
Durability	
Resistance to abrasion	Better
Resistance to erosion	Better
Resistance to sulfate attack	Similar to Type V portland cement concrete[a]
Resistance to frost action	Similar when equivalent air entrainment present

[a] This is applicable to Type K cements made with ASTM Type II or V portland cement.

Construction requirements for a 150-mm, 186,000-m^2 floor slab on grade for a warehouse in the Midwest included minimal joints and cracks and a carrying capacity for heavy loads both fixed and moving.[43] Normal portland cement concrete would have required small placements, checkerboard casting, 50 percent of slab reinforcement cut at saw joints, and saw cuts performed 12 hr after casting. The concrete is generally cast in a 40- by 40-ft (12- by 12-m) section and then saw-cut into 20- by 20-ft (6- by 6-m) sections (Fig. 11–18a). Immediately after cutting, the cuts are cleaned and sealed until the slab has fully cured; then the joints are uncovered and filled with a joint compound. Following this practice the amount of construction and saw-cut joints required for the warehouse would be 33.4 miles (54 km). Instead, with a Type K shrinkage-compensating cement concrete (Table 11–7), 80- by 120-ft (24- by 36-m) sections were cast without any saw cuts, and the amount of joints required totaled only 6.6 miles or 10.5 km (Fig. 11–18b). A reduction in the number of joints by about 80 percent by the use of shrinkage-compensating concrete was considered to be a big advantage from the standpoint of appearance, and construction and maintenance costs of the floor.

Houston's 69th St. Complex wastewater and sludge treatment plant,[44] completed in 1984 (Fig. 11–19), required that shrinkage-compensating concrete be used

---

[43] M. W. Hoffman and E. G. Opbroek, *Concr. Int.*, Vol. 1, No. 3, pp. 19–25, 1979.

[44] "Houston Plant uses Type K Concrete."

TABLE 11-9 SOME APPLICATIONS OF SHRINKAGE-COMPENSATING CONCRETE IN THE UNITED STATES

Type of structure	Owner and location	Pertinent facts	Observations
Airport pavement	Love Field, Dallas, Texas	Taxiway 1 (1969) Taxiway 2 (1972) Each taxiway is in excess of 2 km (1 mile) in length and consists of three 7.6-m (25-ft) lanes reinforced with welded wire fabric providing 0.12% and 0.06% steel in longitudinal and transverse directions.	Existing portland cement pavement has joints spaced 15 m (50 ft) with cracks midway between them. The Type K concrete pavement, which had joints spaced 23 and 38 m (75 and 125 ft), has only occasional cracks between joints.
Parking structure	O'Hare International Airport, Chicago, Illinois	Completed in 1972 10,000 automobile six-level structure 92,000 m^3 (120,000 yd^3) Type K concrete Type K concrete used in combination with post-tensioning for low maintenance design	Decks are divided by columns into bays of multiple pan-and-beam construction. Each bay consists of 46 cm (12-18 in.) deep pans with a relatively thin 11-cm (4½-in.) slab overhead. The shrinkage-compensating concrete section has an excellent inspection rating after 5 years of heavy traffic.
Office building and parking structure	Los Angeles World Trade Center, Los Angeles, California	Completed in January 1974 A six-level modular parking structure supporting the building superstructure and 10-story tower. 11×10^6 kg (12,000 tons) of Type K cement and 41,000 m^3 (53,000 yd^3) Type K concrete Mix designs based on structural requirements ranged from	The six-level parking structure contained 513 precast table-type modules each 15 m^3 (20 yd^3) with cast-in-place slabs pumped into the structure at the site. The precast table modules sit on cast-in-place pedestals into which are embedded the post-tensioning cables. The tower is supported by the park-

TABLE 11-9 Continued

Type of structure	Owner and location	Pertinent facts	Observations
		280 to 350 kgf/m² (4000 to 5000 psi)	ing structure. The structure used natural and lightweight aggregate (seven different mix designs) because of unique structural design.
Cold-storage warehouse	Meijer Frozen Foods, Lansing, Michigan	Completed in October 1975. Temperature range −23 to +4°C (−10 to +40°F). Type K shrinkage-compensating concrete slabs subjected to both drying shrinkage and thermal contraction associated with a cold storage warehouse 8900 m² (96,000 ft²)	This project was used to compare slab design theory with on-site analysis of concrete expansion and shrinkage. Field measurements were made to determine center and edge slab movement under a variety of restraint and temperature conditions. Based on performance to date, shrinkage-compensating concrete can be considered an economic asset for cold storage warehouse design.
Industrial warehouse, slabs on grade	J. C. Penney Co. 1. Lenexa, Kansas	Completed in November 1976. 186,000 m² (2 million ft²) under roof. 24 × 36 m (80 × 120 ft) placements (no sawed joints); 15 cm (6 in.) thick; 4 × 4—W4 × W4 4000-psi concrete (12,000 tons) of Type K cement	A Type I or II portland cement concrete design would call for 28 km (16.7 miles) of construction joints and an additional 28 km (16.7 miles) of sawed joints. Slab size limited to 12 × 12 m (40 × 40 ft) placements. Type K shrinkage-compensating concrete 24 × 36 m (80 × 120 ft) placements allow only 11 km (6.6 miles) of construction joints and no intermediate

TABLE 11-9 Continued

Type of structure	Owner and location	Pertinent facts	Observations
	2. Reno, Nevada	Completed in December 1977 139,000 m² (1.5 million ft²) under roof Design same as Kansas 8×10^6 kg (9000 tons) Type K cement	sawed joints. Slab in excellent condition. Slabs at final inspection were free of any cracks (first placement on 7/6/77). Contractor averaged 1670 m² (18,000 ft²) per day over entire project.

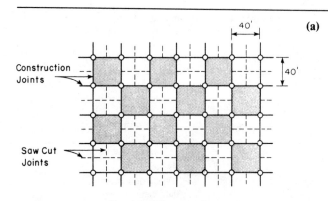

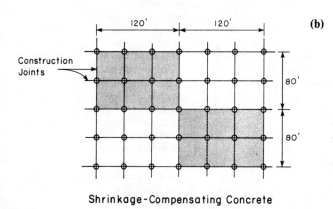

Figure 11-18 Relative proportions of construction joints in concrete slabs made with (a) portland cement, (b) shrinkage-compensating cement. (From M. W. Hoffman and E. G. Opbroek, *Concr. Int.*, Vol. 1, No. 3, pp. 19-25, 1979.)

Figure 11-19 Houston's 69th Street Complex wastewater and sludge treatment facilities: (a) artist's conception of the completed facilities; (b), (c) shrinkage-compensating concrete is used for the construction of reactors, clarifiers, thickeners, digesters, and pumping stations. (Photographs courtesy of Rom Young, Texas Industries, Inc., Dallas, Tex.); part (a) from *Concr. Int.*, Vol. 3, No. 4, p. 42, 1981.)

for the construction of all foundation slabs for the reactors, clarifiers, pump stations, thickeners and digesters, roof slabs, and monolithic beams of reactor trains. The higher unit cost of Type K cement was balanced by the reduced amount of drying shrinkage reinforcement required, larger placement sections, and fewer construction joints and water stops compared with standard portland cement concrete construction. It was concluded that the expected crack-free performance of the structure was worth the extra effort and care needed for concreting with this concrete, even in hot weather.

FIBER-REINFORCED CONCRETE

Definition and Significance

Concrete containing a hydraulic cement, water, fine or fine and coarse aggregate, and discontinuous discrete fibers is called **fiber-reinforced concrete.** It may also contain pozzolans and other admixtures commonly used with conventional concrete. Fibers of various shapes and sizes produced from steel, plastic, glass, and natural materials are being used; however, for most structural and nonstructural purposes, steel fiber is the most commonly used of all the fibers.

LESSONS IN THE HISTORY OF BUILDING MATERIALS

When was the concept of fiber reinforcement first used in building materials? According to Exodus 5:6, Egyptians used straw to reinforce mud bricks. There is evidence that asbestos fiber was used to reinforce clay posts about 5000 years ago.

But what about horneros? Professor Alberto Fava of the University of La Plata in Argentina points out that the hornero is a tiny bird native to Argentina, Chile, Bolivia, and other South American countries; the bird had been painstakingly building straw-reinforced clay nests on treetops since before the advent of man (see photograph).

When was concrete invented? Before you attempt to answer the question, take a look at a conglomerate rock* that is found abundantly in nature.

*A sedimentary rock composed of gravel which has been consolidated and cemented by precipitated silica, iron oxide, or calcium carbonate.

Fiber-Reinforced Concrete

Ordinarily, concrete contains numerous microcracks. As described in Chapter 3, it is the rapid propagation of microcracks under applied stress that is responsible for the low tensile strength of the material. Initially, it was assumed that tensile as well as flexural strengths of concrete can be substantially increased by introducing closely spaced fibers which would obstruct the propagation of microcracks, therefore delaying the onset of tension cracks and increasing the tensile strength of the material. But experimental studies showed that with the volumes and sizes of fibers that could conveniently be incorporated into conventional mortars or concretes, the fiber-reinforced products did not offer a substantial improvement in strength over corresponding mixtures without fibers.

Researchers, however, found considerable improvement in the post-cracking behavior of concretes containing fibers. In other words, *whereas the ultimate tensile strengths did not increase appreciably, the tensile strains at rupture did.* Thus compared to plain concrete, fiber-reinforced concrete is much tougher and more resistant to impact. It is clear from the applications described later in this chapter that the advent of fiber reinforcement has extended the versatility of concrete as a material by providing an effective method of overcoming its characteristic brittleness.

Toughening Mechanism

Typical load-deflection curves for plain concrete and fiber-reinforced concrete are shown in Fig. 11–20a. Plain concrete fails suddenly once the deflection corresponding to the ultimate flexural strength is exceeded; on the other hand, fiber-reinforced concrete continues to sustain considerable loads even at deflections considerably in excess of the fracture deflection of the plain concrete. Examination of fractured specimens (Fig. 11–20b) of fiber-reinforced concrete shows that failure takes place primarily due to fiber pull-out or debonding. Thus unlike plain concrete, a fiber-reinforced concrete specimen does not break immediately after initiation of the first crack (Fig. 11–20c). This has the effect of increasing the work of fracture, which is referred to as toughness and is represented by the area under the load-deflection curve. Explaining the toughening mechanism in fiber-reinforced composites, Shah states:

> The composite will carry increasing loads after the first cracking of the matrix if the pull-out resistance of the fibers at the first crack is greater than the load at first cracking; . . . at the cracked section, the matrix does not resist any tension and the fibers carry the entire load taken by the composite. With an increasing load on the composite, the fibers will tend to transfer the additional stress to the matrix through bond stresses. If these bond stresses do not exceed the bond strength, then there may be additional cracking in the matrix. This process of multiple cracking will continue until either fibers fail or the accumulated local debonding will lead to fiber pull-out.[45]

---

[45] S. P. Shah, "Fiber Reinforced Concrete," in *Handbook of Structural Concrete,* eds. F. K. Kong, R. H. Evans, E. Cohen, and F. Roll, McGraw-Hill Book Company, New York, 1984.

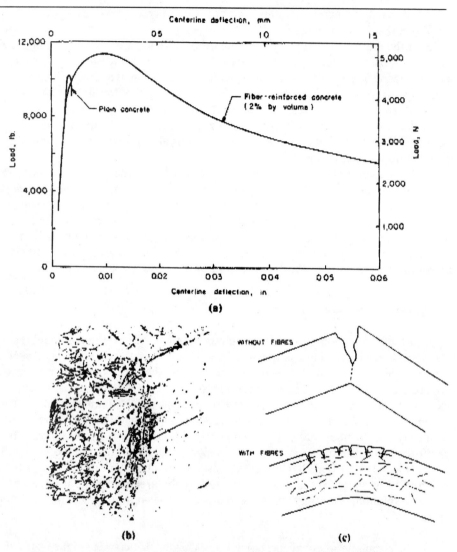

Figure 11–20 (a) Load-deflection behavior of plain and fiber-reinforced concrete; (b) Cross-section of a steel-fiber-reinforced beam after fracture showing that the failure mode is by fiber pullout; (c) mechanism of increase in flexure toughness of concrete with fibers. [(a), From A. C. Hanna, *Report RD049.01P,* Portland Cement Association, Skokie, Ill., 1977; (b), from H. Krenchel, *Fiber Reinforced Concrete,* ACI SP-44, 1974, pp. 45–77; (c), from C. D. Johnston, in *Progress in Concrete Technology,* ed. V. M. Malhotra, CANMET, Ottawa, 1980.]

TABLE 11-10 TYPICAL PROPERTIES OF FIBERS

Type of fiber	Tensile strength (ksi)	Young's modulus (10^3 ksi)	Ultimate elongation (%)	Specific gravity
Acrylic	30–60	0.3	25–45	1.1
Asbestos	80–140	12–20	~0.6	3.2
Cotton	60–100	0.7	3–10	1.5
Glass	150–550	10	1.5–3.5	2.5
Nylon (high tenacity)	110–120	0.6	16–20	1.1
Polyester (high tenacity)	105–125	1.2	11–13	1.4
Polyethylene	~100	0.02–0.06	~10	0.95
Polypropylene	80–110	0.5	~25	0.90
Rayon (high tenacity)	60–90	1.0	10–25	1.5
Rock wool (Scandinavian)	70–110	10–17	~0.6	2.7
Steel	40–400	29	0.5–35	7.8

Source: ACI Committee 544, Report 544.IR-82, *Concr. Int.*, Vol. 4, No. 5, p. 11, 1982.

The data from the tests by Krenchel[46] on both plain and steel fiber-reinforced mortars showed that incorporation of 0.9 and 2 percent fiber by volume of concrete increased the flexural strength by approximately 15 and 30 percent, respectively; however, in both cases the elongation at rupture was 9 to 10 times that of the unreinforced mortar. No visible cracks were ascertained in the tensile zone immediately prior to final rupture; the fine distribution of microcracks showed that fibers acted primarily as micro-reinforcement for crack distribution.

According to the report by ACI Committee 544,[47] the total energy absorbed in fiber debonding as measured by the area under the load-deflection curve before complete separation of a beam is at least 10 to 40 times higher for fiber-reinforced concrete than for plain concrete. The magnitude of improvement in toughness is strongly influenced by fiber concentration and resistance of fibers to pull-out which, in turn, is governed primarily by the fiber **aspect ratio** (length/diameter ratio) and other factors, such as shape or surface texture.

Materials and Mix Proportioning

Fibers. Typical properties of various kinds of fibers are given in Table 11–10, and steel fibers of different shapes and sizes are shown in Fig. 11–21. Round steel fibers have diameters in the range 0.25 to 0.75 mm. Flat steel fibers have cross sections ranging from 0.15 to 0.4 mm thickness by 0.25 to 0.9 mm width. Crimped and deformed steel fibers are available both in full length or crimped at the ends only. To facilitate handling and mixing, fibers collated into bundles of 10 to 30 with water-soluble glue are also available. Typical aspect ratios range from

---

[46] H. Krenchel, *Fiber Reinforced Concrete*, ACI SP-44, 1974, pp. 45–77.

[47] Report ACI 544.1R-82, *Concr. Int.*, Vol. 4, No. 5, pp. 9–30, 1982.

Figure 11–21 Typical fiber types used in concrete: (a) straight, smooth, drawn wire steel fibers; (b) deformed (crimped) wire steel fibers; (c) variable-cross-section steel fibers; (d) glued bundles of steel fibers with crimped ends.

about 30 to 150. Typical glass fibers (chopped strain) have diameters of 0.005 to 0.015 mm, but these fibers may be bonded together to produce glass fiber elements with diameters of 0.013 to 1.3 mm. Since ordinary glass is not durable to chemical attack by portland cement paste, alkali-resistant glass fibers with better durability have been developed.[48] Fibrillated and woven polypropylene fibers are also being used.

General considerations. It is well known that the addition of any type of fibers to plain concrete *reduces the workability*. Regardless of the fiber type, the loss of workability is proportional to the volume concentration of the fibers in concrete. Since fibers impart considerable stability to a fresh concrete mass, the slump cone test is not a good index of workability. For example, introduction of 1.5 volume percent steel or glass fibers to a concrete with 200 mm of slump is likely to reduce the slump of the mixture to about 25 mm, but the placeability of the concrete and its compactability under vibration may still be satisfactory. Therefore, the Vebe test (Chapter 10) is considered more appropriate for evaluating the workability of fiber-reinforced concrete mixtures.

[48] A. J. Majumdar, "Glass Fiber Reinforced Cement and Gypsum Products," *Proc. R. Soc., London*, Vol. A319, pp. 69–78, 1970.

Fiber-Reinforced Concrete

The effects of fiber content and aspect ratio on Vebe time, as found by researchers at the British Building Research Establishment,[49] are shown in Fig. 11–22a. From the standpoint of strengthening and toughening the concrete, it is desirable to have a large fiber aspect ratio as well as a high fiber concentration in concrete; on the other hand, the data in the figure clearly show that increases in these two variables would have the effect of reducing the workability. In fact, data from

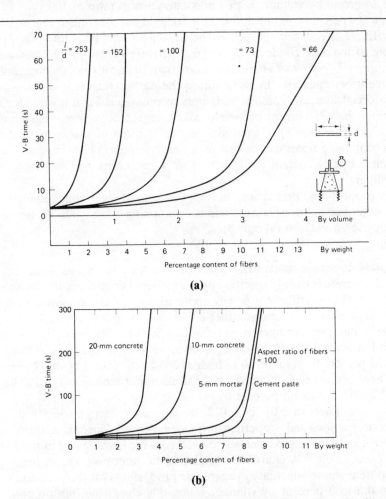

Figure 11–22 (a) Effect of fiber aspect ratio on workability of mortar; (b) effect of aggregate size and fiber content on workability. (From J. Edington, D. J. Hannant, and R. I. T. Williams, BRE Current Paper 69/74, July 1974. By courtesy of Building Research Establishment, U.K., and reproduced by permission of Her Majesty's Stationery Office. Crown Copyright.)

---

[49] J. E. Edington, P. J. Hannant, and R. I. T. Williams, Building Research Establishment, BRE Current Paper CP 69/74, 1974.

the work of Swamy and Mangat[50] show that steel fibers with aspect ratios greater than about 100 tended to produce the phenomenon known as ***curling up*** or ***balling*** at fiber concentrations as low as 1.13 percent by volume. Evidently, a compromise must be reached in selecting the proper amount and the aspect ratio of fibers. This compromise plays an important part in the selection of fibers and the design of fiber-reinforced concrete mixtures. In general, the amount of steel fibers in concrete is limited to about 2 percent by volume, with a maximum aspect ratio of 100.

Even at 2 percent concentration, the workability of concrete or mortar containing steel fibers is reduced sharply as the aggregate size goes up (Fig. 11–22b). According to the *ACI Guide for Specifying, Mixing, Placing, and Finishing Steel Fiber Reinforced Concrete*,[51] aggregates larger than 19 mm are not recommended for use in steel-fiber concrete. In fact, during the earlier stages of development of fiber-reinforced concrete, some investigators recommended that it was desirable to use no more than 25 percent by weight coarse aggregate of 9 mm maximum size (ASTM C 33, Size No. 8). Generally, the requirement of proper workability in mixtures containing fibers can be met by air entrainment, plasticizing admixtures, higher cement paste content (with or without a pozzolan), and the use of glued-together fibers.

Mix proportions, properties, and applications of steel-fiber-reinforced concrete are described below because concretes containing other types of fibers are not commonly used for structural purposes.

Typical concrete mixtures. ACI Committee 544[52] states that steel-fiber-reinforced concrete is usually specified by strength and fiber content. Normally, the flexural strength is specified for paving applications, and compressive strength for structural applications. Typically, either a flexural strength of 700 to 1000 psi (5 to 7 MPa) at 28 days, or a compressive strength of 5000 to 7000 psi (34 to 48 MPa), is specified. For normal-weight concrete, fiber contents from as low as 50 lb/yd^3 or 30 kg/m^3 (0.38 percent by volume) to as high as 365 lb/yd^3 or 220 kg/m^3 (2 percent by volume) have been specified, although the usual upper limit is 160 to 200 lb/yd^3 or 95 to 120 kg/m^3 (1.2 to 1.5 percent by volume).

With steel fibers of 0.01- by 0.022- by 1-in. cross section (0.25 × 0.55 × 25 mm), mix proportions and properties of fiber-reinforced concrete mixture suitable for highways and airport pavements and overlays were investigated at the Portland Cement Association.[53] A chart was developed to determine the increase in the cement content and the decrease in aggregate proportions for the fiber additions in the range 0.5 to 2 percent by volume. Using this chart, the mix proportions in

---

[50] R. N. Swamy and P. S. Mangat, *Cem. Concr. Res.*, Vol. 4, No. 3, pp. 451–65, 1974.

[51] Report ACI 544.3R-84, *J. ACI,* Proc., Vol. 81, No. 2, pp. 140–48, 1984.

[52] Ibid.

[53] A. N. Hanna, *Steel Fiber Reinforced Concrete Properties and Resurfacing Applications,* Portland Cement Association, Skokie, Ill., Report RD049.01P, 1977.

Table 11–11 show how at a given water/cement ratio, the cement paste content has to be increased with a corresponding decrease in the proportion of aggregates to maintain adequate workability when 2 percent steel fibers are added to the plain concrete mixture. The range of proportions recommended by ACI 544 for normal-weight fiber-reinforced mortars and concretes containing entrained air is shown in Table 11–12.

Flexural rather than compressive strengths are generally specified for pavements. It is possible to reduce the cost of paving mixtures by replacing a substantial proportion of the cement by a good-quality fly ash. This also has the effect of improving the needed workability and the 28-day flexural strength. Typical proportions of airfield paving mixtures containing about 30 percent fly ash by weight of the total cementing material (cement + fly ash) are shown in Table 11–13.

TABLE 11–11 COMPARISON OF MIX PROPORTIONS BETWEEN PLAIN CONCRETE AND FIBER-REINFORCED CONCRETE (LB/YD^3)

Material	Plain concrete	Fiber-reinforced concrete[a]
Cement	752	875
Water (water/cement ratio = 0.45)	338	394
Fine aggregate	1440	1282
Coarse aggregate	1150	1024
Fibers (2% by volume)	—	265

[a] The 14-day flexural strength (1150 psi) of the fiber-reinforced concrete was about 20% higher than that for the plain concrete.

Source: Adapted from A. N. Hanna, PCA Report RD 049.01P, Portland Cement Association, Skokie, Ill., 1977.

TABLE 11–12 RANGE OF PROPORTIONS FOR NORMAL-WEIGHT FIBER-REINFORCED CONCRETE

	Mortar	⅜ in. Maximum aggregate	¾ in. Maximum aggregate
Cement (lb/yd^3)	700–1200	600–1000	500–900
Water/cement ratio	0.30–0.45	0.35–0.45	0.40–0.50
Percent of fine to coarse aggregate	100	45–60	45–55
Entrained air content (%)	7–10	4–7	4–6
Fiber content (vol. %)			
Deformed steel fiber	0.5–1.0	0.4–0.9	0.3–0.8
Smooth steel fiber	1.0–2.0	0.9–1.8	0.8–1.6
Glass fiber	2–5	0.3–1.2	

Source: ACI Committee 544, Report 544.IR-82, *Concr. Int.*, Vol. 4, No. 5, p. 16, 1982.

TABLE 11-13 TYPICAL PROPORTIONS FOR AIRFIELD PAVING MIXTURES CONTAINING FLY ASH (LB/YD³)

Material	Mixture 1[a] ⅜ in. maximum aggregate	Mixture 2[a] ¾ in. maximum aggregate
Cement	500	525
Fly ash	235	250
Fine aggregate	1370	1440
Coarse aggregate	1470	1330
Water	255	283
Steel fibers (0.6 to 1.0% by volume)[b]	83–140	83–140

[a] Suitable amounts of water-reducing and/or air-entraining admixtures may be included. Flexural strengths at 28 days ranged from 1045 to 1100 psi.

[b] The lower volume was found to be adequate when crimped-end fibers were used instead of flat steel fibers.

Source: J. ACI., Proc. Vol. 81, No. 2, pp. 140–48, 1984.

Properties

It is difficult to evaluate some of the properties of fiber-reinforced concrete. A guidance on the subject is provided by ACI Report 544.2R.[54] A summary of the essential properties is given below.

Workability. The marked effect of fiber additions on workability of fresh concrete mixtures has already been described. For most applications, typical mortar or concrete mixtures containing fibers possess very low consistencies; however, the placeability and compactability of the concrete is much better than reflected by the low consistency.

Strength. Relatively speaking, the most important contribution of fiber-reinforcement in concrete is not to strength but to the flexural toughness of the material. This can be observed from the data in Fig. 11–23, which is taken from the report of an experimental study by Shah and Rangan. The test results in the tabulated data are from concrete specimens containing a constant amount of fibers (1 percent by volume) of given cross section (0.25 by 0.25 mm) but with aspect ratio varied by increase in fiber length from 6.5 to 25 mm. The authors concluded:

> It can be seen that increasing the length of fibers up to a point increases strength as well as toughness. The increase in toughness is an order of magnitude while that in strength is only mild. Comparison between conventional reinforcement and fiber reinforcement is also shown in the table. The percentage volume of steel was the same in both cases (1%). Conventional reinforcement was a deformed bar of 0.233 in. (6 mm) diameter and was placed with 0.5 in. (12 mm) cover from the tension face of the beam. It can

---

[54] ACI Report 544.2R-78, *J. ACI,* Proc., Vol. 75, No. 7, pp. 283–89, 1978.

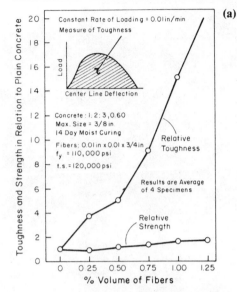

Type of reinforcement	Aspect ratio L/d	Relative strength	Relative toughness
Effect of aspect ratio			
Plain concrete	0	1.00	1.0
	25	1.50	2.0
Random	50	1.60	8.0
fibers	75	1.70	10.5
	100	1.50	8.5
Effect of type of reinforcement			
Conventional tensile bar	—	3.15	—
Random fibers	75	1.00	—

Figure 11–23 Factors affecting properties of fiber-reinforced concrete: (a) influence of increasing fiber volume; (b) influence of increasing the aspect ratio. (From S. P. Shah and B. V. Rangan, *J. ACI,* Proc., Vol. 68, No. 2, p. 128, 1971.)

be seen that the conventional reinforcement gave a maximum flexural load which was more than three times that of fiber reinforced concrete.[55]

Clearly when flexural strength is the main consideration, fiber reinforcement of concrete is not a substitute for conventional reinforcement. From the curve in Fig. 11–23, which is on the effect of volume of fibers on flexural strength and toughness of beams of concrete containing 0.01- by 0.01- by 0.75-in (0.25 × 0.25 × 19 mm) steel fibers, it is obvious that increasing the volume fraction of fibers enhanced both the flexural strength and toughness; whereas the increase in toughness was as much as 20 times for 1.25 percent volume of fibers, the increase in strength was less than twofold.

It seems that the first crack strength (Fig. 11–24a) and the ultimate flexural strength of fiber-reinforced systems (i.e., regardless of whether the matrix is hardened cement paste, mortar, or concrete) can be predicted using a composite material approach:

$$\sigma_c = A\sigma_m (1 - V_f) + BV_f \frac{l}{d}$$

where σ_c and σ_m are strengths of the composite (containing the fiber) and the matrix, respectively; V_f is the volume fraction and l/d the aspect ratio of fibers; and A and B are constants.

[55] S. P. Shah and B. V. Rangan, *J. ACI,* Proc., Vol. 68, No. 2, pp. 126–35, 1971.

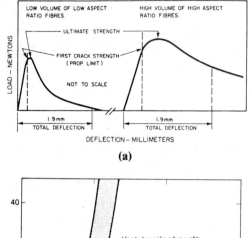

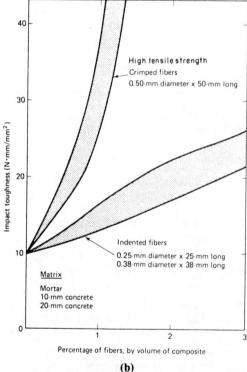

Figure 11–24 (a) Influence of fiber reinforcement on the first crack strength; (b) influence of crimped versus indented fibers on the impact strength. [(a), From C. D. Johnston, in *Progress in Concrete Technology*, ed. V. M. Malhotra, CANMET, Ottawa, 1980; (b), From J. Edington, D. J. Hannant, and R. I. T. Williams, BRE Current Paper 69/74, 1974. By courtesy of Building Research Establishment, U.K., and reproduced by permission of Her Majesty's Stationery Office, Crown Copyright.]

Swamy and his associates at Sheffield University[56] made regression analyses of test data from several investigations of fiber-reinforced pastes, mortars, and concretes with a wide range of mix proportions and fiber geometry to determine the values of the constants. The respective values of A and B were found to be 0.843 and 2.93 for first crack strength, and 0.97 and 3.41 for the ultimate strength of the

[56] R. N. Swamy, P. S. Mangat, and C. V. S. K. Rao, *Fiber Reinforced Concrete,* ACI SP-44, 1974, pp. 1–28.

composite. The coefficient of correlation from the regression analysis was reported to be 0.98.

Toughness and impact resistance. As shown by the data in Fig. 11-23, the greatest advantage in fiber reinforcement of concrete is the improvement in flexural toughness (total energy absorbed in breaking a specimen in flexure). Related to flexural toughness are the impact and fatigue resistance of concrete, which are also increased considerably. Unfortunately, for want of satisfactory tests for impact resistance of fiber-reinforced concrete, it has been difficult for researchers to assess the exact magnitude of improvement. Typical data showing the relative improvement achieved by *the substitution of crimped steel fibers for indented fibers* are shown in Fig. 11-24b.

Even *low-modulus fibers* such as nylon and polypropylene have been found to be very effective in producing precast concrete elements subject to severe impact. According to Johnston,[57] a construction company in the United Kingdom annually produces about 500,000 hollow segmental precast pile units. These units, which are 915 mm long by 280 mm in diameter with a wall thickness of about 50 mm, form the outer casing for conventional reinforced concrete piles subsequently cast within. By employing a concrete with 10-mm aggregate and 0.5 percent by volume of 40-mm-long fibrillated twine of polypropylene, resistance to breakage during pile driving was reduced by 40 percent compared to the steel mesh reinforced units used previously.

In regard to strength,[58] it has been shown that the addition of fiber to conventionally reinforced beams increased *the fatigue life* and decreased the crack width under fatigue loading. In general, a properly designed fiber-reinforced concrete will have a fatigue strength of about 90 percent of the static strength at 2×10^6 cycles when nonreversed loading is used, and about 70 percent when full reversal of load is used.

Elastic modulus, creep, and drying shrinkage. Inclusion of steel fibers in concrete has little effect on the modulus of elasticity, drying shrinkage, and compressive creep. Tensile creep is reduced slightly, but flexural creep can be substantially reduced when very stiff carbon fibers are used. However, in most studies, because of the low volume the fibers simply acted as rigid inclusions in the matrix, without producing much effect on the dimensional stability of the composite.

Durability. Fiber-reinforced concrete is generally made with a high cement content and a low water/cement ratio. When well compacted and cured, concretes containing steel fibers seem to possess excellent durability as long as fibers remain protected by the cement paste. In most environments, especially those containing chloride, surface rusting is inevitable but the fibers in the interior usually remain

---

[57] C. D. Johnston, in *Progress in Concrete Technology,* ed. V. M. Malhotra, CANMET, Ottawa, 1980, pp. 452–504.

[58] Report ACI 544.IR-82, *Concr. Int.,* Vol. 4, No. 5, pp. 9–30, 1982.

uncorroded. Long-term tests of steel-fiber concrete durability at the Battelle Laboratories in Columbus, Ohio, showed minimum corrosion of fibers and no adverse effect after 7 years of exposure to deicing salt.[59]

As stated earlier, ordinary glass fiber cannot be used in portland cement mortars or concretes because of chemical attack by the alkaline cement paste. Zirconia and other alkali-resistant glass fibers possess better durability to alkaline environments, but even these are reported to show a gradual deterioration with time. Similarly, most natural fibers, such as cotton and wool, and many synthetic polymers suffer from lack of durability to the alkaline environment of the portland cement paste.

According to ACI 544,[60] steel-fiber-reinforced concretes have been investigated for durability to surface erosion and cavitation. Tests by the Corps of Engineers suggest that erosion resistance against scour from debris in flowing water is not improved. These tests show that when erosion is due to the gradual wearing away of concrete due to relatively small particles of debris rolling over the surface at low velocity, the quality of aggregate and hardness of the surface determine the rate of erosion (Chapter 5). Inclusion of fibers in concrete did not cause any improvement; in fact, if the use of fibers results in a higher water/cement ratio and a larger paste volume, an increase in wear would occur. However, when loss of material is due to the cavitation impact of large debris, fibrous concrete has proved durable. The largest project of this kind yet undertaken was the repair of the stilling basin of the Tarbella Dam in Pakistan in 1977, with a 500-mm overlay of concrete containing 444 kg/m^3 cement and 73 kg/m^3 fibers.[61]

For the purpose of easy reference, a chart prepared by Johnston on the relative improvements in various properties of plain concrete when 25 to 38 mm straight steel fibers are incorporated, is shown in Fig. 11–25.

Applications

A report of 112 worldwide applications involving fibrous concrete was compiled by ACI Committee 544; Henager[62] summarized some of the large and more interesting projects. Highlights from typical applications included in this summary and in the report by ACI Committee 544[63] are given here.

The first structural use of steel-fiber-reinforced concrete was in 1971 for the production of demountable 3250 mm^2 by 65 mm-thick panels for a parking garage at London's Heathrow Airport. The concrete contained 3 percent by weight of 0.25 mm-diameter by 25 mm-long cold-drawn steel fibers. At the time of the last reported inspection, after 5 years of use, the slabs showed no signs of cracking.

[59] Ibid.
[60] Ibid.
[61] C. D. Johnston, in *Progress in Concrete Technology*.
[62] C. H. Henager, *Concr. Int.*, Vol. 3, No. 9, pp. 13–18, 1981.
[63] Report ACI 544.IR-82, *Concr. Int.*, Vol. 4, No. 5, pp. 9–30, 1982.

Fiber-Reinforced Concrete

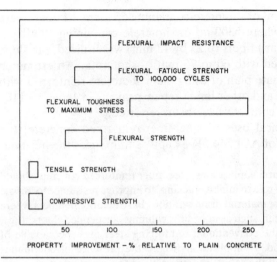

Figure 11–25 Relative improvement in various properties of concrete by fiber reinforcement. [From C. D. Johnston, in *Progress in Concrete Technology*, ed. V. M. Malhotra, CANMET, Ottawa, 1980, p. 502.]

Also in 1971, the U.S. Army Construction Engineering Research Laboratory performed controlled testing of fiber-reinforced concrete (2 percent fiber by volume) runway slabs at Vicksburg, Mississippi. The runway was subjected to C5A aircraft wheel loading (13,600 kg per wheel, over 12 wheels). The experiment compared the performance of a 150-mm-thick fiber-reinforced concrete slab with that of a 250-mm plain concrete slab. With the former, the first crack appeared after 350 loadings compared to 40 for the plain concrete; while the plain concrete was judged to have completely failed after 950 loadings, the fibrous concrete pavement with hairline cracks was serviceable after 8735 loadings.

At McCarran International Airport, Las Vegas, Nevada, in 1976 an existing asphaltic-paved aircraft parking area (63,000 yd^2) was overlaid with 150 mm-thick steel-fiber-reinforced concrete, compared to the 380 mm thickness that would have been required for conventional reinforced concrete. Based on satisfactory performance, in 1979 another pavement overlay (175 mm thick) was slip-formed over a newly constructed base of asphalt on an aggregate foundation. For this, 85 lb/yd^3 (50 kg/m^3) of crimped-end fibers 50 mm long by 0.5 mm in diameter were used; the 28-day flexural strength of concrete was 1045 psi (7 MPa). The use of crimped-end fibers in the second job allowed the use of about one-half the amount of fibers required in the first job (Table 11–13). Both these parking areas support DC-10 and Boeing 747 loads (up to 775,000 lb).

In 1980, steel fiber concrete was used in the construction of a new taxiway (850 m by 23 m by 170 mm) at Cannon International Airport, Reno, Nevada. A total of 5600 m^3 of steel-fiber concrete containing 12 mm-maximum aggregate and 87 lb/yd^3 (52 kg/m^3) of crimped fibers of the type employed at Las Vegas were used. The taxiway was constructed in three 7.5 m-wide strips with 12 m between sawed transverse joints; keyed longitudinal construction joints were placed at 7.5 m intervals.

Two large areas of rock slopes were stabilized with steel fibrous shotcrete, one at a refinery in Sweden, the other at a railroad cut along the Snake River in the state of Washington. In Sweden, 4100 m³ of shotcrete containing 55 kg/m³ of fibers were used. On the Snake River job, where 5300 m³ of shotcrete containing 200 lb/yd³ (120 kg/m³) to 250 lb/yd³ (150 kg/m³) fibers (0.25 by 12 mm) was utilized, the slope was first rock-bolted, then covered with 60 to 75 mm of shotcrete. An estimated $50,000 was saved over the alternate plan of using plain shotcrete reinforced with wire mesh. Several tunnel linings and other lining repair projects in Japan and the United States using steel fibrous concrete have been reported.

Finally, with regard to typical uses of steel-fiber-reinforced concrete, the following statement in the report of ACI 544.3R-84 aptly summarizes the position:

> Generally, when used in structural applications, steel fiber reinforced concrete should only be used in a supplementary role to inhibit cracking, to improve resistance to impact or dynamic loading, and to resist material disintegration. In structural members where flexural or tensile loads will occur, such as in beams, columns, suspended floors, (i.e., floors or slabs not on grade) the conventional reinforcing steel must be capable of supporting the total tensile load. In applications where the presence of continuous reinforcement is not essential to the safety and integrity of the structure, e.g., pavements, overlays, and shotcrete linings, the improvements in flexural strength associated with fibers can be used to reduce section thickness or improve performance or both.[64]

As to the economics of steel-fiber use, on the basis of 1980 prices ($660 per metric ton) Johnston[65] estimated that 1 percent fibers by volume will increase the materials cost of a concrete by $52 per cubic meter. Probably due to the high cost of steel-fiber concrete, for the most part its use has been restricted to overlays rather than for full-depth pavements. Such a composite construction makes sense because the maximum tensile stress occurs at the top.

If it is assumed that the cost of mixing, transporting, and placing concrete does not change by fiber incorporation, the difference in cost between in-place fibrous concrete and plain concrete will not be large. Also, compared to plain concrete, since the thickness of fibrous concrete slab designed for a given load can be substantially reduced, the overall difference in the first cost may turn out to be negligible. Considering the service life, therefore, fibrous concrete would appear to be cost-effective.

CONCRETES CONTAINING POLYMERS

Nomenclature and Significance

Concretes containing polymers can be classified into three categories: **polymer concrete** (PC) is formed by polymerizing a mixture of a monomer and aggregate—there is no other bonding material present; **latex-modified concrete** (LMC), which is also known as **polymer portland cement concrete** (PPCC), is a conventional

---

[64] Report ACI 544.3R-84, *J. ACI,* Proc., Vol. 81, No. 2, pp. 140–48, 1984.
[65] C. D. Johnston, in *Progress in Concrete Technology.*

portland cement concrete which is usually made by replacing a part of the mixing water with a latex (polymer emulsion); and **polymer-impregnated concrete** (PIC) is produced by impregnating or infiltrating a hardened portland cement concrete with a monomer and subsequently polymerizing the monomer in situ.

Both PC and LMC have been in commercial use since the 1950s; PIC was developed later and has been in use since the 1970s. Depending on the materials employed, PC can develop compressive strengths of the order of 20,000 psi (140 MPa) within hours or even minutes and is therefore suitable for emergency concreting jobs in mines, tunnels, and highways. LMC possesses excellent bonding ability to old concrete, and high durability to aggressive solutions; it has therefore been used mainly for overlays in industrial floors, and for rehabilitation of deteriorated bridge decks. In the case of PIC, by effectively sealing the microcracks and capillary pores, it is possible to produce a virtually impermeable product which gives an ultimate strength of the same order as that of PC. PIC has been used for the production of high-strength precast products and for improving the durability of bridge deck surfaces.

Because of the high material cost and cumbersome production technology (except LMC), the use of polymer-containing concretes is very limited. Hence only a brief review of materials, production technology, and properties will be given here. For details, other publications should be consulted.[66]

Polymer Concrete

What is referred to as polymer concrete (PC) is a mixture of aggregates with a polymer as the sole binder. To minimize the amount of the expensive binder, it is very important to achieve the maximum possible dry-packed density of the aggregate. For example, using two different size fractions of 19 mm maximum coarse aggregate and five different size fractions of sand in an investigation carried out at the Building Research Institute of Japan, Ohama[67] attempted to match Fuller's curve for maximum density of the aggregate mixture; the voids in the range of 20 to 25 percent were filled with a 1:1 mixture of unsaturated polyester resin and finely ground limestone. It was important to use dry aggregate because the presence of moisture caused a serious deterioration in the properties of the concrete.

In the Japanese study described above, a peroxide catalyst and an accelerator were included with the monomer for promoting subsequent polymerization in concrete. Two different curing procedures were adapted; thermal curing at 50 to 70°C, or room temperature curing at 20°C. The heat-cured specimens gave about 20,000-psi (140 MPa) compressive strength in 5 hr, whereas the normal-cured specimens gave about 15,000 psi in 7 days. The polyester resins are attractive because of their relatively lower cost compared to other products. Commercial products are available

---

[66] *Polymers in Concrete*, ACI SP-40, 1973; Report by ACI Committee 548, 1977; *Polymers in Concrete*, ACI SP-58, 1978; J. T. Dikeau, in *Progress in Concrete Technology*, ed. V. M. Malhotra, CANMET, Ottawa 1980, pp. 539–82.

[67] *Polymers in Concrete*, ACI SP-40, 1973.

with a variety of formulations, some capable of hardening to 15,000 psi (105 MPa) within a few minutes without any thermal treatment. Epoxy resins are higher in cost but may offer advantages such as adhesion to wet surfaces. The use of styrene monomer, and methyl methacrylate (MMA) with benzoyl peroxide catalyst and an amine promoter, seems to be increasing in PC formulations. Products with increased strength have been obtained by adding to the PC monomer system a silane coupling agent, which increases the interfacial bond between the polymer and aggregates.

The properties of PC are largely dependent on the amount and properties of polymer in the concrete; for example, PC made with MMA is a brittle material that shows a nearly linear stress-strain relationship with high ultimate strength, but the addition of butyl acrylate produces a more ductile material (Figure 11-26a). Typical mechanical properties of a polyester PC and a polymethyl methacrylate PC are shown in Table 11-14. By-product sulfur from oil refineries has a low viscosity at 120°C and has been used successfully for making PC.

Due to good chemical resistance and high initial strength and modulus of elasticity, industrial use of PC has been mainly in overlays and repair jobs. Thermal

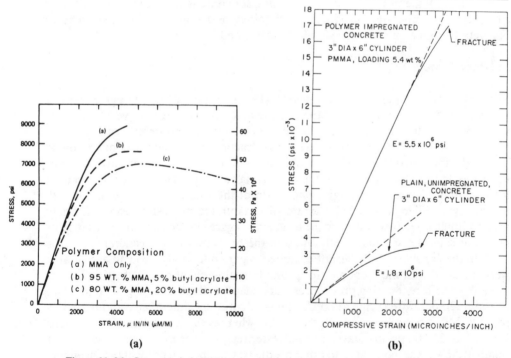

Figure 11-26 Stress-strain behavior of concrete containing polymers: (a) polymer concretes containing different polymer types; (b) polymer-impregnated concrete containing polymethylmethacrylate. [(a), From J. T. Dikeou, in *Progress in Concrete Technology*, ed. V. M. Malhotra, CANMET, Ottawa, 1980; (b), From M. Steinberg, *Polymers in Concrete*, ACI SP-40, 1973, p. 12.]

TABLE 11-14 TYPICAL MECHANICAL PROPERTIES OF CONCRETES CONTAINING POLYMERS (PSI)

	PC		LMC			PIC	
	Polyester	Polymerized MMA	Control		LMC containing styrene butadiene air-cured	Control unimpregnated	MMA impregnated thermal-catalytical polymerization
	1:10	1:15	Moist-cured	Air-cured			
	polymer/aggregate ratio						
Compressive strength	18,000	20,000	5800	4500	4800	5300	18,000
Tensile strength	2,000	1,500	535	310	620	420	1,500
Flexural strength	5,000	3,000	1070	610	1430	740	2,300
Elastic modulus, $\times 10^6$	5	5.5	3.4	—	1.56	3.5	6.2

Source: PIC from ACI SP-40, LMC from ACI Committee 548, and PC from Dikeau, in *Progress in Concrete Technology*, ed. V. M. Malhotra, CANMET, 1980.

and creep characteristics of the material are usually not favorable for structural application of PC. According to Lott et al.,[68] polyester concretes are viscoelastic and will fail under a sustained compressive loading at stress levels greater than 50 percent of the ultimate strength. Sustained loadings at a stress level of 25 percent did not reduce ultimate strength capacity for a loading period of 1000 hr. The researchers recommend, therefore, that polyester concrete be considered for structures with a high ratio of live to dead load and for composite structures in which the polyester concrete may relax during long-term loading.

Latex-Modified Concrete

The materials and the production technology for concrete in LMC are the same as those used in normal portland cement concrete except that latex, which is a colloidal suspension of polymer in water, is used as an admixture. Earlier latexes were based on polyvinyl acetate or polyvinylidene chloride, but these are seldom used now because of the risk of corrosion of steel in concrete in the latter case, and low wet strengths in the former. Elastomeric or rubberlike polymers based on styrene-butadiene and polyacrylate copolymers are more commonly used now.

A latex generally contains about 50 percent by weight of spherical and very small (0.01 to 1 μm in diameter) polymer particles held in suspension in water by surface-active agents. The presence of surface-active agents in the latex tends to incorporate large amounts of entrained air in concrete; therefore, air detraining agents are usually added to commercial latexes. Since 10 to 25 percent polymer (solid basis) by weight of cement is used in typical LMC formulations, the addition of latex provides a large quantity of the needed mixing water in concrete. For reasons given below, the application of LMC is limited to overlays where durability to severe environmental conditions is of primary concern. Therefore, LMC is made with as low an addition of extra mixing water as possible; the spherical polymer molecules and the entrained air associated with the latex usually provide excellent workability. Typically, water/cement ratios are in the range 0.40 to 0.45, and cement contents are of the order of 650 to 700 lb/yd^3 (390 to 420 kg/m^3).

It should be noted that unlike polymerization of monomers by additives and thermal activation, the hardening of a latex takes place by drying or loss of water. There is some internal moisture loss in a fresh LMC mixture because water will be needed for the hydration of portland cement; however, this is not sufficient to develop adequate strength. Consequently, dry curing is mandatory for LMC; the material cured in air is believed to form a continuous and coherent polymer film which coats the cement hydration products, aggregate particles, and even the capillary pores.

The mechanical properties of a typical latex-modifed mortar (sand/cement ratio of 3) containing 20 percent polymer solids by weight and dry-cured at 50 percent RH for 28 days, are compared to those of control portland cement mortars in Table

[68] Ibid.

11-14. It is evident from the data that LMC is better than the control material in tensile and flexural strengths. However, the strength gains are not impressive enough to justify the use of expensive latexes in making LMC products. The most impressive characteristics of LMC are its ability to bond strongly with old concrete, and to resist the entry of water and aggressive solutions. It is believed that the polymer film lining the capillary pores and microcracks does an excellent job in impeding the fluid flow in LMC. These characteristics have made the LMC a popular material for rehabilitation of deteriorated floors, pavements, and bridge decks.

Polymer-Impregnated Concrete

As stated earlier, the concept underlying PIC is simply that if voids are responsible for low strength as well as poor durability of concrete in severe environments, then eliminating them by filling with a polymer should improve the characteristics of the material. In hardened concrete, the void system, consisting of capillary pores and microcracks, is very tortuous. It is difficult for a liquid to penetrate it if the viscosity of the liquid is high and the voids in concrete are not empty (they contain water and air). Therefore, for producing PIC, it is essential not only to select a low-viscosity liquid for penetration but also to dry and evacuate the concrete before subjecting it to the penetration process.

Monomers such as methyl methacrylate (MMA) and styrene are commonly used for penetration because of relatively low viscosity, high boiling point (less loss due to volatilization), and low cost. After penetration, the monomer has to be polymerized in situ. This can be accomplished in one of three ways. A combination of promoter chemicals and catalysts can be used for room-temperature polymerization; but it is not favored because the process is slow and less controllable. Gamma radiation can also induce polymerization at room temperature, but the health hazard associated with it discourages the wide acceptance of this process in field practice. The third method, which is generally employed, consists of using a monomer-catalyst mixture for penetration, and subsequently polymerizing the monomer by heating the concrete to 70 to 90°C with steam, hot water, or infrared heaters.

From the above it should be apparent that the technology of producing PIC will be far more complex than that of conventional concrete. Although techniques have been developed to penetrate hardened concrete in the field, for obvious reasons PIC elements are more conveniently made as precast products in a factory. Typically, the sequence of operation consists of the following steps:

1. **Casting conventional concrete elements:** As will be discussed later, since the quality of concrete before penetration is not important from the standpoint of properties of the end product, no special care is needed in the selection of materials and proportioning of concrete mixtures. Section thickness is generally limited to a maximum of about 150 mm, since it is difficult to fully penetrate thick sections.
2. **Curing the elements:** Following the removal of elements from forms, at ambient

temperatures conventional moist curing for 28 days or even 7 days is adequate because the ultimate properties of PIC do not depend on the prepenetration concrete quality. For fast production schedules, thermal curing techniques may be adopted.

3. **Drying and evacuation:** The time and temperature needed for removal of free water from the capillary pores of moist-cured products depend on the thickness of the element. At the drying temperatures ordinarily used (i.e., 105 to 110°C), it may require 3 to 7 days before free water has been completely removed from a 150- by 300-mm concrete cylinder. Temperatures of the order of 150 to 175°C can accelerate the drying process so that it is complete in 1 to 2 days; the risk of thermal cracking due to rapid heating and cooling may not have much significance if all pores and microcracks are eliminated subsequently by the monomer penetration.

    The dried elements should be evacuated before immersion in monomer if relatively rapid (within 1 hr) and complete penetration is desired. This may not be essential in durability applications, in which case overnight monomer soaking of the dried concrete, without prior evacuation, will result in one-half to three-fourths depth of penetration from the exposed surface.

4. **Soaking the dried concrete in monomer:** The in situ penetration of concrete in the field may be achieved by surface ponding, but precast elements are directly immersed in the monomer-catalyst mixture. Commercial monomers contain inhibitors that prevent premature polymerization during storage; the catalyst serves to overcome the effect of the inhibitor. In the case of MMA, 3 percent by weight of benzoyl peroxide may be used as a catalyst.

    As stated earlier, due to the tortuous void system of hardened concrete, complete penetration of the dried specimens by soaking or infiltration is very difficult. In a study by Sopler et al.,[69] 10-cm cubes of a 0.56-water/cement ratio concrete were moist-cured for 7 days, dried at 150°C for 4 days, and soaked in MMA for periods ranging from 5 min to 48 hr. Inspection of polymerized specimens soaked for 48 hr showed that the total polymer loading was about 4½ percent by weight of concrete, and the middle 3-cm core was still unpenetrated (i.e., only 3½-cm-deep penetration was achieved from each surface). In the first 5 min of exposure about 38 percent of the 48-hr loading was realized; thereafter the rate of penetration slowed down; penetration depths at 100 min, 4 hr, and 8 hr were 2, 2½, and 3 cm, respectively. Consequently, when full penetration is desired, rather than soaking it is necessary to impregnate the monomer under pressure.

5. **Sealing the monomer:** To prevent loss of monomer by evaporation during handling and polymerization, the impregnated elements must be effectively sealed in steel containers or several layers of aluminum foil; in the rehabilitation of bridge decks this has been achieved by covering the surface with sand.

6. **Polymerizing the monomer:** As discussed earlier, thermal-catalytical polymer-

[69] Ibid.

ization is the preferred technique. The time for complete polymerization of the monomer in the sealed elements exposed to steam, hot water or air, or infrared heat at 70 to 80°C may vary from a few to several hours. In the case of a MMA-benzoyl peroxide mixture, no differences in strength were found between specimens polymerized at 70°C with hot air for 16 hr or with hot water for 4 hr.[70] Polymerization in hot water simplifies the precast PIC production process by eliminating the need for sealing.

The mechanical properties of PIC are shown in Table 11–14 and Fig. 11–26b. Various studies have shown that the properties of PIC are not affected by the initial quality of the unimpregnated concrete. For example, Sopler et al.[71] subjected concretes of three different qualities, 3000 psi or 20 MPa (0.83 water/cement ratio), 5500 psi or 38 MPa (0.56 water/cement ratio), and 8500 psi or 59 MPa (0.38 water/cement ratio), to the same process; all concretes achieved similar strength characteristics, although a higher polymer loading was required in the 0.83-water/cement ratio concrete.

As expected, the sealing of microcracks and pores renders the material brittle. The stress-strain curve in compression remains linear up to about 75 percent of ultimate load, and the deviation from linearity is only 10 to 15 percent at failure. Due to the absence of adsorbed water, shrinkage and creep are insignificant.

In spite of high strength, PIC elements will be of little interest for structural use due to the size limitation. There has been considerable interest in applications of PIC where the main consideration is the excellent durability of the material to abrasion, frost action, and attack by strong chemical solutions. The durability characteristics of PIC are due to impermeability and the absence of freezable water. PIC also shows promise for substrate treatment in the rehabilitation of concrete bridge decks. There is evidence to suggest that the overlay systems most commonly in current use for rehabilitation of deteriorated pavements (LMC, PC, or plain high-strength portland cement concrete) may not be wholly compatible with untreated concrete substrate. Cady et al.[72] investigated combinations of several substrate treatments and overlays and concluded that only the MMA (soaked or impregnated) substrate did not suffer from debonding, and showed excellent durability to frost action.

HEAVYWEIGHT CONCRETE FOR RADIATION SHIELDING

Significance

Concrete is commonly used for biological shielding in nuclear power plants, medical units, and atomic research and testing facilities. Other materials can be employed for this purpose, but concrete is usually the most economical and has several other

[70] Ibid.
[71] Ibid.
[72] P. D. Cady, R. E. Weyers, and D. T. Wilson, *Concr. Int.*, Vol. 6, No. 6, pp. 36–44, 1984.

advantages. Massive walls of conventional concrete are being used for shielding purposes. However, where usable space is limited, the reduction in the thickness of the shield is accomplished by the use of heavyweight concrete. ***Heavyweight concretes*** are produced generally by using natural heavyweight aggregates. The concrete unit weights are in the range 210 to 240 lb/ft^3 (3360 to 3840 kg/m^3), which is about 50 percent higher than the unit weight of concrete containing normal-weight aggregates.

Concrete as a Shielding Material

Two types of radiation have to be considered in the design of biological shields. First are ***X-rays and gamma rays,*** which have a high power of penetration but can be absorbed adequately by an appropriate mass of any material. Most materials attenuate these high-energy, high-frequency electromagnetic waves primarily according to the Compton scattering effect, the attenuation efficiency being approximately proportional to the mass of the material in the path of radiation. Since attenuation is not influenced by the type of material, different materials of the same mass have the ability to offer equal protection against X-rays and gamma rays.

The second type of radiation involves ***neutrons,*** which are heavy particles of atomic nuclei and do not carry an electrical charge. Thus neutrons are not affected by the electric field of the surroundings and therefore slow down only on collision with atomic nuclei. According to Polivka and Davis, for an efficient neutron shield it is often desirable to incorporate the following three classes of materials:

> The shield should contain some heavy material, such as iron, whose atomic mass is 56, or elements of higher atomic number. These heavy elements slow down the ***fast neutrons*** by inelastic collisions. It is desirable to have light elements, preferably hydrogen, to slow down further the ***moderately fast neutrons*** through elastic collisions. Hydrogen is particularly effective because it weighs about the same as a neutron. Finally, it is necessary to remove the ***slow-thermal neutrons*** by absorption. Hydrogen is effective in this action but in the process emits a 2.2 million volt gamma ray, which requires considerable shielding itself. Boron, on the other hand, has not only a very high cross section (absorbing power) for neutrons but emits gamma rays of only 0.0478 million volts. For these reasons boron-containing materials are often used in neutron shields. The gamma rays given off during absorption of neutrons are a factor in the required shield thickness.[73]

Commenting on the ***shielding ability of concrete,*** Polivka and Davis state:

> Concrete is an excellent shielding material that possesses the needed characteristics for both neutron and gamma ray attenuation, has satisfactory mechanical properties, and has a relatively low initial as well as maintenance cost. Also the ease of construction makes concrete an especially suitable material for radiation shielding.

---

[73] M. Polivka and H. S. Davis, ASTM STP 169B, 1979, Chap. 26, pp. 420–34.

Heavyweight Concrete for Radiation Shielding

Since concrete is a mixture of hydrogen and other light nuclei, and nuclei of higher atomic number, and can be produced within a relatively wide range of density, it is efficient in absorbing gammas, slowing down fast neutrons, and absorbing resonance and slow neutrons. The hydrogen and oxygen, contained in chemically combined form in the hydrated cement, moderate the neutron flux satisfactorily. The oxygen may also be present in concrete in another form in addition to water; concrete with siliceous aggregate is about one half oxygen.[74]

Materials and Mix Proportions

Except for the heavyweight aggregates (Table 7–3) and some hydrous ores as well as boron minerals, the same materials and proportioning methods are used for producing heavyweight concrete mixtures as are used for conventional normal-weight concrete. For details pertaining to concrete-making materials for biological shielding, the following standard specifications should be consulted: ASTM C 637 (*Specification for Aggregates for Radiation Shielding Concrete*) and ASTM C 638 (*Nomenclature of Constituents of Aggregates for Radiation Shielding Concrete*).

Because of the high density of aggregate particles, segregation of fresh concrete is one of the principal concerns in mix proportioning. From the standpoints of high unit weight and less tendency for segregation, it is desirable that both fine and coarse aggregate be produced from high-density rocks and minerals. Due to the rough shape and texture of crushed aggregate particles, heavyweight concrete mixtures tend to be harsh. To overcome this problem it is customary to use a finer sand, a greater proportion of sand in aggregate than with conventional concrete, and cement contents higher than 600 lb/yd^3 (360 kg/m^3). It should be noted that to get around the problem of segregation, sometimes other than conventional methods of placing, such as the **preplaced aggregate method,** may be employed. In this method, after filling the forms with compacted aggregate coarser than 6 mm, the voids in the aggregate are filled by pumping in a grout mix containing cement, fine sand, pozzolans, and other pumpability aids.

Important Properties

As stated above, the workability of fresh concrete can be a problem. Heavyweight concrete can be pumped or placed by chutes over short distances only because of the tendency of coarse aggregate to segregate. Concretes containing borate ores, such as colemanite and borocalcite, may suffer from slow setting and hardening problems because these minerals are somewhat soluble, and borate solutions are strong retarders of cement hydration. Unit weights of concrete containing barite, magnetite, or ilmenite aggregate are in the range 215 to 235 lb/ft^3 (3450–3760 kg/m^3); when hydrous and boron ores (which are not of high density) are used as partial replacement for heavyweight aggregate, the unit weight of concrete may come down to about 200 to 215 lb/ft^3 (3200–3450 kg/m^3).

[74] Ibid.

Massive shielding walls need not be designed for more than 2000-psi (14 MPa) compressive strength; for structural concrete, strengths of the order of 3000 to 5000 psi (20 to 35 MPa) are sufficient and not difficult to achieve with the high cement contents normally used. Strength is, however, of principal concern in the design of heavyweight concrete mixtures suitable for use in ***prestressed concrete reactor vessels*** (PCRV). These are pressure vessels that operate at higher stress levels and temperatures than conventional structures, and concrete is subject to appreciable thermal and moisture gradients. In such cases, inelastic deformations such as creep and thermal shrinkage should be minimized because they can cause microcracking and loss of prestress. Obviously, the elastic modulus of aggregate and compatibility of coefficients of thermal expansion between aggregate and cement paste should be considered to minimize microcracking.

The reactor vessels are usually designed to operate with concrete temperatures up to 71°C, but higher accidental temperatures and some thermal cycling is expected during the service life. Considerable strength loss can occur when concrete is subjected to wide and frequent fluctuations in temperature; hence PCRV concrete is designed not only for high density but also for high strength. In a study at the Corps of Engineers, Waterways Experiment Station,[75] using 720 to 970 lb/yd^3 (430 to 575 kg/m^3) Type I portland cement, 12 mm or 38 mm-maximum magnetite or ilmenite aggregate, and an 0.30 to 0.35 water/cement ratio, heavyweight concretes (3680 kg/m^3 unit weight) were produced which gave 7600 to 9400-psi (52 to 65 MPa) compressive strength at 7 days, and 9000 to 11,000 psi (62 to 76 MPa) at 28 days.

MASS CONCRETE

Definition and Significance

ACI Committee 116 has defined **mass concrete** as concrete in a massive structure, e.g., a beam, column, pier, lock, or dam where its volume is of such magnitude as to require special means of coping with the generation of heat and subsequent volume change. There is a popular assumption that the composition and properties of mass concrete are of interest only to those who are involved in the design and construction of dams; this definition attempts to correct that erroneous impression because many construction practices developed over a long period in building large concrete dams are applicable to structures far less massive.

Designers and builders of large concrete dams were the first to recognize the significance of temperature rise in concrete due to heat of hydration, and subsequent shrinkage and cracking that occurred on cooling. Cracks parallel to the axis of the dam endanger its structural stability; a monolithic structure (that is essentially free from cracking) will remain in intimate contact with the foundation and abutments and will behave as predicted by the design stress distributions. Concrete piers,

---

[75] K. Mather, *J. ACI*, Proc., Vol. 62, pp. 951–62, 1965.

columns, beams, walls, and foundations for large structures are much smaller than a typical concrete gravity dam. If they are several meters thick and are made of high-strength concrete mixtures (high cement content), the problem of thermal cracking can be as serious as in dams.

ACI Committee 207 has authored comprehensive reports[76] on concrete for dams and other massive structures, which should be consulted for details. Based mainly on these reports, only a brief summary of the general principles and their applications to lean (low-strength) concrete are presented here.

General Considerations

As discussed in Chapter 4,[77] the tensile stress on cooling concrete can, at first, be assumed to be the product of four quantities, $REet$, where t is the temperature drop, e the coefficient of thermal expansion, E the elastic modulus, and R the degree of restraint. Since the temperature drop and the resulting stress do not occur simultaneously, a correction in the calculated stress is necessary to take into account the stress relief due to creep. Thus the product of the quantities listed in the equation, minus the stress relaxed by creep, would determine the actual stress; concrete will crack when the magnitude of stress exceeds the tensile strength of the material (Fig. 4–1). Methods of determination of the tensile strength of massive unreinforced concrete structures are discussed in Chapter 3. The temperature drop, is the easiest to control and has received the most attention in dam building. At the heart of various construction practices for reducing the temperature drop is the recognition that a cost-effective strategy is to restrict the heat of hydration, which is the source of temperature rise in the first place. Basic principles, derived primarily from experience governing the selection of materials, mix proportions, and construction practices for controlling the temperature rise or drop in mass concrete, are discussed below.

Materials and Mix Proportions

Cement. As discussed earlier (Chapter 6), the heat of hydration of a cement is a function of its compound composition and fineness. Portland cements which contain relatively more C_3A and C_3S show higher heats of hydration than do coarser cements with less C_3A and C_3S. For instance, from the adiabatic-temperature-rise curves for a mass concrete containing 223 kg/m^3 of any one of the five types of portland cements (Fig. 4–21), it can be seen that between a normal cement (Type I) and a low-heat cement (Type IV) the difference in temperature rise was 13°C in 7 days and 9°C in 90 days. It should be noted that at this cement content (223 kg/m^3), the total temperature rise was above 30°C even with the low-heat cement.

In the event that temperature rise and the subsequent temperature drop of the order of 30°C is judged too high from the standpoint of thermal cracking, one way

---

[76] Report of ACI Committee 207 on Mass Concrete, *ACI Manual of Concrete Practice*, Part 1, 1991.

[77] For an advanced treatment of the subject, see Chapter 12.

to lower it would be by reducing the cement content of the concrete provided that this can be done without compromising the minimum strength and workability requirements needed for the job. By using several methods, which are described below, it is possible to achieve cement contents as low as 100 kg/m^3 in mass concrete suitable for the interior structure of gravity dams. With such low cement contents, even ASTM Type II portland cement is considered adequate; substitution of 20 percent pozzolan by volume of portland cement produces a further drop in the adiabatic temperature rise (Fig. 4–24).

Admixtures. With cement contents as low as 100 kg/m^3, it is essential to use a low water content to achieve the designed 1-year compressive strength (in the range 13 to 17 MPa) which is normally specified for interior concrete of large gravity structures. Approximately 4 to 8 percent entrained air is routinely incorporated into the concrete mixtures for the purpose of reducing the water content while maintaining the desired workability. Increasingly, water-reducing admixtures are simultaneously being employed for the same purpose. Whereas pozzolans are used primarily as a partial replacement for portland cement to reduce the heat of hydration, most fly ashes when used as pozzolans have the ability to improve the workability of concrete and reduce the water content by 5 to 8 percent.

Aggregate. With concrete mixtures for dams, every possible method of reducing the water content that would permit a corresponding reduction in the cement content (i.e., maintaining a constant water/cement ratio) has to be explored. In this regard, the two cost-effective methods are the choice of the largest possible size of coarse aggregate, and the selection of two or more individual size groups of coarse aggregate that should be combined to produce a gradation approaching maximum density on compaction (minimum void content). Typical coarse aggregate gradation limits for mass concrete and idealized combined grading for 150-mm and 75-mm maximum aggregate are given in the report by ACI Committee 211.[78]

Figure 11–27a, based on the U.S. Bureau of Reclamation's investigations on mass concrete for Grand Coulee Dam, shows the extent of reduction in water content by the use of entrained air and the largest possible size of aggregate. The same data are illustrated in Fig. 11–27b, which shows that at a given water/cement ratio and consistency, as the maximum aggregate size is increased, both the water and the cement contents are reduced.

Aggregate content and mineralogy have a great influence on properties that are important to mass concrete, such as elastic modulus, coefficient of thermal expansion, diffusivity, and strain capacity. Values of instantaneous modulus of elasticity at various ages for a typical mass concrete containing basalt aggregate were 2.3, 3.5, 4.1, and 5.0 × 10^6 psi at 7, 28, 90, and 365 days, respectively;[79] at the same

---

[78] ACI Committee 211, *Concr. Int.,* Vol. 2, No. 12, pp. 67–80, 1980.
[79] Ibid.

Mass Concrete

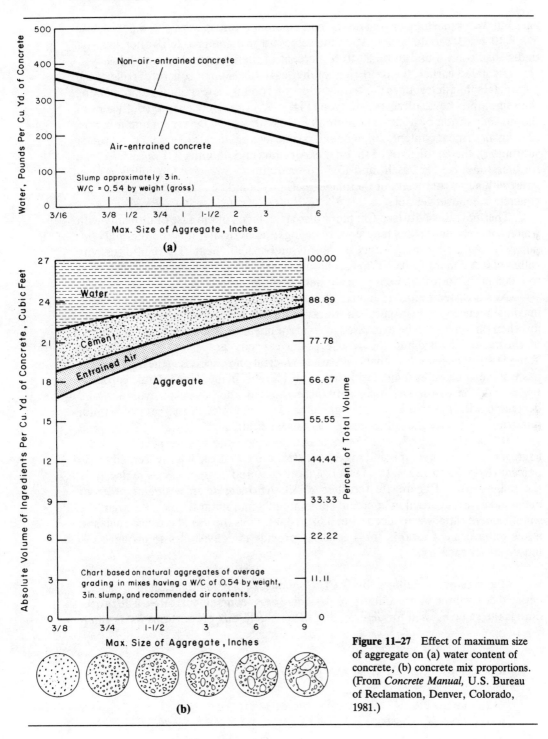

Figure 11–27 Effect of maximum size of aggregate on (a) water content of concrete, (b) concrete mix proportions. (From *Concrete Manual*, U.S. Bureau of Reclamation, Denver, Colorado, 1981.)

ages but with sandstone aggregate the corresponding values were 4.2, 4.5, 5.2, and 5.7×10^6 psi. It should be noted that the values for sustained modulus after 365 days under load were found to be 50 to 60 percent of the instantaneous modulus.[80]

As stated earlier, the coefficient of thermal expansion of concrete is one of the parameters that determines the tensile stress on cooling. Everything else remaining the same, the choice of aggregate type (Fig. 4-25) can decrease the coefficient of thermal expansion by a factor of more than 2. With typical mass concrete mixtures (237 kg/m³ cement content, 30:70 fine to coarse aggregate ratio, and high degree of saturation), the coefficients of thermal expansion (in millionths/°C) were 5.4 to 8.6 for limestone, 8.3 for basalt, and 13.5 for quartzite aggregate.[81] Obviously, aggregates with a low coefficient of thermal expansion should be selected for use in mass concrete whenever feasible.

Thermal diffusivity (see Chapter 4 also) is an index of the ease or difficulty with which concrete undergoes temperature change; numerically, it is the thermal conductivity divided by the product of density and specific heat. Thermal diffusivity values of 0.003 m²/hr (0.032 ft²/hr) for basalt, 0.0047 m²/hr for limestone, and 0.0054 to 0.006 m²/hr for quartzite aggregate have been reported.[82]

As suggested before (Chapter 4), some designers feel that designs based on maximum tensile strain rather than stress are simpler for predicting cracking behavior when the forces can be expressed in terms of linear or volumetric changes. Some of the factors controlling the tensile strain capacity are given by the data in Table 11-15. The data show that compared to mortar and concretes, the neat cement paste of the same water cement ratio has a considerably higher tensile strain capacity. In general, the tensile strain capacity increased with the period of hydration and decreased with the size of coarse aggregate. For example, in the case of natural quartzite aggregate, the values of 28-day tensile strain capacity were 165, 95, and 71×10^{-6} with 4.75, 37.5, and 75 mm maximum aggregate size, respectively. With aggregate of the same type and maximum size, the strain capacity increased by 50 percent (from 95 to 139×10^{-6}) when smooth-textured aggregate was replaced by the rough-textured aggregate (crushed rock). An increase of a similar order of magnitude was recorded in concrete containing 38 mm natural quartzite when the water/cement ratio was reduced from 0.68 to 0.40. Thus the use of crushed rock and a low water/cement ratio in mass concrete provide effective ways to enhance the tensile strain capacity.

Mix design. Reports by ACI Committee 211[83] and Scanlon[84] contain detailed description of procedures for designing mass concrete mixtures. The procedure is the same as used for determining the concrete mix proportions for normal-

---

[80] Ibid.
[81] Report of ACI Committee 207 on Mass Concrete.
[82] Ibid.
[83] ACI Committee 211, *ACI Materials Jour.*, Vol. 89, No. 5, 1991.
[84] J. M. Scanlon, *Mix Proportioning for Mass Concrete*, ACI SP-46, pp. 77–96, 1974.

TABLE 11–15 EFFECT OF MAXIMUM SIZE OF AGGREGATE ON TENSILE STRAIN CAPACITY OF CONCRETE

Mix	Aggregate	Maximum size of aggregate [in. (mm)]	W/C + P	Tensile strain capacity 10^{-6}	
				7 days	28 days
1	Quartzite, natural	3 (75)	0.68	45	71
2	Quartzite, natural	1½ (37.5)	0.68	76	95
3	Quartzite, natural	No. 4 (4.75)	0.68	138	165
4	None (paste)		0.68	310	357
5	Quartzite, natural	1½ (37.5)	0.68	119	139
6	Quartzite, natural	1½ (37.5)	0.40[a]	151	145

[a] All mixes contained 30% fly ash by absolute volume, except mix 6, which had no fly ash.
Source: R. W. Carlson, D. L. Houghton, and M. Polivka, *J. ACI,* Proc., Vol. 76, No. 7, p. 834, 1979.

weight concrete (Chapter 9). Some of the points from Appendix 5 of *ACI Committee 211 Recommended Practice,*[85] are discussed below.

In addition to the largest size of aggregate, determination of the water content should be based on the stiffest possible consistency of fresh concrete that can be adequately mixed, placed, and compacted. Typically, mass concrete slumps in unreinforced structures are of the order of 25 ± 12 mm. If the job-site equipment is inadequate for handling concrete with a stiff consistency, alternative equipment should be sought rather than increasing the water and the cement contents of the concrete mixture. In the case of precooled concrete, the laboratory trial mixtures should also be made at low temperature because less water will be needed to achieve the given consistency at 5 to 10°C than at normal ambient temperatures (20 to 30°C), due to the slower hydration of cement at low temperatures.

Determination of the cement content of mass concrete is guided by the relation between water/cement ratio and strength, which seems to be significantly affected by the aggregate texture (Table 11–16). In a moderate or mild climate, generally a maximum of 0.8-water/cement ratio concrete is permitted for interior of dam and lock walls, and 0.6 for exterior surfaces exposed to water. The maximum compressive stress in gravity dams that are properly designed against overturning and sliding is fairly low; in MPa units it is usually 0.025 to 0.03 times the height of dam in meters.[86] Thus a 100-m-high dam would have a maximum stress of about 3 MPa (450 psi). Stress in arch dams may be as high as 8 MPa (1200 psi). For safety it is recommended that the concrete strength should be four times the maximum stress at 1 year after construction. Gravity dams completed before 1940 (i.e., before the

---

[85] ACI Committee 211, *Concr. Int.,* Vol. 2, No. 12, pp. 67–80, 1980.
[86] W. H. Price, *Concr. Int.,* Vol. 4, No. 10, pp. 36–44, 1982.

TABLE 11-16 APPROXIMATE COMPRESSIVE STRENGTHS OF AIR-ENTRAINED MASS CONCRETE FOR VARIOUS WATER/CEMENT RATIOS[a]

Water/cement ratio by weight	Approximate 28-day compressive strength (f'_c), psi (MPa)	
	Natural aggregate	Crushed aggregate
0.40	4500 (31.0)	5000 (34.5)
0.50	3400 (23.4)	3800 (26.2)
0.60	2700 (18.6)	3100 (21.4)
0.70	2100 (14.5)	2500 (17.2)
0.80	1600 (11.0)	1900 (13.1)

[a] When pozzolan (P) is used, the strength should be determined at 90 days, and the water/cement ratios may be converted to W(C + P) ratios by the use of special equations.

Source: ACI 211, *Concr. Int.,* Vol. 2, No. 12, p. 76, 1980.

use of air-entraining, water-reducing, and pozzolanic admixtures) contained concretes made with approximately 376 lb/yd³ (223 kg/m³) cement content.

Comparison of laboratory specimens under standard curing conditions with cores drilled from high dams containing 223 kg/m³ cement showed that the actual strengths in the structure were considerably above that required: for example, Hoover Dam, 4260 psi (29 MPa); Grand Coulee Dam, 7950 psi (55 MPa); and Shasta Dam, 5100 psi (35 MPa). Even more impressive strength gains were observed in concretes containing pozzolans. Hungry Horse Dam (172 m high and 98 m thick at the base) was the first large dam built by the U.S. Bureau of Reclamation in which less than 223 kg/m³ cement was used. Concrete mixtures for the Hungry Horse Dam, which was completed in 1952, and subsequent large dams such as Flaming Gorge and Glen Canyon, contained 111 kg/m³ Type II portland cement, 56 kg/m³ pozzolan, and air-entraining admixtures.

Generally, a minimum of 3 to 4 percent air is always specified for mass concrete, although in practice 6 to 8 percent air is at times incorporated without any strength loss because the water/cement ratio is substantially reduced. Typically, 35 percent fly ash by volume of total cementing material is used for interior concretes, and 25 percent for exposed concrete. For sands with average fineness (2.6 to 2.8 fineness modulus), approximate coarse aggregate content is 78 to 80 percent by absolute volume of the total aggregate; thus the fine aggregate content is only 20 to 22 percent.

Construction practices for controlling temperature rise. In addition to the reduction of cement content in concrete mixtures, certain construction practices are used to control the temperature rise in massive concrete structures. Again, only a brief description is given below; an excellent and detailed report on cooling and insulating systems for mass concrete has been prepared by ACI 207.[87]

Post-cooling. The first major use of post-cooling of in-place concrete was in the construction of Hoover Dam in the early 1930s. In addition to control of temperature rise, a primary objective of post-cooling was to shrink the columns of concrete composing the dam to a stable volume so that the construction joints could be filled with grout to ensure monolithic action of the dam. Due to the low diffusivity of concrete (0.7 to 0.9 ft^2 or 0.065 to 0.084 m^2 per day), it would have taken more than 100 years for dissipation of 90 percent of the temperature rise if left to natural processes. This is in spite of the fact that a low-heat cement (Type IV) was used in making the concrete mixture, which contained 6-in.-maximum aggregate; also a special method of block construction for efficient heat dissipation was employed.

The cooling was achieved by circulating cold water through thin-wall steel pipes (typically 25 mm in nominal diameter, 1.5 mm in wall thickness) embedded in the concrete. In Hoover Dam, the circulation of cold water was started after the concrete temperature had reached 65°C (i.e., several weeks after the concrete had been placed). Subsequently, for the construction of several large dams the U.S. Bureau of Reclamation followed essentially the same practice, except that circulation of cooling water was started simultaneously with the placement of concrete. Also, pipe spacing and lift thickness are varied to limit the maximum temperature to a pre-designed level in all seasons.

According to the ACI 207 Recommended Practice,[88] during the first few days following placement the rate of cooling or heat removal can be as high as possible because the elastic modulus of concrete is relatively low. The strength and the elastic modulus generally increase rapidly until after the initial peak in concrete temperature has been experienced, which may be some time during the first 15 days following placement. Thereafter cooling should be continued at a rate such that the concrete temperature drop generally does not exceed 1°F (0.6°C) per day. Experience has shown that most mass concretes having average elastic and thermal expansion properties can sustain a temperature drop from 11 to 17°C (20 to 30°F) over approximately a 30-day period following the initial peak. When concrete has become elastic, it is important to have the temperature drop as slowly as possible to allow for stress relaxation by creep; under the slow cooling conditions, concrete can stand a 20°C drop in temperature without cracking.

Precooling. The first use of precooling of concrete materials to reduce the maximum temperature of mass concrete was by the Corps of Engineers during the construction of Norfork Dam in the early 1940s. A part of the mixing water was introduced into the concrete mixture as crushed ice so that the temperature of in-place fresh concrete was limited to about 6°C. Subsequently, combinations of crushed ice,

---

[87] ACI Committee 207, "Cooling and Insulating Systems for Mass Concrete." *Concrete Int.*, Vol. 2, No. 5, pp. 45–64, 1980.

[88] W. H. Price, *Concr. Int.*, Vol. 4, No. 10, pp. 36–44, 1982.

cold mixing water, and cooled aggregates were utilized by Corps of Engineers in the construction of several large concrete gravity dams (60 to 150 m high) to achieve placing temperatures as low as 4.5°C.

According to ACI 207 Recommended Practice,[89] one of the strongest influences on the avoidance of thermal cracking in mass concrete is the control of placing temperature. Generally, the lower the temperature of the concrete when it passes from a plastic state to an elastic state, the less will be the tendency toward cracking. In massive structures, each 10°F (6°C) lowering of the placing temperature below the average air temperature will result in a lowering by about 6°F (3°C) of the maximum temperature the concrete will reach.

To raise the temperature by 1°F, water absorbs 1 Btu/lb heat, whereas cement and aggregates absorb only 0.22 Btu/lb. Therefore, pound for pound, it is more efficient to use chilled water in reducing the temperature of concrete. Of course, the use of ice is most efficient, because ice absorbs 144 Btu/lb heat when it changes to water. For concrete homogeneity, it is important that all the ice in the concrete mixture has melted before the conclusion of mixing. Therefore, flake ice or biscuit-shaped extruded ice is preferable to crushed ice blocks. Cooling the coarse aggregate while enroute to the batch bins by spraying with chilled water may be necessary to supplement the use of ice and cooled mixing water.

Surface Insulation. The purpose of surface insulation is not to restrict the temperature rise, but to regulate the rate of temperature drop so that stress differences due to steep temperature gradients between the concrete surface and the interior are reduced. After the concrete has hardened and acquired considerable elasticity, decreasing ambient temperatures and rising internal temperatures work together to steepen the temperature gradient and the stress differential. Especially in cold climates, it may be desirable to moderate the rate of heat loss from the surface by covering with pads of expanded polystyrene or urethane (k factor of the order of 0.2 to 0.3 Btu-in./hr-ft^2°F).

Application of the Principles

Price[90] reviewed the last 50 years of construction practices in the United States to show the development of strategies for control of cracking in concrete dams. In the construction of Hoover (1935), Grand Coulee (1942), and Shasta (1945) Dams, which contain 2.4, 8.0, and 4.5 million cubic meters of concrete, respectively, ASTM Type IV low-heat portland cement (223 kg/m^3 cement content) was used, and concrete was post-cooled by circulating cold water through the embedded pipes. The heights and scheduling of placements were controlled, and special block construction procedures were devised for more efficient heat dissipation. All three dams remained free of objectionable cracks and leakage. Also in the early 1940s, the Tennessee Valley Authority utilized post-cooling in the construction of Fontana

---

[89] Ibid.
[90] Ibid.

Dam. In all the cases, post-cooling not only reduced the temperature rise, particularly in the base of the dam, which was more vulnerable to cracking due to the restraining effect of the foundation, but also stabilized the columns within the construction period of the dam so that the construction joints between columns could be filled with grout to ensure monolithic action.

Beginning with Norfork Dam (1945), precooling of concrete materials was successfully used to control cracking in Detroit Dam (1953). By limiting the cement content (Type II) to 134 kg/m^3 and the concrete placement temperature to 6 to 10°C, the temperature rise was restricted to 17°C above the mean ambient temperature. No post-cooling of concrete was necessary. Also, the exposed surfaces of concrete were protected from rapid cooling by covering with an insulating material. The complete absence of objectionable cracking in the Detroit Dam blocks, as long as 102 m, pointed the way to future application of these techniques. Thus the trend toward lower peak concrete temperatures began when it was discovered that this allowed the use of larger monolith lengths without harmful consequences. Precooling and post-cooling were used in combination in the construction of several large dams, notably Glen Canyon Dam (1963), Dworshak Dam (1973), and Libby Dam (1975). In every case the temperature rise was limited to 14°C.

Limiting the temperature drop to less than 20°C by precooling concrete-making materials is possible only when, at the same time, the cement content is reduced substantially. Compared to 223 kg/m^3 cement for the Hoover Dam concrete, only 111 kg/m^3 cement and 56 kg/m^3 pozzolan (pumicite) were used for the Glen Canyon Dam concrete, which gave about 3000-psi (20 MPa) compressive strength at 28 days, and 7000 psi (48 MPa) at 360 days. Such a low cement content could be used because a considerable reduction in the water content was achieved by using properly graded aggregates, an air-entraining admixture, and a water-reducing admixture. The data in Table 11-17 illustrate how water content varies with the maximum size of aggregate, fine/coarse aggregate ratio, and the use of admixtures. For example, instead of a 210-kg/m^3 water requirement with a 10-mm-maximum aggregate size and no admixtures present, only 92 kg/m^3 water is needed to achieve the same consistency when 150-mm-maximum aggregate and both water-reducing and air-entraining admixtures were used. According to Price,[91] for a given aggregate and consistency of concrete, about 35 percent reduction in the water content can be accomplished by using entrained air, a water-reducing admixture, a lower sand content, and a lower placing temperature.

Laboratory investigations on mass concrete for the recently constructed Itaipu Dam (Fig. 1–2) showed that the permissible <20°C rise in adiabatic temperature of concrete could be achieved when the cement content was held to about 108 kg/m^3. However, the designed 1-year compressive strength of 14 MPa was not achieved with this cement content even with 13 kg/m^3 pozzolan present, when 38 or 75 mm-maximum aggregate size was used. This is because the water/cement ratio required for the desired consistency was too high. A mixture of 150-, 75-, and 38-mm

---

[91] Ibid.

TABLE 11-17 APPROXIMATE AIR AND WATER CONTENTS PER CUBIC METER FOR CONCRETES CONTAINING DIFFERENT MAXIMUM SIZE AGGREGATE

Maximum size of coarse aggregate [mm (in.)]	Recommended air content (% for full mix)	Sand, percent of total aggregate by solid volume	Non-air-entrained concrete: average water content [kg/m³ (lb/yd³)]	Air-entrained concrete: average water content [kg/m³ (lb/yd³)]	Air-entrained concrete with WRA:[a] average water content [kg/m³ (lb/yd³)]
10 (⅜)	8	60	210 (352)	190 (320)	180 (300)
12.5 (½)	7	50	200 (336)	180 (305)	170 (285)
20 (¾)	6	42	185 (316)	165 (280)	157 (265)
25 (1)	5	37	178 (300)	158 (265)	148 (250)
40 (1½)	4.5	34	165 (280)	145 (245)	135 (230)
50 (2)	4	30	158 (266)	136 (230)	127 (215)
80 (3)	3.5	28	143 (242)	120 (205)	112 (190)
150 (6)	3	24	125 (210)	98 (165)	92 (150)

[a] WRA stands for water-reducing admixture.

Source: W. H. Price, Concr. Int., Vol. 4, No. 10, p. 43, 1982.

maximum aggregates, and incorporation of an air-entraining admixture, caused enough lowering of the water/cement ratio to produce 25 to 50 mm-slump concrete with a 17.5-MPa compressive strength at 90 days. This concrete mixture contained 108 kg/m^3 cement, 13 kg/m^3 fly ash, 85 kg/m^3 water, 580 kg/m^3 sand, and a mixture of 729, 465, and 643 kg/m^3 coarse aggregates with 38, 75, and 150 mm maximum size, respectively. The concrete placement temperature was restricted to 6°C by precooling all coarse aggregate with chilled water, followed by cold air, using some of the mixing water at 5°C and most of it in the form of flake ice.

Roller-Compacted Concrete

Concept and significance. Roller-compacted concrete (RCC) presents a relatively recent development in the construction technology of dams and locks. It is based on the concept that a ***no-slump concrete*** mixture transported, placed, and compacted with the same construction equipment that is used for earth and rockfill dams can meet the design specifications for conventional mass concrete. The materials harden into a concrete with essentially similar physical appearance and properties as those of conventional mass concrete. Gradation controls for aggregates are much more relaxed than those ordinarily required for mass concrete, because the usual relation between water/cement ratio and strength does not apply; in most cases, material finer than No. 200 sieve need not be removed. ACI 207·5R-80 is a detailed report on RCC which contains a description of materials, mix proportions, properties, and design and construction technology. Based on this report[92] and several others,[93] only a brief summary is given here.

From the Japanese experience of construction with RCC, Hirose and Yanagida[94] list several advantages:

- Cement consumption is lower because much leaner concrete can be used.
- Formwork costs are lower because of the layer placement method.
- Pipe cooling is unnecessary because of the low temperature rise.
- Cost of transporting concrete is lower than with the cable crane method because concrete can be hauled by end dump trucks; it is spread by bulldozers and compacted by vibratory rollers.
- Rates of equipment and labor utilization are high because of the higher speed of concrete placement.
- The construction period can be shortened considerably.

Since 1974, 0.5 million cubic meters of RCC have been placed in three dams in Japan;

---

[92] Report of ACI Committee 207, *ACI Materials J.*, Vol. 85, No. 5, pp. 400–45, 1988.

[93] E. Schrader, *Concr. Int.*, Vol. 4, No. 10, pp. 15–25, 1982; T. Hirose and T. Yanagida, *Concr. Int.*, Vol. 6, No. 5, pp. 14–19, 1984; J. E. Oliverson and A. T. Richardson, *Concr. Int.*, Vol. 6, No. 5, pp. 20–28, 1984; E. Schrader and R. McKinnon, *Concr. Int.*, Vol. 6, No. 5, pp. 38–45, 1984.

[94] T. Hirose and T. Yanagida, *Concr. Int.*, Vol. 6, No. 5, pp. 14–19, 1984.

Figure 11-28 Willow Creek Dam, Oregon, the world's first all-roller-compacted concrete structure, shown about 18 weeks after the start of placement. (Photograph courtesy of E. Schrader *Concr. Int.*, Vol. 6, No. 5, 1984.)

two more requiring 1.35 million cubic meters of RCC, including the 103-m-high Tamagawa Dam, are under construction at the present time.[95]

The experience with RCC in the United States has been similar. In 1982, the construction of the world's first all-RCC structure, the Willow Creek Dam (Fig. 11-28), was functionally complete in 1 year after the work was opened for bidding.[96] A saving of $13 million resulted over the alternative rock-filled design, which called for a 3-year schedule. If conventional concrete had been used as a reference, the savings would have been considerably more. The in-place RCC cost averaged about $20 per cubic yard, compared to $65 per cubic yard, estimated for conventional mass concrete. Several RCC dams are being planned; two are under construction, including the Upper Stillwater Dam in Utah, which is 84 m high and will require 1.07 million cubic meters of RCC.

Materials, mix proportions, and properties. RCC differs from conventional concrete principally in its consistency requirement; for effective consolidation, the concrete must be dry to prevent sinking of the vibratory roller equipment but wet enough to permit adequate distribution of the binder mortar throughout the material during the mixing and vibratory compaction operations. The conventional concept of minimizing water/cement ratio to maximize strength does not hold; the best compaction gives the best strength, and the best compaction occurs at the wettest mix that will support an operating vibratory roller. In other words, the consistency requirement plays a major part in the selection of materials and mix proportions.

The aggregates for the Willow Creek Dam concrete contained approximately 25 percent silty, sandy gravel overburden. There was no aggregate washing; in fact, the presence of material passing a No. 200 sieve improved the compactability and increased the strength of the compacted concrete. Although maximum aggregate

---

[95] Ibid.
[96] E. Schrader and R. McKinnon, *Concr. Int.*, Vol. 6, No. 5, pp. 38–45, 1984.

TABLE 11-18 MIX PROPORTIONS (LB/YD^3) AND MATERIAL PROPERTIES OF ROLLER-COMPACTED CONCRETES

	Spillway face	Downstream face	Upstream face	Interior mix
Cement (Type II)	315	175	175	80
Fly ash	135	80	0	32
Water	200	185	185	182
Max aggregate size (in.)	1½	3	3	3
Compressive strength (psi)				
90-day	4450	2850	2310	1060
1-year	6150	4550	3650	1680

Source: E. Schrader, *Concr. Int.*, Vol. 4, No. 10, p. 17, 1982.

size up to 230 mm has been used for RCC (Tarbela Dam, Pakistan), a 75 mm maximum size was preferred for the Willow Creek Dam to minimize the tendency for segregation. RCC concrete mixtures investigated for the Upper Stillwater Dam contained 38 mm maximum size aggregate.[97]

From the standpoint of workability, fly ash is commonly included in RCC mixtures. Probably due to the dry consistency and the low amount of cement paste, the use of either water-reducing or air-entraining admixtures did not provide any benefit with the concrete mixtures for the Willow Creek Dam. Durability to frost action was not required because most of the concrete will never be subjected to wet-dry and freeze-thaw cycles. The exposed upstream face of the structure was made of conventional air-entrained concrete. Mix proportions and material properties for the range of RCC concretes used at the Willow Creek Dam are shown in Table 11-18.

According to Schrader,[98] the Willow Creek Dam is constructed as one monolithic mass with no vertical joints. This is considered a major deviation from conventional design and construction of concrete dams, where vertical joints are typically located every 40 to 60 ft (12 to 18 m). No post-cooling, aggregate chilling, and ice were used. Due to the very low cement content of the interior concrete, the adiabatic temperature rise was only 11°C in 4 weeks.

A problem created by the drier consistency of RCC is the difficulty of bonding fresh concrete to hardened concrete. Investigations have shown that the use of special high-consistency bedding mixtures for starting the new concrete placement is helpful in reducing the tendency to form cold joints. Typically, bedding mixtures contain 330 lb/yd^3 (196 kg/m^3) of cement, 130 lb/yd^3 (77 kg/m^3) of fly ash, and 19 mm maximum size aggregate.

In general, the creep and thermal properties of RCC are within the range of those of conventional mass concrete. Results of permeability tests for the Upper Stillwater Dam concrete indicated a permeability rate of 1.2×10^{-4} m/yr at a

---

[97] J. E. Oliverson and A. T. Richardson, *Concr. Int.*, Vol 6, No. 5, pp. 20–28, 1984.
[98] E. Schrader, *Concr. Int.*, Vol. 4, No. 10, pp. 15–25, 1982.

comparable compressive strength, which is equal to or less than that of conventional mass concrete.[99] This is attributed to the high content of fines furnished by the fly ash and the low ratio of water to cementing materials (0.43 to 0.45).

TEST YOUR KNOWLEDGE

1. (a) Slump loss and floating of coarse aggregates can be major problems with fresh lightweight concrete mixtures. How are they controlled?
   (b) The compressive strength of conventional lightweight concrete is limited to about 7000 psi (48 MPa). How can this be increased?
2. (a) Compared to normal-weight concrete of the same water/cement ratio, a structural lightweight concrete would show higher drying shrinkage but less tendency to crack. Can you explain why?
   (b) In spite of the cellular structure of aggregate, lightweight concretes show less microcracking and excellent durability. Why?
3. (a) In high-rise buildings, what are the advantages of constructing shear walls and columns with high-strength concrete?
   (b) Discuss why the use of water-reducing and pozzolanic admixtures is essential for producing ultra-high-strength concrete.
   (c) Superplasticized concrete is, in general, prone to slump loss. How can this problem be overcome in construction practice?
4. With regard to the microstructure and mechanical properties of high-strength concrete, discuss the conclusions drawn from the Cornell University investigation.
5. What is the significance of superplasticized flowing concrete to the construction industry? What changes may have to be made in the mix proportions for flowing concrete?
6. Explain how the concept works for eliminating drying shrinkage cracking by the use of shrinkage-compensating concrete.
7. Compared to normal portland cement concrete of the same water/cement ratio (e.g., 0.6), shrinkage-compensating concrete generally shows much higher strength. Explain the possible reasons for this behavior.
8. What is the principal advantage of using fiber-reinforced concrete? Explain how the concrete acquires this property.
9. Write a short note on the selection of materials and proportioning of mixtures suitable for use in fiber-reinforced concrete, with special attention to how the requirements of toughness and workability are harmonized.
10. What do you understand by the terms *polymer concrete, latex-modified concrete,* and *polymer-impregnated concrete?* What is the principal consideration in the design of polymer concrete mixtures?
11. Compare the technologies of producing latex-modified and polymer-impregnated concretes. Also, compare the typical mechanical properties and durability characteristics of the two concrete types.

---

[99] J. E. Oliverson and A. T. Richardson, *Concr. Int.,* Vol. 6, No. 5, pp. 20–28, 1984.

12. (a) Discuss the two principal problems in radiation shielding. What are the benefits of using concrete as a shielding material?
    (b) Explain why both high strength and high density are sought in concrete for PCRV (prestressed concrete reactor vessels).
13. Define *mass concrete*. What type of cements and admixtures are used for making mass concrete mixtures? From the standpoint of tensile strain capacity, which factors are important in mix proportioning?
14. Write a brief note on the pros and cons of construction practices for controlling temperature rise in concrete.
15. What is roller-compacted concrete? How do the materials and construction technology for building a dam with roller-compacted concrete differ from conventional mass concrete practice?

SUGGESTIONS FOR FURTHER STUDY

General

MALHOTRA, V. M., ed., *Progress in Concrete Technology,* CANMET, Ottawa, 1980.

Lightweight Concrete

ACI Committee 213, Report 213R-87, "Guide for Structural Lightweight Concrete," ACI Manual of Concrete Practice, Part 1, 1991.
SHORT, A., and W. KINNIBURGH, *Lightweight Concrete,* Applied Science Publishers Ltd., Barking, Essex, U.K., 1978.

High-Strength Concrete

ACI Committee 363, Report 363R-84, "State-of-the-Art Report on High-Strength Concrete," *J. ACI,* Proc., Vol. 81, No. 4. pp. 364–411, 1984.
ACI Manual of Concrete Practice, Part 1, 1991.
Proc. High Strength Concrete Conf., ed. W. Hester, ACI SP 121, 1990.

Superplasticized Concrete

ACI, *Superplasticizers in Concrete,* SP-62 (1979), SP-68 (1981), and SP-119 (1989).
Superplasticizers in Concrete, Proc. Symp., Transportation Research Board, Washington, D.C., Transportation Research Record 720, 1979.

Expansive Cement Concrete

ACI, *Expansive Cement,* SP-38, 1973.
ACI, *Expansive Cement,* SP-64, 1980.

ACI Committee 223, "Standard Practice for the Use of Shrinkage-Compensating Concrete," *J. ACI,* Proc., Vol. 73, No. 6, pp. 319–39, 1976, and ACI Manual of Concrete Practice, Part 1, 1991.

Fiber-Reinforced Concrete

ACI, *Fiber Reinforced Concrete,* SP-44, 1974.

ACI Committee 544, Report 544.1R-82, "State-of-the-Art Report on Fiber Reinforced Concrete," *Concr. Int.,* Vol. 4, No. 5, pp. 9–30, 1982, and ACI Manual of Concrete Practice, Part 5, 1991.

ACI Committee 544, Report 544.3R-84, "Guide for Specifying, Mixing, Placing, and Finishing Steel Fiber Reinforced Concrete," *J. ACI,* Proc., Vol. 81, No. 2, pp. 140–48, 1984.

HANNANT, D. J., *Fibre Cements and Fibre Concretes,* John Wiley & Sons, Ltd., 1978.

NEVILLE, A., ed., *Fiber Reinforced Cements and Concrete,* Proceedings of RILEM Symp., The Construction Press, Ltd., London, 1975.

Concrete Containing Polymers

ACI, *Polymers in Concrete,* SP-40, 1973.

ACI, *Polymers in Concrete,* SP-58, 1978.

The Concrete Society, *Polymers in Concrete,* Proceedings of the International Congress on Polymer Concrete, The Construction Press, Ltd., London, 1975.

Heavyweight Concrete

ACI, *Concrete for Nuclear Reactors,* SP-34 (3 vols.), 1973.

ASTM, *Radiation Effects and Shielding,* STP 169B, 1979, pp. 420–34.

Mass Concrete

ACI Committee 207 Report on Mass Concrete, ACI Manual of Concrete Practice, Part 1, 1991.

ACI Committee 207, "Cooling and Insulating Systems for Mass Concrete," *Concr. Int.,* Vol. 2, No. 5, pp. 45–64, 1980.

ACI Committee 207, Report 207.5R-89, "Roller Compacted Concrete," *ACI Materials J.,* Vol. 85, No. 5, pp. 400–45, 1988.

CHAPTER 12
Advances In Concrete Mechanics

PREVIEW

The theory of composite materials has been extensively used to model the elastic behavior of advanced ceramics, rocks, soils, etc. In this chapter, the theory will be applied to estimate the elastic moduli of concrete, when the elastic moduli and the volume fractions of cement paste and aggregate are known. Concrete is a porous material with a number of microcracks even before load is applied. The **differential scheme** and the **Mori-Tanaka method** allow the computation of the effect of these imperfections on the elastic moduli of concrete. The concept of upper and lower bounds for elastic moduli is discussed, and the **Hashin-Shtrikman bounds** are presented.

In Chapter 4, the concepts of creep and stress relaxation in viscoelastic materials were introduced. In this chapter, a number of rheological models which aid understanding of the underlying mechanisms are presented. Changes in properties of concrete, with time, makes the task of mathematical modelling more complex. Methods for incorporating the age of concrete into mathematical models are discussed. Some viscoelastic formulations, such as the principle of superposition, are also discussed. Finally, **methods of estimating the creep and shrinkage** are illustrated by the use of CEB-FIP and ACI codes as well as the **Bazant-Panula model**.

The concept of thermal shrinkage was introduced in Chapter 4, whereas the technology to reduce cracking due to thermal stresses was discussed in Chapter 11. In this chapter, the finite element method for computing the **temperature distribution in mass concrete** is introduced. Examples of application of this method to a number of practical situations in concrete are also given.

Fracture mechanics of concrete has become a powerful method for studying the behavior of plain and reinforced concrete members in tension. The traditional concept of linear fracture mechanics and its limitations when applied to concrete are discussed in this chapter. The use of the finite element method for cracks in structures is demonstrated. Also presented are nonlinear fracture mechanics models, such as the **fictitious crack model, and the smeared crack model**, together with fracture resistance curves. The importance of size effect is illustrated, using the smeared crack band model.

ELASTIC BEHAVIOR

In this section some fundamental approaches to describe composite materials are introduced. In particular, they can be employed to estimate the elastic moduli of concrete, if the elastic moduli of the concrete phases and their volume fractions are known. In Chapter 4, some relationships between elastic modulus and compressive strength, as recommended by ACI and CEB-FIP were presented. These relationships are empirical and, by and large, were developed from experimental results on normal structural concretes with compressive strength ranging from 3000 to 6000 psi (21 to 42 MPa). The difference between predicted and measured values using this type of formulation can be significant, and special care should be taken when estimating the elastic modulus for different concrete types such as high-strength concrete and mass concrete (see Chapter 11). For instance, with mass concrete the maximum size aggregate and the volume fractions of aggregate and cement paste are quite different from structural concrete. Therefore, the prediction of the elastic modulus using the ACI equation will be unreliable. Now, suppose you are the designer of a large concrete dam performing a preliminary thermal stress analysis. Unfortunately, in most cases, experimental results are not available at this stage of the project, although an estimation of the elastic modulus of concrete is crucial for the prediction of thermal stresses. The solution to this problem consists of obtaining an estimate of the elastic properties using a composite materials formulation that incorporates the elastic moduli and the volume fractions of the cement paste and aggregate, as described later in this chapter.

The two simplest models used to simulate a composite material are shown in Fig. 12–1a, b. In the first model the phases are arranged in a parallel configuration, imposing a condition of uniform strain. This arrangement is often referred to as the **Voigt model**. In the second model, the phases are arranged in a series configuration imposing a condition of uniform stress. In the literature this geometry is known as the Reuss model.

Let us start solving the Voigt model using a simple strength of materials approach. For a first approximation, lateral deformations are neglected. The following equations are obtained:

Equilibrium equation $\qquad \sigma A = \sigma_1 A_1 + \sigma_2 A_2 \qquad$ (12-1)

Compatibility equation $\qquad \epsilon = \epsilon_1 = \epsilon_2 \qquad$ (12-2)

Constitutive relationship $\qquad \sigma = E\epsilon \qquad$ (12-3)

Substituting Eqs. (12-3) into (12-1)

$$E\epsilon A = E_1 \epsilon_1 A_1 + E_2 \epsilon_2 A_2 \qquad (12\text{-}4)$$

Elastic Behavior

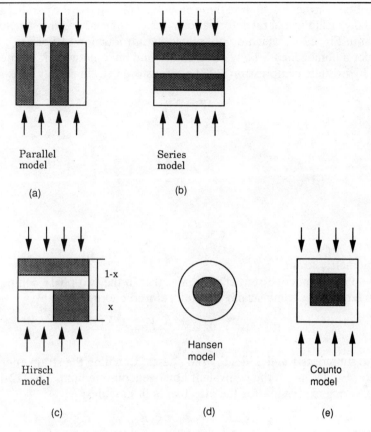

Figure 12–1 Some two-phase models for concrete.

Using the compatibility Eq. (12-2), Eq. (12-4) reduces to:

$$EA = E_1 A_1 + E_2 A_2$$

For composite materials it is more convenient to deal with volume than area, therefore, for unit length:

$$EV = E_1 V_1 + E_2 V_2$$

or

$$E = E_1 c_1 + E_2 c_2 \qquad (12\text{-}5)$$

where $c_i = V_i/V$ is the volume fraction of the ith phase. Using the same approach to solve for the series (Reuss) model:

$$\frac{1}{E} = \frac{c_1}{E_1} + \frac{c_2}{E_2} \qquad (12\text{-}6)$$

To obtain further insight into these models let us rederive the parallel and series models, now including lateral deformations. Instead of studying the structural models shown in Fig. 12–1, which do not allow the introduction of Poisson's ratio, let us consider a homogeneous body of volume V and bulk modulus K, subjected to a uniform hydrostatic compression P. The total stored strain energy W is given by:

$$W = \frac{P^2 V}{2K} \tag{12-7}$$

or,

$$W = \frac{\epsilon^2 K V}{2} \tag{12-8}$$

where

$$\epsilon = \frac{dV}{V} = -\frac{P}{K} \tag{12-9}$$

is the volumetric strain.

The parallel model essentially assumes that in the two-phase composite each phase undergoes the same strain. The total stored energy is

$$W = W_1 + W_2 = \frac{\epsilon^2 K_1 V_1}{2} + \frac{\epsilon^2 K_2 V_2}{2} \tag{12-10}$$

where the subscripts 1 and 2 identify the phases. Equating the strain energy in the composite, Eq. (12-8), to the equivalent homogeneous medium, Eq. (12-10), leads to the following expression for the effective bulk modulus:

$$K = c_1 K_1 + c_2 K_2 \tag{12-11}$$

A similar expression can be obtained for the effective shear modulus G. The effective modulus of elasticity can be calculated from Eq. (12-11) in combination with the following relations from the theory of elasticity:

$$E = \frac{9KG}{3K + G} = 2G(1 + v) = 3K(1 - 2v) \tag{12-12}$$

Therefore, from Eq. (12-11) and (12-12), the effective modulus of elasticity for the parallel model is given by:[1]

$$E = c_1 E_1 + c_2 E_2 + \frac{27 c_1 c_2 (G_1 K_2 - G_2 K_1)^2}{(3K_v + G_v)(3K_1 + G_1)(3K_2 + G_2)} \tag{12-13}$$

where K_v and G_v refer to the values obtained using the Voigt model. For the special case where the two phases have the same Poisson's ratio, Eq. (12-13) reduces to Eq. (12-5), $E = E_1 c_1 + E_2 c_2$, which was obtained neglecting lateral deformations.

The series model assumes that the stress state in each phase will be a uniform

---

[1] G. Grimvall, *Thermophysical Properties of Materials*, North-Holland, 1986.

hydrostatic compression of magnitude P. The total stored energy for the composite is given by:

$$W = W_1 + W_2 = \frac{P^2 V_1}{2K_1} + \frac{P^2 V_2}{2K_2} = \frac{P^2}{2}\left[\frac{V_1}{K_1} + \frac{V_2}{K_2}\right] \qquad (12\text{-}14)$$

The effective bulk modulus can be obtained by equating Eqs. (12-7) and (12-14):

$$\frac{1}{K} = \frac{c_1}{K_1} + \frac{c_2}{K_2} \qquad (12\text{-}15)$$

Using the relationships for elastic modulus given by Eq. (12-12), Eq. (12-15) can be rewritten as

$$\frac{1}{E} = \frac{c_1}{E_1} + \frac{c_2}{E_2} \qquad (12\text{-}16)$$

Note that Eq. (12-16) is the same as Eq. (12-6), which was obtained neglecting lateral deformations.

Neither the Voigt nor the Reuss model can be precisely correct, except in the special case where the moduli of the two materials are equal. This is because the equal-stress assumption satisfies the stress equations of equilibrium, but, in general, gives rise to displacements that are discontinuous at the interfaces between the two phases. Similarly, the equal-strain assumption leads to an admissible strain field, but the resulting stresses are discontinuous.

An important result was obtained by Hill,[2] who, using various energy considerations from the theory of elasticity, showed that the parallel and series assumptions lead to upper and lower bounds on G and K. This result is significant because, given the elastic moduli of the phases and their volume fractions, it allows the determination of the maximum and minimum value that the concrete elastic moduli can have. If the maximum and minimum values are close, the problem would be solved from an engineering point of view. However, when hard inclusions are dispersed in a softer matrix the maximum and minimum values are far apart, as shown in Fig. 12–2. Therefore, there is a strong motivation to establish stricter upper and lower bounds, such as the Hashin-Shtrikman bounds, which will be discussed later, after a brief review of more elaborate models.

Hirsch[3] proposed a **model** (Fig. 12–1c) and a relationship which relates the modulus of elasticity of concrete to the moduli of the two phases (aggregate and matrix), their volume fractions and an empirical constant, x.

$$\frac{1}{E_c} = (1-x)\left[\frac{c}{E_a} + \frac{1-c}{E_m}\right] + x\left[\frac{1}{cE_a + (1-c)E_m}\right] \qquad (12\text{-}17)$$

where

$$c = \frac{V_a}{V_c}$$

---

[2] R. Hill, *Proceedings of the Physical Society of London*, 65A, 349, 1952.
[3] T. J. Hirsch, *ACI Journal*, 59, 427, 1962.

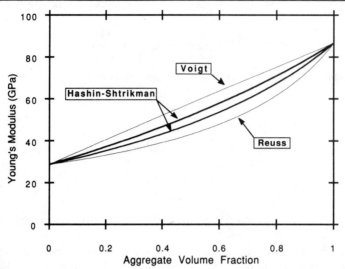

Figure 12-2 Bounds for Young's modulus (elastic moduli of the matrix: $E_m = 28.7$ GPa, $K_m = 20.8$ GPa and elastic moduli of the aggregate: $E_a = 86.7$ GPa, $K_a = 44$ GPa).

In this case x and $(1 - x)$ indicate the relative contributions of the parallel (uniform strain) and series (uniform stress) models. Researchers have reported the use of the Hirsch model to estimate the degree of bonding between the cement paste matrix and aggregate in concrete. It is assumed that perfect bonding would lead to the uniform strain situation, whereas a complete lack of bonding would lead to the uniform stress situation. However, this assumption is not correct because the uniform stress model ($x = 0$) does not imply *no bond* between the phases, and the uniform strain model ($x = 1$) does not indicate a *perfect bond*. It is perhaps more appropriate to view Eq. (12-17) as merely providing a result that lies somewhere between the parallel and the series predictions. For practical application, the value 0.5 for x is often recommended, which gives the arithmetic average of the parallel and series moduli.

Hansen[4] proposed a **model** (Fig. 12-1d) which consists of spherical aggregate located at the center of a spherical mass of matrix, Fig. 12-1. This model was based on a general formulation by Hashin,[5] which, for the particular case when the Poisson's ratio of both phases is equal to 0.2 yields:

$$E = \left[\frac{c_1 E_1 + (1 + c_2)E_2}{(1 + c_2)E_1 + c_1 E_2}\right] E_1 \qquad (12\text{-}18)$$

where phase 1 corresponds to the matrix and phase 2 to the aggregate.

---

[4] T. C. Hansen, *ACI Journal*, Vol. 62, No. 2, pp. 193–216, February 1965.

[5] Z. Hashin, *Journal of Applied Mechanics*, Vol. 29, No. 1, pp. 143–150, March 1962.

Counto[6] considered the case (Fig. 12–1e) where an aggregate prism is placed at the center of a prism of concrete, both having the same ratio of height to area of cross section. By simple strength of materials approach, it can be shown that the modulus of elasticity for the concrete is given by:

$$\frac{1}{E} = \frac{1 - \sqrt{c_2}}{E_1} + \frac{1}{\left(\frac{1 - \sqrt{c_2}}{\sqrt{c_2}}\right)E_1 + E_2} \qquad (12\text{-}19)$$

again, phase 1 corresponds to the matrix and phase 2 to the aggregate.

Let us now return to your original assignment of computing the elastic modulus of concrete. The task has become quite complicated because of the large number of models from which to choose. Selecting the most appropriate model to use may be challenging. Fortunately, for mass concrete, where a large amount of aggregate is present, the predictions are not far apart. Take, for example, a typical lean concrete mixture with 75 percent of aggregate and 25 percent of cement paste by volume. For a given age, assuming the elastic modulus of the cement paste and aggregate (say a quartzite) to be 20 GPa and 45 GPa respectively, the predicted value of the elastic modulus of the concrete from the Voigt model is 38.8 GPa and from the Reuss model 34.3 GPa. No matter how sophisticated or simple the model may be, its prediction should lie inside these bounds. The Hirsch, Hansen and Counto models predict 36.5, 36.4, and 36.2 GPa for the concrete, so, for practical purposes, the three models estimate the same elastic modulus for this particular example. Other cases of estimating elastic moduli may be more problematic.

The models presented so far have limited success in computing the effect of voids, cracks and phase changes (such as water to ice during freezing of the cement paste). Another shortcoming of these methods is that they do not take into account any of the specific geometrical features of the phases, or how the pores and aggregate particles interact with one another under various loading conditions. As a general rule for two-component materials, the effect of the shape of the inclusion is more important when the two components have a vastly different moduli, but is of minor importance when the two components have roughly equal moduli. Hence, when trying to estimate the moduli of a mixture of cement paste and aggregate, we can use models that ignore aggregate shape. When studying the effect of pores and cracks, however, more sophisticated models are needed that explicitly consider the shape of the "inclusions."

As mentioned previously, when estimating the effect of pores or cracks on the elastic moduli, we need models that account for microgeometry. Among the many models that have been proposed for this purpose, two of the most accurate are the **differential scheme** and the **Mori-Tanaka method**. Discussion on the rationale behind these methods, and their detailed predictions, is beyond the scope of this chapter. However, for two important idealized pore shapes, namely the sphere and the "penny-shaped" crack, the results have a relatively simple form, as will now be described.

---

[6] U. J. Counto, *Magazine of Concrete Research,* Vol. 16, No. 48, 964.

If a solid body of modulus E_o and Poisson's ratio v_o contains a volume fraction c of spherical pores, its overall moduli E will be:[7]

Differential method: $\qquad\qquad E = E_o(1 - c)^2 \qquad\qquad$ (12-20)

Mori-Tanaka method: $\qquad\qquad E = E_o(1 - c)/(1 + \alpha c) \qquad\qquad$ (12-21)

where $\alpha = 3(1 + v_o)(13 - 15v_o)/6(7 - 5v_o)$. The parameter α is nearly independent of v_o, and is approximately equal to 1. Therefore, the Mori-Tanaka method essentially predicts $E = E_o(1 - c)/(1 + c)$. The differential method also predicts a slight, but algebraically complicated, dependence of E on v_o. This dependence is negligible for practical purposes, and has been omitted from equation (12-20).

If the body is filled with circular cracks, it is not convenient to quantify their concentration using volume fractions, since a small volume fraction of very thin cracks can cause an appreciable degradation of the moduli. Instead, we use the crack-density parameter Γ, which is defined by $\Gamma = Na^3/V$, where a is the radius of the crack in its plane, and N/V is the number of cracks per unit volume. The effective moduli of a body containing a density Γ of circular cracks will be:

Differential method: $\qquad\qquad E = E_o e^{-16\Gamma/9} \qquad\qquad$ (12-22)

Mori-Tanaka method: $\qquad\qquad E = E_o/(1 + \beta\Gamma) \qquad\qquad$ (12-23)

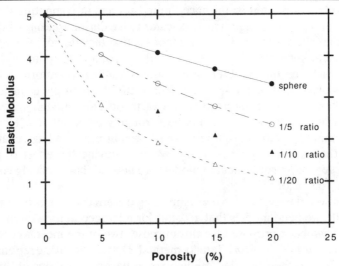

Figure 12-3 Effect of porosity and shape of pores on the elastic modulus of a material.

---

[7] R. W. Zimmerman, *Mechanics of Materials*, Vol. 12, pp. 17-24, 1991.

where $\beta = 16(10 - 3v_o)(1 - v_o^2)/45(2 - v_o)$. For typical values of v_o, β is essentially equal to $16/9 = 1.78$. Again, the differential methods predicts a negligible dependence of E on v_o, which we have omitted from equation (12-20).

More general treatments of the effect of pores on the elastic moduli assume that the pores are oblate spheroids of a certain aspect ratio. The sphere (aspect ratio = 1) and the crack (aspect ratio = 0) are the two extreme cases. Figure 12–3 shows the elastic moduli as a function of porosity, for various pore aspect ratios, as calculated from the Mori-Tanaka model. The detailed equations used to generate this figure can be found in the references given at the end of the chapter.

Hashin-Shtrikman (H-S) Bounds

The Voigt and Reuss models produce an upper and lower bounds for the elastic moduli. However, as shown in Fig. 12–2, the bounds are often far apart, in which case they will be of little use for certain specific cases. For instance, assuming a volume fraction of 0.6 in Fig. 12–2, the upper and lower bounds are 63.9 and 47.9 MPa, respectively. The spread is large and, therefore, of limited use for engineering decisions. Fortunately Hashin and Shtrikman (H-S) developed more stringent bounds for a composite material which is, in a statistical sense, isotropic and homogeneous. The H-S bounds were derived using variational principles of the linear theory of elasticity for multiphase materials of arbitrary phase geometry. For two-phase composites the expressions take the form:

$$K_{low} = K_1 + \frac{c_2}{\dfrac{1}{K_2 - K_1} + \dfrac{3c_1}{3K_1 + 4G_1}}$$

$$K_{up} = K_2 + \frac{c_1}{\dfrac{1}{K_1 - K_2} + \dfrac{3c_2}{3K_2 + 4G_2}} \quad (12\text{-}24)$$

$$G_{low} = G_1 + \frac{c_2}{\dfrac{1}{G_2 - G_1} + \dfrac{6(K_1 + 2G_1)c_1}{5G_1(3K_1 + 4G_1)}}$$

$$G_{up} = G_2 + \frac{c_1}{\dfrac{1}{G_1 - G_2} + \dfrac{6(K_2 + 2G_2)c_2}{5G_2(3K_2 + 4G_2)}} \quad (12\text{-}25)$$

where K and G are the bulk and shear moduli, respectively. Here $K_2 > K_1$; $G_2 > G_1$. K_{up} and G_{up} refer to the upper bounds and K_{low} and G_{low} to the lower bounds.

Figure 12–2 shows that the H-S bounds are inside the Voigt-Reuss bounds. Taking the previous example, for a volume fraction of 0.6, the H-S bounds give 58.4 and 54.0 MPa. The range is significantly narrower than that obtained from the Voigt-Reuss bounds.

TRANSPORT PROPERTIES

This section has concentrated on the various methods of estimating elastic modulus, however, other important properties can also be predicted using the theorems of composite materials. Consider the following relationships, which have the same mathematical structure:

Electrical conduction: $\quad j = \sigma E$
Thermal conduction: $\quad Q = -\kappa \nabla T$
Dielectric displacement: $\quad D = \epsilon E$
Magnetic induction: $\quad B = \mu H$
Diffusion: $\quad Q = -D \nabla c$

For each of these five transport relationships, the flux vector is related to the driving force vector by a second-order physical property tensor, i.e., a 3×3 matrix $(\sigma, \kappa, \epsilon, \mu, D)$. For isotropic materials, the electrical conductivity σ, the thermal conductivity κ, the dielectric constant ϵ, the magnetic susceptibility μ, and the diffusion constant D reduce to a single constant. It should be noted that the elastic moduli is a fourth order tensor and, even for isotropic materials, contains two independent constants. Any model that can predict, say, diffusion constant D from the individual phases properties, will also be able to predict σ, κ, ϵ, and μ.

Hashin and Shtrikman derived the following bounds for transport constants. For thermal conductivity ($\kappa_2 > \kappa_1$), we have, for the 3-dimensional case:

Upper bound:
$$\kappa_u = \kappa_2 + \frac{c_1}{\dfrac{1}{\kappa_1 - \kappa_2} + \dfrac{c_2}{3\kappa_2}}$$

Lower bound:
$$\kappa_1 = \kappa_1 + \frac{c_2}{\dfrac{1}{\kappa_2 - \kappa_1} + \dfrac{c_1}{3\kappa_1}}$$

The number 3 in the denominator should be replaced by 2 and 1 for two-dimensional or one-dimensional cases, respectively. Similar equations apply for the other transport constants.

VISCOELASTICITY

The one-dimensional viscoelastic behavior of concrete may be studied from two experiments: (a) the creep test, where the stress is kept constant and the increase

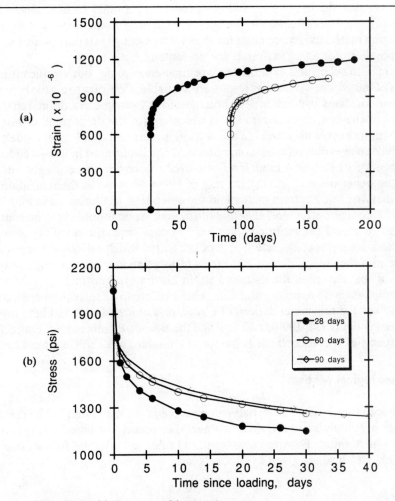

Figure 12–4 (a) Creep test; (b) relaxation test.

(a) creep tests with a constant stress of 2100 psi loaded at 28 and 90 days; (b) relaxation tests performed at 28, 60, and 90 days. All concrete specimens had the same composition. The original data are from K. Thomas and D. Pirtz, "Experimental and Theoretical Studies on Stress Relaxation of Concrete under Uniaxial Compressive Strain," Report No. SESM 81/05, University of California at Berkeley, 1981, and D. Pirtz, K. Thomas and P. J. M. Monteiro, Proceedings of the ACI Journal, Vol. 83, 432, 1986.

in strain over time is recorded, and (b) the relaxation test, where the strain is kept constant and the decrease in stress over time is recorded. Figure 12–4 shows experimental results from creep and relaxation tests. As it can be seen in Fig. 12–4, the creep response is a function of the duration of loading and the age of concrete when the load was applied: the longer the concrete is under load, the greater will be the deformation, and the greater the age of loading, the lower will be the deformation. This behavior classifies concrete as an aging viscoelastic material, which is to be expected because the hydration reactions proceed with time. In fact, most of the mechanical properties of concrete are age-dependent. The mathematical formulation for aging materials is more complex than for nonaging materials; in this section basic expressions for aging materials are presented.

Creep and relaxation experiments are time-consuming, but significant information is obtained when such tests results are available. Contrary to elastic behavior, where two constants can describe a homogeneous isotropic elastic material, for viscoelastic behavior, an evolution law is necessary for the description of how the stress or strain changes over time. In this section, we present rheological models that can produce such evolution laws, some practical equations used in design codes are also discussed. Rheological models will be used to provide some insight into the viscoelastic behavior, such as why the rate of stress decrease in the relaxation test is faster than the rate of strain increase in the creep test, as illustrated in Fig. 12–4.

In real concrete structures the state of stress or strain is unlikely to be constant over time. To model the more complex loading conditions the principle of superposition and integral representations will be presented which will allow computation of the strain if the creep function and stress history are known, or computation of the stress if the relaxation function and strain history are known.

Finally, when no experimental data, either on creep or relaxation are known, one has to use the recommendations of a code or a model. Discussed here are the CEB model codes (1978, 1990), ACI-209 and the Bazant-Panula model. For technological aspects of the viscoelastic behavior of concrete, refer to Chapter 4.

Basic Rheological Models

The behavior of viscoelastic materials can be successfully estimated by the creation of rheological models based on two fundamental elements: the linear spring and the linear viscous dashpot. For the linear spring (Table 12–1(a)), the relationship between stress and strain is given by Hooke's law:

$$\sigma(t) = E\epsilon(t) \quad (12\text{-}24)$$

The response of the spring to the stress is immediate. During a creep test, where the stress, σ_o, is kept constant, the strain will be σ_o/E, constant over time. Similarly, for a relaxation test, where the strain, ϵ_o, is kept constant the stress will be $\epsilon_o E$, constant over time.

The viscous dashpot can be visualized as a piston displacing a viscous fluid in a cylinder with a perforated bottom. Using Newton's law of viscosity:

$$\dot{\epsilon}(t) = \frac{\sigma(t)}{\eta} \quad (12\text{-}25)$$

where

$\dot{\epsilon} = \dfrac{d\epsilon}{dt}$ = the strain rate

η = the viscosity coefficient

Viscoelasticity

TABLE 12-1

Name	Representation	Creep	Relaxation
		σ vs t: constant at σ_0	ξ vs t: constant at ξ_0
(a) Spring	spring, E	ξ vs t: constant at σ_0/E	σ vs t: constant at $E\xi_0$
(b) Dashpot	dashpot, η	ξ vs t: linear increase	σ vs t: impulse spike
(c) Maxwell	spring E in series with dashpot η	ξ vs t: linear increase from nonzero intercept	σ vs t: exponential decay from $\sigma = E\xi_0$
(d) Kelvin	spring and dashpot in parallel	ξ vs t: exponential rise to asymptote	σ vs t: impulse spike
(e) Standard Solid	spring in series with Kelvin element	ξ vs t: exponential rise to asymptote $\dfrac{\sigma}{E_\infty}$	σ vs t: exponential decay

Newton's law states that the strain rate is proportional to the stress; hence, for the creep experiment, the dashpot will deform at a constant rate, as shown in Table 12–1b. However, for a relaxation experiment, with the application of an instantaneous constant strain, the stress will become instantaneously infinite as indicated in Table 12–1b.

Complex formulations can be obtained by combining springs and dashpots in different configurations. One of the simplest combinations consists in assembling one spring and one dashpot in series or in parallel. The **Maxwell model** comprises a linear spring and a linear viscous dashpot connected in series, as shown in Table 12–1c. The following equations apply:

Equilibrium equation $\quad\quad\quad \sigma_E(t) = \sigma_\eta(t) = \sigma(t)$ $\quad\quad\quad$ (12-26)

Compatibility equation $\quad\quad\quad \epsilon(t) = \epsilon_E(t) + \epsilon_\eta(t)$ $\quad\quad\quad$ (12-27)

Constitutive relationship (spring) $\quad\quad \sigma_E(t) = E\epsilon_E(t)$ $\quad\quad\quad$ (12-28)

$\quad\quad\quad\quad\quad$ (dashpot) $\quad\quad \sigma_\eta(t) = \eta\dot{\epsilon}_\eta(t)$ $\quad\quad\quad$ (12-29)

Differentiating Eqs. (12-27) and (12-28) with respect to time t and using Eqs. (12-26) and (12-29):

$$\dot{\epsilon}(t) = \frac{\dot{\sigma}(t)}{E} + \frac{\sigma(t)}{\eta} \quad\quad\quad (12\text{-}30)$$

Note that for a rigid spring ($E = \infty$) the model reduces to a Newtonian fluid; likewise, if the dashpot becomes rigid ($\eta = \infty$) the model reduces to a Hookean spring. The response of the Maxwell model to various kinds of time dependent stress or strain patterns can be determined by solving Eq. (12-30). For instance, consider again a creep test, with the initial conditions $\sigma = \sigma_o$ at $t = 0$. From integration of Eq. (12-30), we obtain:

$$\epsilon(t) = \frac{\sigma_o}{E} + \frac{\sigma_o}{\eta} t \quad\quad\quad (12\text{-}31)$$

The model predicts that the strain increases without bounds. This is a characteristic of many fluids and, for this reason, a material described by Eq. (12-30) is known as a Maxwell fluid. If after the creep experiment, the system is unloaded at time t_1, the elastic strain σ_o/E in the spring recovers instantaneously, whilst permanent strain $(\sigma_o/\eta)t_1$ remains in the dashpot.

In a relaxation experiment, where the strain, ϵ_o, is constant, the model predicts from integration of Eq. (12-30):

$$\sigma(t) = E\epsilon_o e^{-Et/\eta} \quad\quad\quad (12\text{-}32)$$

The ratio $T = \eta/E$ is called the **relaxation time**, and it helps to characterize the viscoelastic response of the material. A small relaxation time indicates that the relaxation process will be fast. In the extreme case of a purely viscous fluid, $E = \infty$, Eq. (12-32) would indicate an infinitely fast stress relaxation, $T = 0$; while for an elastic spring, $\eta = \infty$, the stress would not relax at all, since $T = \infty$.

Viscoelasticity

The **Kelvin model** combines a linear spring and a dashpot in parallel as shown in Table 12–1d. The following equations apply:

Equilibrium equation $\quad\quad\quad\quad \sigma(t) = \sigma_E(t) + \sigma_\eta(t) \quad\quad\quad$ (12-33)

Compatibility equation $\quad\quad\quad \epsilon(t) = \epsilon_E(t) = \epsilon_\eta(t) \quad\quad\quad\;$ (12-34)

Constitutive relationship (spring) $\quad \sigma_E(t) = E\epsilon_E(t) \quad\quad\quad\quad\quad$ (12-35)

$\quad\quad\quad\quad\quad$ (dashpot) $\quad\quad \sigma_\eta(t) = \eta\dot{\epsilon}_\eta(t) \quad\quad\quad\quad\quad$ (12-36)

Resulting in the differential equation

$$\sigma(t) = E\epsilon(t) + \eta\dot{\epsilon}(t) \quad\quad\quad (12\text{-}37)$$

Note that the model reduces to a Hookean spring if $\eta = 0$, and to a Newtonian fluid if $E = 0$. Equation (12-37) may be used to predict strain if the stress history is given, or to predict stress if the strain history is given. For instance, for the creep experiment, integrating Eq. (12-37) with the boundary condition $\sigma = \sigma_o$ at time $t_o = 0$ yields:

$$\epsilon(t) = \frac{\sigma_o}{E}(1 - e^{-Et/\eta}) \quad\quad\quad (12\text{-}38)$$

In Eq. (12-38), the strain increases at a decreasing rate and has an asymptotic value of σ_o/E, as shown in Table 12–1d. During the creep test, the stress is initially carried by the dashpot and, as time goes by, the stress is transferred to the spring. Analogous to the relaxation time, it is convenient to define the **retardation time** $\theta = \eta/E$. A small retardation time indicates that the creep process will be fast. In the extreme case of an elastic spring ($\eta = 0$), the final strains would be obtained instantaneously, since $\theta = 0$.

The Kelvin model requires an infinite stress to produce the instantaneous strain necessary for the relaxation test, which makes it physically impossible to perform.

The Maxwell and Kelvin models have significant limitations in representing the behavior of most viscoelastic materials. As discussed before, the Maxwell model shows a constant strain rate under constant stress, which may be adequate for fluids, but not for solids. The Kelvin model cannot predict a time-dependent relaxation and does not show a permanent deformation upon unloading.

The next step in complexity is the **standard solid** model, where a spring is connected in series with a Kelvin element as shown in Table 12–1e.

Assuming ϵ_1 and ϵ_2 to be the strain in the spring and Kelvin elements respectively, the total strain, for the standard solid, is given by:

$$\epsilon = \epsilon_1 + \epsilon_2 \quad\quad\quad (12\text{-}39)$$

Since the stress in the spring and the Kelvin element is the same, the stress can be determined from Eq. (12-37)

$$\sigma(t) = E_2\epsilon_2(t) + \eta\frac{\partial\dot{\epsilon}_2(t)}{\partial t} \quad\quad\quad (12\text{-}40)$$

where $\partial/\partial t$ is a differential operator that may be handled as a algebraic entity,

$$\sigma(t) = \epsilon_2(t)\left(E_2 + \eta\frac{\partial}{\partial t}\right) \qquad (12\text{-}41)$$

leading to

$$\epsilon_2(t) = \frac{\sigma(t)}{\left(E_2 + \eta\dfrac{\partial}{\partial t}\right)} \qquad (12\text{-}42)$$

Therefore, from Eq. (12-39) the strain for the standard solid is,

$$\epsilon(t) = \frac{\sigma(t)}{E_1} + \frac{\sigma(t)}{\left(E_2 + \eta\dfrac{\partial}{\partial t}\right)} \qquad (12\text{-}43)$$

or,

$$E_1\epsilon(t)\left(E_2 + \eta\frac{\partial}{\partial t}\right) = E_1\sigma(t) + \sigma(t)\left(E_2 + \eta\frac{\partial}{\partial t}\right) \qquad (12\text{-}44)$$

which leads to the differential equation

$$\eta E_1\dot\epsilon(t) + E_1 E_2\epsilon(t) = \eta\dot\sigma(t) + (E_1 + E_2)\sigma(t) \qquad (12\text{-}45)$$

Equation (12-45) can be integrated for an arbitrary stress history,

$$\epsilon(t) = \frac{\sigma(t)}{E_1} + \frac{1}{\eta}\int_0^t \sigma(\tau)e^{-E_2(t-\tau)/\eta}\,d\tau \qquad (12\text{-}46)$$

For the particular case of the creep experiment, Eq. (12-46) reduces to:

$$\epsilon(t) = \frac{\sigma_o}{E_1} + \frac{\sigma_o}{E_2}[1 - e^{-E_2 t/\eta}] \qquad (12\text{-}47)$$

which can be rewritten as:

$$\epsilon(t) = \sigma_o\left[\frac{E_1 + E_2}{E_1 E_2} - \frac{1}{E_2}e^{-E_2 t/\eta}\right] \qquad (12\text{-}48)$$

Equation (12-48) indicates that the strain is proportional to σ_o, changing from σ_o/E_1 at $t = 0$, to σ_o/E_∞ at $t = \infty$. E_∞ is called the asymptotic modulus and is given by:

$$E_\infty = \frac{E_1 E_2}{E_1 + E_2} \qquad (12\text{-}49)$$

Therefore, during the creep test, the elastic modulus of the standard solid model $E_c(t)$ reduces from the initial value E_1 to its asymptotic value E_∞, according to the following evolution law:

$$\frac{1}{E_c(t)} = \frac{\epsilon(t)}{\sigma_o} = \frac{E_1 + E_2}{E_1 E_2} - \frac{1}{E_2}e^{-E_2 t/\eta} \qquad (12\text{-}50)$$

Viscoelasticity

We will now integrate Eq. (12-45) for an arbitrary strain history.

$$\sigma(t) = \epsilon(t)E_\infty + (E_1 - E_\infty)\int_0^t e^{-(E_1 + E_2)(t - \tau)/\eta}\dot{\epsilon}(\tau)d\tau \qquad (12\text{-}51)$$

For the particular case of relaxation experiment the stress evolution is given by:

$$\sigma(t) = \epsilon_o[E_\infty + (E_1 - E_\infty)e^{-(E_1 + E_2)t/\eta}] \qquad (12\text{-}52)$$

Equation (12-52) indicates that the stress is proportional to ϵ_o, changing from $E_1\epsilon_o$ at $t = 0$, up to $E_\infty\epsilon_o$ at $t = \infty$. Therefore, during the relaxation test, the elastic modulus $E_r(t)$ reduces from the initial value E_1 to its asymptotic value E_∞, according to the following law:

$$E_r(t) = [E_\infty + (E_1 - E_\infty)e^{-(E_1 + E_2)t/\eta}] \qquad (12\text{-}53)$$

Even though both creep and relaxation may be understood as a decrease in elastic modulus over time from E_1 to its asymptotic value E_∞, Eqs. (12-50) and (12-53) have different rates of decrease. In the relaxation test, the decrease in the elastic modulus occurs at a significantly faster rate than in the creep test. As an example, let us take the following values for concrete: $E_1 = 35$ GPa, $E_2 = 18$ GPa, $T(E_2/\eta) = 1/300$ days. Figure 12–5 illustrates the faster reduction of elastic modulus during relaxation than for the creep test.

Example 1

The testing of materials is usually performed either by controlling the stress or strain rate. Study the response of a standard solid model loaded under these conditions. Solve the problem analytically and then expand the discussion for instantaneous, slow, and

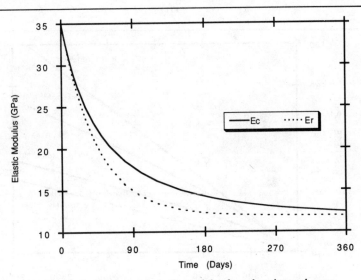

Figure 12–5 Decrease of elastic modulus for relaxation and creep.

medium stress and strain rates; assume the following properties for the standard solid: $E_1 = 35$ GPa, $E_2 = 18$ GPa, $T = 1$ min

(*A*) Test with a constant stress rate (*v*): The stress increases linearly with time, according to:

$$\sigma(t) = vt \tag{12-54}$$

The strain in the standard solid model is obtained by combining Eqs. (12-46) and (12-54),

$$\epsilon(t) = \frac{vt}{E_1} + \frac{v}{\eta} \int_0^t \tau e^{-E_2(t-\tau)/\eta} d\tau \tag{12-55}$$

which leads to

$$\epsilon(t) = \frac{vt}{E_\infty} - \frac{v\eta}{E_2^2}(1 - e^{-E_2 t/\eta}) \tag{12-56}$$

Figure 12–6 presents the stress [Eq. (12-54)] in function of strain [Eq. 12-56] using the given material properties. It shows that the stress-strain diagram is strongly influenced by the rate of loading. Note that the stress-strain diagram may be non-linear, a common feature for viscoelastic materials where the strain at a given time is influenced by the entire stress history. This phenomenon will be presented in more detail in the following sections. The stress-strain relationships in Fig. 12–6 are bounded by the very slow and very fast rates. The latter gives the upper bound and physically corresponds to the linear spring (E_1) absorbing all the stress, because the Kelvin element has no time to deform. For very slow rates, the standard solid models responds with the asymptotic modulus E_∞ and physically corresponds to the linear spring (E_1) in series with the spring from the Kelvin element (E_2), the dashpot does not contribute to the stiffness of the system.

(*B*) *Test with constant strain rate*: The strain increases with time, according to:

$$\epsilon(t) = vt \tag{12-57}$$

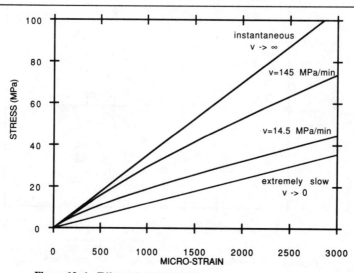

Figure 12–6 Effect of stress rate on the stress-strain diagram.

Viscoelasticity

The stress in the model is obtained by combining Eqs. (12-51) and (12-57),

$$\sigma(t) = vtE_\infty + (E_1 - E_\infty) \int_0^t ve^{-(E_1 + E_2)(t - \tau)/\eta} d\tau \tag{12-58}$$

which leads to

$$\sigma(t) = vtE_\infty + (E_1 - E_\infty)v \frac{\eta}{(E_1 + E_2)}(1 - e^{-(E_1 + E_2)t/\eta}) \tag{12-59}$$

Figure 12–7 shows the stress [Eq. (12-59)] in function of strain [Eq. (12-57)] with the specified concrete properties.

Example 2

Study the response of a viscoelastic material subjected to a cyclic strain $\epsilon(t) = \epsilon_0 \cos wt$, where ϵ_0 is the strain amplitude and w the frequency. Write explicit equations for the Maxwell and Kelvin models.

For a linear elastic spring, the stress will be in phase with the cyclic strain [see Eq. (12-5)], that is,

$$\sigma(t) = E\epsilon(t) = E\epsilon_o \cos wt \tag{12-60}$$

For a Newtonian fluid the stress will lead the strain by $\pi/2$ [see Eq. (12-29)]:

$$\sigma(t) = \eta\dot\epsilon(t) = -\eta w\epsilon_o \sin wt = \eta w\epsilon_o \cos(wt + \delta) \tag{12-61}$$

where

$$\delta = \frac{\pi}{2}$$

For a viscoelastic material, the phase difference between stress and strain ranges from 0 to $\pi/2$. A convenient way of representing oscillatory strain is by using the expression:

$$e^{iwt} = \cos wt + i \sin wt \tag{12-62}$$

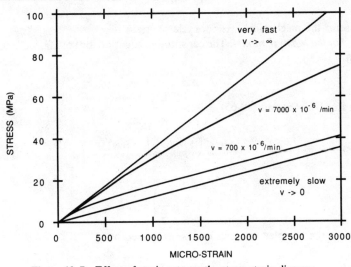

Figure 12–7 Effect of strain rate on the stress-strain diagram.

Taking the real part of the expression, the strain equation can be rewritten

$$\epsilon(t) = \epsilon_o e^{iwt} \tag{12-63}$$

The stress oscillates with the same frequency w, but leads the strain by a phase angle δ

$$\sigma(t) = \sigma_o e^{i(wt + \delta)} \tag{12-64}$$

which can be rewritten as

$$\sigma(t) = \sigma_o e^{i\delta} e^{iwt} = \sigma^* e^{iwt} \tag{12-65}$$

where σ^* is the complex stress amplitude, given by:

$$\sigma^* = \sigma_o e^{i\delta} = \sigma_o(\cos \delta + i \sin \delta) \tag{12-66}$$

A complex modulus, E^*, can be defined as

$$E^* = \frac{\sigma^*}{\epsilon_o} = \frac{\sigma_o(\cos \delta + i \sin \delta)}{\epsilon_o} = E_1 + i E_2 \tag{12-67}$$

where E_1, the storage modulus, is in phase with the strain, and is given by

$$E_1 = \frac{\sigma_o}{\epsilon_o} \cos \delta \tag{12-68}$$

E_2, the loss modulus, is the imaginary part, and is given by:

$$E_2 = \frac{\sigma_o}{\epsilon_o} \sin \delta, \tag{12-69}$$

and the magnitude of the complex modulus is given by:

$$|E^*| = \sqrt{E_1^2 + E_2^2} \tag{12-70}$$

It should be noted that

$$\tan \delta = \frac{E_2}{E_1} \tag{12-71}$$

represents the mechanical loss per cycle of strain.

For the Maxwell model: The constitutive equation for the Maxwell model is given by Eq. (12-30)

$$\sigma + \frac{\eta}{E}\dot{\sigma} = \eta \dot{\epsilon} \tag{12-72}$$

Using Eqs. (12-63) and (12-65)

$$\sigma_o e^{i\delta}\left(1 + iw\frac{\eta}{E}\right) = iw\epsilon_o \eta \tag{12-73}$$

or

$$\sigma^*\left(1 + iw\frac{\eta}{E}\right) = iw\epsilon_o \eta \tag{12-74}$$

Therefore the complex modulus can be expressed by

$$E^* = \frac{\sigma^*}{\epsilon_o} = \frac{iw\eta}{1 + \frac{iw\eta}{E}} \tag{12-75}$$

Viscoelasticity

Separating the real and imaginary parts we find

$$E^* = \frac{\eta^2 w^2/E}{1 + \eta^2 w^2/E^2} + i\frac{\eta w}{1 + \eta^2 w^2/E^2} \qquad (12\text{-}76)$$

hence the magnitude of the complex modulus is given by:

$$|E^*| = w\eta\left(1 + \frac{\eta^2 w^2}{E^2}\right)^{-1/2} \qquad (12\text{-}77)$$

and

$$\tan\delta = \frac{E_2}{E_1} = \frac{E}{w\eta} \qquad (12\text{-}78)$$

If we take the material constants from the previous example we can plot the magnitude of complex modulus against the angular frequency in Fig. 12–8, it can be seen that for very high frequencies the dynamic modulus approaches the spring constant E, and for the low-frequencies the dynamic modulus approaches zero.

For the Kelvin model: The constitutive equation for the Kelvin model is given by Eq. (12-37)

$$\sigma(t) = E\epsilon(t) + \eta\dot\epsilon(t) \qquad (12\text{-}79)$$

Using equations (12–64) and (12–79)

$$\sigma_o e^{i\delta} = \epsilon_o(E + iw\eta) \qquad (12\text{-}80)$$

Therefore the complex modulus is expressed by

$$E^* = \frac{\sigma^*}{\epsilon_o} = E + iw\eta \qquad (12\text{-}81)$$

and the magnitude of the complex modulus by

$$|E^*| = (E^2 + w^2\eta^2)^{1/2} \qquad (12\text{-}82)$$

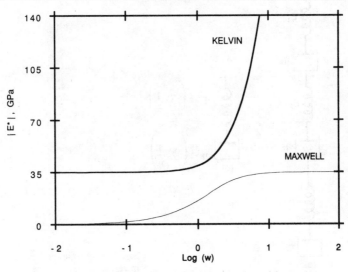

Figure 12–8 Complex elastic modulus in function of frequency.

The mechanical loss for the model is

$$\tan \delta = \frac{\eta}{E} w \qquad (12\text{-}83)$$

Again taking the material constants from the previous example the results for the Kelvin model can be plotted in Fig. 12–8. It can be seen that for low frequencies the dynamic modulus is given by the spring constant E, while for high frequencies the stiffness increases.

The fact that there is a significantly different response for the Maxwell and Kelvin models under oscillatory stress points to the advantage of performing such a test to assess which model is most representative for a specific material.

Generalized Rheological Models

The modelling of viscoelastic behavior can be improved by combining a large number of springs and dashpots in series or in parallel. By adding many elements, several relaxation times can be obtained, which is typical of complex materials such as concrete.

When we decide to generalize the Maxwell model we have the following choices: connect the units in series or in parallel. Let us start by studying the response when the units are connected in series, as shown in Fig. 12–9. The constitutive equation has the form:

$$\dot{\epsilon}(t) = \dot{\sigma}(t) \sum_{i=1}^{n} \frac{1}{E_i} + \sigma(t) \sum_{i=1}^{n} \frac{1}{\eta_i} \qquad (12\text{-}84)$$

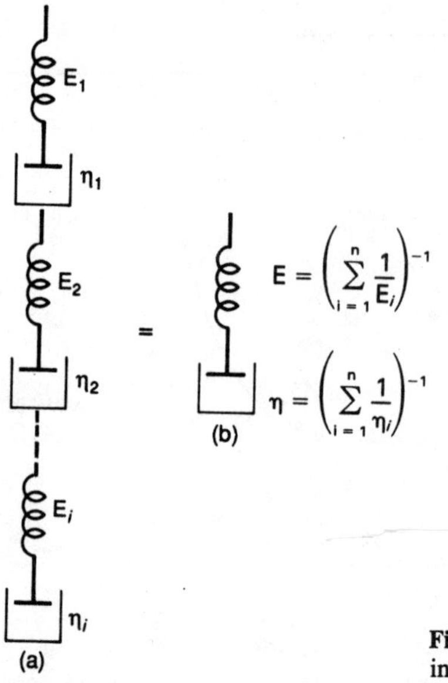

Figure 12–9 Generalized Maxwell model in series.

Viscoelasticity

Where n is the number of elements. The equation is equivalent to Eq. (12-30) consequently the chain of elements is identical to a single Maxwell element as shown in Fig. 12–9b. Therefore not much was accomplished by connecting the units in series. Let us now analyse the response when the units are connected in parallel, as shown in Fig. 12–10b.

The strain in each unit of a generalized Maxwell model in parallel is given by:

$$\frac{\partial}{\partial t}\epsilon_i(t) = \left\{\frac{1}{E_i}\frac{\partial}{\partial t} + \frac{1}{\eta_i}\right\}\sigma_i(t) \qquad (12\text{-}85)$$

The stress for the generalized model is given by:

$$\sigma(t) = \left\{\sum_{i=1}^{n} \frac{\partial/\partial t}{\frac{1}{E_i}\frac{\partial}{\partial t} + \frac{1}{\eta_i}}\right\}\epsilon(t) \qquad (12\text{-}86)$$

and the relaxation function for the generalized Maxwell model is:

$$E(t - \tau) = \sum_{i=1}^{n} E_i\{\exp - (t - \tau)/T_i\} \qquad (12\text{-}87)$$

which indicates that the response of the material depends on a distribution of relaxation times. This formulation is useful in modelling complex viscoelastic materials.

When generalizing the Kelvin model the same question arises: should we

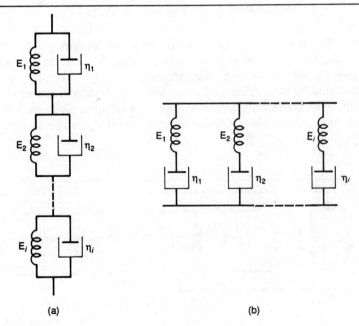

Figure 12–10 (a) Generalized Kelvin model in series and (b) Generalized Maxwell model in parallel.

connect the units in series or parallel? Let us start with the units connected in parallel, as shown in Fig. 12–11. The constitutive equation for the model has the form:

$$\sigma(t) = \epsilon(t) \sum_{i=1}^{n} E_i + \dot{\epsilon}(t) \sum_{i=1}^{n} \eta_i \quad (12\text{-}88)$$

which has the same form as a Kelvin element shown in Fig. 12–11b.

Consider a generalized Kelvin model in series, Fig. 12–10a. The stress in each unit is given by:

$$\sigma_i(t) = \left(E_i + \eta_i \frac{\partial}{\partial t} \right) \epsilon_i(t) \quad (12\text{-}89)$$

The strain for the generalized model is given by:

$$\epsilon(t) = \sum_{i=1}^{n} \left\{ \frac{1}{E_i + \eta_i \dfrac{\partial}{\partial t}} \right\} \sigma(t) \quad (12\text{-}90)$$

Equations (12-85) and (12-90) are differential equations of the general form:

$$\sum_{i=1}^{h} p_i \frac{d^i \sigma}{dt^i} = \sum_{i=1}^{k} q_i \frac{d^i \epsilon}{dt^i} \quad (12\text{-}91)$$

The specific creep function for the generalized Kelvin model in series is:

$$\Phi(t - \tau) = \sum_{i=1}^{n} \frac{1}{E_i} \{ 1 - \exp(-t - \tau)/\theta_i \} \quad (12\text{-}92)$$

To be useful for modelling the material's response the spring constants E_i and the dashpot constants η_i should vary over a large range. Sometimes to model a fluid or a solid it is convenient to take some limiting value for the spring or dashpot constant. It should be noted that a Maxwell model with infinite spring constant or a Kelvin model with zero spring constant becomes a dashpot. Conversely a Maxwell model with infinite viscosity or a Kelvin model with zero viscosity results in a spring.

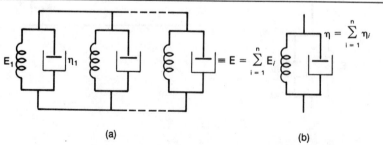

(a) (b)

Figure 12–11 Generalized Kelvin model in parallel.

Viscoelasticity

Time-Variable Rheological Models

Concrete changes its mechanical properties with time due to hydration reaction. However, in the models presented so far, the elastic modulus E and the viscosity coefficient η are constant over time. Consequently they have limited success in modelling the complex response of concrete. To include aging of concrete, we will now study how the differential equations for the basic elements—spring and dashpot—change when their mechanical properties change with time.

Consider a linear spring with elastic modulus varying in time. Hooke's law can be expressed in two forms:

$$\sigma(t) = E(t)\epsilon(t) \tag{12-93}$$

and

$$\dot{\sigma}(t) = E(t)\dot{\epsilon}(t) \tag{12-94}$$

The equations are not equivalent. In the solid mechanics literature, a body following Eq. (12-93) is said to be elastic, whereas a body following Eq. (12-94) is said to be hypoelastic.

A linear viscous dashpot with viscosity coefficient varying in time is expressed unequivocally by:

$$\sigma(t) = \eta(t)\dot{\epsilon}(t) \tag{12-95}$$

Let us reconstruct the previous models (Maxwell, Kelvin, standard-solid, generalized) for aging materials such as concrete.

The equations for a Maxwell element with a hypoelastic spring or with an elastic spring are given by:

$$\dot{\epsilon}(t) = \frac{\dot{\sigma}(t)}{E(t)} + \frac{\sigma(t)}{\eta(t)} \tag{12-96}$$

$$\dot{\epsilon}(t) = \frac{\dot{\sigma}(t)}{E(t)} + \left(\frac{1}{\eta(t)} + \frac{d}{dt}\frac{1}{E(t)}\right)\sigma(t) \tag{12-97}$$

Note that Eqs. (12-96) and (12-97) may be expressed as:

$$\dot{\epsilon}(t) = q_o(t)\dot{\sigma}(t) + q_1(t)\sigma(t) \tag{12-98}$$

where $q_o(t), q_1(t)$ are independent functions of time. Equation (12-98) is able to represent the constitutive law for the Maxwell model with either an elastic or a hypoelastic spring.

Dischinger[8] used the aging Maxwell element, Eq. (12-96), to derive the so-called rate-of-creep method. The specific creep function $\Phi(t, \tau)$, that is the strain per unit stress at time t for the stress applied at age τ is given by:

$$\Phi(t, \tau) = \frac{1}{E(\tau)} + \int_\tau^t \frac{d\tau'}{\eta(\tau')} \tag{12-99}$$

[8] F. Dischinger, *Der Bauingenieur*, Vol. 18:487–520, 539–62, 595–621, 1937.

The creep coefficient $\varphi(t, \tau)$ representing the ratio between the creep strain and the initial elastic deformation is

$$\varphi(t, \tau) = E(\tau) \int_\tau^t \frac{d\tau'}{\eta(\tau')} \qquad (12\text{-}100)$$

Equation (12-96) can be expressed in function of the creep coefficient as

$$\frac{\partial \epsilon}{\partial \varphi} = \frac{1}{E(t)} \frac{\partial \sigma}{\partial \varphi} + \frac{\sigma}{E(\tau)} \qquad (12\text{-}101)$$

The Dischinger formulation implies that the creep curves are parallel for all ages. Experimental results do not indicate that is assumption is valid, as it can be seen in Fig. 12–4a, where the creep curves are not parallel. Usually this method substantially underestimates the creep for stresses applied at ages greater than τ.

A Kelvin element with an elastic spring is described by:

$$\sigma(t) = E(t)\epsilon(t) + \eta(t)\dot{\epsilon}(t) \qquad (12\text{-}102)$$

and for a hypoelastic spring,

$$\dot{\sigma}(t) = [E(t) + \dot{\eta}(t)]\dot{\epsilon}(t) + \eta(t)\ddot{\epsilon}(t) \qquad (12\text{-}103)$$

Equations (12-102) and (12-103) are not equivalent.

Let us now consider the standard solid. Previously, for nonaging materials, we solved the model for the Kelvin element in series with a spring. The same differential equation would have been obtained for a Maxwell element in parallel with a spring. For aging materials, the number of combinations for the standard solid greatly increases, as indicated in Table 12–2.[9]

with the notation:

E = elastic spring
H = hypoelastic spring
Ke, Kh = Kelvin element with elastic and hypoelastic spring, respectively
M = Maxwell element (Note Eq. (12-98) satisfies both springs)
—, // = series and parallel configurations

As an example, let us solve case (a) which was previously analyzed for a nonaging material. In this model we have an elastic spring with an elastic modulus $E_1(t)$ in series with a Kelvin model with an elastic modulus $E_2(t)$ and a dashpot of viscosity $\eta(t)$.

TABLE 12–2

(a) E—Ke	(b) E—Kh
(c) H—Kh	(d) H—Ke
(e) M//E	(f) M//H

[9] J. Lubliner, *Nuclear Engineering and Design*, Vol. 4, 287, 1966.

Viscoelasticity

Let ϵ_1 and ϵ_2 denote the strains of the spring and of the Kelvin element, respectively, then:

$$\epsilon(t) = \epsilon_1(t) + \epsilon_2(t) \qquad \epsilon_1(t) = \sigma(t)/E_1(t) \tag{12-104}$$

$$E_2(t)\epsilon_2(t) + \eta(t)\dot{\epsilon}_2(t) = \sigma(t) \tag{12-105}$$

Eliminating ϵ_1 and ϵ_2, we obtain:

$$E_2(t)\epsilon(t) + \eta(t)\dot{\epsilon}(t) = \left[1 + \frac{E_2(t)}{E_1(t)} + \eta \frac{d}{dt}\frac{1}{E_1(t)}\right]\sigma(t) \\ + \frac{\eta(t)}{E_1(t)}\dot{\sigma}(t) \tag{12-106}$$

These aging models can be generalized to obtain the following differential constitutive equation:

$$\left[\frac{d^n}{dt^n} + p_1(t)\frac{d^{n-1}}{dt^{n-1}} + \cdots + p_n(t)\right] \\ \epsilon(t) = \left[q_o(t)\frac{d^n}{dt^n} + q_1(t)\frac{d^{n-1}}{dt^{n-1}} + \cdots + q_n(t)\right]\sigma(t) \tag{12-107}$$

It should be mentioned that models having two or more Maxwell elements in parallel, or Kelvin elements in series, will not, in general, lead to a differential equation, but rather to an integro-differential equation.

Superposition Principle and Integral Representation

In the lifetime of a concrete structure, it is unlikely that the load will be kept constant as in a creep test, nor, will the strain be kept constant, as in a relaxation test. In order to estimate the strain at a given time from a known stress history further assumptions are necessary. McHenry made a significant contribution by postulating the principle of superposition, which states:[10]

> The strains produced in concrete at any time t by a stress increment at any time t_o are independent of the effects of any stress applied either earlier or later than t_o. The stresses which approach the ultimate strength are excluded.

Experimental results indicate that the principle of superposition works well for sealed concrete specimens, that is for basic creep. When creep is associated with drying shrinkage other methods should be used.

The principle of superposition may also be formulated as "the effect of sum of causes is equal to sum of effects of each of these causes."[11] Consider $\epsilon_1(\tau)$ and $\epsilon_2(\tau)$,

---

[10] D. A. McHenry, *ASTM Proc.*, Vol. 43 (1943).
[11] M. G. Sharma, *Viscoelasticity and Mechanical Properties of Polymers*, University Park, Pennsylvania, 1964.

the strains resulting from the stress history $\sigma_1(\tau)$ and $\sigma_2(\tau)$ respectively. For a *linear* viscoelastic material we can simply add the two stress histories,

$$\sigma(\tau) = \sigma_1(\tau) + \sigma_2(\tau) \tag{12-108}$$

Using the principle of superposition, the following strain history is obtained:

$$\epsilon(\tau) = \epsilon_1(\tau) + \epsilon_2(\tau) \tag{12-109}$$

We will show next that using the principle of superposition and a known creep function, the strain at any time can be determined for a given stress history. For a creep test we may write the strain $\epsilon(t)$ as a function of the stress σ_o, time t and age of loading τ, that is,

$$\epsilon(t) = \Phi(\sigma_o, t, \tau) \tag{12-110}$$

In the linear range Eq. (12-110) may be written as

$$\epsilon(t) = \sigma_o \Phi(t, \tau) \tag{12-111}$$

where $\Phi(t, \tau)$ is the specific creep function.

The graph in Fig. 12–12 shows an arbitrary stress changing with time. Breaking the stress history up into small intervals we have:

$$\sigma(t) \cong \sum_{i=0}^{n} \Delta\sigma(\tau_i), \qquad \tau_n = t \tag{12-112}$$

Using Eq. (12-111), the strain history is given by

$$\epsilon(t) \cong \sum_{i=0}^{n} \Delta\sigma(\tau_i)\Phi(t, \tau) \tag{12-113}$$

and, in the limit:

$$\epsilon(t) = \int_{\tau_o}^{t} \Phi(t, \tau) d\sigma(\tau) \tag{12-114}$$

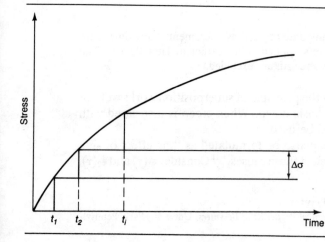

Figure 12–12

Viscoelasticity

Equation (12-114) is often referred to as the hereditary, or Volterra, integral. It shows that at time t, the strain $\epsilon(t)$ not only depends on the stress $\sigma(t)$, but rather on the whole stress history. Integrating equation (12-114) by parts we obtain

$$\epsilon(t) = \frac{\sigma(t)}{E(t)} - \int_{\tau_0}^{t} \sigma(\tau) \frac{\partial \Phi(t,\tau)}{\partial \tau} d\tau \qquad (12\text{-}115)$$

where $E(t) = 1/\Phi(t,t)$.

Suppose that the objective, now, is to compute the stress for a given strain history and relaxation function $E(t,\tau)$. Equations analogous to Eqs. (12-113) and (12-115) can be formulated.

$$\sigma(t) = \int_{\tau_0}^{t} E(t,\tau) \dot{\epsilon}(\tau) d\tau \qquad (12\text{-}116)$$

$$\sigma(t) = E(t)\epsilon(t) - \int_{\tau_0}^{t} \epsilon(\tau) \frac{\partial E(t,\tau)}{\partial \tau} d\tau \qquad (12\text{-}117)$$

where $E(t) = E(t,t)$.

Mathematical Expressions for Creep

As we mentioned before, creep tests are time-consuming, and special care needs to be taken to select a creep function which best fits the experimental results. From relatively short-time creep experiments the selected creep function also needs to be able to predict the long-term deformation. In the early days, when the curve fitting was done manually, researchers had to use intuition and experience to select simple and well-behaved functions. Today curve-fitting can be performed on almost any personal computer, hence the number and degree of sophistication of the functions has increased significantly. Before we present some functions for creep of concrete commonly used in structural analysis, we will make the following general statements regarding to the specific creep function $\Phi(t,\tau)$. You may consider it as a guideline in case you feel the need to introduce another creep function.

1. For a given age of loading τ, the creep function is a monotonic increasing function of time t.

$$\frac{\partial \Phi(t,\tau)}{\partial t} \geq 0 \qquad (12\text{-}118)$$

2. However, the rate of creep increment is always negative.

$$\frac{\partial^2 \Phi(t,\tau)}{\partial t^2} \leq 0 \qquad (12\text{-}119)$$

3. The aging of concrete causes a decrease in creep as the age of loading τ increases. For a given value of load duration $(t - \tau)$ due to aging of concrete,

$$\left[\frac{\partial \Phi(t,\tau)}{\partial \tau}\right]_{(t-\tau)} \leq 0 \qquad (12\text{-}120)$$

4. Creep has an asymptotic value

$$\lim_{t \to \infty} \Phi(t, \tau) \leq M \qquad (12\text{-}121)$$

Current codes differ in this topic. ACI-209 gives a finite final creep value while CEB-78 does not. The issue for most practical cases is not relevant, for instance, since according to CEB-78 the difference between 50-year and 100-year basic creep values is only about 8 percent.

In many structural models, the function $\Phi(t, \tau)$ is separated into instantaneous and delayed components.

$$\Phi(t, \tau) = \frac{1}{E(\tau)} + C(t, \tau) \qquad (12\text{-}122)$$

and to take aging of the concrete into account, the specific creep function $C(t, \tau)$ is further separated into:

$$C(t, \tau) = F(\tau) f(t - \tau) \qquad (12\text{-}123)$$

By writing $C(t, \tau)$ in this fashion, we indicate that concrete, at a given time t, should recollect not only the actions to which it was subjected since time τ, given by the function $f(t - \tau)$, but also its own material state at time τ, given by the function $F(\tau)$. Therefore, the function $F(\tau)$ characterizes the aging of concrete. The following expressions for $F(\tau)$ and $f(t - \tau)$ have been traditionally used for fitting short term experimental data, with the objective of predicting the long-term deformation.

Expressions for $f(t - \tau)$

1. *Logarithmic expression*: The U.S. Bureau of Reclamation[12] proposed the logarithmic expression for its projects dealing with mass concrete. When the stress/strength ratios do not exceed 0.40 the following equation is used.

$$f(t - \tau) = a + b \log[1 + (t - \tau)] \qquad (12\text{-}124)$$

The constants a and b are easily obtained when the creep data are plotted semi-logarithmically. The equation was originally developed for modeling basic creep of large dams and the duration of load $(t - \tau)$ is measured in days. The expression is unbounded and usually overestimates the later creep.

2. *Power expression*: The general expression is given by

$$f(t - \tau) = a(t - \tau)^m \qquad (12\text{-}125)$$

The constants a, m can be easily obtained on a log-log plot, where the power expression gives a straight line. The expression is able to capture the early creep well but overestimates the later creep. It also gives unbounded results.

---

[12] U.S. Bureau of Reclamation, "Creep of Concrete Under High Intensity Loading," Concrete Laboratory Report No. C-820, Denver, Colorado, 1956.

Viscoelasticity

3. *Hyperbolic expression*: Ross[13] proposed the following hyperbolic expression:

$$f(t - \tau) = \frac{(t - \tau)}{a + b(t - \tau)} \quad (12\text{-}126)$$

This expression provides a limiting value for creep, $1/b$. It usually underestimates early creep but provides good agreement for late creep. ACI code uses this formulation for creep evolution.

4. *Exponential expression*: The exponential expression provides a limiting value for creep, in its simplest formulation it is given by:

$$f(t - \tau) = a(1 - e^{-b(t - \tau)}) \quad (12\text{-}127)$$

It does not provide a good fit for experimental values. For numerical analysis more terms are usually incorporated.

Expressions for $F(\tau)$

$F(\tau)$ takes into account the aging of concrete, therefore it should be monotonically decreasing. While expressions for $f(t - \tau)$ have been developed during the last 60 years, expressions for $F(\tau)$ are much more recent. Among the expressions, we cite:

1. *Power law*:

$$F(\tau) = a + b\tau^{-c} \quad (12\text{-}128)$$

2. *Exponential*:

$$F(\tau) = a + be^{-c\tau} \quad (12\text{-}129)$$

Methods for Predicting Creep and Shrinkage

When experimental data are not available, the designer has to rely on a relevant code, which usually represents the consensus among researchers and practitioners. Unfortunately, the CEB-FIP model code 1978 proved to be somewhat controversial; however the new CEB-FIP model code 1990 should resolve many of the criticisms of the previous model. In this section, we present both the 78 and 90 CEB-FIP models as well as the recommendations of ACI-209 and the Bazant-Panula model.

The creep function $\Phi(t, t_o)$ representing the strain at time t for a constant unit stress acting from time t_o[14] is given by:

$$\Phi(t, t_o) = \frac{\epsilon(t, t_o)}{\sigma_o} = \frac{1}{E_c(t_o)} + C(t, t_o) \quad (12\text{-}130)$$

---

[13] A. D. Ross, *The Structural Engineer*, 15, No. 8, pp. 314–26, 1937.

[14] In codes, it is common to refer to the age of loading as t_o, instead of τ which is often used in mechanics. From now on, to be consisted with the code nomenclature we will use t_o as the age of loading.

In the prediction models two types of creep coefficient exist:

1. Creep coefficient representing the ratio between creep strain at time t and initial strain at time t_o. This definition is used in the ACI and Bazant-Panula models.

$$\varphi_o(t, t_o) = \frac{\epsilon_c(t, t_o)}{\sigma_o/E_c(t_o)} \quad (12\text{-}131)$$

Therefore Eq. (12-130) may be written as

$$\Phi(t, t_o) = \frac{1}{E_c(t_o)}[1 + \varphi_o(t, t_o)] \quad (12\text{-}132)$$

2. Creep coefficient representing the ratio between the creep strain at time t and the initial strain for a stress applied at 28 days. This definition is used in the CEB-78 model.

$$\varphi_{28}(t, t_o) = \frac{\epsilon_c(t, t_o)}{\sigma_o/E_{c28}} \quad (12\text{-}133)$$

Therefore Eq. (12-130) may be written as

$$\Phi(t, t_o) = \frac{1}{E_c(t_o)} + \frac{\varphi_{28}(t, t_o)}{E_{c28}} \quad (12\text{-}134)$$

CEB 1990. This method estimates creep and shrinkage for structural concretes in the range of 12 MPa to 80 MPa in the linear domain, that is, for compressive stresses $\sigma_c(t_o)$ not exceeding $0.4 f_{cm}(t_o)$ at the age of loading t_o. In this case, the total strain at time t, $\epsilon_c(t)$ may be subdivided into:

$$\epsilon(t) = \epsilon_{ci}(t) + \epsilon_{cc}(t) + \epsilon_{cs}(t) + \epsilon_{cT}(t) = \epsilon_{c\sigma}(t) + \epsilon_{cn}(t) \quad (12\text{-}135)$$

where
$\epsilon_{c\sigma}(t) = \epsilon_{ci}(t) + \epsilon_{cc}(t)$
$\epsilon_{cn}(t) = \epsilon_{cs}(t) + \epsilon_{cT}(t)$
$\epsilon_{ci}(t_o) = $ initial strain at loading
$\epsilon_{cc}(t) = $ creep strain
$\epsilon_{cs}(t) = $ shrinkage strain
$\epsilon_{cT}(t) = $ thermal strain
$\epsilon_{c\sigma}(t) = $ stress dependent strain
$\epsilon_{cn}(t) = $ stress independent strain

The creep strain, $\epsilon_{cc}(t, t_o)$, is given by:

$$\epsilon_{cc}(t, t_o) = \frac{\sigma_c(t_o)}{E_c}\varphi(t, t_o) \quad (12\text{-}136)$$

where
$\varphi(t, t_o) = $ creep coefficient
$E_c = $ 28-day modulus of elasticity

Table 12-3 indicates the parameters necessary to compute the creep coefficient

Viscoelasticity

TABLE 12-3

$$\varphi(t,t_o) = \phi_o \beta_c(t - t_o) \qquad \phi_o = \phi_{RH} \beta(f_{cm})(\beta t_o)$$

$$h = \frac{2A_c}{u} \qquad \phi_{RH} = 1 + \frac{1 - RH/100}{0.46(h_o/100)^3}$$

$$\beta(f_{cm}) = \frac{5.3}{(f_{cm}/f_{cmo})^{0.5}} \qquad \beta(t_o) = \frac{1}{0.1 + (t_o/t_i)^{0.20}}$$

$$\beta_c(t - t_o) = \left[\frac{(t - t_o)/t_1}{\beta_H + (t - t_o)/t_1}\right]^{0.3} \qquad \beta_H = 150\left[1 + \left(1.2\frac{RH}{100}\right)^{18}\right]\frac{h}{100} + 250 \leq 1500$$

where t and t_o = measured in days
t_1 = 1 day
f_{cm} = 28-day compressive strength, in MPa
f_{cmo} = 10 MPa
RH = relative humidity, in percent
A_c = cross-section of the member
u = perimeter of the member in contact with the atmosphere

The development of creep with time, β_c, is hyperbolic, therefore giving an asymptotic value of strain as $t \rightarrow \infty$. The effect of type of cement may be considered by modifying the age of loading to, as

$$t_o = t_{o,T}\left(\frac{9}{2 + t_{o,T}^{1/2}} + 1\right)^\alpha \geq 0.5 \text{ days} \qquad (12\text{-}137)$$

and

$$t_{o,T} = \sum_{i=1}^{n} \Delta t_i \exp -\left[\frac{4000}{273 + T(\Delta t_i)/T_o} - 13.65\right] \qquad (12\text{-}138)$$

where α = -1 for slowly hardening cements, 0 for normal or rapid hardening cement, 1 for rapid hardening high strength cements
$T(\Delta t_i)$ = temperature, in C, during the time period Δt_i
Δt_i = number of days with temperature T
T_o = 1°C

Shrinkage

The total shrinkage $\epsilon_{cs}(t,t_s)$ can be computed from the equations shown in Table 12–4:

where t = age of concrete (days)
t_s = age of concrete (days) at the beginning of the shrinkage
t_1 = 1 day
h_o = 100 mm
f_{cm} = mean compressive strength of concrete at the age of 28 days [MPa]
f_{cmo} = 10 MPa

TABLE 12-4

$$\epsilon_{cs}(t,t_s) = \epsilon_{cso}\beta_s(t - t_s) \qquad \epsilon_{cso} = \epsilon_s(f_{cm})\beta_{RH}$$

$$\epsilon_s(f_{cm}) = [160 + 10\beta_{sc}(9 - f_{cm}/f_{cmo})] \times 10^{-6} \qquad \beta_s(t - t_s) = \left[\frac{(t - t_s)/t_i}{350(h/h_o)^2 + (t - t_s)/t_i}\right]^{0.5}$$

β_{sc} = coefficient (4 for slowly hardening cements, 5 for normal or rapid hardening cements, 8 for rapid hardening high strength cements
$\beta_{RH} = -1.55[1 - (RH/100)^3]$ for $40\% \leq RH < 99\%$;
$\beta_{RH} = 0.25$ for $RH \geq 99\%$

CEB 1978. According to CEB 1978, the creep function $\Phi(t, t_o)$ is separated into the following additive components:

1. Initial strain:

$$\frac{1}{E_c} = \frac{\beta_i(t_o)}{E_{c28}} \qquad (12\text{-}139)$$

When the value of E_c is not available from experimental results it is recommended to increase, by 25 percent the value obtained from CEB-FIP model code on the basis of the mean compressive strength $f_{cm}(t_o)$.

$$E_c = 1.25 x \sqrt{f_{cm}(t_o)} \quad \text{in MPa} \qquad (12\text{-}140)$$

2. Rapid initial strain developing in the first day after load application:

$$\frac{\beta_a(t, t_o)}{E_{c28}} = \frac{\beta_{a1}(t_o) - \beta_{a2}(t - t_o)}{E_{c28}} \qquad (12\text{-}141)$$

This term is significant for loading at early ages. Note that it is always considered as part of the creep deformability and not related to the "initial" or "elastic" term.

3. Delayed elastic strain:

$$\frac{\varphi_d \beta_d(t - t_o)}{E_{c28}} \qquad (12\text{-}142)$$

This term represents the recoverable part of the delayed deformation; it is assumed to be independent of aging.

4. Unrecoverable time-dependent strain (flow, delayed plastic strain):

$$\frac{\varphi_d[\beta_f(t) - \beta_f(t_o)]}{E_{c28}} \qquad (12\text{-}143)$$

This term is correlated to the environmental parameters (e.g., relative humidity), dimensions of the specimen, and consistency of the concrete. The effect

Viscoelasticity

of temperature on the rate of hydration or aging is taken in account by a change of time scale, according to:

$$t = \frac{\alpha}{30} \sum_0^{t_m} \{[T(t_m) + 10]\Delta t_m\} \tag{12-144}$$

where α = coefficient (1 for normally and slowly hardening cements, 2 for rapid-hardening cements, 3 for rapid-hardening high strength cements).
T = mean daily temperature of the concrete in degrees C.
Δt_m = number of days when the mean daily temperature has assumed the value T.

Leading to the nondimensional creep function

$$E_{c28}\Phi(t,t_o) = \beta_i(t_o) + \beta_a(t_o) + 0.4\beta_d(t - t_o) + \varphi_f[\beta_f(t) - \beta_f(t_o)] \tag{12-145}$$

The analytical expressions for the model parameters are given in Table 12–5.

ACI 209.   The creep coefficient $\varphi(t, t_o)$ is defined as

$$\varphi = \frac{(t - t_o)^{0.6}}{10 + (t - t_o)^{0.6}} \varphi(\infty, t_o) \tag{12-146}$$

where (t, t_o) = time since application of load
$\varphi(\infty, t_o)$ = ultimate creep coefficient given by

$$\varphi(\infty, t_o) = 2.35 \, k_1 k_2 k_3 k_4 k_5 k_6 \tag{12-147}$$

TABLE 12–5 ANALYTICAL EXPRESSIONS[15]

$\beta_c(t_o) = \left[\dfrac{t_o}{t_o + 47}\right]^{\frac{1}{2.45}}$ $\beta_i(t_o) = 0.875 \left[\dfrac{t_o + 47}{t_o}\right]^{\frac{1}{7.35}}$

$\beta_a(t_o) = 0.8\left[1 - \left(\dfrac{t_o}{t_o + 47}\right)^{\frac{1}{2.45}}\right]$ $\beta_f(t) = \left[\dfrac{t}{t + K_1(h_o)}\right]^{K_2(h_o)}$

$K_1(h_o) = e^{\left[\frac{5.02}{h_o} + \ln(6.95h_o^{1.25})\right]}$ $K_2(h_o) = e^{\left[0.00144h_o - \frac{1.1}{h_o} - \ln(1.005h_o^{0.2954})\right]}$

$\varphi_{f1} = 4.45 - 0.035RH$ $\varphi_{f2} = e^{\left[4.4 \times 10^{-5}h_o + \frac{0.357}{h_o} - \ln\left(\frac{h_o^{0.1667}}{2.6}\right)\right]}$

$\beta_s(t) = \left[\dfrac{t}{t + K_3(h_o)}\right]^{K_4(h_o)}$ $\beta_d(t - t_o) = \left[\dfrac{t - t_o}{t - t_o + 328}\right]^{\frac{1}{4.2}}$

$K_3(h_o) = 11.8h_o + 16$ $K_4(h_o) = e^{\left[-0.00257h_o + \frac{0.32}{h_o} + \ln(0.22h_o^{0.4})\right]}$

$\epsilon_{s1} = (0.000775RH^3 - 0.1565RH^2 + 11.0325RH - 303.25) \times 10^{-5}$

$\epsilon_{s2} = e^{\left[0.00174h_o - \frac{0.32}{h_o} - \ln\left(\frac{h_o^{0.251}}{1.9}\right)\right]}$

---

[15] *Structural Effects of Time-dependent Behaviour of Concrete,* CEB Manual No. 142, 1984.

At loading ages greater than 7 days for moist cured concrete and greater than 1–3 days for steam cured concrete

$$k_1 = 1.25 t_o^{-0.118} \quad \text{for moist cured concrete}$$

$$k_1 = 1.13 t_o^{-0.095} \quad \text{for steam cured concrete}$$

Coefficients k_4, k_5 and k_6 are all related to the concrete composition

$k_4 = 0.82 + 0.00264s$
$s =$ slump of concrete (mm)
$k_5 = 0.88 + 0.024f$
$f =$ ratio of fine aggregate to total aggregate by weight in percent
$k_6 = 0.46 + 0.09a$
$a =$ air content (percent). k_6 should not be less than 1.

k_2, the humidity coefficient is given by:

$$k_2 = 1.27 - 0.006 RH \quad (RH > 40\%)$$

where RH is the relative humidity in percent.

The member thickness coefficient, k_3, can be computed by two methods:

1. *Average-thickness method* for average thickness less than 150 mm:

$$k_3 = 1.14 - 0.023h \quad \text{for } (t - t_o) < 1 \text{ year}$$

$$k_3 = 1.10 - 0.017h \quad \text{for } (t - t_o) < 1 \text{ year}$$

where h is the average thickness in mm.

2. *Volume-surface ratio method*:

$$k_3 = \frac{2}{3}[1 + 1.13 \exp(-0.0213 V/S)]$$

where V/S is the volume/surface ratio (mm).

Bazant-Panula Method

Only the simplified version of the method will be presented here. The refined formulation can be found in the original publication. In this model total creep is separated into basic and drying creep. The basic creep function is given by:

$$\Phi_b(t, t_o) = \frac{1}{E_o} + C_o(t, t_o) = \frac{1}{E_o} + \frac{\varphi_1}{E_o}(t_o^{-m} + \alpha)(t - t_o)^n \qquad (12\text{-}148)$$

E_o, asymptotic modulus, is a material parameter. It is not an actual elastic modulus for any load duration.

The five materials parameters of Eq. (12-148) are given by the following relations:

$$\frac{1}{E_o} = 0.09 + \frac{0.465}{f_{cm\,28}} \qquad \alpha = 0.05$$

$$\varphi_1 = 0.3 + 15 f_{cm\,28}^{-1.2} \qquad m = 0.28 + \frac{1}{f_{cm\,28}^2}$$

$$n = 0.115 + 0.00013 f_{cm\,28}^{3/4}$$

where $f_{cm\,28}$ is in ksi and $1/E_o$ is in 10^{-6} per psi.

The only input is the 28-day mean cylinder strength. In the refined formulation, the material parameters are corrected both to the strength $f_{cm\,28}$ and to several composition parameters.

Equation (12-148) also computes the static and dynamic modulus. The load duration $(t - t_o)$ for the static modulus is approximately 0.1 days, and for the dynamic modulus, 10^{-7} days.

The total creep function is given by:

$$\Phi(t, t_o) = \Phi_o(t, t_o) + \frac{\varphi_{ds}(t, t_o, t_s)}{E_o} \qquad (12\text{-}149)$$

The drying creep coefficient $\varphi_{ds}(t, t_o, t_s)$ is given by a series of equations relating concrete composition, thickness of the member, and environmental conditions.

TEMPERATURE DISTRIBUTION IN MASS CONCRETE

The problems created by the temperature rise in mass concrete due to the exothermic reactions of cement were presented in Chapter 4. There, a simple equation relating the tensile stress in concrete to its coefficient of thermal expansion, elastic modulus, creep coefficient, degree of restraint and temperature change was introduced. The major difficulty is how to compute the temperature distribution in complex geometries, and how to incorporate the incremental construction into the analysis. One of the challenges in mass concrete design is to maximize the thickness of the layer of concrete without causing cracks due to thermal stresses and also to optimize the time before next layer can be placed. The designer is under pressure from the contractor, who wants large layers to be placed in rapid succession to speed the concrete construction. For large projects major economic savings can be achieved when the size and placement of the lifts are perfectly orchestrated; the penalty for not doing this is to have the construction crew waiting for the next placement or to be forced to repair, or even demolish, an overly thick layer that cracked due to thermal stresses.

In this section we present an introduction to the finite element method, which is the most powerful tool in computing temperature distributions in solid materials.

After deriving the fundamental equations, a series of simulations of concrete constructions will be analyzed using this method. The technology of selecting materials, mix proportions, and construction practices has been presented in Chapter 11.

Heat Transfer Analysis

The fundamental equation governing the distribution of temperature in a solid subjected to internal heat generation was developed by Fourier. Consider a parallelepiped representing a volumetric element of a material, with conductivity coefficient k (Fig. 12–13). The change in heat flux in the x-direction is given by the equation:

$$\frac{\partial}{\partial x}\left(k\frac{\partial T}{\partial x}\right) dx\, dy\, dz \qquad (12\text{-}150)$$

where T is the temperature.

Similarly for the y and z directions:

$$\frac{\partial}{\partial y}\left(k\frac{\partial T}{\partial y}\right) dy\, dz\, dx \qquad (12\text{-}151)$$

and

$$\frac{\partial}{\partial z}\left(k\frac{\partial T}{\partial z}\right) dz\, dx\, dy \qquad (12\text{-}152)$$

Addition of the flux variation in the three directions, Eqs. (12-150)–(12-152) determines the amount of heat introduced in the interior of the element per unit time:

$$\left\{\frac{\partial}{\partial x}\left(k\frac{\partial T}{\partial x}\right) + \frac{\partial}{\partial y}\left(k\frac{\partial T}{\partial y}\right) + \frac{\partial}{\partial z}\left(k\frac{\partial T}{\partial z}\right)\right\} dx\, dy\, dz \qquad (12\text{-}153)$$

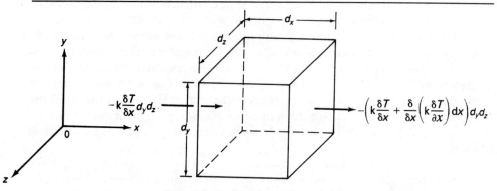

Figure 12–13 Heat flux in the x-direction.

Temperature Distribution in Mass Concrete

In the above derivation the material was considered isotropic. Considering it also homogeneous, Eq. (12-153) becomes:

$$k\left(\frac{\partial^2 T}{\partial x^2} + \frac{\partial^2 T}{\partial y^2} + \frac{\partial^2 T}{\partial z^2}\right) dx\, dy\, dz \qquad (12\text{-}154)$$

For a material with mass density ρ and specific heat c, the increase of internal energy in the element is given by:

$$\rho c\, dx\, dy\, dz\, \frac{\partial T}{\partial t} \qquad (12\text{-}155)$$

where t is the time.

When there is no heat generation in the material, equating Eqs. (12-154) and (12-155):

$$k\left(\frac{\partial^2 T}{\partial x^2} + \frac{\partial^2 T}{\partial y^2} + \frac{\partial^2 T}{\partial z^2}\right) = \rho c\, \frac{\partial T}{\partial t} \qquad (12\text{-}156)$$

Equation (12-156) may be rewritten as:

$$\kappa \nabla^2 T = \dot{T} \qquad (12\text{-}157)$$

where
$$\dot{T} = \frac{\partial T}{\partial t}$$

$$\nabla^2 T = \left(\frac{\partial^2 T}{\partial x^2} + \frac{\partial^2 T}{\partial y^2} + \frac{\partial^2 T}{\partial z^2}\right)$$

$$\kappa = \frac{K}{c\rho} = \text{thermal diffusivity}$$

Now consider the case when there is heat generation inside the material. Equation (12-154) when added to the quantity of heat generated in the interior of the element per unit of time—$w\,dxdydz$—can be equated with the increase of internal energy in the element. Therefore, the Fourier equation is obtained:

$$k\left(\frac{\partial^2 T}{\partial x^2} + \frac{\partial^2 T}{\partial y^2} + \frac{\partial^2 T}{\partial z^2}\right) + w = \rho c\, \frac{\partial T}{\partial t} \qquad (12\text{-}158)$$

or

$$k\nabla^2 T + w = \rho c \dot{T} \qquad (12\text{-}159)$$

In the steady-state, T and w are not function of time, Eq. (12-159) becomes:

$$k\nabla^2 T + w = 0 \qquad (12\text{-}160)$$

Equation (12-157) applies for any isotropic homogeneous material. We will concentrate the discussion for the problem of temperature distribution in mass concrete. In this case, the heat generation rate, w, is associated to the adiabatical temperature rise. In Chapter 4, the various factors affecting the temperature rise were presented; here we will show how to incorporate it into the Fourier equation. For a concrete with a density ρ and a cement content β (kg/m³), the relationship between the adiabatic temperature rise T_a and the heat of hydration Q_h is given by:

$$T_a = \frac{\beta}{c\rho} Q_h \tag{12-161}$$

The heat of hydration Q_h is obtained per unit mass of cement, therefore the factor β/ρ must be used to calculate the heat of hydration per unit mass of concrete. The heat generation rate w is related to the heat of hydration by the following equation:

$$w = \beta \frac{dQ_h}{dt} \tag{12-162}$$

and using Eq. (12-161)

$$w = \rho c \frac{dT_a}{dt} \tag{12-163}$$

In order to determine a unique solution to the Fourier Eq. (12-159), adequate initial and boundary conditions must be given. They should be compatible with the physical conditions of the particular problem.

Initial Condition

The initial condition must be defined by prescribing the temperature distribution throughout the body at time zero as a known function of x, y, and z.

$$T(x, y, z, t = 0) = f(x, y, z) \tag{12-164}$$

Boundary Conditions

I. Prescribed temperature boundary. The temperature existing on a portion of the boundary of the body Γ_t must be given as

$$T(x, y, z, t) = f(x, y, z, t) \qquad x, y, z \quad \text{on } \Gamma_t \tag{12-165}$$

This condition is also known as Dirichlet or essential boundary condition. In mass concrete, this condition may exist in the concrete-water contact, where the convection is small, making the temperature of the concrete that is in contact with water the same as that of the water.

II. Prescribed heat flow boundary. A prescribed heat flow boundary condition can be expressed as:

$$k \frac{\partial T}{\partial n}(x, y, z, t) = q_n(x, y, z, t); \qquad x, y, z \quad \text{on } \Gamma_q \tag{12-166}$$

where q_n is the given amount of heat flow at point (x, y, z), and n is the outward normal to the surface.

III. Convection boundary condition. The rate of heat transfer across a boundary layer is given by:

$$k\frac{\partial T}{\partial n}(x,y,z,t) = h(T_e - T_s)^N; \quad x,y,z \text{ on } \Gamma_h \quad (12\text{-}167)$$

where h is the heat transfer coefficient, T_e is the known temperature of the external environment, T_s is the surface temperature of the solid, and Γ_h is the portion of the boundary surface undergoing convective heat transfer.

For a linear convection boundary condition, $N = 1$, and Eq. (12-167) becomes:

$$k\frac{\partial T}{\partial n}(x,y,z,t) = h(T_e - T_s) = g(x,y,z,t) - hT_s \quad (12\text{-}168)$$

where $g(x,y,z,t) = hT_e$.

IV. Radiation boundary condition. Heat transfer by radiation between boundary condition surface Γ_r and its surroundings can be expressed by:

$$q_r(x,y,z,t) = V\sigma\left[\frac{1}{\frac{1}{\epsilon_r} + \frac{1}{\epsilon_s} - 1}\right][T_r^4 - T_s^4]; \quad x,y,z \text{ on } \Gamma_r \quad (12\text{-}169)$$

where V is the radiation view factor, σ the Stefan-Boltzmann constant, ϵ_r the emissivity of the external radiation source, ϵ_s the emissivity of the surface, and T_r and T_s are the absolute temperature of the radiation source and the surface, respectively.

Finite Element Formulation

The finite element method is a powerful tool in approximately solving thermal problems. The method is completely general with respect to geometry, material properties and arbitrary boundary conditions. Complex bodies of arbitrary shape including several different anisotropic material can be easily represented.

For mass concrete structures, the significant boundary conditions are cases I and III, that is, the prescribed temperature and the convection boundary conditions. There are many approaches to present a finite element formulation for temperature distribution in mass concrete, here we will follow the one suggested by Souza Lima et al.[16] To become familiar with the method, we start with the steady-state case, moving then to the transient-state case. The objective is to solve the Fourier equation given the necessary initial and boundary conditions. Consider a body with the two different boundary conditions: Γ_t where the temperature is prescribed and Γ_h where there is a convection boundary condition (see Fig. 12–14).

For a point P in Γ_t (steady-state case)

$$T = f(P) \quad (12\text{-}170)$$

[16] V. M. Souza Lima, D. Zagottis, J. C. André, XI National Conference on Large Dams, Ceará, Brazil, Theme I, 1, 1976.

and for a point P in Γ_h (steady-state case)

$$k\frac{\partial T}{\partial n} = g(P) - hT \qquad (12\text{-}171)$$

Consider a continuous and differentiable function Φ in the domain shown in Fig. 12–14 with the condition $\Phi = 0$ along Γ_t. No condition on Φ is imposed along Γ_h. Φ is often referred as "weighting function" and is relevant to note that it is, and will remain, arbitrary.

Multiplying both sides of Eq. (12-160) by Φ,

$$k\Phi\nabla^2 T = -\Phi w \qquad (12\text{-}172)$$

Integrating the above equation in domain V,

$$k\int_V \Phi\nabla^2 T\, dV = -\int_V \Phi w\, dV \qquad (12\text{-}173)$$

Using the divergence theorem in the left side of Eq. (12-173):

$$k\int_V \Phi\nabla^2 T\, dV = -k\int_V \left(\frac{\partial\Phi}{\partial x}\frac{\partial T}{\partial x} + \frac{\partial\Phi}{\partial y}\frac{\partial T}{\partial y} + \frac{\partial\Phi}{\partial z}\frac{\partial T}{\partial z}\right) dV + k\int_\Gamma \Phi\frac{\partial T}{\partial n}\, dS \qquad (12\text{-}174)$$

Since $\Phi = 0$ along Γ_t and using the boundary conditions, Eq. (12-168):

$$k\int_\Gamma \Phi\frac{\partial T}{\partial n}\, dS = k\int_{\Gamma_h} \Phi\frac{\partial T}{\partial n}\, dS = \int_{\Gamma_h} \Phi g\, dS - h\int_{\Gamma_h} \Phi T\, dS \qquad (12\text{-}175)$$

Introducing Eqs. (12-174) and (12-175) into Eq. (12-173):

$$k\int_V \left(\frac{\partial\Phi}{\partial x}\frac{\partial T}{\partial x} + \frac{\partial\Phi}{\partial y}\frac{\partial T}{\partial y} + \frac{\partial\Phi}{\partial z}\frac{\partial T}{\partial z}\right) dV + h\int_{\Gamma_h} \Phi T\, dS = \int_V \Phi w\, dV \\ + \int_{\Gamma_h} \Phi g\, dS \qquad (12\text{-}176)$$

Equation (12-176) is very useful to solve the steady-state Fourier heat equation. Consider a set of n functions Φ_i with $\Phi_i = 0$ on Γ_t. Thus any temperature field satisfying the boundary conditions on Γ_h also satisfies Eq. (12-176) for each of the functions Φ_i.

Equation (12-176) may be used instead of Eqs. (12-160), (12-170), and (12-171) to solve approximately for T in the following manner:

$$T = \Phi_o + \Sigma c_i \Phi_i \qquad (12\text{-}177)$$

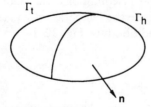

Figure 12–14

where c_i are unknown constants and Φ_o is any smooth function satisfying the boundary condition on Φ_t. Of course the above may not satisfy Eq. (12-160) at every point in the body. However, substituting the expression for T given by Eq. (12-177) into Eq. (12-176) a system of linear equations is obtained that allows the determination of the coefficients c_i. By increasing the number of coefficients in Eq. (12-177) a better approximation to the solution is obtained. Φ_i are referred as **interpolation functions** and they are almost invariably polynomials.

In the finite element method of analysis, the continuum is idealized by an assemblage of discrete elements or subregions. These elements may be of variable size and shape, and are interconnected by a finite number of nodal points P_i. The interpolation functions Φ_i should be chosen so that the coefficients c_i be numerically equal to the temperature, T, at n nodal points P_i, previously chosen in the domain. In order for the equality $c_i = T(P_i)$ to be true at the nodal points P_i the following conditions should be obeyed: $\Phi_i(P_i) = 1$, $\Phi_j(P_i) = 0$ ($j \neq i$) and $\Phi_o(P_i) = 0$ (see Fig. 12–15).

It is convenient to introduce a matrix formulation: $\{T\}$ a vector of n elements with values $T(P_i)$, and $\{w\}$ a vector of n elements with the values:

$$\int_V \Phi_i w \, dV + \int_{\Gamma_h} \Phi_i g \, dS - w_{oi} \tag{12-178}$$

where w_{oi} is the value of the first term of Eq. (12-173), for $\Phi = \Phi_i$ and replacing T by the function Φ_o.

For the steady-state case, this notation leads to:

$$[K]\{T\} = \{w\} \tag{12-179}$$

where $[K]$ is the conductivity matrix ($n \times n$) with values:

$$[K] = K_{ij} = \int_V \nabla^T \Phi_i k \nabla^T \Phi_j \, dV \tag{12-180}$$

The heat transfer in mass concrete is additionally complicated, because it involves the solution of the transient case and a change of geometry during construction. The solution can be found by introducing an incremental calculation of the linear transient problem. In the transient case, the Fourier equation is given by

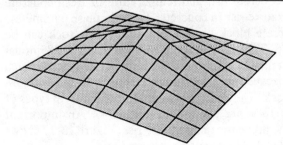

Figure 12–15 Typical trial function.

Eq. (12-159), which differs from the steady-state case by the term $\rho c \dot{T}$ and by the fact that w is a function of time. Using the divergence theorem:

$$k \int_V \left(\frac{\partial \Phi}{\partial x} \frac{\partial T}{\partial x} + \frac{\partial \Phi}{\partial y} \frac{\partial T}{\partial y} + \frac{\partial \Phi}{\partial z} \frac{\partial T}{\partial z} \right) dV + h \int_{\Gamma_h} \Phi T \, dS =$$
$$\int_V \Phi w \, dV + \int_{\Gamma_h} \Phi g \, dS - \rho c \int_V \Phi \dot{T} \, dV \qquad (12\text{-}181)$$

Or, using the matrix notation:

$$[k]\{T\} = \{w\} - [c]\{\dot{T}\} \qquad (12\text{-}182)$$

where $[c]$ is the capacity matrix $(n \times n)$ with values:

$$c_{ij} = \rho c \int_V \Phi_i \Phi_j \, dV \qquad (12\text{-}183)$$

To integrate Eq. (12-182), an incremental method is usually employed. Taking small interval Δt.

$$\{\dot{T}\} = \frac{1}{\Delta T}[\{T(t)\} - \{T(t - \Delta t)\}] \qquad (12\text{-}184)$$

Using Eq. (12-184) into Eq. (12-182)

$$\left([k] + \frac{1}{\Delta T}[c]\right)\{T(t)\} = \{w\} + \frac{1}{\Delta T}[c]\{T(t - \Delta t)\} \qquad (12\text{-}185)$$

Starting from a known initial temperature distribution, we proceed stepwise. Equation (12-185) allows the determination of $\{T(\Delta t)\}$ for the first step. The new temperature known, we proceed to the next step, giving a new increment Δt and continuing the process until the distribution of temperatures over the period of time of interest is known.

Examples of Application

Consider some practical problems that a concrete technologist has to face when studying thermal stresses in mass concrete. For example, how the type of aggregate, amount of pozzolan, size of the concrete lift, and temperature of fresh concrete would affect the maximum temperature rise in concrete. To study these parameters a finite element model of a concrete block placed on a foundation rock can be developed, as shown in Fig. 12–16. The finite element mesh is made of 385 nodal points and 344 elements. Note that the size of elements in the concrete block is much smaller than in the foundation, because we are mainly interested in temperature distribution inside the concrete block. The material properties for different types of concrete and for the foundation rock are shown in Table 12–6. An important parameter in thermal analysis is the adiabatic temperature rise. Figure 12–17 shows the assumed values for different levels of pozzolan replacements.

Temperature Distribution in Mass Concrete

TABLE 12-6 PROPERTIES OF CONCRETE AND FOUNDATION ROCK

	Properties of concrete made with different aggregates			Foundation rock
	Basalt	Granite	Gravel	
Thermal conductivity (kcal/m.h. C)	1.740	2.367	3.960	2.800
Specific heat (kcal.kg. C)	0.24	0.23	0.22	0.20
Thermal diffusivity (m²/hr)	0.0029	0.0042	0.0075	0.0050
Density (kg/m³)	2500	2450	2400	2800
Cement consumption (kg/m³)	315	315	315	—

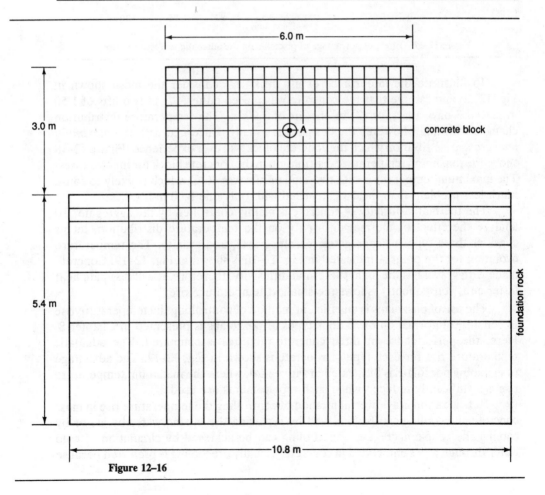

Figure 12–16

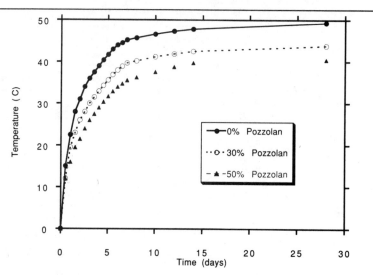

Figure 12-17 Effect of percentage of pozzolan on the adiabatic temperature rise.

To illustrate the importance of lift thickness, consider the mesh shown in Fig. 12-16 and assume that the concrete was placed either: (1) in two lifts of 1.50 m placed 3 days apart, or (2) in one lift of 3.00 m. The temperature distribution changes with time, however the designer is usually concerned with the *maximum temperature* distribution which the concrete block will ever experience. Figure 12-18 shows the maximum temperature distribution in the concrete block for the two cases. The maximum temperature with two 1.50 m-lifts was 46°C, which is likely to cause much less problem than the 56°C reached with only one 3.00 m-lift.

The thermal diffusivity of concrete is mainly controlled by the aggregate. To analyze the effect of thermal diffusivity on the temperature distribution, let us consider three types of aggregates: basalt, gravel, and granite. The temperature evolution for the point A indicated in Fig. 12-16 is shown in Fig. 12-19. Concrete made with gravel has the highest thermal diffusivity, therefore, it will dissipate heat faster and, consequently, shows the smallest temperature rise.

The use of pozzolans is an efficient method of controlling the temperature rise in concrete. Information on various types of pozzolans is presented in Chapter 8. Here, the performance of three concrete mixtures is compared. The adiabatic temperature rise for each type of concrete is shown in Fig. 12-17. The advantage of including pozzolans is illustrated in Fig. 12-20, where the maximum temperature rise is significantly reduced when such replacements are used.

Refrigeration is a powerful method of controlling the temperature rise in mass concrete. Precooling can be achieved by replacing the mixing water by ice or by cooling the coarse aggregate. Postcooling can be achieved by circulation of cold water through pipes embedded in concrete. Usually precooling is preferred because

Temperature Distribution in Mass Concrete

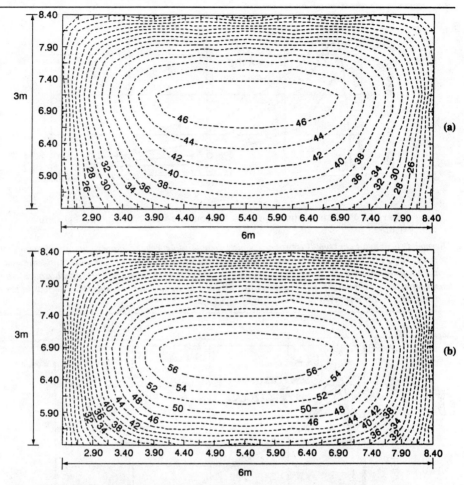

Figure 12–18 Maximum temperature distribution in the concrete block.

The size of the concrete lift is an important parameter in the temperature distribution in mass concrete. Thick lifts are attractive for fast construction, but high temperatures are usually generated in the concrete. Smaller lifts generate much lower temperatures, however they may create problems with the construction scheduling. It is the responsibility of the engineer to establish the optimal thickness of the lifts. For this, a thermal analysis is usually performed using the finite element method. As an illustration, a thermal analysis was conducted for the finite element mesh shown in Fig. 12–16, with two conditions: (a) two concrete lifts of 1.50 m placed 3 days apart, and (b) one lift of 3.00 m. The temperature distribution in the concrete is shown above. For case (a) the maximum temperature in the concrete is 46°C, which is much lower than the 56°C for case (b). For this analysis, the temperature of the fresh concrete was assumed to be 17°C, the aggregate was assumed to be granite, and no pozzolan was used.

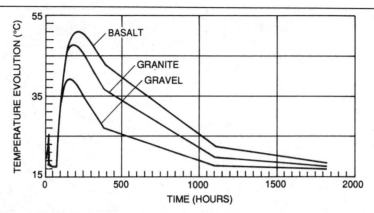

Figure 12–19 Temperature evolution for concrete with different thermal diffusivities.

Thermal diffusivity of concrete is a property which greatly influences the temperature distribution within the mass. Higher thermal diffusivity leads to faster heat loss which results in a lower maximum temperature. It may not always be desirable to have a rapid dissipation of heat, because the concrete may not have enough tensile strength at earlier ages. For this study two lifts of 1.50 m each was considered. The reason for the first temperature drop is given in the caption of Fig. 12–20.

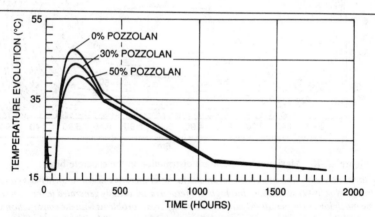

Figure 12–20 Influence of pozzolan on the temperature of concrete.

Pozzolans can significantly reduce the temperature inside mass concrete. The plot above shows the temperature evolution for point A of Fig. 12–16. Placement consisted of two concrete lifts of 1.50 m 3 days apart. Point A is at the top of the first lift, so there is an initial temperature increase, followed by a quick heat loss to the ambient temperature (17°C) until the next lift is placed. The temperature then increases up to a maximum, after which the block starts to cool.

Temperature Distribution in Mass Concrete

it is more economical and it does not involve extra construction work such as embedding pipes, pumping cold water, and eventually regrouting the pipes. The importance of the temperature of fresh concrete is shown in Fig. 12–21. When the concrete is placed at 25°C, the maximum temperature is 52°C, compared to 42°C when the concrete is placed at 10°C.

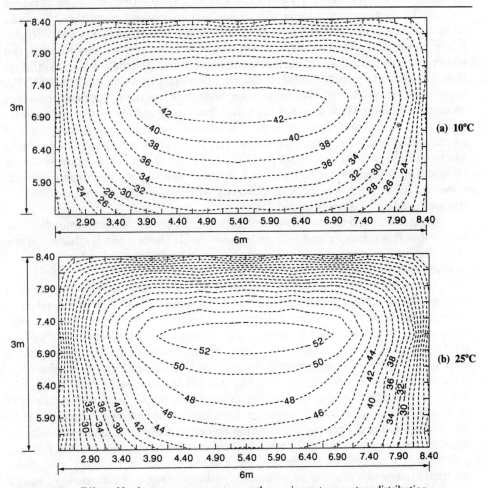

Figure 12–21 Effect of fresh concrete temperature on the maximum temperature distribution.

One of the most effective methods of controlling the temperature rise in mass concrete is by lowering the temperature of fresh concrete. A simple method is to use ice instead of mixing water, or precooling the aggregate. Unlike change in the size of lift, modifications in the temperature of fresh concrete do not affect construction scheduling. In this finite element analysis, three different temperatures of fresh concrete are assumed: 10°C, 17°C and 25°C. The temperature distribution for fresh concrete placed at a temperature of 17°C is shown in Fig. 12–18a. The concrete was placed in two 1.50 m lifts each, with granite as the coarse aggregate, and the ambient temperature was 17°C.

FRACTURE MECHANICS

Fracture mechanics for concrete can be a useful tool for the designer because of the insight it provides on size effects, that is, how the size of a structural element will affect the ultimate load capacity. Fracture mechanics also provides powerful criteria for the prediction of crack propagation. For instance, consider a case where you are responsible for determining if a given crack in a large structure, such as a concrete dam, will propagate catastrophically under certain loading conditions. You can adopt a strength criteria which predicts that a crack will propagate when the stresses reach the ultimate tensile strength of the material. However, for sharp cracks, the theory of linear elasticity predicts that the stresses at the tip of the crack go to infinity, therefore, this theory predicts that the crack will propagate no matter how small the applied stress; which is unlikely. Fracture mechanics, on the other hand, provides an energy criterion which does not have such drawbacks and allows for more precise predictions of the stability of the crack. The application of this energy criterion can be particularly useful when using traditional finite element methods to study cracks where mesh sensitivity becomes a problem. Figure 12–22 shows an example where the result is greatly affected by the size of the mesh when a strength criterion is used, however, little mesh sensitivity is observed when an energy criterion based on fracture mechanics is employed.

Considering the advantages of using fracture mechanics for concrete, you may expect that it is a mature and well understood field. Unfortunately, this is not the case. The development of fracture mechanics of concrete was slow when compared to other structural materials.

Linear elastic fracture mechanics theory was developed in 1920, but not until 1961 was the first experimental research in concrete performed. Fracture mechanics was used successfully in design for metallic and brittle materials early on; however, comparatively, few applications were found for concrete. This trend continued up until the middle 1970s when, finally, major advances were made. The contributions were based on the development of nonlinear fracture mechanics models, where the structure and behavior of concrete could be taken in account. In the 1980s and 1990s, intensive research has been performed and applications of fracture mechanics in design of beams, anchorage, and large dams are becoming more common. Still, when compared to the previous continuum theories covered in this chapter (elasticity, viscoelasticity, and thermal problems), fracture mechanics is not yet as mature, and this will be reflected in its presentation. The text is at an introductory level where we tried to present a fair, but simplified, exposition of some of the existing fracture mechanics models for concrete.

Linear Elastic Fracture Mechanics

We will start our presentation with the work of A. A. Griffith,[17] who is often regarded as the founder of fracture mechanics. His original interest was on the effect

---

[17] A. A. Griffith, "The phenomena of rupture and flow in solids," Philosophical Transactions, Royal Society of London, series A 221, 163–198 (1920).

Fracture Mechanics

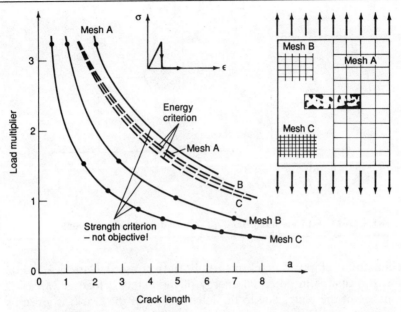

Figure 12-22 Example of mesh sensitivity (From Z. P. Bazant and L. Cedolin, *Stability of Structures*, New York, Oxford University Press, 1991, p. 917).

of surface treatment in the strength of solids. It had been observed, experimentally, that small imperfections have a much less damaging effect on the material properties than the large imperfections. This was theoretically puzzling because the fracture criteria used at that time would predict that if the imperfections were geometrically similar, the stress concentrations caused by the imperfections should be the same, so that the effect on the strength is the same, no matter what the size of the imperfection is. Griffith tackled this dilemma by developing a new criterion for fracture prediction. He suggested an energy balance approach, in contrast to the simplistic strength approach, based not only on the potential energy of the external loads and on the stored elastic strain energy, but also on another energy term: the **surface energy**. This surface energy, γ, is associated with the creation of fresh surface during the fracture process. Griffith applied his method to a crack of length $2a$ in an infinite plate of unit thickness. Figure 12-23a,b shows that when the crack is extended *under constant load*, the change in potential energy of the external load due to crack growth is $P\Delta x$ and the increase in strain energy is $\frac{1}{2} P\Delta x$. In other words, the decrease in potential energy of the external load is twice the increase in strain energy. During crack extension there is an increase of surface energy $4a\gamma$ (remember, the crack length is $2a$, and both the upper and lower surface of the crack should be included). Griffith used a result obtained by Inglis,[18] that the change in strain

---

[18] C. E. Inglis, "Stresses in a plate due to the presence of cracks and sharp corners," Trans. Inst., Naval Architects, 55, 219–241 (1913).

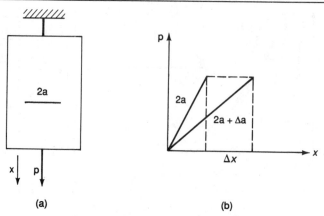

Figure 12–23 (a) Plate with crack $2a$; (b) Load-displacement diagram.

energy due to an elliptical crack in a uniformly stressed plate is $\pi a^2 \sigma^2/E$, and therefore, the change in potential energy of the external load is $2\pi a^2 \sigma^2/E$. The energy change of the plate, due to the introduction of the crack, is given by:

$$U_{\text{cracked}} - U_{\text{uncracked}} = -\frac{2\pi a^2 \sigma^2}{E} + \frac{\pi a^2 \sigma^2}{E} + 4a\gamma \qquad (12\text{-}186)$$

Minimizing the energy in relation to the crack length,

$$\frac{\partial}{\partial a}\left(-\frac{\pi a^2 \sigma^2}{E} + 4a\gamma\right) = 0 \qquad (12\text{-}187)$$

gives the critical stress (for plane stress):

$$\sigma = \sqrt{\frac{2E\gamma}{\pi a}} \qquad (12\text{-}188)$$

This equation is significant because it relates the size of the imperfection ($2a$) to the tensile strength of the material. It predicts that small imperfections are less damaging than large imperfections, as observed experimentally. Even though Griffith's work was ignored for several years, it paved the way for the creation and development of the mature field of linear fracture mechanics.

A problem with Griffith's approach was that the surface energy obtained from his equation was found to be orders of magnitude higher than the one obtained using thermodynamical tests unrelated to fracture. The reason is that the energy required for crack extension exceeds the thermodynamical value because the dissipative processes associated to the fracture propagation absorb a significant amount of energy. Irwin[19] proposed that instead of using the thermodynamic surface energy, you should measure the characteristic surface energy of a material in a fracture test. He introduced the quantity G_c as the work required to produce a unit increase in

---

[19] G. R. Irwin, Trans ASME, J. Applied Mechanics, Vol. 24, pp. 361–364 (1957).

Fracture Mechanics

crack area. G_c is also referred to as the **critical energy release rate**. G_c is determined experimentally, normally using simple specimen configuration. Once G_c for a given material is known, and we assume that it is a material property, we have a powerful method of determining if a given crack will or will not propagate under any other loading condition. The process is quite simple: the energy release per unit increase crack area, G, is computed; if the energy release rate is lower than the critical energy release rate ($G < G_c$), the crack is stable, and, conversely, if $G > G_c$, the crack propagates. In the particular case when the energy release rate is equal to the critical energy release rate ($G = G_c$) a metastable equilibrium is obtained.

The following analysis helps to compute the value of G_c. Considering the plate, shown in Fig. 12–23, with thickness B, we can express the energy released by crack growth Δa as:

$$GB\Delta a = P\Delta x - \Delta U_e \qquad (12\text{-}189)$$

Where ΔU_e is the change in elastic energy due to crack growth Δa. In the limit:

$$GB = P\frac{dx}{da} - \frac{dU_e}{da} \qquad (12\text{-}190)$$

Introducing the compliance $c = x/P$, the strain energy U_e is given by:

$$\Delta U_e = \frac{cP^2}{2} \qquad (12\text{-}191)$$

Equation (12-190) becomes:

$$GB = P\frac{d(cP)}{da} - \frac{d(cP^2/2)}{da} \qquad (12\text{-}192)$$

or,

$$G = \frac{P^2}{2B}\frac{dc}{da} \qquad (12\text{-}193)$$

When, for a given specimen configuration, the compliance versus crack length has been obtained, the critical energy release rate, G_c, can be determined by recording the load at fracture.

Example 1

Compute the energy release rate for the double cantilever beam shown in Fig. 12–24. Also study the stability of the crack in its own plane under (a) load control, and (b) displacement control. Shear deflections may be ignored.

The deflection of each cantilever can be easily found using simple beam theory:

$$\frac{\delta}{2} = \frac{Pa^3}{3EI} \qquad (12\text{-}194)$$

where E is the elastic modulus and I is the moment of inertia,

$$I = \frac{1}{12}b\left(\frac{h}{2}\right)^3$$

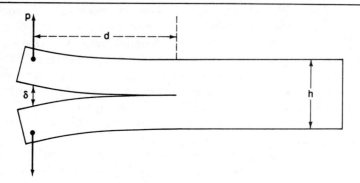

Figure 12-24 Double cantilever beam, with thickness B.

The compliance is given by:

$$c = \frac{\delta}{P} = \frac{2a^3}{3EI} \qquad (12\text{-}195)$$

Therefore the energy release rate is given by

$$G = \frac{P^2}{2B}\frac{dc}{da} = \frac{P^2 a^2}{BEI} \qquad (12\text{-}196)$$

Stability criteria: A crack is stable if the derivative of the strain energy rate, with respect to crack length, is negative, in other words,

$$\frac{1}{G}\frac{\partial G}{\partial a} < 0 \qquad (12\text{-}197)$$

1. *For load control*:

$$\frac{\partial G}{\partial a} = \frac{2P^2 a}{BEI} \qquad (12\text{-}198)$$

$(1/G)(\partial G/\partial a) = 2/a$ is a positive number so the crack will propagate in an unstable way.

2. *For displacement control*: Combining Eqs. (12-196) and (12-194), the energy release rate can be expressed in terms of the deflection:

$$G = \frac{9EI\delta^2}{4a^4 B} \qquad (12\text{-}199)$$

and

$$\frac{\partial G}{\partial a} = -\frac{9EI\delta^2}{a^5 B} \qquad (12\text{-}200)$$

$(1/G)(\partial G/\partial a) = -4/a$ is a negative number, therefore, the crack will propagate in a stable manner.

Fracture Mechanics

Let us now analyze what happens to the stress field near the tip of a crack for the three configurations shown in Fig. 12–25. The three types of relative movements of two crack surfaces are classified as (a) **mode I**: Opening or tensile mode, (b) **mode II**: Sliding or in-plane shear mode, and (c) **mode III**: Tearing or antiplane shear mode.

Most practical design situations and failures are associated with Mode I. The stresses at the tip of the crack for this mode are given by (see Fig. 12–26):

$$\sigma_y = \frac{K_I}{\sqrt{2\pi r}} \cos\frac{\varphi}{2}\left[1 + \sin\frac{\varphi}{2}\sin\frac{3\varphi}{2}\right] \quad (12\text{-}201)$$

$$\sigma_x = \frac{K_I}{\sqrt{2\pi r}} \cos\frac{\varphi}{2}\left[1 - \sin\frac{\varphi}{2}\sin\frac{3\varphi}{2}\right] \quad (12\text{-}202)$$

$$\tau_{xy} = \frac{K_I}{\sqrt{2\pi r}}\left[\sin\frac{\varphi}{2}\cos\frac{\varphi}{2}\cos\frac{3\varphi}{2}\right] \quad (12\text{-}203)$$

K_I is called **stress-intensity** factor for mode I. Dimensional analysis of Eqs. (12-201)–(12-203) indicates that the stress-intensity factor must be linearly related to stress and to the square root of a characteristic length. If we assume that this characteristic length is associated with the crack length, we have:

$$K_I = \sigma\sqrt{a}\, f(g) \quad (12\text{-}204)$$

where $f(g)$ is a function that depends on the specimen and crack geometry.

The stress-intensity factor can be computed for a variety of crack shape configurations. Suppose we measure the value of the stress at fracture in a given test. Using Eq. (12-204) we determine the **critical stress intensity factor**, K_c, or, as it is usually called in the literature, the **fracture toughness**. If we make the assumption that K_c is a material property (as we did for critical energy rate G_c) we have another powerful tool of predicting critical combinations of stress and crack length for other configurations of mode I.

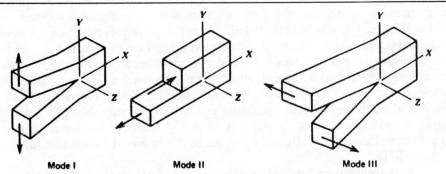

Figure 12–25 Basic modes of loading (From R. W. Hertzberg, *Deformation and Fracture Mechanics of Engineering Materials*, New York, John Wiley & Sons, 1976, p. 262).

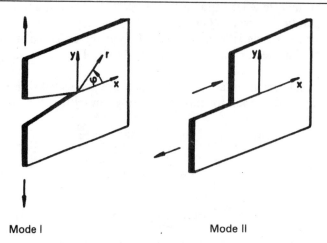

Figure 12–26 Coordinate system at the tip of the crack. (From R. Piltner, *Spezielle finite Elemente mit Löchern, Ecken und Rissen, unter Verwendung von analytischen Teillösungen*, VDI-Verlag Dusseldorf, n. 96, 1982).

Irwin showed that the energy release rate and the stress intensity factor approaches are equivalent. For linear elastic behavior, considering only mode I and plane stress condition:

$$G_I = \frac{K_I^2}{E} \qquad (12\text{-}205)$$

Finite Elements for Cracking Problems[20]

The concept of finite elements was introduced in the last section for the determination of temperature distribution in mass concrete. Here we will discuss how finite elements can be used to determine the stress intensity factors K_I and K_{II} for complex geometries. As we mentioned before, the finite element method is a powerful tool for the numerical treatment of partial differential equations. For elasticity problems with complex geometric boundaries it is usually impossible to find an exact solution for the displacements and stresses. In order to construct an approximate solution, the domain under consideration is divided into subdomains that are called finite elements. For every finite element, linearly independent basis functions similar to those used for the heat transfer problems can be used in order to approximate the displacement field. Restricting attention to plane problems, the nodal jth values of a finite element are usually chosen to be the displacement components u_j and y_j (Fig. 12–27).

The basis functions for standard displacement finite elements consist of shape functions which are multiplied by the unknown nodal values. By coupling the finite

---

[20] This section is more advanced and it may be skipped without loss of continuity.

Fracture Mechanics

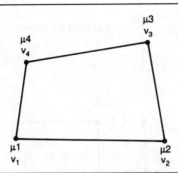

Figure 12-27 Quadrilateral 4-node element.

elements we "glue" the pieces of our solution together. The unknowns in the finite element solution are the nodal values. For the unknown nodal values we require that the potential energy of the system is minimized, and, by doing this, we set up the system of equations for the unknowns.

The stresses in standard displacement elements are finite. Therefore, standard displacement elements are not appropriate to approximate the stress singularities at the crack tip. For crack problems special finite elements are needed which include the crack tip singularities in the trial functions. Moreover we would like to couple crack elements with standard displacement elements for which polynomials are used as approximation functions.

In order to couple crack elements with standard displacement elements we have to make sure that the displacements along the edges of adjacent elements are compatible. The procedure of coupling a crack element with a standard displacement element is illustrated in Fig. 12-28.

For the crack element, displacement trial functions need to be used which have the following properties:

The linearly independent trial functions satisfy the equilibrium equations.

The trial functions satisfy the stress-free boundary conditions on the crack surface.

The trial functions for the displacements contain terms that are proportional to $\sqrt{r}$ so that the associated stresses are proportional to $1/\sqrt{r}$.

These conditions ensure that the correct form of the singular stress function terms will be used in the finite element approach. From the finite element analysis we get the coefficients of the singular stress functions for the crack tip. Apart from a factor, the coefficients of the singular stress functions are the stress intensity factors.

Two points need to be explained in more details: (1) How can we systematically construct linearly independent trial functions for the displacements and stresses with the properties listed above; (2) How can we compute a stiffness matrix for a crack element when the stresses are singular at one point.

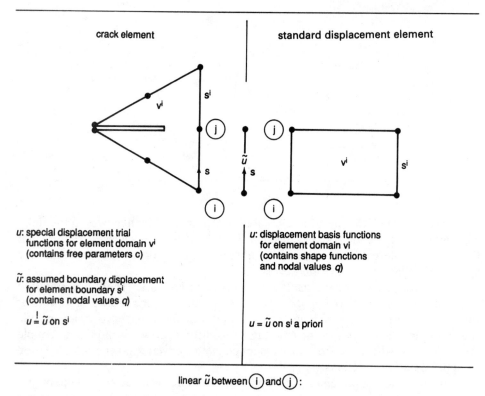

Figure 12-28 Illustration of the coupling between a crack element and a standard displacement element.

For the construction of linearly independent trial functions a representation of the displacements and stresses in terms of arbitrary functions is helpful. Using two complex functions, $\phi(z)$ and $\Psi(z)$, the displacements and stresses can be written in the form[21]:

$$2\mu u = Re[\kappa\phi(z) - z\overline{\phi'(z)} - \overline{\Psi(z)}]$$
$$2\mu v = Im[\kappa\phi(z) - z\overline{\phi'(z)} - \overline{\Psi(z)}]$$

[21] N. I. Muskhelishvili, *Some Basic Problems of the Mathematical Theory of Elasticity*, Noordhoff, Groningen, Holland, 1953.

Fracture Mechanics

$$\sigma_x = Re[2\phi'(z) - \bar{z}\phi''(z) - \Psi'(z)] \qquad (12\text{-}206)$$

$$\sigma_y = Re[2\phi'(z) + \bar{z}\phi''(z) + \Psi'(z)]$$

$$\tau_{xy} = Im[\bar{z}\phi''(z) + \Psi'(z)]$$

where $z = x + iy$, ϕ' denotes differentiation with respect to z, $2\mu = E/(1 + v)$ and $\kappa = (3 - 4v)$ for plane strain and $\kappa = (3 - v)/(1 + v)$ for plane stress. The advantage of using Eq. (12-206) is that for any choice of the functions ϕ and Ψ the equilibrium equations are automatically satisfied. For our crack element we need functions ϕ and Ψ, which ensure the satisfaction of the stress free boundary conditions on the crack surfaces. These functions may be represented in the form of a power series as:

$$\phi(z) = \sum_{j=0}^{N} a_j \zeta^j \qquad (12\text{-}207)$$

and

$$\Psi(z) = -\sum_{j=0}^{N} \left[\bar{a}_j(-1)^j + \frac{j}{2}a_j \right] \zeta^j \qquad (12\text{-}208)$$

where $a_j = \alpha_j + i\beta_j$ and $\zeta = \sqrt{z}$.

Substituting Eqs. (12-207) and (12-208) into Eq. (12-206) gives us the linearly independent real trial functions for the crack element. It should be noted that the terms with the index $j = 1$ give the singular stress terms for mode I and mode II. For example, for σ_x we obtain the singular terms in the form:

$$\sigma_x = \frac{1}{\sqrt{r}} \cos\frac{\varphi}{2}\left[1 - \sin\frac{\varphi}{2}\sin\frac{3\varphi}{2}\right]\alpha_1 + \frac{1}{\sqrt{r}} \sin\frac{\varphi}{2}\left[2 + \cos\frac{\varphi}{2}\cos\frac{3\varphi}{2}\right]\beta_1 \qquad (12\text{-}209)$$

The stress intensity factors K_I and K_{II} can be calculated from

$$K_I = \sqrt{2\pi}\alpha_1 \qquad (12\text{-}210)$$

$$K_{II} = -\sqrt{2\pi}\beta_1 \qquad (12\text{-}211)$$

Collecting the unknown coefficients α_j and β_j into a vector $\underline{c}$, the displacements for our crack element can be written in matrix notation as

$$\underline{u} = \underline{U}\,\underline{c} + \underline{u}_p = \underline{u}_h + \underline{u}_p \qquad (12\text{-}212)$$

where $\underline{u}_p$ is a particular solution involving no unknown coefficients. A particular solution can be used if we want to take nonhomogeneous stress boundary conditions on a crack surface into account (for example, constant pressure on the crack). Only the homogeneous solution $\underline{u}_h$ involves unknown coefficients. Since the unknowns in the vector $\underline{c}$ are not associated with finite element nodal values, we have to relate the vector $\underline{c}$ somehow to the vector of nodal displacements $\underline{q}$. The vector $\underline{q}$ contains the nodal values u_j, v_j of the chosen element nodes. In Fig. 12–28 a linear variation of the boundary displacements, $[\bar{u}\bar{v}]^T = \bar{\underline{u}}$, is assumed between nodes i and j. If we want to couple the crack element with linear standard displacement elements, $\bar{u}, \bar{v}$

are chosen linear between two nodes. If the crack element shall be coupled with quadratic standard elements, a quadratic variation of the boundary displacement $\bar{u}$ of the crack element is chosen.

In the first step for the evaluation of a crack element stiffness matrix the vector $\underline{c}$ of the displacement field $\underline{u}$ for the domain V^i of the crack elements is calculated such that an optimal agreement between $\underline{u}$ and $\underline{\bar{u}}$ is achieved along the boundary of the crack element. This gives us a relationship of the form:

$$\underline{c} = \underline{G}\underline{q} + \underline{g} \tag{12-213}$$

so that the unknowns $\underline{c}$ can be eliminated and only the nodal displacements $\underline{q}$ will remain as unknowns of the crack element.

For the evaluation of the crack element stiffness matrix the following displacement functional is used:

$$\Pi_H^i = \Pi^i + \int_{S_u^i} \underline{T}^T(\underline{\bar{u}} - \underline{u})dS^i \tag{12-214}$$

where

$$\Pi^i = \int_{V_i} \left[\frac{1}{2}(\underline{u}^T\underline{D}^T)\underline{E}(\underline{D}\underline{u}) - \underline{u}^T\underline{\bar{f}}\right]dV^i - \int_{S_u^i} \underline{u}^T\underline{\bar{T}}\,dS^i \tag{12-215}$$

Where V_i denotes the domain of the finite element. S_i is the boundary of the element and $\underline{T}$ are the tractions along the element boundary. Using the decomposition of the displacements and tractions in the form $\underline{u} = \underline{u}_h + \underline{u}_p$ and $\underline{T} = \underline{T}_h + \underline{T}_p$ we can simplify the variational formulation. Since the displacement field for the crack element is constructed such that the governing partial differential equations (Navier-equation in matrix notation):

$$\underline{D}^T\,\underline{E}\,\underline{D}\,\underline{u} = -\underline{\bar{f}} \quad \text{in } V \tag{12-216}$$

are satisfied a priori (30), can be simplified to an expression with boundary integrals:

$$\Pi^i = \int_{S_u^i} \frac{1}{2}\underline{u}_h^T\underline{T}_h\,dS^i + \int_{S_u^i} \underline{u}_h^T\underline{T}_p\,dS^i - \int_{S_u^i} \underline{u}_h^T\underline{\bar{T}}\,dS^i \tag{12-217}$$
$$+ \text{ terms without } \underline{u}_h \text{ and } \underline{T}_h$$

Using Eqs. (12-214), (12-215), and (12-217) the stiffness matrix of a crack element can be obtained by evaluations of the boundary integrals along the element boundary.[22,23] It is important that the boundary conditions on the crack surface are satisfied a priori so that all integrals along the crack surface vanish. Therefore, during the calculations of the stiffness matrix we do not need to evaluate stresses at the crack tip and although the stress singularities are included in the model, the evaluation of the stiffness coefficients will not be "polluted" from the presence of the singularities.

[22] R. Piltner, *Int. J. Numer. Methods Eng.*, 21, 1471, 1985.
[23] R. Piltner, in *Local Effects in the Analysis of Structures*, Elsevier, Amsterdam, 299, 1985.

Fracture Mechanics

Different shapes for the crack elements can be chosen (triangles, quadrangles, or other shapes). The location of the crack can be inside the finite elements or the crack can have its opening at a point on the element boundary.

As an example, consider a cracked plate under uniform tension (Fig. 12–29a), where the stress intensity factors were calculated numerically. The finite element model consists of one triangular crack element and 70 quadrilateral (linear 4-node) displacement elements (Fig. 12–29b). Only one half of the symmetric plate is discretised. The results from the finite element analysis are $K_I/p = 0.7915$ and $K_{II}/p = -0.1 \times 10^{-15}$ (for $x_o = 1.0705, L = 0.4109, L/x_o = 0.3838, y_o/x_o = 1$). These results are in very good agreement with the exact solution given by O. L. Bowie[24] ($K_I/p = 0.793, K_{II}/p = 0$).

Concrete Fracture Mechanics

The first experimental research on fracture mechanics of concrete was performed by Kaplan in 1961. Subsequent workers studied the affects of various parameters on K_c and G_c. Experimental studies indicated that the fracture toughness increases with increasing aggregate volume, increasing the dimensions of the maximum size aggregate, and increasing roughness of the aggregate. As expected, the toughness decreases with increasing w/c ratio and increasing air content. One of the problems encountered in the early research was the fact that the value of the fracture toughness K_c, instead of being a material property, was strongly influenced by the size of the specimen tested. It soon became apparent that fracture mechanics measurements should not be made on small concrete specimens.

Let us study the following case proposed by Cedolin[25] to analyze what happens to the ultimate stress when we change the dimensions of a cracked plate shown in Fig. 12–30b. The stress intensity for this configuration is given by $K = p\sqrt{\pi a} f(a/b)$ where $f(a/b)$ is a correction factor for the geometry.

Th critical stress, p_c, associated with the fracture toughness K_c is given by:

$$p_c = \frac{K_c}{\sqrt{\pi a}\, f(a/b)} \qquad (12\text{-}218)$$

This relationship is shown in Fig. 12–30c. Instead of the fracture mechanics criteria, let us now analyze the strength criteria. The average tensile stress, f_t, in the plane that contains the crack will vary because the crack dimensions affect the net section of the specimen. The relationship is given by:

$$p_t 2b = f_t(2b - 2a) \qquad (12\text{-}219)$$

or

$$p_t = f_t\left(1 - \frac{a}{b}\right) \qquad (12\text{-}220)$$

---

[24] O. L. Bowie, *J. Appl. Mech.* 31, p. 208–212, 1964.
[25] L. Cedolin, "Introduction to Fracture Mechanics of Concrete," *Il Cemento*, 283 (1986).

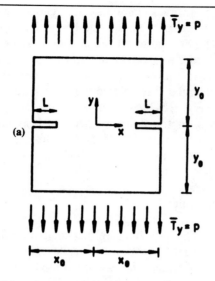

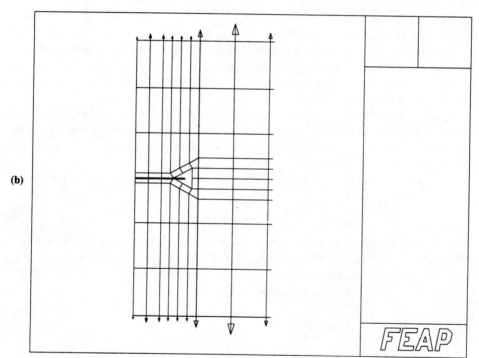

Figure 12-29 (a) Dimensions of the plate; (b) finite element discretization;

Fracture Mechanics

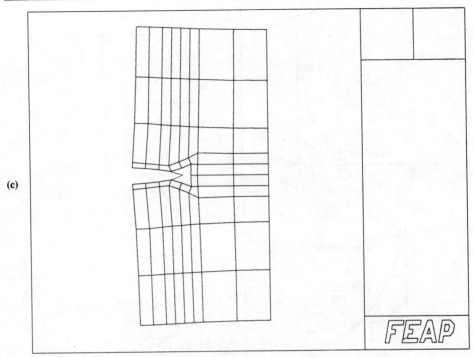

(c)

Figure 12–29 (c) deformed configuration. (Courtesy of R. Piltner)
The finite element model consisted of one triangular crack element and 70 quadrilateral displacement elements. The finite element program FEAP was used for the computations.

which is also shown in Fig. 12–30c. Therefore, as we can see in Fig. 12–30c, for a small crack, the strength criteria dominates and we cannot infer fracture mechanics properties.

It is also fruitful to study the case of geometrically similar plates (a/b constant) and varying b. Equation (12-218) may be rewritten as:

$$p_c = \frac{K_c}{\sqrt{b} f^*(a/b)} \qquad (12\text{-}221)$$

where $f^*(a/b) = \sqrt{\pi a/b} f(a/b)$. Since (a/b) is constant, when Equation (12-221) is plotted as function of b in a logarithmic scale it gives a straight line with slope 1/2 (Fig. 12–30d). Equation (12-220) is also plotted in Fig. 12–30d, and because a/b is constant it yields a straight line with zero slope. Again, we conclude that for small specimen sizes the strength criteria dominates and fracture mechanics properties cannot be inferred.

The ratio between the fracture mechanics criteria (Equation (31)) and the strength criteria (Equation (12-220)) is given by:

$$\frac{p_c}{p_t} = \frac{K_c}{f_t \sqrt{b}(1 - a/b) f^*(a/b)} \qquad (12\text{-}222)$$

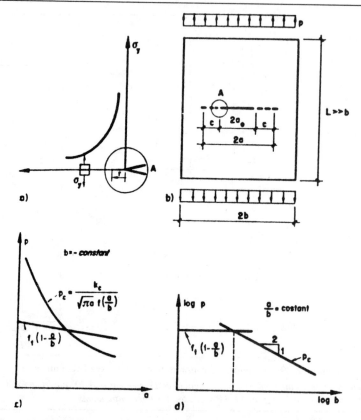

Figure 12-30 (a) Variation of σ_y at the crack tip in an elastic body; (b) cracked plate under tension; (c) comparison between ultimate values of applied tension, calculated according to fracture mechanics and tensile strength; (d) effect of plate width for geometrically similar plates. (From L. Cedolin, "Introduction to Fracture Mechanics of Concrete," *El Cemento*, 1986, No. 4, page 285.)

It is convenient to define a **brittleness number**, $s = K_c/f_t\sqrt{b}$, to characterize the nature of the collapse; the lower the brittleness number the more brittle the behavior of the specimen will be. Fracture will occur for a small brittleness number, that is, for materials with a comparatively low fracture toughness, high tensile strengths, and large specimens. The brittleness number characterizes the nature of the collapse for one-dimensional problems; for beams or slabs in flexure, additional information on the slenderness is necessary. It should be noted that the physical dimensions of the tensile strength $[FL^{-2}]$ and fracture toughness $[FL^{-3/2}]$ are different, however, the brittleness number is dimensionless.

The brittleness number can also be expressed as a function of elastic modulus, E, and energy release rate, G, instead of the fracture toughness K_c: $s = \sqrt{EG}/(f_t\sqrt{b})$. This number helps to explain the experimental results, which show that

Fracture Mechanics

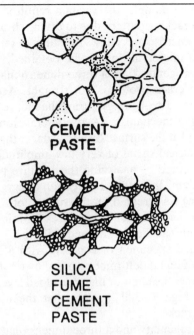

Figure 12–31 Structure of crack front in ordinary cement paste and in silica fume cement paste. (From H. H. Bache, *Fracture Mechanics in Design of Concrete and Concrete Structures*, in *Fracture Toughness and Fracture Energy of Concrete*, edited by F. H. Wittmann, Amsterdam, Elsevier Science Publishers B.V., 1986, p. 582.)

concretes made with high-strength silica fume cement paste usually have more fine microcracks than normal strength concrete (Fig. 12–31). In the high-strength matrix, the tensile strength can be 2–5 times greater than the normal-strength matrix, however, the increase in fracture energy or elastic modulus is not as much. Consequently, high-strength matrix has a much lower brittleness number and is more susceptible to the development of cracks.

Fracture Process Zone

In concrete, it has been observed that microcracks develop ahead of the crack tip, creating a **fracture process zone** that originates from strain localization. The characterization of this zone is of fundamental importance in the development of modern nonlinear fracture mechanics for concrete. The experimental characterization is challenging, and new methods have recently been proposed.

It is desirable to determine, among other parameters, the position of the crack tip, the profile of the crack opening, and the overall state of microcracking ahead of the crack tip. Optical microscopy can provide useful information, but resolution is limited (in the order of 10 μm). Scanning electron microscopy has a much better resolution, but in traditional models, the vacuum required for operation induces significant changes in the cracking pattern due to drying shrinkage. The new generation of scanning electron microscopes allows the study of saturated specimens, and meaningful information of microcracking can be obtained.

Due to concrete heterogeneity and a three-dimensional stress state along the crack front, generally the crack profile is not straight. **Acoustic emissions** (AE) resulting from the sudden release of energy during the failure process provide useful information on the cracking mechanism. Acoustic emissions are transient elastic waves which can be detected at the surface by a transducer that converts an acoustic-pressure pulse into an electrical signal of very low amplitude.

Another powerful method of analyzing the fracture process zone is **optical interferometry** with laser light. In a study by Cedolin et al.[26] a reference grid (with a density of 1000 lines/mm) was created on the surface of the concrete specimen. When a load is applied to the specimen it produces a Moirè fringe pattern, from which the extensional strain can be determined. Figure 12–32a shows the fringe configurations for a notched specimen of concrete with a 1.2 cm MSA. In Fig. 12–32b, the contour lines of equal deformation are shown for an intermediate load level, while Fig. 12–32d shows the contour lines for a load level close to failure. Note that the contour lines in Fig. 12–32d indicate that the microcracked zone has propagated ahead of the crack.

Due to concrete heterogeneity and a three-dimensional stress state along the crack front, generally the crack profile is not straight.

The additional elongation in the fracture zone can be estimated by introducing the additional strains ϵ_w over the length of the fracture zone (Fig. 12–33).

$$w = \int \epsilon_w \, dx \qquad (12\text{-}223)$$

Unfortunately, the real strain distribution is often very hard to incorporate in an analytical model. Therefore, simplified models have been proposed. Bazant and co-workers developed the **smeared crack band model**, where the entire fracture zone is represented by a band of microcracked material with width w_c. The model assumes a linear stress-strain relationship with slope E_c up to the tensile strength f_t and a strain-softening relationship with slope E_t. The area enclosed by the diagram in Fig. 12–34 represents the fracture energy G_f, given by:

$$G_f = w_c \int_0^{\epsilon_o} \sigma \, d\epsilon_f = \frac{1}{2} w_c f_t^2 \left(\frac{1}{E_c} - \frac{1}{E_t} \right) \qquad (12\text{-}224)$$

This method proved to be very successful when used with the finite element method. We will use the smeared crack band model to exemplify the importance of size effect.

Further simplification is obtained when the fracture process zone is modelled as a "tied crack" (Fig. 12–33), that is, a crack with a width w and a specified stress-elongation (σ-w) relationship. Since the aim is to replace the real fracture

---

[26] L. Cedolin, S. D. Poli, and I. Iori, *Journal of Engineering Mechanics*, Vol. 113, 431, 1987.

Intermediate Load Level

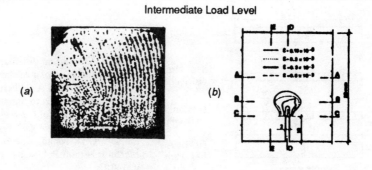

Load Level Close to Failure

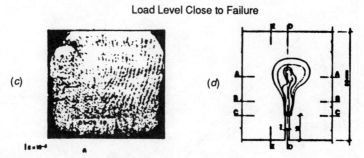

Figure 12–32 (a) and (c) fringe configuration; (b) and (d) equal strain contours and developing crack. (From L. Cedolin et al., "Tensile Behavior of Concrete," *Journal of Engineering Mechanics*, Vol. 113, p. 439, 1987.)

process zone by an equivalent fictitious tied crack, this representation has been called **the fictitious crack model**. The development of this model is presented in detail in the following section.

Fictitious Crack Model

The fictitious crack model was created and further developed by Hillerborg, Petersson, and co-workers. One of the objectives of the model is to capture the complex nature of concrete in tension. The amount of microcracking in concrete, which is in tension, is small before the peak stress is reached, therefore, the deformation ϵ along the specimen can be assumed to be uniform, and the total elongation Δl of the specimen can be expressed in terms of the length of the specimen l (Fig. 12–35a).

$$\Delta l = l\epsilon \tag{12-225}$$

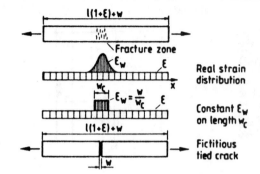

Figure 12-33 Strain distribution during fracture, and two possible simplifying assumptions. (From A. Hillerborg, "Numerical Methods to Simulate Softening and Fracture of Concrete," in *Fracture Mechanics of Concrete*, G. C. Sih and A. Di Tommaso, editors, Martinus Nijhoff Publishers, 1985, p. 148.)

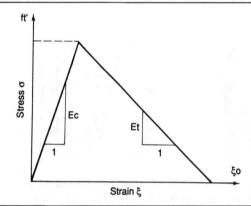

Figure 12-34 Stress-strain relationship for the smeared crack band model.

A localized fracture zone starts to develop just after the peak load is reached. In the model, this zone is assumed to form simultaneously across an entire cross section. As the total elongation increases, the stress decreases and the region outside the fracture zone experiences an unloading, while inside the fracture zone, there is softening. The fracture zone remains localized and does not spread along the specimen, this is called **strain localization**, somewhat akin to that seen in plasticity. Beyond the peak stress, the total elongation of the specimen is the sum of the uniform deformation outside the fracture zone and the additional localized deformation w existing in the fracture zone, as shown in Fig. 12–35b.

$$\Delta l = l\epsilon + w \quad (12\text{-}226)$$

As illustrated in Fig. 12–35c, to characterize the mechanical behavior of concrete in tension, two relationships are needed: (1) a stress-strain (σ-ϵ) relationship for the region outside the fracture zone, and (2) a stress-elongation (σ-w) relationship for the fracture zone. Note that in the σ-ϵ diagram, the horizontal axis is given by the strain, which is nondimensional, while for the σ-w diagram, the

Fracture Mechanics

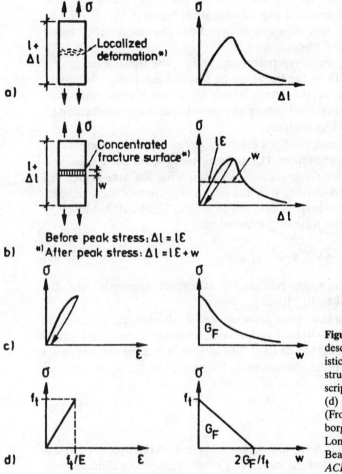

Figure 12-35 Fictitious crack model description of tensile fracture: (a) Realistic structural behavior; (b) model of structural behavior; (c) model for description of properties of material; and (d) simplified properties of material. (From P. J. Gustafsson and A. Hillerborg, "Sensitivity in Shear Strength of Longitudinally Reinforced Concrete Beams to Fracture Energy of Concrete," *ACI Structural Journal*, 1988, p. 287).

horizontal axis is given by the elongation, which has units of length. The curves shown in Fig. 12-35c may be influenced by the rate of loading and temperature, but they are assumed to be independent of the shape and size of the specimen. Figure 12-35d shows simplified stress-strain and stress-elongation relationships. There is no fundamental reason to choose linear or bilinear relationships with the exception that they are numerically simple and seem to satisfy experimental results rather well. It should be mentioned that other researchers preferred to use a nonlinear stress-elongation (σ-w) relationship.

The fracture energy, G_f, is equal to the area under the stress-elongation curve.

$$G_f = \int_0^\infty \sigma(w)\,dw \qquad (12\text{-}227)$$

Figure 12–36a shows typical experimental stress-elongation curves for different concrete mixes. The results presented in Fig. 12–36a are redrawn in Fig. 12–36b to show that, even for concretes with different composition, the normalized stress-elongation curves have the same shape.

For very large specimens with deep preexisting cracks, the fracture energy G_f corresponds to the parameter G_c of the linear elastic fracture mechanics. While its measurement is fairly easy to make, the determination of the σ-w relationship is not. Therefore, formulations, based on the fracture energy, such as the one indicated in Fig. 12–35, is usually preferred in analysis.

The fracture energy of concrete G_f is generally determined experimentally, according to RILEM Recommendation TC-50 FMC using a notched specimen loaded in flexure. The value for G_f is obtained by computing the area under the load-deflection relationship and dividing it by the net cross-section of the specimen above the notch. When experimental data is not available, CEB-FIP model code 1990 recommends the use of the following expression:

$$G_f = \alpha_f (f_{cm}/f_{cmo})^{0.7} \quad (12\text{-}228)$$

where α_f is a coefficient, which depends on the maximum aggregate size d_{max} (Table 12–7), and f_{cmo} is equal to 10 MPa.

It is interesting to analyze how the stress-strain and stress-elongation curves are related. The slope of the stress-strain diagram is E, and the slope of the stress-deformation curve is proportional to $f_t/(G_f/f_t)$. The ratio between the two slopes has units of length, and it is called the **characteristic length** (l_{ch}) of the material:

$$l_{ch} = \frac{EG_f}{f_t^2} \quad (12\text{-}229)$$

The characteristic length is often considered to be a material property, and it gives a measure of the brittleness of the material. Cement paste has a characteristic length in the range 5–15 mm, mortar in the range 100–200 mm, and concrete 200–400 mm. Compared to normal-strength concrete, high-strength concretes and light-weight aggregate concrete have lower characteristic lengths.

Designers realized the importance of the stress-strain and stress-elongation relationships of concrete in tension. For instance, the CEB-FIP model code 1990 recommends the following stress-strain relationships for uniaxial tension (Fig. 12–37).

$$\sigma_{ct} = E_c \epsilon_{ct} \quad \text{for } \sigma_{ct} \leq 0.9 f_{ctm} \quad (12\text{-}230)$$

$$\sigma_{ct} = f_{ctm} - \frac{0.1 f_{ctm}}{0.00015 - (0.9 f_{ctm}/E_c)} (0.00015 - \epsilon_{ct}) \quad (12\text{-}231)$$

$$\text{for } 0.9 f_{ctm} < \sigma_{ct} \leq f_{ctm}$$

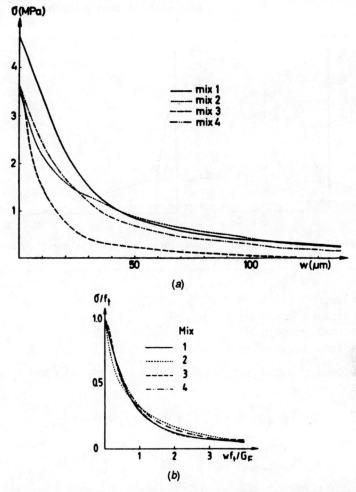

Figure 12–36 (a) σ-*w* curves for four concrete mixes. (From P. Petersson, "Crack Growth and Development of Fracture Zones in Plain Concrete and Similar Materials," Report TVBM-1006, Lund, Sweden, 1981, p. 167); (b) the curves from (a) are redrawn to show that their shape is similar. (From A. Hillerborg, "Numerical Methods to Simulate Softening and Fracture of Concrete," in *Fracture Mechanics of Concrete*, G. C. Shih and A. DiTommaso, eds., Martinus Nijhoff Publishers, 1985, p. 152).

TABLE 12–7 COEFFICIENT α_f AS FUNCTION OF THE MAXIMUM AGGREGATE SIZE d_{max}

d_{max} (mm)	α_f (Nmm/mm²)
8	0.02
16	0.03
32	0.05

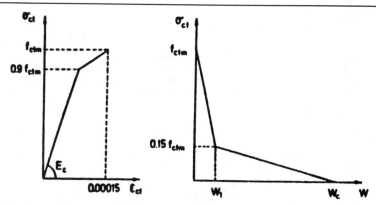

Figure 12-37 Stress-strain and stress-elongation for concrete in uniaxial tension. (From CEB-FIP Model Code 1990).

where E_c = tangent modulus of elasticity in MPa
f_{ctm} = tensile stress in MPa
σ_{ct} = tensile stress in MPa
ϵ_{ct} = tensile strain

For the cracked section, the following bilinear stress-crack opening relation is recommended:

$$\sigma_{ct} = f_{ctm}\left(1 - 0.85\frac{w}{w_1}\right) \quad \text{for } 0.15f_{ctm} \leq \sigma_{ct} \leq f_{ctm} \qquad (12\text{-}232)$$

$$\sigma_{ct} = \frac{0.15f_{ctm}}{w_c - w_1}(w_c - w) \quad \text{for } 0 \leq \sigma_{ct} \leq 0.15f_{ctm} \qquad (12\text{-}233)$$

and

$$w_1 = \frac{2G_f}{f_{ctm}} - 0.15\, w_c \quad \text{and} \quad w_c = \beta_F \frac{G_f}{f_{ctm}} \qquad (12\text{-}234)$$

where w_1 = crack opening (mm)
w_c = crack opening (mm) for $\sigma_{ct} = 0$
G_f = fracture energy [Nm/m²]
β_F = coefficient given in Table 12.8

TABLE 12-8 CRACK OPENING AT $\sigma_{ct} = 0$

d_{max} (mm)	β_F
8	8
16	7
32	5

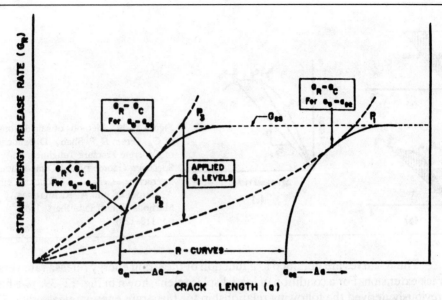

Figure 12-38 Typical *R*-curve. (From S. P. Shah, "Dependence of Concrete Fracture Toughness on Specimen Geometry and Composition," in *Fracture Mechanics of Concrete*, eds., A. Carpinteri and A. R. Ingraffea, Martinus Nijhoff Publishers, 1984, pp. 118-120).

Fracture Resistance Curves

In many practical applications, it is useful to develop fracture resistance curves, or *R*-curves, which indicate the energy release rate as a function of the extension of a crack from a notch. Different approaches for the development of *R*-curves have been proposed; here we follow the one suggested by Shah.[27] Figure 12-38 shows schematic *R*-curves for concrete, where the strain energy release rate at each crack initiation (G_R) is plotted as a function of the actual crack extension. Note that the strain energy release rate (G_R) increases with increasing crack extension, until it reaches the maximum value (G_{ss}), at which point it is not affected by further crack extension. Consider, now, the strain energy release rate caused by increasing loads P_1, P_2, and P_3, shown as dashed lines in Fig. 12-39. When a strain energy release rate curve becomes tangent to the resistance curve, it will cause the propagation of a specified notch length. In linear elastic fracture mechanics, the critical strain energy release rate, G_C, is a material property, but in this case, it is a function of the initial notch depth. If the process zone is accounted for, the R-curves appear to be independent of the specimen geometry for a given range of specimen sizes.

[27] S. P. Shah, in *Fracture Mechanics of Concrete*, A. Carpinteri and A. R. Ingraffea, eds., Netherlands, Martinus Nijhoff Publishers, 1984, pp. 111-135.

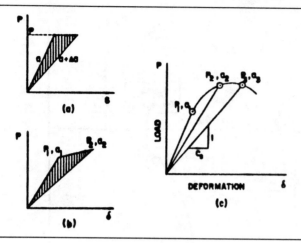

Figure 12-39 Methods of calculations of G_R. (From S. P. Shah, "Dependence of Concrete Fracture Toughness on Specimen Geometry and Composition," in *Fracture Mechanics of Concrete*, eds., A. Carpinteri and A. R. Ingraffea, Martinus Nijhoff Publishers, 1984, pp. 118–120).

The R-curves are obtained as a function of the strain energy release rate versus crack extension. For a condition of load displacement shown in Fig. 12–39a, we have previously derived the following relationship for the strain energy release rate [Eq. (12-157)]:

$$G = \frac{P^2}{2B} \frac{dc}{da} \quad (12\text{-}235)$$

Considering, now, the case where the crack propagation occurs with a rising load-deflection curve (Fig. 12–39b), Eq. (12-235) becomes:

$$G = \frac{P_1 P_2}{2B} \frac{dc_s}{da} \quad (12\text{-}236)$$

where P_1 and P_2 are two consecutive, neighboring loads, and c_s is the secant compliance.

A limitation to the secant compliance method is that it assumes the material to be elastic with no permanent deformation upon unloading. To avoid this, an unloading-reloading technique, where the secant compliance is replaced by the reloading compliance, c_R, has been suggested (Fig. 12–40). Equation (12-236) can be rewritten as:

$$G = \frac{P_1 P_2}{2B} \frac{dc_r}{da} \quad (12\text{-}237)$$

Equation (12-227), however, does not incorporate the inelastic energy absorbed during crack growth (area OADC, Fig. 12–40b). This can be taken into account by the modified energy release rate, G_R:

$$G_r = \frac{P_1 P_2}{2B} \left[\frac{dc_r}{da} + \left(\frac{P_1 + P_2}{P_1 P_2} \right) \frac{d\delta_p}{da} \right] \quad (12\text{-}238)$$

where δp is the permanent deformation, as shown in Fig. 12–40b.

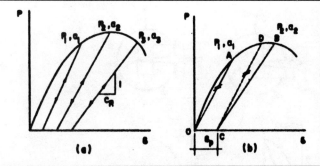

Figure 12-40 Compliance measurements during slow crack growth. (From S. P. Shah, "Dependence of Concrete Fracture Toughness on Specimen Geometry and Composition," in *Fracture Mechanics of Concrete*, eds., A. Carpinteri and A. R. Ingraffea, Martinus Nijhoff Publishers, 1984, pp. 118–120).

Two-Parameter Fracture Model

Jenq and Shah[28] proposed a model based in two-parameter that does not require the post-peak constitutive relationship. The two parameters are (1) the critical stress intensity factor (K_{ICs}), calculated at the tip of the effective crack, and (2) the elastic critical crack tip opening displacement (CTODc). Based on their experimental results, it was concluded that the two parameters are size independent. One of the advantages of this method is that the two parameters can be calculated using linear elastic fracture mechanics.

Figure 12–41 shows the relationships between the load P, and the crack mouth opening displacement (CMOD). Figure 12–41a shows a linear response for smaller loads (up to 50% of the maximum load), as the load increases (Fig. 12–41b), inelastic displacement and slow crack growth start to develop until it reaches a critical point (Fig. 12–41c), when the crack tip opening displacement reaches a critical value $K_I = K_{ICS}$.

Size Effect

One of the most important consequences of fracture mechanics is that the size of the structure affects the ultimate failure load, as shown in Fig. 12–42b. In this section we will discuss how to incorporate nonlinear fracture mechanics to study the importance of structural size. As an illustration, we will present an example of how to develop a simplified size effect law using the smeared crack band model. Bazant[29] applied this model for various loads and crack configurations. Let us consider the simple case of a crack with length $2a$ in a plate of thickness b, width $2d$ and length $2L$. The width of the crack band is w_c as indicated in Fig. 12–42a. Applying a stress σ to the plate originates a uniform strain energy density of $\sigma^2/2E$ before cracking.

---

[28] Y. Jenq and S. P. Shah, "Two Parameter Fracture Model for Concrete," *Journal of Engineering Mechanics*, Vol. III, No. 10, 127 (1985).

[29] Z. P. Bazant, in *Fracture Mechanics of Concrete*, G. C. Sih and A. DiTommaso editors, Martinus Nijhoff Publishers. Netherlands, pp. 1–94, 1985.

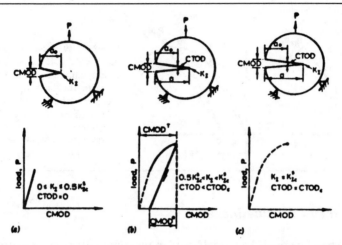

Figure 12–41 Fracture resistance stages of plain concrete: (a) $K_1 < 0.5K_k^s$; (b) nonlinear range; (c) critical point, $K_1 = K_k^s$. (From Y. Jenq and S. P. Shah, "Two Parameter Fracture Model for Concrete," *Journal of Engineering Mechanics*, Vol. 10, No. 10, 1985, p. 229).

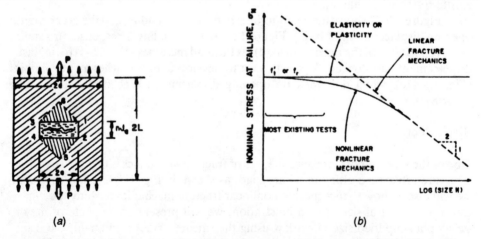

Figure 12–42 (a) Approximate analysis of energy release; (b) structural size effect. (From Z. P. Bazant, "Mechanics of fracture and progressive cracking in concrete structures," in *Fracture Mechanics of Concrete*, G. C. Sih and A. DiTommaso eds., Martinus Nijhoff Publishers, 1985, pp. 42, 44).

Fracture Mechanics

After the crack band is formed, it will relieve strain energy both from the area 1342, and from the uncracked areas 163 and 245, as shown in Fig. 12–42a. Let us further assume that the lines 25, 45, 16, and 36 have a fixed slope k_1. These are strong assumptions and they make the computation of the energy release due to the crack band approximate in nature. The energy release is given by:

$$W = W_1 + W_2 \tag{12-239}$$

where W_1 is the energy released by the uncracked areas 163 and 245 given by $2k_1 a^2 b(\sigma^2/2E)$, and W_2 is the energy released in the area 1342 given by $2w_c ab(\sigma^2/2E)$.

Considering fixed top and bottom conditions, the work of the load on the boundary is zero. Therefore, the potential energy release rate, $\partial W/\partial a$, is given by:

$$\frac{\partial W}{\partial a} = 2(2k_1 a + w_c) \frac{b\sigma^2}{2E} \tag{12-240}$$

In the critical stage, the energy criterion for the crack band extension is given by

$$2G_f b - \frac{\partial W}{\partial a} = 0 \tag{12-241}$$

Where G_f is the fracture energy of the smeared crack band model (Equation (12-224)). Combining Eqs. (12-240) and (12-241) yields:

$$\sigma = \left[\frac{2EG_f}{2k_1 a + w_c}\right]^{1/2} \tag{12-242}$$

which can be rewritten as:

$$\sigma = \frac{Af_t^*}{\sqrt{1+\lambda}} \qquad \lambda = \frac{d}{d_o} \tag{12-243}$$

where $d_o = (w_c/2k_1)(d/a)$ is a constant because d/a is kept constant for geometrically similar plates and $Af_t^* = (2EG_f/w_c)^{1/2}$.

Equation (12-243) represents the size affect law. For small structures, $\lambda \ll 1$ and Eq. (12-243) predicts a constant strength, independent of size. For large structures λ is much larger than unity so Eq. (12-243) is inversely proportional to the square root of d, as predicted by linear elastic fracture mechanics. Finally, Eq. (12-243) is also able to provide a smooth transition between the strength and linear elastic fracture mechanics criteria, as shown in Fig. 12–42b.

Another aspect of the size effect is the qualitative behavior of the structure in terms of softening response, which may change dramatically when considering different sizes. For instance, two geometrically similar structures may be characterized by a "ductile," or snap-through response for a small size, and a "brittle," or snap-back response for a larger size. This is a natural consequence of the mismatch in elastic energy stored per unit volume, and the fracture energy dissipated per unit area. This problem is particularly important when laboratory data must be used in structural prediction.

TEST YOUR KNOWLEDGE

1. Suppose the objective of a laboratory experiment is to measure the effect of freezing on the elastic modulus of concrete. You take two identical concrete samples from the fog-room, and then you test one sample in the saturated condition and the other, in a frozen condition (say $-20°C$). Assume that the freezing was done carefully and, therefore, did not generate microcracks. Which concrete will have a higher elastic modulus: saturated-concrete or frozen concrete?

2. A 5000 psi concrete is made with limestone aggregate. Suppose you replace 50 percent of the aggregate with solid steel balls (about the same size as the aggregates). Is the compressive strength going to increase? What about the elastic modulus? Please justify your answer.

3. A series of experiments on identical specimens of Maxwell material were performed such that in each experiment the strain rate was held constant. Sketch a family of stress-strain curves corresponding to three different strain rates: very slow, moderate, and very fast. For each case determine $E(0)$. Discuss the implications of results in practical applications.

4. Study the response of a standard-solid material subjected to a cyclic strain $\epsilon(t) = \epsilon_o \cos wt$, where ϵ_o is the strain amplitude and w the frequency.

5. Consider the following conditions for a 5000 psi compressive strength concrete (*justify* your answers):
   (a) The basic creep of the concrete at 28 days is 300×10^{-6} under a compressive load of 1000 psi. Can you estimate the basic creep of the same concrete at 90 days under a compressive load of 1000 psi?
   (b) The basic creep of the concrete at 28 days is 300×10^{-6} under a compressive load of 1000 psi. Can you estimate the basic creep of the same concrete at 28 days under a compressive load of 1500 psi?
   (c) The basic creep of the concrete at 90 days is 1200×10^{-6} under a compressive load of 4000 psi. Can you estimate the basic creep of the same concrete at 90 days under a compressive load of 1000 psi?
   (d) The basic creep of the concrete at 28 days is 300×10^{-6} under a compressive load of 1000 psi. The drying shrinkage (50 percent R.H.) at 28 days is 100×10^{-6}. Can you estimate the creep of the same concrete at 28 days under a compressive load of 1500 psi and exposed to 50 percent R.H.?
   (e) The basic creep of the concrete at 28 days is 300×10^{-6} under a compressive load of 1000 psi. The drying shrinkage (50 percent R.H.) at 28 days is 100×10^{-6}. Can you estimate the basic creep of the same concrete at 28 days under a compressive load of 4000 psi?

6. A Burgers model is made by connecting a Maxwell and a Kelvin model in series. Suppose that a Burgers material is maintained under a constant stress until time t_1 and then unloaded. Draw the graph of strain versus time.

7. Assume that a mass concrete structure should not have a temperature difference greater than 13°C. Given the following conditions: adiabatic temperature rise, 42°C; ambient temperature, 23°C; temperature losses, 15°C. Determine the maximum temperature of fresh concrete to avoid cracking.

8. Compute the energy release rate for the double cantilever beam when loaded by end moments.

9. Show for plane stress that the energy release rate G is equal to

$$G = \frac{K_I^2}{E} + \frac{K_{II}^2}{E} + \frac{K_{III}^2}{2S}$$

where S is the shear modulus.

10. Compare, critically, the advantages and limitations of the various techniques used for the determination of the fracture process zone in concrete.

SUGGESTIONS FOR FURTHER STUDY

Elastic Behavior

CHRISTENSEN, R. M., "A Critical Evaluation for a Class of Micromechanics Models," *J. Mech. Phys. Solids*, Vol. 18, No. 3, 379–404, 1990.

CHRISTENSEN, R. M., *Mechanics of Composite Materials*, John Wiley & Sons, 1976.

HENDRIKS, M. A. N., "Identification of Elastic Properties by a Numerical-Experimental Method," *Heron*, Delft Univ. of Technology, The Netherlands, Vol. 36, No. 2, 1991.

Viscoelasticity

CREUSS, G. J., *Viscoelasticity—Basic Theory and Applications to Concrete Structures*, Springer-Verlag, 1986.

FLUGGE, W., *Viscoelasticity*, Springer-Verlag, New York, 1975.

GILBERT, R. I., *Time Effects in Concrete Structures*, Elsevier Science Publishers, 1988.

NEVILLE, A. M., W. H. DILGER, and J. J. BROOKS, *Creep of Plain and Structural Concrete*, Longman Inc., New York, 1983.

RUSCH, H., JUNGWIRTH D., and HILSDORF H. K., *Creep and Shrinkage; Their Effect on the Behavior of Concrete Structures*, Springer-Verlag, New York, 1986.

Thermal Stresses in Mass Concrete

WILSON, E., "The Determination of Temperatures within Mass Concrete Structures," Report No. UCB/SESM-68-17, University of California at Berkeley, 1968.

POLIVKA, R. M., and WILSON, E., "DOT/DETECT: Finite Element Analysis of Nonlinear Heat Transfer Problems," Report No. UCB/SESM-76/2, University of California at Berkeley, 1976.

ACKER, P., and REGOURD, M., "Physicochemical Mechanisms of Concrete Cracking," in *Materials Science of Concrete II*, Skalny, J. and Mindess, S., eds., The American Ceramic Society, 1991.

Fracture Mechanics of Concrete

Analysis of Concrete Structures by Fracture Mechanics, L. Elfgren and S. P. Shah, eds., Chapman and Hall, 1991.

Fracture Mechanics: Application to Concrete, V. C. Li, and Z. P. Bazant, eds., American Concrete Institute SP-118, 1989.

Fracture of Concrete and Rock: Recent Developments, S. P. Shah, S. E. Swartz, and B. Barr, eds., Elsevier Applied Science, 1989.

SLUYS, L. J., and DE BORST, R., "Rate-dependent Modeling of Concrete Fracture," *Heron*, Delft Univ. of Technology, The Netherlands, Vol. 36, No. 2, 1991.

WHITTMAN, F. H., ed., *Fracture Mechanics of Concrete*, Elsevier Applied Science Publishers, Barking, Essex, U.K., 1983.

CHAPTER

13

The Future of Concrete

PREVIEW

For a variety of reasons discussed in Chapter 1, concrete is the most widely used construction material today. What about the future? In "Concrete for the Year 2000," C. E. Kesler says: "Concrete, as a construction material, has been important in the past, is more useful now, and is confidently forecast to be indispensable in the future."[1] The forecast is borne out if we apply to concrete the time-honored rules of the marketplace: that is, demand, supply, and economical and technical advantages over the alternative structural materials, such as lumber and steel. Furthermore, it is shown in this chapter that compared to other building materials the use of concrete is not only more energy-efficient but also ecologically beneficial. Large-scale application of the principles of material science to concrete production technology offers the hope that in the future the general product will be considerably superior in strength, elasticity, and toughness to the one available today.

FUTURE DEMAND FOR STRUCTURAL MATERIALS

From the standpoint of industrial development, the world can be divided into two parts: one where the process of industrialization-urbanization began more than 100 years ago, and one where it started essentially after the end of World War II. It seems that in the foreseeable future both parts of the world will require large amounts of structural materials.

[1] C. E. Kesler, in *Progress in Concrete Technology*, ed. V. M. Malhotra, CANMET, Ottawa, 1980, pp. 1–23.

525

In the developed world, gigantic construction programs are being planned for metropolitan areas not only for new construction but also for rehabilitation or replacement of existing structures, for example, **buildings** for home, office, and industrial use; **transit systems** (highways, railroads, bridges, harbors, airports, etc.) for transporting people and goods; and **water and sewage-handling facilities** such as pipelines, storage tanks, and treatment plants. Today's structures, meant for use by a large number of people in most of the well-known metropolitan areas of the world, are bigger and more complex; therefore, they require massive foundations, beams, columns, and piers, for which reinforced or prestressed concrete generally offers technical and economical superiority over lumber and steel. Commenting on the future of concrete in North America, B. C. Gerwick, Jr., says:

> The trend to urbanization, including satellite cities, will continue. Underground structures and deep foundations will require large volumes of concrete construction. Multi-story buildings will increasingly utilize concrete, with emphasis on design for ductility under earthquake, wind, and other lateral and accidental forces. Rapid transit systems will have to be expanded and new ones undertaken....
>
> As the 20th century has progressed, concrete construction has increasingly become the dominant building material in North America. Shear volume however, is not the only measure of success. It is encouraging to note its rapidly increasing acceptance as the material of choice for use under extreme load and severe conditions in the newer, more sophisticated and more demanding applications.[2]

Even more rapid escalation of population in the metropolitan areas of developing countries (Table 13–1) ensures the future of concrete there. In 1950, only 7 urban areas held more than 5 million people: New York, London, Paris, Rhine-Ruhr complex, Tokyo-Yokohama, Shanghai, and Buenos Aires. According to a United Nations report, in 1980, 34 world centers contained this number; by the year 2000 there will be 60. The data in the table show that 17 of 20 cities with a population over 11 million will be in the developing countries. The upsurge in urban population has already caused sprawling slums, massive transport problems, and an acute shortage of housing, water, and sewage facilities. A report in *Time* magazine (August 6, 1984) says that in Mexico City more than 2 million of the 17 million inhabitants have no running water in their homes; more than 3 million have no sewage facilities. To remedy the situation, an elaborate system of canals and pipelines for water supply is already under construction. Similarly, in Cairo (12 million) more than one-third of residential buildings are not connected to any sewage system. In Calcutta (10 million), 40 percent of homes are over 75 years old, and 20 percent of them are classified as unsafe. The experience of urban growth rate in Asian and South American countries shows that leisurely development is a luxury of the past, and structural materials such as concrete will be needed for a long time and in large quantities.

---

[2] B. C. Gerwick, Jr., *Concr. Int.*, Vol. 6, No. 2, pp. 36–40, 1984.

TABLE 13-1 LARGEST METROPOLITAN AREAS OF THE WORLD

1980 population (millions)		2000 population (millions)	
1. Tokyo-Yokohama	17.0	Mexico City	26.3
2. New York	15.6	São Paulo	24.0
3. Mexico City	15.0	Tokyo-Yokohama	17.1
4. São Paulo	12.8	Calcutta	16.6
5. Shanghai	11.8	Bombay	16.0
6. Buenos Aires	10.1	New York	15.5
7. London	10.0	Seoul	13.5
8. Calcutta	9.5	Shanghai	13.5
9. Los Angeles	9.5	Rio de Janeiro	13.3
10. Rhine-Ruhr	9.3	New Delhi	13.3
11. Rio de Janeiro	9.2	Buenos Aires	13.2
12. Beijing	9.1	Cairo	13.2
13. Paris	8.8	Jakarta	12.8
14. Bombay	8.5	Baghdad	12.8
15. Seoul	8.5	Tehran	12.7
16. Moscow	8.2	Karachi	12.2
17. Osaka-Kobe	8.0	Istanbul	11.9
18. Tianjin	7.7	Los Angeles	11.3
19. Cairo	7.3	Dhaka	11.2
20. Chicago	6.8	Manila	11.1

FUTURE SUPPLY OF CONCRETE

The concrete needed for the construction programs of both the developed and the developing parts of the world will be readily available. Portland cement and aggregate are the essential raw materials for concrete. Essentially unlimited supplies of aggregate such as sand, gravel, and crushed rock are available almost everywhere in the world. The main raw materials for making portland cement are limestone, clay, and fossil fuel in the form of coal, oil, or natural gas. In most parts of the world, limestone and clay occur in abundance. As discussed below, compared to steel the production of cement requires less energy, and the production of concrete far less. Coal is generally available in most countries, and the type or quality of coal is not a problem in the manufacture of cement.

ADVANTAGES OF CONCRETE OVER STEEL STRUCTURES

Due to the high cost of lumber in urban areas and the massive size of needed structural elements, it is usually steel that competes with concrete. In the future, the choice of steel versus concrete as a construction material will be increasingly in favor of concrete because it will be governed by considerations listed in Fig. 13–1, which are discussed below in detail.

Engineering Considerations

The following arguments presented by Gjerde,[3] which were probably instrumental in the selection of prestressed concrete gravity platforms instead of steel jacket structures for many offshore oil fields in the North Sea, amply demonstrate the desirable engineering characteristics of concrete.

Maintenance. Concrete does not corrode, needs no surface treatment, and its strength increases with time; therefore, concrete structures require essentially no maintenance. Steel structures, on the other hand, are susceptible to rather heavy corrosion in an offshore environment, require costly surface treatment and other methods of protection, and entail considerable maintenance and repair.

Fire resistance. An offshore fire could easily reach temperatures at which steel structures would be permanently damaged, thus causing a major threat to human safety and investment (e.g., total investment in the Stratfjord A platform was close to $1 billion). According to Gjerde, the fire resistance of concrete is perhaps the most important single aspect of offshore safety and, at the same time, the area in which the advantages of concrete are most evident. Since an adequate concrete cover on reinforcement or tendons is required for assuming structural integrity in

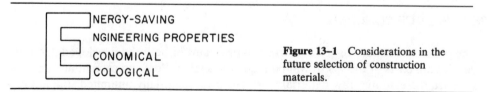

Figure 13–1 Considerations in the future selection of construction materials.

---

[3]T. Gjerde, *Nordisk Betong* (*Stockholm*), No. 2–4, pp. 95–96, 1982.

reinforced and prestressed concrete structures, protection against failure due to excessive heat is normally provided at the same time.

Resistance to cyclic loading. The fatigue strength of steel structures is greatly influenced by local stress fields in welded joints, corrosion pitting, and sudden changes in geometry, such as from thin web to thick frame connections. In most codes of practice, the allowable concrete stresses are limited to about 50 percent of the ultimate strength; thus the fatigue strength of concrete is generally not a problem.

In the Ekofisk oil field on March 27, 1980, Alexander L. Kielland, a 4-year-old steel platform of the Pentagon type, suddenly heeled over, turned upside down within about 20 min, and sank with 212 workers on board; 123 lives were lost. According to a report issued by the inquiry commission, one of the bracings holding a column fractured, and this was followed in rapid succession by failures (due to overloading) of the five other bracings which connected the column to the platform. The fracture in the bracing that initiated the structural failure was judged to be a fatigue fracture.

Vibration damping. In addition to human comfort, a good damping effect with minimal vibration in structures and machinery foundations on a production platform is very important for the success of operations. Due to lower self-weight, platform decks of steel show a lower damping effect against vibrations and dynamic loading than do corresponding concrete decks.

Control of deflections. Gjerde cites Leonhardt's observation that compared with steel girders of the same slenderness, the deflection of prestressed concrete girders is only about 35 percent. Also, by prestressing it is possible to give a girder a positive camber (upward deflection) under self-weight, and zero camber for the total payload.

Explosion resistance. Owing mainly to the very high elastic limit of the tendons commonly used in prestressed concrete beams, their explosion resistance is better than that of normal steel girders. An FIP report dealing with the behavior of floating concrete structures says: "Considering explosions, fires, sabotage, and missiles, structures of reinforced concrete imply less residual risks than alternative materials."

Resistance to cryogenic temperatures. Of immediate interest in North America is the construction of Arctic marine structures for exploration and production of oil off the Alaskan and Canadian coasts. Compared to the North Sea, the presence of floating icebergs and sheet ice offers a unique challenge for the construction material. According to Gerwick:

> An overriding criterion for the design of marine structures for the Arctic is that of high local pressures, which may reach almost 6000 kips over an area 5 ft by 5 ft in size. Typical

steel designs suffer in the lack of stress distribution between stiffeners, whereas concrete shells and slabs suitably prestressed and confined with heavy reinforcing steel are admirably suited to resist the punching shear from such ice impact.[4]

Another aspect favoring concrete is its ductile behavior under impact at subzero temperatures. Normal structural steel becomes brittle at low temperatures and loses its impact resistance. On the other hand, successful experience with prestressed concrete tanks for the storage of LNG (liquefied natural gas) at temperatures as low as $-260°F$ ($-162°C$) has opened the opportunities for expanding the use of concrete under cryogenic conditions. It seems that prestressed concrete is the only economically feasible material which is safe for use under ambient- as well as low-temperature conditions.

Economic Considerations

The principal ingredient in concrete, the aggregate, costs only $3 to $5 per ton; the cost of portland cement varies between $40 and $60 per ton (1984 U.S. prices). Depending on the cement and admixture contents, the delivered price of normal-weight concrete at the job site is usually in the range $40 to $80 per cubic yard, or $20 to $40 per ton. Plain concrete is, therefore, a relatively inexpensive material.

Considering the lower strength/weight ratio and added costs of reinforcing steel and labor, the difference between the direct cost of a steel versus a reinforced or prestressed concrete substructure may not turn out to be large. With the advent of high-strength concrete, however, the cost-effectiveness is shifting in favor of concrete. For example, a report on a cost analysis of offshore platforms in the Norwegian continental shelf shows that during the period 1973–1979, whereas the relative cost of a concrete platform has almost doubled, the cost of steel platform has risen from 100 percent to about 240 percent. However, equally important to the direct platform cost are the maintenance and inspection costs capitalized over the platform's operational life. In this respect, concrete definitely has an advantage over steel.

Furthermore, as the increasing use of high-strength concrete in prestressed and reinforced elements for high-rise structures shows (Chapter 11), in this era of high interest rates and the high cost of large projects, cash flow considerations are very important to the buyer. Concrete gravity platforms can be installed in a matter of days and be ready for drilling in a few months. Steel platforms, on the other hand, because of pile installation and deck completion at sea, may take almost a year.

Energy Considerations

Many studies have shown that concrete structures are more energy-conserving than steel structures. On the basis of studies at the Tampere University of Technology,

---

[4] B. C. Gerwick, Jr., *Proc. Star Symp.*, *Society of Naval Architects and Marine Engineers*, New York, April 1984.

Finland,[5] depending on the cement content, the energy content of 1 m³ of concrete may be calculated as follows:

Item	Energy content charged to concrete	
	kwh/m³	kwh/ton
Cement, 250 to 500 kg/m³ concrete	330–660	137–275
Aggregate, 1750 to 1950 kg/m³ concrete	20	8
Production and handling concrete	90	37
Total	440–770	182–320

In these calculations the energy content of cement is assumed to be 1300 kwh/ton or 1.3 kwh/kg. The energy content of steel (about 8000 kwh/ton) is six times as high as that of cement. Due to the relatively large amount of aggregate, which contributes only 8 kwh/ton, the energy content of plain concrete amounts to 1/25 to 1/40 that of steel. Because of the presence of steel, the energy content of reinforced or prestressed concrete rises to relatively high levels. Depending on the amount of cement and steel, reinforced concrete elements will have 800 to 3200 kwh/m³, and prestressed concrete elements 700 to 1700 kwh/m³ energy contents. These data include heat treatment for precast elements (50 kwh/m³) and transportation energy for finished elements (100 kwh/m³).

Oscar Beijer of the Swedish Cement and Concrete Research Institute[6] also showed that the steel reinforcement and the cement make major contributions to the energy content of concrete, whereas the aggregate component and the transport of concrete require only a little energy. The comparative energy consumption data in Fig. 13-2a are shown in terms of liters of oil. The author also made a comparison of the energy consumption of a 1 m-high column resisting a 1000-ton load and constructed of unreinforced concrete, brick (a single-wythe wall), or steel (Fig. 13-2b). The concrete column clearly required much less energy.

According to Kesler,[7] even though dead weight is a significant factor in concrete flexural members, the energy required for a concrete beam in place may be only one-fourth to one-sixth that of a comparable steel beam. Another energy-saving aspect associated with concrete is the high thermal lag of the material. In what can be termed as comparable structures, concrete structures will require 10 to 35 percent less energy for heating and cooling. Kesler also points out that in addition to the contributions of concrete in conserving energy, the role of concrete in the exploration and production of new energy sources should not be overlooked. Offshore platforms, storage tanks, and terminals made of concrete are playing an important part in the development of new sources of oil and gas.

[5] M. Alasalmi, and H. Kukko, Publ. 2, Nordic Concrete Research, Oslo, 1983, pp. 8–15.
[6] O. Beijer, *J. ACI*, Proc., Vol. 72, No. 11, pp. 598–600, 1975; translated by A. E. Fiorato of PCA.
[7] C. E. Kesler, in *Progress in Concrete Technology*.

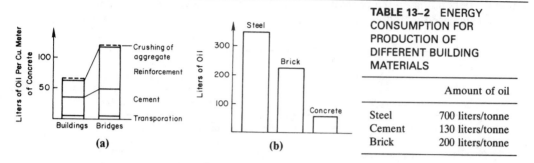

Figure 13-2 Energy efficiency of concrete compared to other building materials. (a) Average energy consumption per cubic meter of concrete in the finished structure; (b) energy consumption of one meter high column resisting a 1000-tonne load and constructed of various materials. (From O. Beijer, translated by A. E. Fiorato, *J. ACI*, Proc., Vol. 72, No. 11, p. 599, 1975).

Ecological Considerations

Designers of structures generally evaluate and select materials on the basis of their engineering properties and cost. Increasingly, the properties influencing human health are being taken into consideration, such as any toxic fumes or radiation associated with the production and use of a material. In the future, the users will also have to consider ecological properties; for instance, the desoiling or deforestation required to obtain the raw materials, energy and water consumption, and pollution, as well as waste created during the manufacturing process.

Based on 1987 costs in the Netherlands expressed per unit of strength or elastic modulus, Kreijger compared the ecological properties of steel, glass, burnt-clay or sand-lime brick masonry, reinforced concrete, and wood (Table 13-3). From the data, typical **ecological profile** for elastic modulus of these materials, expressed in units per volume per N mm-2, is plotted in Fig. 13-3. The plot clearly shows that among the common materials of construction, reinforced concrete is the most **environmental friendly**. In his concluding remarks, the author suggests

> Because it is people who determined how materials are used in the society, each designer, in making his or her choice of building materials, is also responsible for the **ecological and social consequences of that choice**.

From technical, economic, and ecological considerations, there is no better home than concrete for millions of tons of pozzolanic and cementitious by-products (i.e., fly ash and blast-furnace slag). Stockpiling of these by-products on land causes air pollution, whereas dumping into ponds and streams releases toxic metals that are normally present in small amounts. Even low-value applications such as land-fill and the use of granular materials for subgrades and highway shoulders are ultimately

[8] P. C. Kreijger, *Materials and Structures*, Vol. 20, No. 118, 1987, pp. 248-54.

TABLE 13-3 ECOLOGICAL PROPERTIES PER UNIT OF VOLUME MATERIAL, GIVEN PER N mm^{-2} STRENGTH AND PER N mm^{-2} MODULUS OF ELASTICITY. (REF. 8)

Properties	Steel (Fe 360)	Glass	Brickwork	Sand lime brick work	Reinforced concrete	Wood
Costs in (Fl m^{-3})						
per σ	72	270	78.7	53.3	25.9	71.4
per E	0.082	0.12	0.12	0.08	0.013	0.091
Energy (MJ m^{-3})						
per σ	983	1867	1467	653	444	143
per E	1.12	0.86	2.2	0.98	0.21	0.18
Water (1 m^{-3})						
per σ	1788	—	160	133	47	—
per E	2.04	—	0.24	0.20	0.023	—
Pollution SO$_2$ (kg m^{-3})						
per σ	0.058	0.107	0.24	0.053	0.074	+
per E (10^4)	0.67	0.49	3.6	0.80	0.36	+
Dust/soot (kg m^{-3})						
per σ	0.163	+	0.12	0.12	0.074	+
per E (10^4)	1.85	+	1.8	1.8	0.36	+
Desoiling/deforestation in m^2 m^{-3}						
per σ	0.16	—	0.19	0.027	0.074	357
per E (10^4)	1.9	—	2.8	0.40	0.36	0.455
Labour (mh m^{-3})						
per σ	0.34	—	0.50	0.18	0.26	—
per E (10^4)	3.9	—	7.6	2.7	1.23	—

hazardous to human health because toxic metals will find their way into groundwater. On the other hand, incorporation of fly ash and slag as components of blended portland cements or as mineral admixtures in concrete presents a relatively inexpensive way of proper disposal of the toxic elements present. Work in Japan by Tashiro and Oba[9] has shown that many toxic metals can be chemically bonded in the hydration products of portland cement.

As described earlier (Chapter 8), besides the technical advantages of replacing a part of the portland cement in concrete with fly ash or blast-furnace slag, the mineral admixtures offer an easy way of conserving the cement-making raw materials and energy sources. Conservation of natural resources is, after all, ecological. The energy to produce fly ash or slag comes free to concrete. For instance, replacing 30 percent of portland cement in concrete with fly ash will reduce the energy content of the cementing material by almost the same amount. Granulated blast-furnace slag may be used to replace even larger amounts of cement in concrete. It is mainly from the standpoint of resource and energy conservation that the U.S. Environmental Protection Agency has issued guidelines prohibiting specifications that discriminate against the use of fly ash in construction projects funded wholly or partially by the federal government.

[9] C. Tashiro and J. Oba, *Cem. Concr. Res.*, Vol. 9, No. 2, pp. 253–58, 1979.

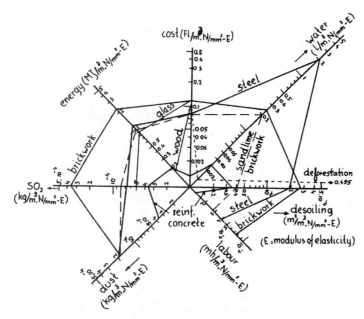

Figure 13-3 Ecological profile for some materials, properties expressed in units per volume per Nmm^{-2} modulus of elasticity. [From P. C. Kreijger, *Materials and Structures*, Vol. 20, No. 118, 1987, p. 253]

Materials with the smallest ecological profile should be preferred for use. Obviously the smallest profile will correspond with the material which, for a given engineering property, causes the least deforestation and desoiling, consumes the least energy and water, and gives out the least dust and SO$_2$ emissions for its manufacture. When ability to consume as raw materials large volumes of waste products generated by other industries is added to their ecological profile, it is portland cement concrete which emerges as the most environment-friendly among all the construction materials.

A BETTER PRODUCT IN THE FUTURE

Although as a structural material concrete generally has a history of satisfactory performance, it is expected that even a better product will be available in the future owing to overall improvements in elastic modulus, flexural strength, tensile strength, impact strength, and permeability.

It is now well known that ordinary concrete is weak in these properties due to certain deficiencies in the microstructure, such as the existence of numerous voids and microcracks in the transition zone, the presence of a low-strength solid component in the hydrated cement paste (i.e., crystalline calcium hydroxide), and heterogeneous distribution of the phases in the composite material. The science and technology to overcome these deficiencies are already available and, in fact, are being effectively used for producing high-strength and high-durability concretes. A wider dissemination of the principles that underlie the production of high-strength concrete, and their application to the concrete practice in general, will make it possible that even for ordinary structural purposes a better product will be available in the future.

Whereas the technology of materials produced either by solidification of high-temperature melts or by sintering of fine particles is not directly applicable to low-cost structural materials (i.e., hydraulic cementing products such as concrete), nevertheless the same scientific principles that govern the production of high-strength metals or ceramics are at work when a high-strength concrete is made. For instance, a reduction of the water content in a concrete mixture decreases the porosity of both the matrix and the transition zone and thus has a strengthening effect. Again, the presence of a pozzolan in a hydrating cement paste can lead to the processes of pore-size and grain-size refinement (Fig. 6–14), which have the effect of reducing both the size and the volume of voids, microcracks, and calcium hydroxide crystals, thus causing a substantial improvement in strength and impermeability. Since the use of water-reducing and mineral admixtures in concrete is generally cost effective, an awareness among the producers of concrete that application of these admixtures can help produce a more homogeneous and better product should result in their being routinely used for normal concrete production. Some of the practices that will be increasingly employed by the industry and will result in better concrete are as follows:

- A better control of the bleeding tendency in concrete mixtures will be sought through proper aggregate grading, and the use of water-reducing and mineral admixtures (e.g., fly ash or finely ground natural pozzolans or slags). In this regard, the use of even small amounts of silica fume (that is, 5 to 10 percent by weight of the cementitious material) may be sufficient, owing to the very high surface area of the material. In developing countries, where silica fume may not be economically available, the use of rice husk ash to reduce bleeding in fresh concrete mixtures, and also to serve as a highly active pozzolan, will result in similar benefits to the product.
- In addition to mineral admixtures, the application of air-entraining agents and superplasticizers will become more widespread for improving the workability, especially the compactability of concrete mixtures.
- Fiber reinforcement of concrete that is subject to cyclic or impact loads will be commonly practiced. For developing countries, the use of natural organic fibers (such as sisal fiber and rice straw) presents interesting possibilities provided that the chemical deterioration of the fiber is prevented by using a cement matrix that is far less alkaline than portland cement paste.
- The use of centralized and high-speed concrete mixers instead of truck mixing will help in the production of more homogeneous concrete than is generally available today.

As structures of tomorrow become larger and more complex, the materials of construction will be required to meet more demanding standards of performance than in force today. To meet these standards, today's products will have to be improved in strength, dimensional stability, durability, and overall reliability. For instance, in Japan the construction of a $10 billion project, The Trans-Tokyo Highway, which is scheduled to be completed by 1996, will include two 10-km tunnels at one end, a 5-km bridge at the other end, and two man-made islands in the middle. Like the offshore concrete platforms in the North Sea, the size and scope of the structures are far beyond the conventional structures because they will be built on very soft soils and under deep water. Therefore, advanced construction technologies and highly durable materials will be needed to meet the more demanding performance standards.

Research is underway to define and develop high-performance concrete mixtures. There seems to be a general agreement that, in addition to high workability, a high-performance concrete should show high dimensional stability, high strength, and high durability in service.[10] With numerous materials, especially pozzolanic and chemical admixtures, and many procedures for concrete mix proportioning, it is not easy to design high-performance concrete mixtures. A simple, step-by-step procedure has been suggested by Mehta and Aitcin.[11]

CONCLUDING REMARKS

Concrete is the dominant structural material today. In the future, in selecting the materials for construction, engineers will consider not only the technical and economical attractiveness of materials but also the ecological and energy-conserving implications of their use. Judged by the predictable parameters, the future of concrete should be even better because it seems to have the strongest overall qualifications for structural use. There is, however, one unpredictable factor, which is best expressed in the words of Gerwick:

> Does concrete really offer an opportunity for the future or just "cast-in-concrete" entombments of crude materialism that, like the stone pyramids of Egypt, will eventually serve only as dismal monuments of the past? The long-term answer is, of course, tied up in our overall societal aspirations and goals, and in our ability to develop local, regional, and national, and worldwide political institutions to match the needs of a modern technological age.[12]

---

[10] A. Mor, *Concrete Construction*, Vol. 37, No. 5, 1992.
[11] P. K. Mehta and P. C. Aitcin, *Cement, Concrete, and Aggregates*, Vol. 12, No. 2, pp. 70–78, 1990.
[12] B. C. Gerwick, Jr., *Concr. Int.*, Vol. 6, No. 2, pp. 36–40, 1984.

In the Hindu mythology, there are several gods. Siva is an ascetic god who is always ready to help the needy. Once the gods collectively decided to dewater an ocean to obtain the pot of nectar of immortality which, it was rumored, lay at the ocean floor. However, in the process a stream of poison was released and it started destroying the whole world. When no other god showed any willingness to handle the poison, Siva volunteered. He drank the whole stream of poison which left a permanent blue mark on his throat.

Cement industry is increasingly recycling hazardous organic wastes as fuel for making clinker, thus recovering their energy value and helping conserve fossil fuels. Reportedly, portland cement clinker is also a safe sink for a variety of toxic elements present in chemical wastes. Furthermore, since portland cement concrete is able to incorporate safely millions of tons of fly ash, slag, and other waste by-products which contain toxic metals, shall we call concrete Lord Siva of the construction materials' world?

Index

Abram's rule, 47
Abrasion
  control, 123
  test methods, 125
Accelerated strength testing, 345
Acid rain, 152
Acid resistance
  calcium aluminate cement, 222
  portland blast-furnace slag cement, 214
  portland cement, 142–44
  portland pozzolan cement, 212
Adiabatic temperature rise
  influence of cement content, 107
  influence of cement type, 107, 429
  influence of pozzolans, 107, 430
Admixtures
  accelerating, 268
  air entraining, 133–136, 259, 262
  applications, 256, 252, 268, 271, 281
  blast-furnace slag, 278
  calcined clay, 276
  calcium chloride, 268
  classification, 257, 273, 286
  condensed silica fume, 279, 372, 380
  definition, 10, 256
  diatomaceous earth, 276
  fly ash, 277
  plasticizing, 256–60, 262–63
  pozzolans (natural), 272
  retarding, 268, 271
  rice husk ash, 280
  set controlling (mechanism), 266
  specifications, 257
  superplasticizers, 263, 372–73
  surfactants, 259
  volcanic glasses, 274
  water-reducing, 259–63
  zeolites, tuff, 276
Adsorbed water, 29, 89
AFm, AFt, 193
Aggregate
  alkali-reactive, 155–156, 233
  classifications, 227, 232–34
  definition, 8
  deleterious substances, 250
  expanded slag, 237
  ferrophosphorous, 236
  fly ash, 237
  foamed slag, 237
  gravel, 8, 239, 232
  graywacke, 229, 232
  heavyweight, 235
  lightweight, 231, 235, 240
  municipal waste, 237
  natural rocks and minerals, 237
  perlite, 234
  production
    lightweight (expanded shale), 240
    natural sand and gravel, 238
  properties and significance

Index

abrasion resistance, 244
absorption capacity, 242
apparent specific gravity, 242
bulk density, 227, 242
crushing strength, 244
elastic modulus, 244
fineness modulus, 247
grading, 245, 294
maximum size, 245
shape, 21, 247
soundness, 244
sphericity, 250
surface moisture, 242
texture, 21, 247, 432
recycled concrete, 237
sedimentary rocks, 228, 232
slag, 236, 237
specifications, 246–47, 252–53
sulfide and sulfate minerals, 231
test methods, 252
Air entrainment
admixtures, 259, 262
cements, 201
effect on frost resistance, 133–34
effect on strength, 48, 136
void size and spacing, 26, 133
Alite, 187, 204
Alkali-aggregate reaction
alkali-reactive aggregates, 154–55
case histories, 157
cement composition, 154, 214
control of expansion, 158
forms of damage, 153, 155
low-alkali cement, 155
mechanism of expansion, 155
sources of alkali, 154
Alkalies in portland cement, 189
Alkali-silica reaction. *See* Alkali-aggregate reaction
Ammonium salt solutions, effect on concrete, 154
Anhydrite, 231
Aphthitalite, 189
Autoclave expansion test, 205–06

Barium ores, 236
Batching. *See* Fresh concrete
Belite, 187, 204
Biaxial behavior, 73–76
Blaine, fineness, 206
Blast-furnace slag
as admixture, 278
as aggregate, 236
mineralogical composition, 207
Blast-furnace slag cements. *See* Cement types
Bleeding
causes and control, 332
definition and significance, 330
influence on strength, 22, 331
measurement, 331
Blended portland cements, 207
Bogue equations, 185
Bond, aggregate-cement paste. *See* Transition zone
Brucite, 169, 172–73
Bulking of sand, 244
Bull-float, 319

C_3A, C_4AF, $C_4A_3\bar{S}$, C_3S, C_2S, 182
Calcium aluminate cement
composition, 220
curing, 221–23
refractory use, 223
strength retrogression, 221
Calcium aluminates
CA, 220, 223
C_3A, 186, 189–90, 193, 203
Calcium chloride, 286–70
Calcium ferroaluminates, Fss, C_4AF, 187, 193, 204
Calcium formate, 267
Calcium hydroxide, 25·
Calcium oxide, 159, 189
Calcium silicate hydrate, 24–26, 196
Calcium silicates, C_3S, C_2S, 187, 196, 204
Calcium sulfates, 180–81
Calcium sulfoaluminate, 189
Calcium sulfoaluminate hydrates, 26
Capillary voids (pores) in cement paste
calculation of the amount, 32
origin and size, 27
significance, 29–32
size distribution plots, 28, 35
Capillary water, 29, 33
Carbon dioxide, effect on concrete, 145, 163, 169–71
Cathodic protection, 167
Cavitation, 123
Cement fineness, 190–91, 202–03, 206
Cement paste. *See* Hydrated cement paste
Cement soundness, 189, 205–06
Cement specifications, 201–05
Cement strength, 206
Cement types
calcium aluminate, 220–24
colored, 209, 219

Cement types (*Cont.*)
  expansive, 208, 214
  gypsum, 180–81
  high iron, 217
  hydraulic, 8, 180–81
  jet set, 208, 217
  lime, 180–81
  nonhydraulic, 180
  oilwell, 208, 217
  portland, 9, 180–205
  portland blast-furnace slag, 207–14
  portland pozzolan, 208–14
  regulated-set, 208, 216
  self-stressing, 214
  shrinkage-compensating, 214
  very high early strength, 208, 217
  white, 209, 219–20
Chalcedony, 230
Chemically combined water, 29
Chert, 229, 232
Chloride, corrosion, 163
Chord modulus of elasticity, 83
Ciment fondu, 220
Clinker
  in hydrated cement paste, 26
  in portland cement, 180–82
Coarse aggregate
  definition, 8, 227
  grading requirement, 246
Cold-weather concreting
  recommended practice, 337
  temperature control, 339
Compacting factor test, 323–24
Compaction (consolidation), 317
Composite models
  differential scheme, 451, 452
  Hansen, 450
  Hashin-Shtrickman, 453
  Hirsch, 449
  Mori-Tanaka, 451, 452
  Reuss, 447
  Voigt, 446, 447
Compressive strength. *See* Strength
Concrete
  advantages over steel, 528
  applications, 2–8, 526
  classifications, 10
  components, 8
  definition, 8
  future product, 534
  high-performance, 535–536
  principal characteristics, 11–14
  ready-mixed, 311

  special types, 357–444
  structure (microstructure), 18
  world consumption rate, 2
  world demand and supply, 525–528
Condensed silica fume, 279, 372, 380
Consistency, 322
Core tests, 348
Corrosion of steel in concrete
  bridge decks, 161, 164
  carbonation, 163
  case histories, 163, 171, 174–76
  cathodic protection, 167
  chloride threshold, 163
  concrete cover, 166
  corrosion cells, 161
  cracking-corrosion cycle, 175
  electrical resistivity, 163
  electrochemical reactions, 161–63
  measures for control, 165–67
  mechanism of expansion, 161–63
  passivity of steel, 163
  permissible crack widths, 166
  protective coatings and overlays, 167
Cost of concrete, 54, 368, 530
Cracking (microcracking)
  chemical causes, 143
  influence of extensibility, 109
  influence of shrinkage and creep, 79
  physical causes, 122
  plastic shrinkage, 332–34
  stable and unstable cracks, 62–63
Creep
  basic creep, 90
  creep coefficient, 90
  definition, 13, 78
  drying creep, 90
  factors affecting
    admixtures, 97
    aggregate, 93–94
    cement content and type, 95–96
    curing history, 100
    mix water content, 96
    relative humidity, 97
    size of member, 98
    strength, 97
    stress intensity, 100
    temperature, 100
    time after loading, 97
  mechanisms, 89
  relationship to cracking, 80
  relationship to drying shrinkage, 89
  significance, 13, 78
  specific creep, 90

Index

Critical aggregate size, 132
Critical degree of saturation, 136–37
Critical energy release rate, 497
Critical stress, 63
Critical stress intensity factor, 499
Cryogenic behavior, 529
Crystallization pressure of salts, 126
Crystal structure and reactivity
  calcium ferroaluminate (C_4AF), 186–87
  calcium oxide, 189
  calcium silicates (C_3S, C_2S), 186–87
  magnesium oxide, 189
  tricalcium aluminate, 186–87
C-S-H, 24
Curing
  definition, 56
  effect on strength, 56–60
  methods, 320
  significance, 320

Darby, 319
D-cracking, 127–28, 133
Deicing salts action, 138
Diatomaceous earth, 276
Dicalcium silicate (C_2S), 185, 187, 197
Dolomite, 229, 233
Drying shrinkage
  definition, 13, 79
  factors affecting
    admixtures, 97
    aggregate, 93–94
    cement content, 96
    exposure time, 97
    relative humidity, 97
    size of member, 98
    water content, 96
  mechanism, 89
  relationship to cracking, 79
  relationship to creep, 89
  reversible shrinkage, 90
  significance, 13, 78
Ductility, 12
Durability
  aggregates, 244, 250
  concrete, 113–77
  definition, 13, 114
  relationship to permeability, 115
  role of water, 114–15
  significance, 114
Durability factor, 137
Dynamic modulus of elasticity, 83

Early-age behavior. See Fresh concrete
Ecological benefits, 532–534

Effective absorption, 243
Efflorescence, 53, 144
Elastic modulus
  chord modulus, 83
  computation from strength, 85
  dynamic modulus, 83
  factors affecting
    aggregate type, 86
    curing age, 87
    moisture condition, 87
    rate of loading, 88
    temperature, 140
    transition zone, 87
  flexural modulus, 83
  models. See Composite models
  secant modulus, 82–83
  static modulus, 83, 84
  steel, 13
  tangent modulus, 83
Electron micrographs
  condensed silica fume, 280
  ettringite, monosulfate, 24
  fly ash, 278
  hydrated cement paste, 20, 185
  rice husk ash, 280
Energy requirement, 530–532
Entrained air. See Air entrainment
Erosion, 123
Ettringite
  composition, morphology, 24, 26
  formation in cement paste, 194, 215
  formation in sulfate attack, 147
  mechanism of expansion, 147
Expanded clay and shale, 235, 240
Expanded slag aggregate, 237
Expansive cement concrete. See Shrinkage-
    compensating concrete
Expansive cements
  self-stressing, 214
  shrinkage-compensating, 214
  Type K, 214
  Type M, Type O, Type S, 215–16
Expansive phenomena in concrete
  alkali-aggregate reaction, 154
  corrosion of steel in concrete, 160
  frost action, 127
  hydration of calcium oxide, 159
  hydration of magnesium oxide, 159
  sulfate attack, 146
Extensibility, 109

False set, 194–95
Feldspar, 230

Fiber-reinforced concrete
  applications (significance), 404, 416
  aspect ratio of fibers, 407, 413
  creep, 415
  definition, 404
  drying shrinkage, 415
  durability, 415
  elastic modulus, 415
  fiber types and properties, 407-08
  first crack strength, 407-08, 414
  impact resistance, 414-15, 417
  load-deflection curves, 406, 414
  mix proportions, 410-12
  strength, 412, 417
  toughening mechanism, 405
  toughness, 415, 417
Final set, 198, 200, 206, 335
Fine aggregate, 8, 227, 247-48
Fineness modulus, 247
Fineness of cement. See Cement fineness
Finishing. See Fresh concrete
Flash set, 195
Finite element formulation
  for cracking problems, 500-505
  for temp. distribution, 485-488
Flexural strength, 68-69
Flint, 229, 232
Floating. See Fresh concrete
Flowing concrete. See High-workability concrete
Fly ash
  as an aggregate, 237
  ASTM classification, 273
  composition, 272, 277
  definition, 277
  effect on properties of concrete, 281-85
  particle characteristics, 273, 278
  plerospheres and cenospheres, 278
  as a pozzolan, 277, 287
  reactivity, 277
Foamed slag. See Aggregate
Formwork removal, 320
Fracture Mechanics
  finite element formulation, 500-505
  for concrete, 505-522
  fracture process zone
    fictitious crack model, 511-516
    methods of determining, 509-511
    smeared crack band, 510-511, 519-521
  fracture resistance curves, 517-518
  linear elastic, 494-500
  size effect, 519-521
  two-parameter fracture model, 519

Free calcium oxide (lime), 161, 189
Freeze-thaw resistance. See Frost action on concrete
Fresh concrete
  batching, 311
  bleeding, 22, 330-31
  compaction, 317
  consistency, 322-30
  curing, 320
  finishing, 317
  floating, 319
  mixing, 311
  placing, 317
  plastic shrinkage, 332-34
  quality control, 344-54
  retempering, 327
  revibration, 318
  screeding, 318
  segregation, 330
  setting time, 335-36
  slump cone test, 323-24
  slump loss, 326-27
  temperature, 337-44
  transporting, 313-17
  Vebe test, 323-25
  vibration, 317-18
  workability, 322-30
Frost action on aggregate, 132-33
Frost action on cement paste, 129-31
Frost action on concrete
  definition, 127
  durability factor, 135
  factors controlling
    aggregate, 134
    air content and spacing, 133-34
    curing, 135
    degree of saturation, 136
  forms of damage, 127-28
  methods of evaluating, 137
Fss, 187
Future of concrete, 525-536

Gamma rays, 426
Gel pores, 24
Gel/space ratio (effect on strength), 45
Gradation
  coarse aggregate, 246
  curves, 249
  fine aggregate, 247
Grain-size refinement, 211
Gravel, 9, 229, 232
Graywacke, 148, 229, 232

Grout, 9
Gypsum
    in aggregate, 231
    in cement, 190
    composition, 190
    role in cement, 193–95
    role in sulfate attack, 147–150

Hardening, 193, 199
hcp, 18
Heat of hydration
    calcium aluminate cement, 223
    factors affecting
        cement composition, 197–201
        cement fineness, 202–04
        pozzolanic materials, 107, 222, 281–82, 430
        temperature, 203
    method of estimation, 199
    portland blast-furnace slag cement, 213
    portland cement, 198–99, 201–03
    portland pozzolan cement, 212
    significance, 78–79
Heat transfer equations, 482–485
Heavyweight aggregates, 235–36
Heavyweight concrete, 425–28
Hemihydrate, 181, 190
High-alumina cement. See Calcium aluminate cement
High-early strength cements, 203, 205. See also Rapid setting and hardening cements
High-performance concrete, 535–536
High-strength concrete
    abrasion and resistance, 381
    applications and significance, 367, 376–77
    compressive strength, 367
    definition, 367
    durability, 381, 375–76
    flexural strength, 375
    materials, 369–73
    mix proportions, 369–72
    modulus of elasticity, 375
    superplasticizer application, 372–73
    tensile strength, 375
High-workability concrete
    applications and significance, 381–83
    definition, 381
    mix proportions, 384
    properties, 384–86
Hot-weather concreting
    definition, 340
    significance, 341
    temperature control method, 341–44

Hydrated (portland) cement paste
    bulk density, 27
    creep, 34
    dimensional stability, 32–34
    disjoining pressure, 34
    drying shrinkage, 32–34
    durability, 34–36
    electron micrographs, 20, 24
    Feldman-Sereda model, 25, 30
    forms of solids present, 23–26
    forms of voids present, 26–29
    forms of water present, 28–30
    microstructure, 20, 22–29
    permeability, 34–35
    pore-size distribution, 26–28, 35
    Powers-Brunauer model, 25
    strength, 30–33
    structure-property relations, 30–36
Hydration of portland cement, 190–97
Hydration reactions of aluminates, 193–95
Hydration reactions of silicates, 196–97
Hydraulic cements, 180
Hydraulic pressure, 129
Hydrophilic and hydrophobic, 259

Igneous rocks for aggregate, 233–34
Impact strength, 64
Impregnation with polymers, 523–25
Initial set
    cement, 198, 200, 206
    concrete, 335
Initial tangent modulus, 83
Interlayer space in C-S-H, 25
Interlayer water in C-S-H, 29–30
Iron blast-furnace slag, 9
Iron ores (heavyweight aggregate), 235

Jet set cement, 208, 217

Laitence, 125, 331
Langbeinite, 189
Latex-modified concrete
    definition, 418
    materials, 422
    properties, 421–22
Leaching of cement paste, 142–44, 170
Le Chatelier's test on soundness, 205
Lightweight aggregates. See Aggregates
Lightweight concrete. See Structural lightweight concrete
Lignosulfonate, 261, 266, 286
Lime cements, 180

Limestone, 180, 229, 232
Low heat portland cement, 203

Macrostructure, 17
Magnesium oxide, 159, 189
Magnesium salts solutions, effect on concrete, 145
Map cracking, 157
Marcasite, 231
Mass concrete
  definition, 428
  general considerations, 429
  measures for temperature control
    admixtures, 430, 438
    aggregate, 430, 437–39
    cement type, 429
    mix proportions, 432
    post-cooling, 435–36
    pre-cooling, 435–37
    surface insulation, 436
  roller-compacted concrete, 439–42
Maturity concept, 339
Maturity meters, 348
Microcracking
  effect on elastic behavior, 80–82
  effect on permeability, 39
  stable and unstable cracks, 63, 82
Microsilica. *See* Condensed silica fume
Microstructure
  definition, 17
  hydrated cement paste, 22–30
  portland cement clinker, 185
  transition zone, 37–38
Mineral admixtures, 258, 273
Mixing of concrete, 311–12
Mixing water
  effect of quality on strength, 53
  effect of temperature on slump, 342
  evaluation method, 55
  use of seawater, 55
Mix proportioning (designing)
  ACI procedure, 296–302
  considerations and principles, 291–94
  objectives, 290
  sample computation, 302
  strength overdesign, 293, 305
Mix proportions
  fiber-reinforced concrete, 411
  high-strength concrete, 11, 372
  high-workability concrete, 384
  low and moderate-strength concretes, 11
  roller-compacted concrete, 439
  shrinkage-compensating concrete, 395

Modified portland cements, 207
Modulus of elasticity. *See* Elastic modulus
Modulus of rupture, 68–69
Mohr rupture diagram, 72
Monosulfate hydrate, 26, 147, 193
Mortar, 9
Multiaxial strength, 73–76
Municipal-waste aggregate, 237

NDT, 346
Neutron radiation, 426
Nondestructive tests
  maturity meters, 348
  penetration resistance techniques, 347
  pullout tests, 349
  radiographic methods, 348
  surface hardness methods, 341
  ultrasonic pulse velocity, 348
Normal distribution, 350
Nuclear shielding concrete, 425–27

Oil-well cements, 208, 217
Oil-well platform concrete, 389–91
Opal, 230
Oriented water, 116
Osmotic pressure, 129
Oven-dry aggregate, 242
Overlays of concrete, 166

Particle size
  cement hydration products, 25–26
  condensed silica fume, 273, 279–80
  fly ash, 273, 277–78
  portland cements, 189–90, 278
  rice husk ash, 273, 280
  sand and silt, 232
PCRV, 428
Penetration resistance, 335, 347
Periclase, 160, 189
Perlite, 234
Permeability
  aggregate, 119
  cement paste, 34–35, 118
  concrete, 120
  Darcy's equation, 118
  factors controlling, 121
Phosphates, 266, 268
Placing of concrete, 317–20
Plaster of paris, 190
Plastic shrinkage, 332–33
Plerosphere, 278
Poisson's ratio, 66, 84–85
Polymer concrete, 419–22

Index

Polymer-impregnated concrete
  definition, 418
  materials, 423
  production, 423
  properties, 420–21, 425
Polymer portland cement concrete, 418
Population of metropolitan areas, 527
Pore-size distribution, 26, 28, 35
Pore-size refinement, 210
Porosity
  aggregate, 241–42
  cement paste, 26–27
  effect on permeability, 33, 200
  effect on strength, 33, 200
Portland blast-furnace slag cement. *See* Cement types
Portland cement
  Bogue equations, 185
  chemical composition, 24, 182
  compound composition, 184–86, 204–05
  compound structure and reactivity, 186–204
  definition, 180
  fineness, 190–91, 202–03, 206
  hydration, 190–205
  manufacturing process, 180–84
  specifications and tests, 201–06
  types, 201–05
Portlandite, 25
Portland pozzolan cement. *See* Cement types
Potential compound composition, 186
Pozzolan
  application, 281
  definition, 207
  types (*see* Admixtures)
Pozzolanic reaction, 209
Preplaced aggregate concrete, 427
Proportioning. *See* Mix proportions
Pulvarized fuel ash, 277
Pull-out test, 347
Pumice, 234
Pyrite, pyrrohotite, 231

Quality assurance
  accelerated strength tests, 345
  control charts, 349–50
  core tests, 348
  in situ and nondestructive tests, 346–48
Quartz, 230
Quartzite, 229
Quick set, 194–95

Radiation shielding concrete
  mix proportions, 427
  preplaced aggregate method, 427
  prestressed concrete reactor vessels, 428
  properties, 427–28
  shielding ability of concrete, 426
  types of radiation, 426
Rapid setting and hardening cements, 216–17
Ready-mixed concrete, 311
Recycled-concrete aggregate, 237–39
Regulated-set cement, 208, 216
Retarding admixtures, 268, 271
Retempering, 341
Revibration, 318
Rice husk ash (rice hull ash), 273, 280
Roller-compacted concrete, 439–442

Salt crystallization pressure, 126
Sand (definition and grading), 9, 247–48
Sandstone, 229, 232
Santorin earth, 274–76
Saturated surface dry condition, 242
Scaling, 127–28, 136
Schmidt rebound hammer, 347
Screeding, 318
Seawater
  chemical attack on concrete, 167–76
  composition, 167
  suitability as mixing water, 55
Secant elastic modulus, 83–85
Sedimentary rocks for aggregate, 232
Segregation, 330–32
Self-stressing cement, 214
Setting of cement paste
  false, flash, and quick set, 194–95
  initial and final set, 198, 200, 206
Setting of concrete
  control, 335
  definitions, 334
  effect of admixtures, 266–68, 271, 336
  effect of temperature, 336
  initial and final set, 335
  method of testing, 335
  significance, 334
Shear-bond failure, 22
Shear strength, 72
Shotcreting, 9
Shrinkage
  drying (*see* Drying shrinkage)
  plastic, 332–33
  thermal (*see* Thermal shrinkage)
Shrinkage-compensating concrete
  applications, 398–403
  concept, 392–93
  creep, 399

Shrinkage-compensating concrete (*Cont.*)
  curing, 397
  definition, 392
  drying shrinkage, 393, 396–97
  durability, 398–99
  effect of w/c on expansion, 396–98
  mix proportions, 394–95
  plastic shrinkage, 396
  slump loss, 395
  strength, 396–97
  workability, 395
Sieve analysis of aggregate, 347–49
Silica fume. *See* Condensed silica fume
Slag
  cements, 207–214
  expanded or foamed, 237
  iron blast-furnace, 9, 278–79
Slip-formed concrete, 6, 346, 381, 391
Slump cone test, 323–24
Slump loss in concrete, 326–30
Solid-space ratio, 31–33, 45
Solid-state hydration, 192
Soundness, 205–06, 244
Spacing factor of entrained air, 27, 133
Special hydraulic cements, 205–24. *See also* Cement types
Specifications
  aggregate, 246–47, 251
  air-entraining admixtures, 257
  blended portland cements, 207
  chemical admixtures, 257–58
  expansive cements, 214
  fly ash and natural pozzolans, 258
  ground slag, 258
  portland cement, 201–06
Specific heat, 340
Specific surface area
  condensed silica fume, 273
  C-S-H, 24
  fly ash, 273
  ground slag, 273, 279
  portland cement, 190–91, 206
  rice husk ash, 273
Sphericity, 250
Splitting tension strength, 66–69
SSD, 242–43
Standard specifications. *See* Specifications
Standard test methods. *See* Test methods
Stiffening of cement paste, 199
Strain
  in compression, 13, 63, 73
  definition, 11–12
  localization, 512

  tensile strain capacity, 109–11
  in tension, 70–73
Strength
  biaxial, 73–76
  compressive (uniaxial), 62–65
  definition, 11, 43
  factors affecting, 61
    admixtures, 56, 262, 265, 284–85
    aggregate, 51–55, 70
    air entrainment, 48–49, 136
    cement content, 48–49
    cement type, 50–51
    curing age, 50, 56–58
    cyclic loads, 64–65
    humidity, 57
    impact loads, 64
    lime leaching, 143–44, 170
    loading rate, 61, 64–65
    mixing water quality, 53–56
    moisture state of concrete, 61
    specimen geometry, 60–61
    stress type, 62–76
    sustained load, 63–64
    temperature, 59–60
    transition zone, 46
    water/cement ratio, 47–51
  flexural, 67–69
  multiaxial, 73–75
  relationship of compressive to tensile and flexural, 68–69
  shear, 72
  tensile, 66–69
Stress, 11
Stress intensity factor, 499
Stress relaxation, 80, 90–91 models. *See* Viscoelasticity
Stress-strain curves
  aggregate, 81
  cement paste, 81
  concrete, 63–65, 69, 74, 81–84
  steel, 13
Structural lightweight concrete
  abrasion resistance, 365
  application and significance, 358–59, 365
  creep, 363–64
  definition, 358–59
  drying shrinkage, 363–64
  durability, 363–65
  elastic modulus, 363
  mix proportioning, 359–61
  strength, 361–63
  thermal conductivity, 365
  thermal expansion, 365

Index

unit weight, 361
workability, 361
Structure (microstructure) of concrete, 17–19
Structures of concrete, photographs
  Baha'i Temple, Wilmette, Illinois, 9
  California aqueduct, 4
  Candlestick Park Stadium, San Francisco, 8
  Central Arizona pipeline, 2
  CN communication tower, Toronto, 346
  Country club reservoir, Oakland, 5
  Ekofisk platform, North Sea, 390
  Floating terminal, Valdez, Alaska, 383
  Fort Peck Dam, Montana, 149
  Fountain of Time, Chicago, 7
  Itaipu Dam, Brazil, 3
  Kinzua Dam, Pennsylvania, 382
  Raffles City building, Singapore, 387
  San Mateo-Hayward Bridge, California, 164
  Stratfjord B offshore platform, 6
  Texas Commerce Tower, Houston, 379
  Tjorn Bridge, Sweden, 380
  Trump Tower building, New York, 388
  Val-de-la-Mare dam, U.K., 158
  Waste treatment plant, Houston, 403
  Water Tower Place, Chicago, 377
  Willow Creek Dam, Oregon, 440
Sulfate attack
  case histories, 148–152
  chemical reactions, 147–48
  factors affecting
    admixtures, 151, 153, 270, 283
    cement content, 150–52
    cement type, 150, 212–13
    permeability, 151
    sulfate concentration, 153
  measures for control, 152–54
  mechanism, 147
  sources of sulfate, 146–47, 152
  types (forms) of damage, 147
Sulfate resisting cement, 203–04
Sulfates in portland cement, 189–90
Sulfides and sulfate aggregate, 231
Superplasticized concrete, 272–74, 381–82
Superplasticizing admixtures, 263–65
Surface area. *See* Specific surface area
Surface energy, 495
Surface moisture, 242

Tangent modulus of elasticity, 82
Temperature effects
  on aggregate, 132–33, 139–40
  on cement paste, 129–31, 139
  on concrete, 133–38, 140, 337–44

Tensile strain, 73–75
Tensile strain capacity, 109–11, 433
Tensile strength, 69–73
Test methods
  abrasion-erosion, 125–26
  accelerated strength, 344–45
  aggregate, 252–53
  bleeding of fresh concrete, 332
  compressive strength
    cement (mortar), 206
    concrete, 62
  consistency, 323–25
  elastic modulus, 84
  fineness of cement, 206
  fineness modulus of aggregate, 249
  flexural strength, 66–68
  mixing water quality, 55
  nondestructive, 346–48
  penetration resistance, 335, 347
  pull-out, 347
  setting time
    cement, 206
    concrete, 335
  soundness
    aggregate, 244
    cement, 205–06
  surface hardness, 347
  tensile strength, 66–68
  ultrasonic pulse velocity, 348
Thermal conductivity, 107, 109, 365
Thermal diffusivity, 109, 110
Thermal expansion coefficient, 108
Thermal shrinkage
  control, 101–103, 435–437
  factors affecting, 103–108
  significance, 13, 79, 108, 428–29
Thermal stress
  factors affecting
    degree of restraining, 103
    placement temperature, 103
    precooling of fresh concrete, 103
    temperature change, 103
    see also adiabatic temperature rise
  importance, 109–111
Through-solution hydration, 191
Time of set. *See* Setting of cement paste; Setting of concrete
Tobermorite gel, 24, 196
Topochemical hydration, 192
Toughness, 12, 415
Trans-Tokyo Highway, 536
Transition zone
  definition, 18

Transition zone (*Cont.*)
    development, 37
    microstructure-properties, 37–39
    significance to concrete, 36–40, 46, 88, 121
Transport properties, 454
Transporting concrete, 313–16
Tricalcium aluminate. *See* Calcium aluminates
Tricalcium silicate. *See* Calcium silicates
Triethanolamine, 266, 268
Truck mixing, 312, 314

Ultrasonic pulse velocity, 348
Uniaxial compression behavior
    axial and lateral strains, 63
    crack propagation, 63–64
    cyclic loads, 65
    elastic moduli, 80—88
    failure mode, 22, 46
    impact load, 64
    sustained load, 63
Units of measurement
    conversion factors, 15
    English and metric system, 15
    International system (SI), 15
Unit weight
    aggregate, 235–36
    concrete, 10, 236, 358–59, 427–28

Van der Waals force, 30, 37
Vebe test, 323, 325
Vermiculite, 235
Very high early strength cement, 208, 216
Vibration, 318–19
Vicat apparatus, 200, 206
Viscoelasticity
    mathematical expressions, 473–475
    methods for predicting creep and shrinkage
        ACI 209, 479, 480
        Bazant-Panula, 480–481
        CEB 1978, 478–479
        CEB 1990, 476–478
    relaxation time, 458
    retardation time, 459
    rheological models
        basic models, 456–459
        Dischinger, 469
        generalized models, 466–471
        Kevin, 459
        Maxwell, 458
        standard solid, 459
    superposition principle, 471
    See also Creep and Relaxation
Void in hydrated cement paste
    capillary, 25–28
    entrained air, 26–28, 133–34
    entrapped air, 27–28
    interlayer, 24–27
Volcanic glass, 234–274
Volcanic tuff, 234, 276

Water
    agent of deterioration, 115–16
    in calcium silicate hydrate, 23, 29
    density, 117
    forms in cement paste, 28–30
    molecular structure, 116
    oriented structure, 116
    quality for concrete mixing, 55
    surface tension, 117, 260
Water/cement ratio
    determination in mix design, 299
    effect on abrasion resistance, 124
    effect on frost resistance, 135
    effect on permeability, 33, 118–21
    relationship to strength, 33, 47–51
Water-reducing admixtures, 259–60, 262–61
Watertightness, 34
White cement, 9, 209, 219–20
Windsor probe, 347
Winter concreting, 337, 339
Workability, 322–30

X-ray diffraction analysis, 150, 185

Zeolite, 276